Shuqin Lou, Chunling Yang
Digital Electronic Circuits

Information and Computer Engineering

Volume 4

Already published in the series

Volume 3
Baolong Guo, Signals and Systems, 2018
ISBN 978-3-11-059541-3, e-ISBN 978-3-11-059390-7,
e-ISBN (EPUB) 978-3-11-059296-2

Volume 2
Jie Yang, Congfeng Liu, Random Signal Analysis, 2018
ISBN 978-3-11-059536-9, e-ISBN 978-3-11-059380-8,
e-ISBN (EPUB) 978-3-11-059297-9

Volume 1
Beija Ning, Analog Electronic Circuit, 2018
ISBN 978-3-11-059540-6, e-ISBN 978-3-11-059386-0,
e-ISBN (EPUB) 978-3-11-059319-8

Shuqin Lou, Chunling Yang

Digital Electronic Circuits

Principles and Practices

DE GRUYTER  Science Press Beijing

Authors
Shuqin Lou
Beijing Jiaotong University

Chunling Yang
Harbin Institute of Technology

ISBN 978-3-11-061466-4
e-ISBN (PDF) 978-3-11-061491-6
e-ISBN (EPUB) 978-3-11-061493-0
ISSN 2570-1614

Library of Congress Control Number: 2018953871

Bibliographic information published by the Deutsche Nationalbibliothek
The Deutsche Nationalbibliothek lists this publication in the Deutsche Nationalbibliografie; detailed bibliographic data are available on the Internet at http://dnb.dnb.de.

© 2019 Walter de Gruyter GmbH, Berlin/Boston, Science Press
Typesetting: Integra Software Services Pvt. Ltd.
Printing and binding: CPI books GmbH, Leck
Cover image: Prill/iStock/Getty Images Plus

www.degruyter.com

Preface

With the rapid development of digital technology, digital electronic circuits are widely used in computers, communications systems, navigation and guidance systems, medical instrumentations, consumer and industrial electrical systems. It is essential for engineers and students to fully understand both the fundamentals and the implementation and application principles of digital circuits, thus enabling them to master the most appropriate and effective technique to satisfy their technical needs. This textbook aims to cover all three aspects related to the teaching of digital circuits: *digital principles, digital electronics*, and *digital design*. It teaches the fundamental principles of digital circuits or systems and covers thoroughly both traditional and modern methods of applying digital design and development techniques. This textbook is appropriate for a compulsory course of digital electronic circuits or systems in the electronic and electrical engineering, computer science and engineering, mechanical and electronic engineering, instrumentation engineer, and others majorly related to electronics; it also can be a self-study reference book for digital designers and engineers who want to design and build real digital circuits and systems.

This book covers three aspects of digital circuits: digital principles, digital electronics, and digital design. It is organized based on the idea that students must grasp the fundamental knowledge, but at the same time to understand how circuits work in the real world. Hence, the "principles and practices" theme are both adopted. It is grounded on basic principles of digital design that do not change with technology, such as combinational logic, sequential logic, and state machine. A fundamental grounding on the basic concepts of digital circuits and systems is reinforced by an abundance of illustrations, examples, applications, and exercises.

The rapid development of integrated circuits technology results in an exponential increase of the integrated density; logic packages in a tiny integrated chip are evolved a dozen to more than millions of transistors. To keep up with the development of integrated circuits technology, this book also introduces the modern design methods using electronic design automation (EDA). These include how to use the hardware description language (HDL) describing the circuit and how to implement designs with programmable logic devices (PLDs) and large-scale integrated (LSI) circuit. We also believe that students should be involved with projects using the state-of-the-art design tools and hardware solutions.

In each chapter of this textbook, there is an introduction and the objectives at the beginning and summary as well as key terms at the end. The treatment of each new topic or device typically starts with the principle of operation and then the explanations of applications. The internal circuit analysis and calculation of a digital circuit are replaced by an emphasis on the introduction to the characteristics and parameters of the circuit. Self-test and problems are arranged in every chapter to help students gain the ability of analyzing and designing digital circuits and systems.

https://doi.org/10.1515/9783110614916-201

The problems are arranged in sequence, from simple to complex. Especially, the in-depth problems are chosen to reinforce the material without simply repeating the principles. This provides with a wide choice of student assignments and also enhances students to demonstrate comprehension of the principles by applying them to different situations. The most difficult goal in this textbook is set to help students adapt to the inevitable changes in the future. As a new technology emerges, you would be able to design circuits with it, utilizing the fundamental concepts and methods introduced in this book.

Content organization and chapter overview

This textbook includes 12 chapters covering three major topics. These are the principle and application of small- and medium-scale digital integrated circuits, large-scale digital integrated circuits, and analog-digital mixed circuits.

Small- and medium-scale digital integrated circuits include logic gate circuits, combinational logic circuits, flip-flops and related device, and sequential logic circuits. The logic gate circuits are composed of transistors, multiple logic gates constitute combinational logic circuits, flip-flops, and combinational logic circuits constitute sequential logic circuits. Their mathematical basis is logical algebra, which is also known as Boolean algebra.

Large-scale digital integrated circuits covered in this book include memories and PLDs. Memories, including read-only memory (ROM) and random access memory (RAM), are used to store data, programs, tables, etc. Programmable logic devices (PLDs) include simple PLDs, complex PLDs, and field programmable gate arrays (FPGAs).

Actually, lots of electronic products are composed of analog circuits and digital circuits, which are analog-digital mixed circuits. Analog-to-digital converters (ADC) and digital-to-analog converters (DAC) are essential interface circuits between analog signal and digital signal.

The contents of the 12 chapters are briefly described in the following.

Chapter 1 provides a broad overview of digital electronic circuit, including the introduction of basic concepts, advantages of digital technology, the classification and the typical package of digital integrated circuits, and the EDA technology. The key terms listed at the end of the chapter are to help learners easily and clearly grasp the basic concepts in digital circuits.

Chapters 2 and 3 are the fundamental parts that will enable students to prepare the fundamental theory and methods for further learning the core contents of digital circuits. Chapter 2 introduces number representations, number conversions, binary arithmetic, and the most commonly used binary codes. Chapter 3 begins with the basic logical operations and logic gates, and it continues with the fundamentals of Boolean algebra, including Boolean theory, axioms, and theorems that can be applied to digital circuits. Then the expression methods of Boolean function, especially standard sum-of-product and product-of-sum expression, are introduced. Finally, the simplification method of Boolean expression with Boolean algebra and Karnaugh map are discussed. These provide theoretical fundamental for further analyzing and designing digital circuit.

Chapter 4 is one of the core contents of digital circuits, which introduces analysis and design methods of gate-level and block-level combinational logic circuits. This chapter introduces some functional modules of digital integrated circuits, such as multiplexers, decoders, adders, and comparators, as well as their applications in constructing the combinational logic circuits and systems. In the last part, the commonly used hardware description language, Verilog hardware description language (HDL), and the typical Verilog HDL descriptions of commonly used logic modules are introduced, which is fundamental to PLDs.

Chapters 5–8 are about sequential logic circuits, which are also the core contents of digital circuits. Chapter 5 covers the basic devices for constructing sequential signal and sequential circuits, which include different types of flip-flops for the applications of memory and multivibrators for pulse generation and transformation. Chapter 6 begins with the basic concepts and structural features of sequential logic circuits and continues with the general analysis and design method of sequential circuits, focusing especially on the analysis and design of synchronous sequential circuits. This chapter ends with the state minimization to further simplify the design of sequential circuits. Chapter 7 focuses on counters, which are very widely used sequential modules in digital systems. Counters are usually manufactured as separate integrated circuits and also incorporated as a part of larger integrated circuits. This chapter introduces asynchronous counters, synchronous counters, cascaded counters, and counter integrated modules and the related applications. Chapter 8 introduces registers and shift registers, which are another kind of commonly used sequential circuits in digital system. The contents include registers, shift registers, and the applications of shift registers.

Chapters 9 and 10 are the contents involving large-scale integrated circuits. Chapter 9 deals with semiconductor memories for storing large amount of data. The contents involve RAM, ROM, and special types of semiconductor memories. This chapter begins with the concepts, the memory cell organization, the basic operations and applications of ROM, and continues with RAM, including static RAMs (SRAMs) and dynamic RAMs (DRAMs). At the end, two methods for expanding the capacity of memory, word-length expansion and word-capacity expansion, are introduced to construct the larger capacity of memory. Chapter 10 covers PLDs including SPLDs, CPLDs, and FPGAs. The programming process of PLDs is briefly introduced at the end of the chapter.

Chapter 11 deals with the interface circuits between analog circuits and digital circuits. The conversion accuracy and speed are two main parameters for ADC and DAC. This chapter introduces the basic concepts and operating principles of DACs and

ADCs. Commonly used DACs include binary-weighted-input DAC and R/2R ladder DAC. Commonly used ADCs include flash ADC, successive-approximation ADC, dual-slope ADC, and sigma-delta ADC. Several typical integrated DACs and ADCs chips and their applications are introduced at the end.

Chapter 12 covers the integrated circuit technologies for designing and manufacturing logic gates. Two kinds of integrated circuit technologies are introduced. One is bipolar technology, and the typical logic family is transistor-transistor logic (TTL). Another is MOSFET (metal oxide semiconductor field effect transistor) or unipolar technology and the typical logic family is CMOS (complementary MOS) logic. The interface between TTL and CMOS-integrated circuits are discussed at the end of this chapter.

This textbook also includes three appendices. Appendix 1 is a brief introduction to Quartus II, which is commonly used EDA software for designing large-scale digital integrated circuits and PLDs. Appendix 2 provides the information of Altera DE2 development and education board for implementing the design of PLDs. Appendix 3 lists the abbreviations in this book.

To the instructors

This textbook covers a comprehensive range of topics, beginning with basic digital concepts and progressing through number systems, Boolean algebra and logic simplification, combinational logic circuit, sequential logic circuit, programmable logic device, semiconductor memory, ADC and DAC, and integrated circuit technologies.

Owing to the time limitations of course and the difference of the major, instructors can change the sequence for certain topics as they appear in the textbook. Chapter 12 is intended to be used as a flexible chapter, which can be covered in whole or in part at any point in the text, or it can be omitted without affecting any other topics. If you want to include Chapter 12 in your lecture, the suggestion points are to put this chapter after Chapter 3 or arrange it at the end of the course without affecting the other contents. If the students already learn the knowledge of number system and analog-to-digital conversion, you can omit Chapter 11 and part of content in Chapter 2. If you want to emphasize the content of pulse generation and pulse transform, you can split one shot, astable multivibrator, Schmitt trigger, and 555 timer from Chapter 5 into separated chapters.

To the students

For students, if you want to master the knowledge of digital circuit in this textbook. You should pay an attention to the following suggestion.

Firstly, logical algebra is the mathematical basis of digital circuit. A basic knowledge about logic algebra is indispensable for learning digital circuits. Secondly, you should master logic function and the external characteristics of digital integrated circuits. Thirdly, digital electronic circuit is a course with strong application background. Strengthening practical training is essential part for this course. Finally, EDA is the up-to-date development of electronic technology. Programmable logic devices are currently the most frequently used logic devices, and the design of programmable logic devices needs to be realized through EDA technology. Examples of Verilog HDL for some logic circuits are provided in Chapter 4. The readers can write their own Verilog HDL codes for other logic circuits accordingly. In addition, Appendix I shows the guidance of programmable logic device design by using software Quartus II. Function simulation can be carried out for digital circuits by using Quartus II, and the digital circuit simulation capabilities of students can be trained and strengthened. Students should actively learn the new knowledge through simulation experiments.

Acknowledgment

Many people have contributed to the publication of this textbook. The authors of this textbook are Shuqin Lou, Chunling Yang, Lei Kang, Xin Wang, Yan Shen, and Mu Li from Beijing Jiaotong University, Harbin Institute of Technology and Hebei University of Technology, China. Among them, Shuqin Lou and Chunling Yang are in charge of the whole planning, writting, and modification of the contents in this book. Lei Kang, Xin Wang, Yan Shen, and Mu Li wrote several main chapters. Zhongqiu Hua, Shuang Bai, and Xinzhi Sheng helped us polish the language. Some students, Wan Zhang, Shibo Yan, Tongtong Zhao, Zijuan Tang, Minqing Wang, Zhen Xing, Ailun Zhang, Yuxin Zhong, Xidan Liu, and Yanan Wu, helped us prepare all figure files. Our overseas students, including James Konchellah, Rueben Kawaka, Joseph Mwanzia, Daniel Muoki, Dadson Wamuiga, Simon Gituba, Edwin Thuo, Issa Abdirizak, Loise Wanjiku, Christopher Kipamet, Clinton Munene, and Bilal Maddihani, helped us proofread the book. We would like to acknowledge the efforts of Professor Zhenggang Lian who reviewed contents of this textbook and gave us a lot of suggestions.

A writing project of this book requires conscientious and professional editorial support. We thank the editors from De Gruyter Press of Germany and Science Press of China for their help to make this publication a success. Miss Pan Sisi and other editors dedicated a lot of energy to the book. We wish to express our heartfelt gratitude to the editorial and publication team from De Gruyter Press and Science Press for supporting this edition of our book.

This textbook might still contain errors and omissions due to the limitation of the authors' ability and language. The authors sincerely hope readers kindly enough to provide feedback on any errors discovered in this textbook to email (lousq@163.com) and help us improve the quality of this textbook.

Authors
13/03/2018

Contents

About the Authors

Shuqin Lou is a professor of electronic and information engineering at Beijing Jiaotong University. Since 1989, she has been teaching courses of digital circuit and analog circuit for the sophomore and junior undergraduate students whose major is electronic and information Engineering. Her teaching has been honored with the Outstanding Teacher Award in 2001 and the Distinguished Teacher Award in 2008 in Beijing Jiaotong University. She earned one first prize and two second prizes of Beijing teaching achievement award. Since 2001, she has published seven textbooks. One of these books was honored by the excellent book award of Beijing Municipal Government in 2011. In 2010, she was granted the project of National bilingual course of digital electronic technology funded by the Ministry of Education of China. In 2014, she finished the translation version of "Digital fundamental: a system approach" written by Thomas L. Floyd and was honored with excellent translator award from Hua Zhang Press in Beijing for her excellent work. Her research interests cover the areas of telecommunication, especially optical communication. She has been holding more than 20 projects, publishing over 250 papers, and getting 14 patents. She has also organized more than 10 conferences on her research areas.

Chunling Yang is a professor of electrical engineering at Harbin Institute of Technology. She is the director of Electronic Technology Research Association of Universities in northeast of China. She was honored with the Teacher Distinguished Award of Hei Longjiang Province in 2017, and Baosteel's Teacher Award in 2008. She earned one first prize and two second prizes of National Teaching Achievement, two first prizes and two second prizes of Teaching Achievement of Hei Longjiang Province. She has published six textbooks, two of which were selected as National 11th Five-Year Planned Textbooks and one of which was selected as National 12th Five-Year Planned Textbooks. Her research interests cover the areas of FPGA design and signal monitoring and processing. She has been the holding more than 10 projects, publishing 68 papers, and getting 6 patents in China. She has won one third prize for Progress in Science and one second prize of Natural Science Award of Hei Longjiang Province.

https://doi.org/10.1515/9783110614916-202

1 Introduction to digital electronic circuit

1.1 Introduction

Electronics is a branch of physics, engineering, and technology that deals with circuits consisting of components that control the flow of electricity. Circuits and components can be divided into two groups: analog and digital. A particular device may consist of circuitry that has analog or digital or a combination of these. Digital electronics or digital electronic circuits operate on digital signals. In the early days, applications of digital electronic circuits were focused on computer systems. Now digital electronics has been applied in a wide range of systems, such as telecommunication systems, military systems, medical systems, control systems, and consumer electronics. This chapter provides a broad overview of digital electronic circuits, including a brief introduction to the basic concepts of digital circuits, their commonly used devices, and technology of electronic design automation (EDA).

The objectives of this chapter are to
- Explain the differences between digital and analog quantities
- Describe the representation of digital quantities
- Explain the classification of digital circuits
- State the advantages of digital over analog
- Explain the characteristics of the commonly used hardware description languages (HDLs)
- Define EDA
- Describe the design and programming process of programmable logic device (PLD)

1.2 Introductory basic concepts of digital electronic circuit

Electronic systems can be divided into two broad categories: digital and analog. Digital circuits are electric circuits that deal with the digital signals that have a number of discrete voltage levels. To most engineers, the terms "digital circuit," "digital system," and "logic" are interchangeable in the context of digital circuits, while analog circuits involve quantities with continuous values. This section introduces some basic concepts about digital circuits.

The objectives of this section are to
- Explain the differences between digital and analog quantities
- Define binary digits
- Describe how to represent voltage levels by bits
- Explain the advantages of digital circuits over analog circuits

https://doi.org/10.1515/9783110614916-001

1.2.1 Analog and digital

An analog quantity is the one having continuous values in time. A digital quantity is the one having a discrete set of values. In the natural world, most of the physical parameters, such as temperature, pressure, and strain, are analog quantities. These physical parameters can be converted into continuous electronic signals, voltage, or current, by the specific sensor so that they can be processed using the circuit. An analog signal refers to a signal that changes its value continuously over time. The typical analog signal is sinusoidal wave or sound wave, as shown in Figure 1.2.1(a). The term analog signal usually refers to electronic signals; however, mechanical, pneumatic, hydraulic, human speech, and other systems may also convey or be considered analog signals.

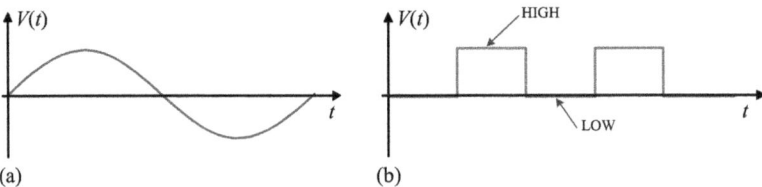

(a)   (b)

Figure 1.2.1: Comparison between (a) analog signal and (b) digital signal.

A digital signal refers to an electrical signal that has a sequence of discrete values; at any given time it can only take one of a finite number of values [1, 2]. This contrasts with an analog signal, which represents continuous values; at any given time it represents a real number within a continuous range of values. In digital circuit, a digital signal is a pulse train, that is, a sequence of fixed width electrical pulses. Figure 1.2.1(b) shows a typical digital signal that varies between low and high voltage levels, in which the high voltage level conveys a binary 1 and the low voltage level conveys a binary 0. This kind of digital signal is also called as a logic signal or a binary signal.

1.2.2 Binary digits, logic levels, and digital waveforms

1. Binary digits
Most digital circuits use a binary system that only has two digits, 1 and 0, which can represent two voltage levels. A *binary digit* is called a *bit*. Often logic "0" will be a lower voltage and referred to as "LOW" while logic "1" is referred to as "HIGH." This is called *positive logic* and is used throughout this book.

Of course, you can use logic "0" representing a HIGH and logic "1" representing a LOW. This logic system is called *negative logic*.

In digital systems, a combination of 1s and 0s is called *codes*, which are used to represent numbers, symbols, alphabetic characters, and other types of information.

2. Logic levels

The voltage used to represent a 1 or a 0 are called *logic level*. In a practical digital circuit, a HIGH level can be any voltage level between a specified minimum value and a specified maximum value. Likewise, a LOW can be any voltage level between a specified minimum and a specified maximum. There is no overlap between the accepted range of HIGH and LOW levels.

3. Digital waveforms

Digital waveform consists of voltage levels that change back and forth between the HIGH and LOW levels. As shown in Figure 1.2.2, a positive-going pulse is generated when the voltage goes from its normally LOW level to its HIGH level. The negative-going pulse is formed when the voltage goes from its normally HIGH level to its LOW level.

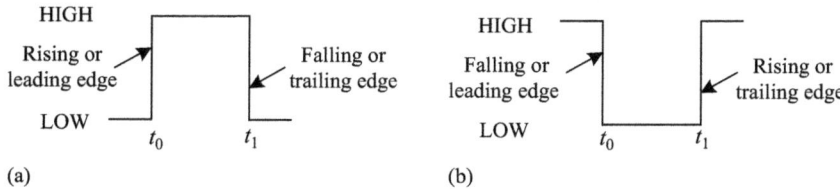

Figure 1.2.2: Waveforms of positive-going (a) and negative-going (b) pulses.

Binary information, handled by digital systems, appears as digital waveforms. A digital waveform is made up of a series of pulses, sometimes called *pulse trains* [3]. When the waveform is HIGH, a binary 1 is present; when the waveform is LOW, a binary 0 is present. Each bit in a sequence occupies a defined time interval called a *bit time*. In digital systems, signal waveforms are synchronized with a basic timing waveform called the *clock*, as shown in Figure 1.2.3. The waveform of the clock is a period of pulse trains in which the pulse period equals to a bit time. Binary data is indicated by the level in the waveform. During each bit time of the clock, waveform of

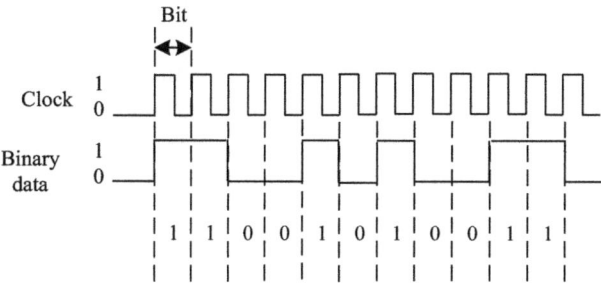

Figure 1.2.3: Binary data indicated by a pulse train synchronized with a clock pulse.

binary data is either HIGH or LOW, in which HIGHs and LOWs represent a sequence of binary digits (bits). Binary data is represented by a group of several bits. Notice that the clock waveform itself does not carry information.

1.2.3 Characteristics of digital circuit

A digital circuit is a circuit that takes digital signals as inputs, processes them, and outputs the processed digital signals. Compared with analog circuits, digital circuits have some distinguished advantages.

Since a digital signal is a signal in which discrete steps are used to represent information, active components in digital circuits typically have one signal level when turned on, and another signal level when turned off. In general, a component in digital circuits is only switched on or off. For example, transistors in digital circuits operate either in saturation region or in cutoff region. While transistors in analog circuits operate in active region, their outputs are susceptible to several factors such as temperature, power supply voltage, and component aging. Therefore, an advantage of digital circuits when compared to analog circuits is that signals represented digitally can be transmitted without degradation due to noise [4].

Information storage can be designed easily in digital systems than in analog ones. The noise immunity of digital systems permits data to be stored and retrieved without degradation. In an analog system, noise from aging and wear degrades the stored information. In a digital system, as long as the total noise is below a certain level, the information can be recovered perfectly.

Digital circuits are the most common physical representation of Boolean algebra [5]. The design of digital circuits is a logical design that does not require designers to have very strong mathematical background, whereas the analog circuit design requires the calculation of the model in order to understand and study the internal characteristics and the operating principle of the circuit. In a digital system, a more precise representation of a signal can be obtained by using more binary digits to represent it. While this requires more digital circuits to process the signals, each digit is handled by the same kind of hardware, resulting in an easily scalable system. In an analog system, additional resolution requires fundamental improvements in the linearity and noise characteristics of each step of the signal chain.

Digital circuits are easy to be integrated, and they are low cost and small in size [6]. The integration level of digital circuits is generally higher than that of analog circuits. In addition, digital circuits are programmable. Computer language can be used to design some digital circuits to achieve corresponding logic functions. Computer-controlled digital systems can be controlled by software, allowing new functions to be added without changing hardware. Often this can be done outside the factory by updating the product's software. So, the product's design errors can be corrected after the product is in the customer's hands.

1.3 Digital integrated circuits and typical packages

Digital electronic circuits are usually made from large assemblies of logic gates to implement various logic functions, which mainly involve combinational logic functions and sequential logic functions. All the logic elements and functions in a digital circuit are available in an integrated circuit (IC) form. A monolithic IC is an electronic circuit that is constructed entirely on a single small chip of silicon. All the components that make up the circuit – transistors, diodes, resistors, and capacitors – are an integral part of that single chip. Digital ICs are divided into two broad categories: fixed-functional logic and programmable logic. This section briefly introduces fixed-function devices, programmable logic devices (PLDs), and their typical packages.

The objectives of this section are to
- Explain the difference between fixed-function logic devices and PLDs
- Recognize the IC packages
- Explain the complexity of ICs
- Describe the IC technology

1.3.1 Fixed-function logic devices

The fixed-function logic devices refer to digital ICs that their logic functions are set by the manufacturer and cannot be altered. They include standard chips and custom-designed chips.

Standard chips refer to ICs that integrate some commonly used logic function circuits into the chips of silicon [7]. Combinational logic circuits and sequential logic circuits are two types of digital circuits. In the combinational logic circuit, its outputs are only determined by its current inputs. Combinational logic ICs include different gates, magnitude comparators, encoders, decoders, multiplexers, demultiplexers, and other logic function devices. In the sequential logic circuit, its outputs depend not only on the current inputs but also on the past inputs. Sequential logic ICs include latches, flip-flops, registers, and counters. These chips have functions and specifications in line with recognized standards. Designers can use these chips to design circuits that perform the desired functions. The advantages of using standard chips are their ease to use and ready availability. They can be bought off-the-shelf. However, their fixed simple functionality results in many chips that should be used to implement a complex functionality on a printed circuit board (PCB). This causes a requirement for more space, more components, and more wires that make volume larger and reliability lower. An example of commonly used standard chips is 7,400 series.

Custom-designed chips refer to ICs that are designed to meet the specific requirement, also known as *application-specific ICs* (ASICs) [7]. The distinguished advantage of custom-designed ICs is that they are optimized for implementing a particular task so that higher performance can be achieved with higher integration level. Although the production cost of these ICs is high, the allocated cost of each chip can be obviously reduced by mass production. In addition, ASICs can integrate multiple chips onto a single chip, reducing the size and cost of products.

1.3.2 Programmable logic devices

Before the 1980s, standard ICs were usually chosen to design logical circuits. However, advances in very large scale integrated technology made possible the design of special chips, which can be configured by a user to implement different logic circuits. These chips are known as PLDs.

A PLD is a type of IC that starts as a "blank slate" and into which a logic design can be programmed [3]. The logic design in the PLD can be changed repeatedly. The designer first designs the prototype of a product; then the performance of the design can be evaluated and problems could be found during the subsequent hardware testing; finally, the product can be updated by adding new functionality and reprogramming the PLD. Compared with the fixed-function logic devices, PLDs have the following advantages. One advantage is that PLD uses much less board space for the equivalent amount of logic. Another advantage is that designs with PLD can be changed without rewiring or replacing components. Also, a logic design can be generally implemented faster and less cost with programmable logic than with fixed-function ICs. PLDs can be used to implement complex, large-scale logic circuits that cannot be implemented with typical standard ICs. PLDs, also known as semicustom-specific ICs, have made the design of products much easier. Various types of PLDs are available, ranging from small devices that can replace a few fixed-function devices to complex high-density devices that can replace thousands of fixed-function devices.

The most popular types of PLDs are simple PLDs (SPLDs), complex PLDs (CPLDs), and field-programmable gate arrays (FPGAs). Various types of PLDs have different internal architecture. Among them, FPGAs have the highest *gate count*, which can implement much larger designs than SPLDs and CPLDs. Today there are millions of transistors in an FPGA chip. Some famous PLD manufacturers include Altera Inc., Xilinx Inc., Lattice Semiconductor, Atmel, Actel, Cypress, Lucent, and QuickLogic. The PLD Cyclone II produced by Altera Corporation is shown in Figure 1.3.1.

1.3.3 Typical IC packages

A package of IC refers to IC shape and pin arrangement [6]. According to their mounting methods in PCBs, packages can be divided into two types: through-hole

Figure 1.3.1: The programmable logic device "Cyclone II" produced by Altera Corporation.

and surface-mount. With a through-hole package, pins of a chip are inserted into the through-hole from one side of the PCB and soldered on the other side. The pins of the chip can be directly inserted into breadboards for experiments and thus through-hole package is widely used in laboratory. Most packages of ICs use surface-mount technology (SMT), which solders the pins of a chip directly on the circuit board without through-holes, so that the other side of the circuit board can be used for other circuits. SMT package can save the space of the circuit board. In addition, the pins of the surface-mount package are arranged more closely. In particular, surface-mount package is generally used in large-scale ICs due to the large amount of pins.

1. Through-hole package

Typical through-hole package is dual in-line package (DIP), which has two packaging materials: plastic and ceramic. The pin center distance is 2.54 mm, and the number of pins typically ranges from 6 to 64. The package width is typically 15.2 mm, as shown in Figure 1.3.2.

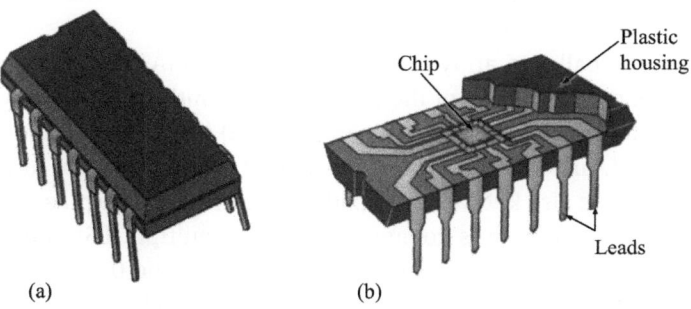

(a)   (b)

Figure 1.3.2: Dual in-line package: (a) outline and (b) cross-sectional diagram.

2. SMT package

There are many types of SMT packages, which are categorized according to the integration level and the number of pins of ICs, as shown in Figure 1.3.3.

(a) (b)

(c) (d)

Figure 1.3.3: Typical SMT package configurations: (a) SOIC; (b) PLCC; (c) BGA; and (d) QFP.

Small outline IC (SOIC) package is used for the chips with a small number of pins. The pins are distributed on both sides of the package like "gull wing," as shown in Figure 1.3.3(a).

Plastic leaded chip carrier (PLCC) package is wrapped around and under the edge of the package, and arranged on the four sides of the package serving as electrodes. PLCC package is a high-speed and high-frequency IC package. The outline of the package is square surrounded by leads, and its dimensions are smaller than the DIP package. PLCC package has advantages such as small size and high reliability, suitable for PCB SMT, as shown in Figure 1.3.3(b).

Figure 1.3.3(c) shows the ball grid array (BGA) package. In a BGA package, the leads are replaced by pads on the bottom of the package, each initially with a tiny solder ball stuck to it; meanwhile, a large-scale integration (LSI) chip is mounted on the top of the printed board, and then the chip is sealed by a molding resin or a potting method. This package is also known as the pad array carrier, whose pins can be more than 200. Thus, this package is commonly used by multipin LSI chips. Typical pitches of the lead of a BGA package are 1.27, 1.0, 0.8, 0.65, and 0.5 mm.

A quad flat package (QFP) is a surface-mount package with "gull wing" leads extending from each of the four sides of the package, as shown in Figure 1.3.3(d). The commonly used materials of QFP include ceramic, metal, and plastic. From the quantitative point of view, plastic packaging accouts for the vast majority. When the material is not specifically required, the plastic QFP is mainly used. Typical pitches of pins include 1.0, 0.8, 0.65, 0.5, 0.4 and 0.3 mm. The 0.65 mm pin pitch has the maximum 304 pins. According to the thickness of package body, this type of package can be further divided into QFP (from 2.0 to 3.6 mm thick), low-profile QFP (1.4 mm thick), thin QFP (1.0 mm thick), and so on.

1.3.4 Complexity classifications for digital ICs

The complexity of digital ICs has various definitions from different sources. Here, the complexity of digital ICs is defined in terms of the number of equivalent gate circuits on a single chip. According to the complexity, digital ICs can be divided into five categories: small-scale integration (SSI), medium-scale integration (MSI), medium-scale integration (MSI), LSI, very large scale integration (VLSI), and ultra-large-scale integration (ULSI) [3].

SSI refers to the digital ICs in which there are up to ten equivalent gate circuits on a single chip, which include basic gates and flip-flops. MSI is a type of ICs in which there are a number of equivalent logic gates from 10 to 100 on a chip, which involves combinational logic function ICs, such as adders, comparators, decoders, and encoders, and sequential logic function ICs including counters and registers. LSI refers to ICs with complexities from more than 100 to 10,000 equivalent gates per chip, which can be used to build memories. VLSI is a kind of ICs in which there are a number of equivalent logic gates from more than 10,000 to 100,000 on a chip. ULSI is a classification of ICs with the complexities of more than 100,000 equivalent gates per chip, which can be used to construct very large memories, larger single-chip computers, and larger microprocessors.

1.3.5 Integrated circuit technology

Logic gates are basic logic elements in digital circuits. Logic gate can be constructed with transistors, including metal oxide semiconductor (MOS) field-effect transistors and bipolar junction transistors (BJT). Bipolar and CMOS (complementary MOS) are two typical digital IC technologies. BiCMOS uses a combination of both CMOS and bipolar.

1.4 Introductory EDA

Design is a key issue in developing digital circuits to satisfy the requirement of various applications. For a small-scale digital circuit, logic gates can be used to design digital logic circuit; thus, this design method is called as gate-level design. For medium- or large-scale digital circuit, digital integration modules can be used to design the required function circuits, so this design method is called as block-level design. For larger scale logic circuit, the system level of hierarchy design is introduced. Early designs were truly handcrafted. But with the rapid development of semiconductor technology, the complex digital systems, for example, microprocessor, contains more than millions of logic elements. Obviously, the handcrafted design method is not appropriate when more than a million logic elements have to be created and assembled. As a result, designers have increasingly adhered to rigid design methodologies and strategies that are more amenable to design automation. EDA is introduced into design practice to save costly engineering effort. This section briefly introduces EDA. HDLs and design process of PLDs are covered.

The objectives of this section are to
- Define EDA
- Explain HDLs
- Describe design and programming process of PLDs

1.4.1 Electronic design automation

EDA is a term for a category of software tools that help to design electronic systems with the aid of a computer [8]. It is also referred to as electronic computer-aided design (ECAD). Integrating electronic technology, computer technology, and intelligent technology, EDA uses a computer as a design workbench mainly supporting PLD design, IC design, electronic circuit design, and PCB design. Design tools include simulation at various complexity levels, design verification, layout generation, and design synthesis. Popular names in the EDA software world are National Instruments (Multisim), Cadence (ORCAD), Altium (Protel), LabCentre Electronics (Proteus), Altera (Quartus II), and Xilinx (ISE).

For the design of digital ICs, EDA tools provide different styles for design entry, which include not only the commonly used entry styles, for example, schematic entry, waveform entry, and state machine, but also the HDL entry.

A large-scale digital system can be implemented with one or several PLDs, using HDL to complete the system behavior-level design, and finally to generate the final target device through synthesizers and adapters. This design method is called high-level electronic design method.

1.4.2 Hardware description language

Due to the exploding complexity of digital electronic circuits since the 1970s (see Moore's law), circuit designers needed digital logic descriptions to be performed at a high level without being tied to a specific electronic technology, such as CMOS or BJT.

HDL was created to implement register transfer-level abstraction, a model of the data flow and timing of a circuit [9].

In electronics, an HDL is a specialized computer language used to describe the structure and behavior of electronic circuits, and most commonly, digital logic circuits. An HDL looks much like a programming language such as C; it is a textual description consisting of expressions, statements, and control structures. One important difference between most programming languages and HDLs is that HDLs explicitly include the notion of time. An HDL enables a precise, formal description of an electronic circuit that allows for the automated analysis and simulation of an electronic circuit. It also allows for the synthesis of an HDL description into a netlist (a specification of physical electronic components and how they are connected together), which can then be placed and routed to produce the set of masks used to create an IC.

Compared with the traditional gate-level description, HDLs form an integral part of EDA systems, especially for complex circuits, such as ASICs, microprocessors, and PLDs. Many HDLs are available, but VHDL and Verilog are by far the most popular ones. Most CAD tools available in the market support these languages. VHDL stands for "very high speed integrated circuit hardware description language." Both VHDL and Verilog are officially endorsed IEEE (Institute of Electrical and Electronics Engineers) standards. Take a 32-bit adder, for example, gate-level description of the adder needs to use 500–1,000 logic gates, whereas VHDL needs only one statement "$A = B + C$."

1. VHDL

In the 1980s, the rapid advances in IC technology necessitated a need to standardize design practices. In 1983, VHDL was developed under the very high speed IC (VHSIC) program of the U.S. Department of Defense, which was originally intended to serve as a language to document descriptions of complex digital circuits. In 1987, IEEE adopted VHDL as the HDL standard (IEEE STD-1076) [10]. It was revised in 1993 as the standard VHDL-93 [11]. VHDL includes multiple design levels, such as system behavior level, register transfer level, and logic gate level, and supports mixed description methods consisting of structural description, data-flow description, and behavioral description. The entire large-scale digital IC design process can be completed with VHDL. VHDL also has several advantages. VHDL has a wide range description capability. This enables the designers to focus on improving the system functions and debugging, instead of emphasizing the physical implementations. VHDL can describe complex control logic with simple codes, which is very flexible and the results can be easily saved and reused. VHDL design does not rely on a specific device. Now VHDL already becomes a standard language, applicable for majority of EDA manufacturers with good portability.

2. Verilog HDL

Verilog HDL and a simulator were released by Gateway Design Automation (GDA) in 1983. In 1989, Cadence acquired GDA, and Verilog HDL became an intellectual property of Cadence. In 1990, Cadence separated the HDL from its simulator (Verilog-XL) and released the HDL into the public domain. Verilog HDL is guarded by the Open-Verilog International Organization, now part of Accellera Organization. In 1995, IEEE adopted Verilog HDL as standard 1364 (Verilog HDL 1364–1995). In 2001, IEEE issued the Verilog HDL standard (Verilog HDL 1364–2001) and revised in 2005 [12, 13].

Many syntaxes of Verilog HDL and C language are similar. For example, Verilog HDL also has an "if-then-else" structure statement, a "for" statement, an "int" variable type, a function, and so forth. However, Verilog HDL is fundamentally an HDL, which is essentially different from the C language. First of all, C language program is executed sequentially, that is, the next statement can't be executed until the execution of the current statement is completed. And Verilog statements are executed in parallel, that is, multiple branches (multiple statements) of the circuit may be executed at the same time. So beginners tend to be confused by concepts, leading to the failure of the design.

Second, the hardware design language has a concept of timing sequence, that is, there is always a delay existing from the input to the output in a hardware circuit. However, this concept doesn't exist in C language. In addition, C language has been used for a long time, and it has perfect compiling and debugging environment and powerful input and output functions. As for Verilog HDL, there are many restrictions, such as rigid syntax rules, poor debugging function, and incomplete error messages. Moreover, the codes of Verilog HDL should be developed with a hardware point of view. This requires that designer must have background knowledge of digital circuits.

3. Comparison between VHDL and Verilog HDL

Both VHDL and Verilog HDL have strong capabilities of hardware description, and each of them has its own characteristics. It is generally believed that Verilog HDL is slightly inferior to VHDL in system-level abstraction and stronger than VHDL in gate-level descriptions. Meanwhile, as Verilog has powerful gate-level description, the bottom layer of the VHDL is supported with device library described by Verilog HDL. It is much easier to learn Verilog HDL than to learn VHDL. This is mainly because Verilog HDL is similar to C language. Therefore, for most designers, the choice of Verilog HDL or VHDL may be more dependent on the habit and work environment. This book will introduce the basic syntax of Verilog HDL and some logic circuits with Verilog HDL description in Chapter 4.

4. Comparison between software language compiler and HDL synthesizer

The HDL program can be synthesized through synthesizers. Then a netlist file is generated, which can be downloaded into devices to realize the corresponding logic circuits. The software language compiles the program into the instruction or data codes of the central processing unit (CPU) through the compiler. HDL synthesizers and software program compilers are nothing more than a "translator" that translates high-level design expressions into low-level expressions, but they have many essential differences. The comparison of compilers and synthesizers is shown in Figure 1.4.1 [6].

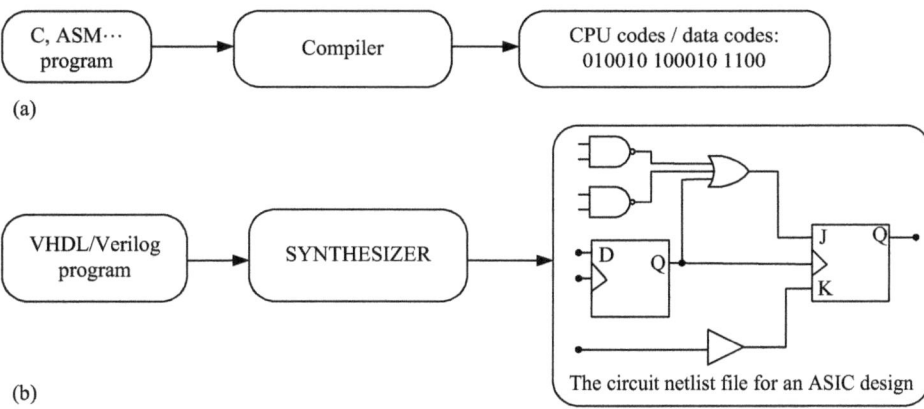

Figure 1.4.1: Comparison between functions of compiler and synthesizer: (a) software language design and (b) hardware description language design.

The compiler translates the software program into machine codes based on a specific CPU, and the CPU executes the corresponding function sequentially. This code is limited to this CPU and cannot be transplanted; the machine codes neither represent the hardware structure nor change the CPU hardware structure. It can only passively be used for the specified hardware circuit.

On the contrary, the "translation" target of a synthesizer is to generate the netlist file for describing the bottom-layer circuit structure. The circuit structure that satisfies the original design function does not depend on any particular hardware environment and therefore can be independently stored and easily transplanted to any common hardware environment, for example, from PLDs to ASICs. In addition, a synthesizer is active and creative in the process of converting the circuit function described with HDL into a specific circuit structure netlist. It doesn't translate mechanically and literally, but choose the optimum way to implement a circuit structure according to the design library, technology library, and various preset constraints. For the same VHDL description, the synthesizer can use different circuit structures to realize the same function.

1.4.3 Design and programming process of PLD

A PLD can be thought of as a "blank slate" on which you implement a specified system design by using a certain process. The implementation of programmable logic requires both hardware and software so that PLDs can be programmed to perform specified logic functions by the designer. The software development package installed on a computer is to implement a circuit design. The computer must be interfaced with a development board or programming fixture containing the device.

PLD manufacturers have their own EDA software package, which integrates all function modules within an EDA environment to facilitate the development of PLDs. The EDA software packages provided by Altera and Xilinx are shown in Figure 1.4.2.

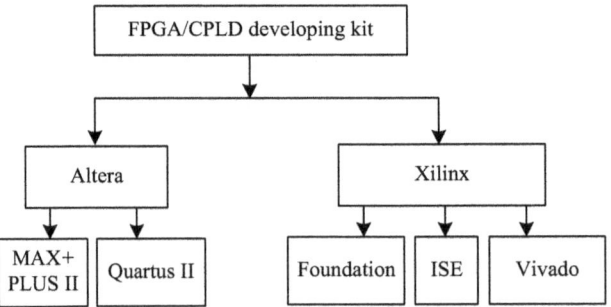

Figure 1.4.2: PLD corporations and their corresponding software packages.

For instance, Altera's MAX+PLUS II and Quartus II design tools include both VHDL and Verilog HDL synthesizers to fully support the design process by using VHDL and Verilog HDL. Quartus II is an upgraded version of MAX+PLUS II for programming the following devices: low-cost FPGA devices (Cyclone series), high-density FPGA devices (Stratix series), structured ASIC devices (Hardcopy series), and CPLD devices (MAX series).

Logic circuit design process implemented by PLDs is shown in Figure 1.4.3. First, according to the selected PLD, download the manufacturer's corresponding design tool; then, input the corresponding designs by using HDL, schematic, or a mixture of HDL and schematic; finally, complete the design of PLD through synthesis, simulation, and downloading.

Figure 1.4.4 shows the Altera development board DE2, whose core PLD is Cyclone II. The design can be downloaded to the PLD of the DE2 board, and the validation of the design can be carried out by inputs and outputs on the board.

The design kits of PLD include Altera's MAX+PLUS II and Quartus II software, Xilinx's foundation and ISE software. With these software products and PLD programming boards, you can program PLD on the board and test the functionality of PLDs. If you do not have PLD programming board, you can use these software products for digital logic circuit design and simulation.

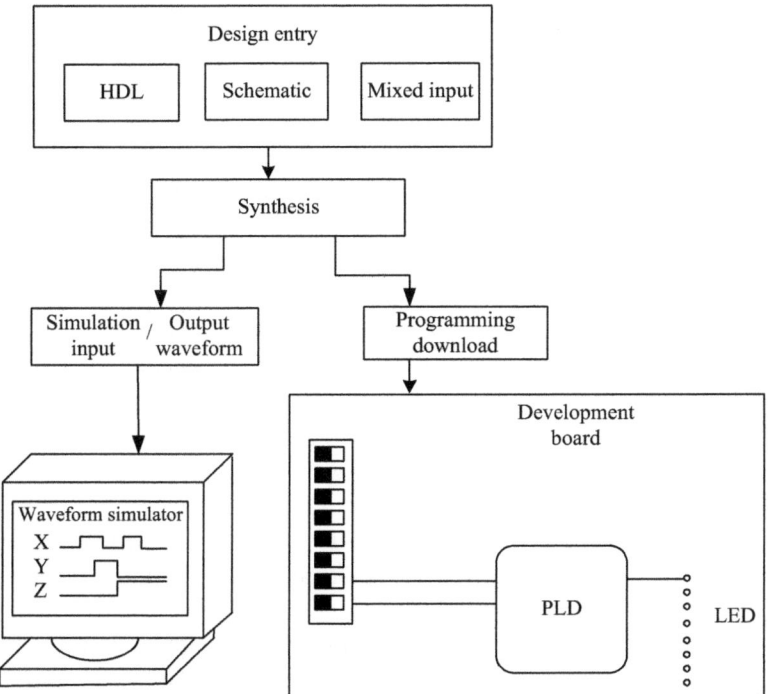

Figure 1.4.3: Design process of using PLD to realize basic logic circuits.

Figure 1.4.4: DE2 development and education board manufactured by Altera.

1.5 Summary

1. Analog and digital quantities are two basic ways of representing the numerical value of physical quantities. An analog quantity is one having continuous values. A digital quantity is one having a discrete set of values.
2. Most digital circuits use a binary system that only has two digits, 1 and 0, which can represent two voltage levels. A binary digit is called a bit.

3. The voltages used to represent a 1 or a 0 are called logic level. In a practical digital circuit, a HIGH level and a LOW level correspond to a certain voltage range. There is no overlap between the accepted range of HIGH levels and LOW level.
4. Digital waveform is one way of representing binary information, which consists of voltage levels that change back and forth between the HIGH and LOW levels.
5. Digital ICs are divided into two broad categories: fixed-functional logic and programmable logic.
6. The fixed-function logic devices have the fixed logic functions that cannot be altered.
7. PLDs are ICs into which logic designs can be programmed.
8. Two types of IC packages are through-hole mounted and surface mounted.
9. According to the complexity, digital ICs can be divided into five categories: SSI, MSI, LSI, VLSI, and ULSI.
10. EDA, also called as ECAD, is a category of software tools that help to design electronic systems with the aid of computers.
11. HDL is a specialized computer language used to describe the structure and behavior of electronic circuits. Currently, VHDL and Verilog HDL are most commonly used HDL for digital circuit design.
12. The implementation of programmable logic requires both hardware and software. The software development package installed on a computer is to implement a circuit design. The computer must be interfaced with a development board or programming fixture containing the device.

Key terms

Analog quantity: The quantity having continuous values
Digital quantity: The quantity having a discrete set of values
Analog signal: A signal that has continuous values
Digital signal: An electrical signal that has a sequence of discrete values
Analog circuit: An electronic circuit involving quantities with continuous values
Digital circuit: An electric circuit based on a number of discrete voltage levels
Bit: A binary digit being either 1 or 0
Code: The combinations of 1s and 0s used to represent numbers, symbols, alphabetic characters, and other types of information
Logic level: The voltage used to represent a 1 or a 0
Digital waveform: A waveform describing voltage level which changes back and forth between the HIGH and LOW level
Clock: The basic timing signal in a digital system; a periodic waveform used to synchronize operation
Pulse: A sudden change from one level to another, followed after a time, by a sudden change back to the original level

Pulse train: A series of pulses

Complexity: The number of equivalent gate circuits on a single chip

Integrated circuit (IC): A type of circuits in which all components are integrated on a single small chip of silicon

Fixed-function logic: A type of digital ICs whose logic functions are set by the manufacturer and cannot be altered

Programmable logic: A type of digital IC that starts as a "blank slate" and into which a logic design can be programmed

Standard integrated circuit: A type of digital ICs that integrate some commonly used logic function circuit into the chips of silicon

Custom-designed integrated circuit: A type of digital ICs that is designed for special applications; also called as ASIC

Electronic design automation (EDA): A category of software tools that help to design electronic systems with the aid of computers; also called as ECAD

Hardware description language (HDL): A specialized computer language used to describe the structure and behavior of electronic circuits, and most commonly, digital logic circuits

2 Number systems and codes

2.1 Introduction

Digital signals are generally represented by a series of binary digits. Binary number system and digital codes are fundamental to digital circuits and digital systems. First, this chapter introduces number systems including binary, decimal, octal, and hexadecimal, and conversion between different number systems. Then binary arithmetic operations with binary numbers are introduced. Finally, digital codes such as binary-coded decimal (BCD), excess-3 code, the Gray code, and the American Standard Code for Information Interchange (ASCII) code are covered.

The objectives of this chapter are to
– Describe binary, decimal, hexadecimal, and octal number system
– Convert number representation between different number systems
– Apply arithmetic operations to binary numbers
– Express signed binary numbers in sign-magnitude, one's complement, two's complement
– Describe BCD, excess-3 code, and Gray code.

2.2 Number systems

*Number system*s are ways to represent the quantities. You are familiar with the decimal number system since you use decimal number in your daily lives. However, a digital circuit and a digital system belong to a two-valued system. A binary number system is the most efficient way to represent quantities. Hexadecimal and octal number systems are used primarily as a compact way of writing binary number since long binary numbers are difficult to read and write. In this section, we will begin with the structure of decimal number system that you are familiar with. This will help you more easily understand the structure of a binary number system, a hexadecimal system, and an octal system. Then, the structure of a binary number system, a hexadecimal system, and an octal system are introduced, because the same quantity can be represented by a decimal number, or an equivalent binary number, hexadecimal number, and octal number. Conversion between different number systems is introduced at the end of this section.

The objectives of this section are to
– Determine the representation of binary, decimal, hexadecimal, and octal number system
– Explain the weighted system and determine the weight of each digit in different systems
– Conversion between different number systems

https://doi.org/10.1515/9783110614916-002

2.2.1 Decimal number system

A *decimal number system* is the most commonly and broadly used number system in our daily life. It contains 10 digits, 0–9, representing a certain quantity. If you express a quantity greater than 9, you can use two or more digits. The position of each digit indicates the magnitude of the quantity [14]. For example, a decimal number, 325, can be expressed by their respective positions as follows:

$$325 = 3 \times \boxed{10^2} + 2 \times \boxed{10^1} + 5 \times \boxed{10^0}$$

$$\text{Weight} \quad \text{Weight} \quad \text{Weight}$$

where the digit 3 represents the quantity of 300, the digit 2 expresses the quantity of 20; and the digit 5 indicates the quantity of five.

The position of each digit in a decimal number represents the magnitude of the quantity assigned a *weight* of 10^i, where 10 is the base of a decimal number and i is the position of each digit relative to the decimal point in a decimal number. For the whole number part, i is a positive number increasing from right to left starting with 0; for fractional numbers part, i is negative number decreasing from left to right starting with −1. The weight structure of the decimal number is illustrated as follows.

$$\text{Weight:} \quad \cdots 10^2 10^1 10^0 . 10^{-1} 10^{-2} 10^{-3} \cdots$$
$$\underset{\text{Decimal point}}{\uparrow}$$
$$\text{Position } i: \quad \cdots 2 \quad 1 \quad 0 \quad -1 \quad -2 \quad -3 \cdots$$

Any decimal number can be represented by the sum of digits after each digit has been multiplied by its weight as shown in eq. (2.2.1).

$$[N]_{10} = K_{n-1} \times 10^{n-1} + \cdots + K_1 \times 10^1 + K_0 \times 10^0 + K_{-1} \times 10^{-1} + \cdots + K_{-m} \times 10^{-m}$$
$$= \sum_{i=-m}^{n-1} K_i \times 10^i \tag{2.2.1}$$

where K_i represents a digit (0–9) in the decimal number, 10^i is the corresponding weight, and the subscript 10 or D illustrates that N is a decimal number.

For instance,

$$(3281)_{10} = 3 \times 10^3 + 2 \times 10^2 + 8 \times 10^1 + 1 \times 10^0$$

$$(209.04)_D = 2 \times 10^2 + 0 \times 10^1 + 9 \times 10^0 + 0 \times 10^{-1} + 4 \times 10^{-2}$$

2.2.2 Binary number system

A *binary number system* is another way to represent quantities. It is fundamental to digital circuit and digital system. Same as the decimal system, the binary system with its two digits (0 and 1) is a *base-two system* [14]. The position of each digit in a binary number represents its weight. The *weight*s in a binary number are based on powers of two represented by 2^i, where 2 is the base of a binary number and i is the position of each digit in the binary number. For the whole number part, i is a positive number increasing for right to left starting with 0; for the fractional number part, i is a negative number, which decreases from left to right starting with –1. The weight structure of the binary number is as follows

$$\text{Weight:} \quad \cdots 2^2 2^1 2^0 . 2^{-1} 2^{-2} 2^{-3} \cdots$$

$$\uparrow \underline{\hspace{4cm}} \text{Binary point}$$

$$\text{Position } i: \quad \cdots 2\ 1\ 0\ \ -1\ -2\ -3 \cdots$$

Any binary number can be represented as a sum of digits after each digit has been multiplied by its weight as shown in eq. (2.2.2):

$$\begin{aligned} [N]_2 &= K_{n-1} \times 2^{n-1} + \cdots + K_1 \times 2^1 + K_0 \times 2^0 + K_{-1} \times 2^{-1} + \cdots + K_{-m} \times 2^{-m} \\ &= \sum_{i=-m}^{n-1} K_i \times 2^i \end{aligned} \tag{2.2.2}$$

where K_i represents a digit (0 or 1) in the binary number, 2^i is the corresponding weight, and the subscript 2 or B illustrates that N is a binary number.

The decimal value of any binary number can be obtained by eq. (2.2.2).

For instance,

$$(101.01)_B = 1 \times 2^2 + 0 \times 2^1 + 1 \times 2^0 + 0 \times 2^{-1} + 1 \times 2^{-2} = (5.25)_{10}$$

From eqs. (2.2.1) and (2.2.2), we can deduce that any arbitrary number in other number systems can be represented as a sum of the digits after each digit has been multiplied by its weight as shown in eq. (2.2.3):

$$[N]_R = \sum_{i=-m}^{n-1} K_i \times R^i \tag{2.2.3}$$

where the definitions of i, m, and n are the same as in eq. (2.2.1); R is the base of the number system; K_i denotes the digit (0 ~ R–1) in the ith position; and R^i is the corresponding weight of K_i.

2.2.3 Hexadecimal and octal number systems

A number represented in a binary number system usually involves many bits, which are difficult to record and memorize. Consequently, hexadecimal and octal numbers are commonly chosen to represent a binary number in a digital system.

A *hexadecimal number system* has 16 digits, 0–9 and A–F, where A–F denotes the decimal numbers 10–15, respectively; 16 is the base of the hexadecimal number system. The weight of each digit in a hexadecimal number is 16^i. Any hexadecimal number can be represented as a sum of the digits after each digit has been multiplied by its weight as shown in eq. (2.2.4). The subscript 16 or H illustrates that N is a hexadecimal number.

$$[N]_{16} = \sum_{i=-m}^{n-1} K_i \times 16^i \tag{2.2.4}$$

The decimal value of any hexadecimal number can be obtained by eq. (2.2.4). For instance,

$$(D8.A)_H = 13 \times 16^1 + 8 \times 16^0 + 10 \times 16^{-1} = (216.625)_{10}$$

An octal number system has eight digits, 0–7. Here, 8 is the base of an octal number, and the weight of each digits is 8^i. Any octal number can be represented as a sum of the digits after each digit has been multiplied by its weight as shown in eq. (2.2.5). The subscript 8 or O illustrates that N is an octal number.

$$[N]_8 = \sum_{i=-m}^{n-1} K_i \times 8^i \tag{2.2.5}$$

The decimal value of any octal number can be obtained by eq. (2.2.5). For instance,

$$(207.04)_O = 2 \times 8^2 + 0 \times 8^1 + 7 \times 8^0 + 0 \times 8^{-1} + 4 \times 8^{-2} = (135.0625)_{10}$$

Table 2.2.1 lists the corresponding relationship between binary, octal, decimal, and hexadecimal numbers. Table 2.2.1 shows that each hexadecimal digit

Table 2.2.1: Equivalent relation of decimal, binary, hexadecimal, and octal numbers.

Decimal	Binary	Octal	Hexadecimal	Decimal	Binary	Octal	Hexadecimal
0	0000	00	0	8	1000	10	8
1	0001	01	1	9	1001	11	9
2	0010	02	2	10	1010	12	A
3	0011	03	3	11	1011	13	B
4	0100	04	4	12	1100	14	C
5	0101	05	5	13	1101	15	D
6	0110	06	6	14	1110	16	E
7	0111	07	7	15	1111	17	F

corresponds to a four-bit binary number and each octal digit is corresponding to a three-bit binary number.

2.3 Conversion between number systems

The conversion between number systems is to transform a number from one number system to another without changing its value. Essentially, the conversion process is the conversion of the weights between different number systems. This section introduces the conversion between different number systems.

The objectives of this section are to
- Convert from a decimal number to binary and from binary to decimal
- Covert from a binary number to hexadecimal and from hexadecimal to binary
- Covert from a binary number to octal and from octal to binary

2.3.1 Conversion between decimal and binary

1. Conversion from binary to decimal
As mentioned in Section 2.2, the conversion from a binary number to a decimal number can be achieved directly by applying eq. (2.2.2).

Example 2.1 Convert the binary number $(1011.01)_2$ to decimal.

Solution

$$(1011.01)_2 = 1 \times 2^3 + 0 \times 2^2 + 1 \times 2^1 + 1 \times 2^0 + 0 \times 2^{-1} + 1 \times 2^{-2} = (11.25)_{10}$$

---

2. Conversion from decimal to binary
The process of converting a decimal number to its equivalent binary number can be divided into whole number conversion and fractional number conversion, respectively.

(1) Whole number conversion
A systematic method of converting a whole number from decimal to binary is the *repeated division-by-2* process.

For a binary whole number, its equivalent decimal number is deduced by

$$[N]_{10} = b_n \times 2^n + b_{n-1} \times 2^{n-1} + \cdots + b_1 \times 2^1 + b_0 \times 2^0 \qquad (2.3.1)$$

where b_n is the digit at the nth position with respect to the binary point, and 2^n is the corresponding weight of b_n.

If both sides of eq. (2.3.1) are divided by the base of 2, the result is

$$\frac{1}{2}[N]_{10} = b_n \times 2^{n-1} + b_{n-1} \times 2^{n-2} + \cdots + b_1 \times 2^0 + \frac{b_0}{2} \qquad (2.3.2)$$

The remainder is b_0, which is the least significant bit (LSB) in the binary number, and the quotient is

$$b_n \times 2^{n-1} + b_{n-1} \times 2^{n-2} + \cdots + b_1 \qquad (2.3.3)$$

If you divide the quotient by the base of 2, the following result is

$$b_n \times 2^{n-2} + b_{n-1} \times 2^{n-3} + \cdots + b_2 + \frac{b_1}{2} \qquad (2.3.4)$$

The remainder is b_1 and the quotient is

$$b_n \times 2^{n-2} + b_{n-1} \times 2^{n-3} + \cdots + b_2 \qquad (2.3.5)$$

The rest can be done in the same manner. Until the quotient is 0, the last reminder is the most significant bit (MSB) of the binary number.

Therefore, the process of converting a decimal number to its equivalent binary can be implemented by the *repeated division-by-2* method. *First, you begin with dividing the decimal number by 2 and the first remainder to be produced is the LSB in the binary number. Then divide each resulting quotient by 2 until the quotient is 0. The last remainder to be produced is the MSB in the binary number.* The procedure is illustrated by the following example.

Consequently, if we divide eq. (2.3.4) by R, b_1 can also be obtained as the remainder of the division. Similarly, the coefficients b_2–b_n can be obtained successively by continuous division by the base R and keeping the corresponding remainders of each division until the quotient of the division is zero. Thus, the conversion of the integer part of a decimal number into the integer part of any number system is achieved.

(2) Fractional part

For the fractional part of any number in a binary, octal, or hexadecimal system, the fractional part of its equivalent decimal number is obtained by

$$[N]_{10} = b_{-1} \times R^{-1} + b_{-2} \times R^{-2} + \cdots + b_{-(m-1)} \times R^{-(m-1)} + b_{-m} \times R^{-m} \qquad (2.3.6)$$

If both sides of the equation are multiplied by the base R (R can be 2, 8, 16, and any other number), we can get

$$R \times [N]_{10} = b_{-1} + b_{-2} \times R^{-1} + \cdots + b_{-(m-1)} \times R^{-(m-2)} + b_{-m} \times R^{-(m-1)} \qquad (2.3.7)$$

where b_{-1} is the integer part of the product. Likewise, if eq. (2.3.7) is continuously multiplied by the base R, then the coefficients b_{-2}, b_{-3}, ..., b_{-m} can be obtained step-by-step. This multiplication process would not be ceased until the rest of the fractional part is zero or the required precision is reached.

Example 2.2 Convert a decimal number $(342.6875)_{10}$ into its equivalent binary, octal, and hexadecimal numbers.

Solution

2	342		
2	171	$\cdots$	0
2	85	$\cdots$	1
2	42	$\cdots$	1
2	21	$\cdots$	0
2	10	$\cdots$	1
2	5	$\cdots$	0
2	2	$\cdots$	1
2	1	$\cdots$	0
	0	$\cdots$	1

8	342		
8	42	$\cdots$	6
8	5	$\cdots$	2
	0	$\cdots$	5

16	342		
16	21	$\cdots$	6
16	1	$\cdots$	5
	0	$\cdots$	1

The integer part is: $(342)_{10} = (101010110)_2 = (526)_8 = (156)_{16}$
The fractional part is: $(0.6875)_{10} = (0.1011)_2 = (0.54)_8 = (0.B)_{16}$
Thus, $(342.6875)_{10} = (101010110.1011)_2 = (526.54)_8 = (156.B)_{16}$.

2.3.2 Conversion between binary and hexadecimal

1. Conversion from binary system into hexadecimal system

A four-bit binary number can represent 16 different decimal values from 0 to 15. A one-bit hexadecimal number also has 16 different digits (0–9 and A–F), corresponding to 16 different decimal values too. Thus, a one-bit hexadecimal number can be represented by a four-bit binary number. The conversion from binary to hexadecimal typically begins with grouping the bits into sets of four starting at the radix point (starting at the LSB for integer part, while the MSB for the fractional part), adding zeros as needed to fill out the groups. Then, assign to each group with the equivalent hexadecimal digit.

Example 2.3 Convert a binary number $(10110100111100.01001)_2$ into its corresponding hexadecimal number.

Solution

$$(0010 \quad 1101 \quad 0011 \quad 1100 \,.\, 0100 \quad 1000)_2$$
$$\downarrow \qquad \downarrow \qquad \downarrow \qquad \downarrow \qquad \downarrow \qquad \downarrow$$
$$2 \qquad D \qquad 3 \qquad C \,.\, 4 \qquad 8$$

Thus, $(10110100111100.01001)_2 = (2D3C.48)_{16}$.

---

2. Conversion from hexadecimal system into binary system

Every bit of a hexadecimal number can be represented by a four-bit binary number. Hence, the conversion from a hexadecimal number into its equivalent binary number is typically realized by converting bit-by-bit the hexadecimal number using the reverse process shown in Example 2.3.

Example 2.4 Convert a hexadecimal number $(4FB.CA)_{16}$ into its corresponding binary number.

Solution

$$\begin{array}{ccccc} (4 & F & B \,. & C & A)_{16} \\ \downarrow & \downarrow & \downarrow & \downarrow & \downarrow \\ 0100 & 1111 & 1011 \,.\, & 1100 & 1010 \end{array}$$

Thus, $(4FB.CA)_{16} = (010011111011.11001010)_2$.

2.3.3 Conversion between binary and octal

The conversion between a binary and octal system is similar to the conversion between binary and hexadecimal system.

1. Conversion from binary system into octal system

Similarly, the conversion from binary into octal system typically begins with grouping the bits into sets of three starting at the radix point (starting at the LSB for the integer part while the MSB for the fractional part), adding zeros as needed to fill out the groups. Then, assign to each group the corresponding equivalent octal digit.

Example 2.5 Convert a binary number $(1111010010.01)_2$ into its corresponding octal number.

Solution

$$\begin{array}{ccccc} (001 & 111 & 010 & 010 \,. & 010)_2 \\ \downarrow & \downarrow & \downarrow & \downarrow & \downarrow \\ 1 & 7 & 2 & 2 \,. & 2 \end{array}$$

Thus, $(1111010010.01)_2 = (1722.2)_8$.

---

2. Conversion from octal system into binary system

Likewise, the conversion: from an octal number to its equivalent binary number is typically realized by bit-by-bit converting the number using the reverse process shown in Example 2.5.

Example 2.6 Convert an octal number $(6407.2)_8$ into its corresponding decimal number.

Solution

$$(6 \quad 4 \quad 0 \quad 7 \; . \quad 2)_8$$
$$\downarrow \quad \downarrow \quad \downarrow \quad \downarrow \quad \quad \downarrow$$
$$110 \quad 100 \quad 000 \quad 111 \; . \quad 010$$

Thus, $(6407.2)_8 = (110100000111.010)_2$.

2.4 Binary arithmetic operations

2.4.1 Basic arithmetic operations

Basic arithmetic operations of binary numbers are performed primarily in the same manner as the arithmetic operations of decimal numbers [15]. For the addition of two decimal numbers, a carry of 1 is produced when the sum is 10, while for the addition of two binary numbers, a carry of 1 is produced when the sum is 2. Likewise, for the subtraction of two decimal numbers, a borrow of 1 represents 10, while for binary numbers, a borrow of 1 represents 2.

The four cases of the addition of two binary digits in any position are as follows:

$$
\begin{array}{cccc}
0 & 0 & 1 & 1 \\
+0 & +1 & +0 & +1 \\
\hline
0 & 1 & 1 & 10
\end{array}
$$

where the first three cases are the same as the addition of decimal digits and the last case produces a carry of 1.

The four cases of the subtraction of two binary digits in any position are as follows:

$$
\begin{array}{cccc}
0 & 0 & 1 & 10 \\
-0 & -1 & -0 & -1 \\
\hline
0 & 1 & 1 & 01
\end{array}
$$

where the first three cases are the same as the subtraction of decimal digits and the last case produces a borrow of 1. As negative numbers cannot be represented by unsigned numbers, the minuend should be greater than the subtrahend.

Likewise, the multiplication and division of binary numbers are the same as the multiplication and division of decimal numbers except that only "0" and "1" are involved in the operations.

Example 2.7 Suppose $X = (1100)_2$ and $Y = (0101)_2$, calculate $X + Y$, $X - Y$, $X \times Y$, and $X \div Y$.

Solution

$$
\begin{array}{r}
1100 \\
\times\ 0101 \\
\hline
1100 \\
0000 \\
1100 \\
\hline
111100
\end{array}
$$

$$
\begin{array}{r}
1100 \\
+\ 0101 \\
\hline
10001
\end{array}
\qquad
\begin{array}{r}
1100 \\
-\ 0101 \\
\hline
0111
\end{array}
$$

$$
\begin{array}{r}
10.011\cdots \\
101\overline{)1100} \\
101 \\
\hline
001000 \\
101 \\
\hline
0110 \\
101 \\
\hline
01\cdots
\end{array}
$$

$$X + Y = (1100)_2 + (0101)_2 = (10001)_2$$
$$X - Y = (1100)_2 - (0101)_2 = (0111)_2$$
$$X \times Y = (1100)_2 \times (0101)_2 = (111100)_2$$
$$X \div Y = (1100)_2 \div (0101)_2 = (10.011)_2$$

From this example, we can find that the multiplication of two binary numbers can be realized by the combination of shifting and addition operations and the division can be realized by the combination of shifting and subtraction operations.

2.4.2 Representation of signed binary numbers

Numbers discussed above are unsigned numbers. For signed numbers, three forms of representation, including sign-magnitude representation, one's complement representation, and two's complement representation, are generally used. In sign-magnitude representation scheme, the left-most position is a sign bit (0 denotes positive and 1 denotes negative), and the rest of the bits indicate the magnitude or absolute value of the number. This representation scheme is straightforward, but its circuit implementation is more complex than the other two forms of representation. Negative numbers are typically represented by one's complement and two's complement forms, and two's complement representation is more widely utilized in the arithmetic operations of binary numbers in computers.

1. Two's complement representation

For an n-bit binary number whose true form is N, its two's complement form is defined as

$$[N]_{\text{two's complement}} = 2^n - N$$

For instance, if $N = 1001$, then $[1001]_{\text{two's complement}} = 2^4 - 1001 = 10000 - 1001 = 0111$

2. One's complement representation

For an n-bit binary number whose true form is N, its one's complement form is defined as

$$[N]_{\text{one's complement}} = (2^n - 1) - N$$

For instance, if $N = 1001$, then $[1001]_{\text{one's complement}} = (2^4-1)-1001 = 1111-1001 = 0110$

From earlier examples, we can find that one's complement of a number can be directly obtained by changing each 0 to a 1 and each 1 to a 0 bit by bit. In addition, from the definitions of one's complement and two's complement representations, the two's complement of a number can be derived by adding a 1 to its one's complement.

2.5 Codes

Coding is a process of using a specific group of characters, alphabets, symbols, or numbers to represent specific information. The digital data are represented, stored, and transmitted as a group of binary bits. This group of binary bits is also called as the *binary code*. The binary codes represent numbers as well as alphanumeric letters. They are widely used for analyzing and designing digital circuit since only 0 and 1 are being used, which can be implemented easily. Many specialized codes are used in a digital system. The codes introduced in this section are BCD code, Gray code, and ASCII code.

The objectives of this section are to
– Express decimal number in BCD
– Convert the representation between decimal number and BCD
– Explain the advantage of Gray code
– Explain the ASCII code

2.5.1 Binary-coded-decimal

BCD is a way to express each of decimal digits with a binary code[16]. The BCD code presented decimal digits, which are often used in daily life, provides an excellent interface to binary system. In the BCD, each decimal digit is represented by a four-bit binary number. Totally they can represent 16 numbers (0000–1111). But in BCD code, only 10 of these are used. The remaining six code combinations are invalid in BCD. The commonly used BCD codes include 8421BCD code, 5421BCD code, 2421BCD code, and excess-3 code, as listed in Table 2.5.1. Some codes in Table 2.5.1, such as 8421, 5421, and 2421 codes, are weighted codes, while some codes, such as excess-3 code, are nonweighted codes. In a weighted code, each digit position has a weight and the sum of all digits multiplied by their weight represents its corresponding decimal number. In a nonweighted code, no specific weights are assigned to bit position [17].

Table 2.5.1: Commonly used BCD codes.

Decimal	8421 BCD	5421 BCD	2421 BCD	BCD2421 *	Excess-3 code
0	0 0 0 0	0 0 0 0	0 0 0 0	0 0 0 0	0 0 1 1
1	0 0 0 1	0 0 0 1	0 0 0 1	0 0 0 1	0 1 0 0
2	0 0 1 0	0 0 1 0	0 0 1 0	0 0 1 0	0 1 0 1
3	0 0 1 1	0 0 1 1	0 0 1 1	0 0 1 1	0 1 1 0
4	0 1 0 0	0 1 0 0	0 1 0 0	0 1 0 0	0 1 1 1
5	0 1 0 1	1 0 0 0	0 1 0 1	1 0 1 1	1 0 0 0
6	0 1 1 0	1 0 0 1	0 1 1 0	1 1 0 0	1 0 0 1
7	0 1 1 1	1 0 1 0	0 1 1 1	1 1 0 1	1 0 1 0
8	1 0 0 0	1 0 1 1	1 1 1 0	1 1 1 0	1 0 1 1
9	1 0 0 1	1 1 0 0	1 1 1 1	1 1 1 1	1 1 0 0

1. The 8421 BCD code

The *8421 BCD code* is the most commonly used BCD code. It only uses the first ten binary numbers (0000–1001) to represent ten decimal digits (0–9), respectively; the remaining six numbers (1010–1111) are invalid in the 8421 BCD code. The 8421 BCD code belongs to the weighted code and its weights from the MSB to the LSB are $2^3(8)$, $2^2(4)$, $2^1(2)$, and $2^0(1)$. The designation 8421 refers to the binary weight of the four bits $(2^3, 2^2, 2^1, 2^0)$. The 8421 code is the predominant BCD, and when we refer to BCD, we always mean the 8421 BCD code unless otherwise stated.

2. The 5421 and 2421 BCD codes

The *5421* and *2421 BCD codes* are weighted codes too, and their weights from MSB to LSB are 5, 4, 2, 1 and 2, 4, 2, 1, respectively. For the 5421 and 2421 BCD codes, one decimal digit may be represented by different binary numbers. For instance, 5 can be represented by either 1011 or 0101 in the 2421 BCD code; likewise, 5 can be represented by either 1000 or 0101 in the 5421 BCD code. However, the 5421 and 2421 BCD codes listed in Table 2.5.1 have been generally accepted, and other forms are no longer used.

In addition, it can be observed that in BCD2421*, the code for decimal 0 is the complement of the code for decimal 9; this also holds true for the codes for decimal 1 and 8, 2 and 7, 3 and 6, and 4 and 5. This property is called the nine's complement of a decimal number, that is, bitwise complementation of a code will produce the nine's complement of the decimal number, which makes hardware implementation of arithmetic operations much simpler in digital systems.

3. The excess-3 Code

The *excess-3 code* is a nonweighted code used to express decimal numbers. The Excess-3 code is derived from the 8421 BCD code adding $(0011)_2$ or $(3)_{10}$ to each code in 8421 BCD. When the addition of two excess-3 codes produces a carry, the carry signal can be directly obtained from the MSB. In addition, excess-3 code also has the property of the nine's complement of the decimal number, and has been commonly used in the arithmetic operation circuitry of BCD codes.

There are only ten codes in the BCD system, and so it is very easy to convert between decimal number and BCD. To convert any decimal number in BCD, simply replace each decimal digit with the corresponding four-bit binary code. To convert a BCD number to a decimal, you can break the code into groups of four bits, starting from the LSB, and then write the corresponding decimal digit represented by each four-bit group.

Example 2.8 Covert the following decimal numbers to the 8421 BCD codes and the excess-3 codes, respectively.

(a) 15   (b) 276

Solution

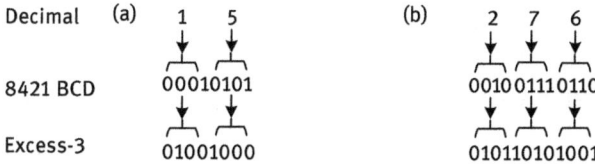

Example 2.9 Covert the following the 8421 BCD codes to decimal numbers.

(a) 10010100   (b) 000110000110

Solution

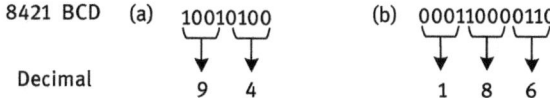

2.5.2 Gray code

The Gray code is the nonweighted code and it is not arithmetic code. This means that there are no specific weights assigned to the bit position. It has a very special feature that only one bit will change each time the decimal number is incremented as shown in Table 2.5.2. As only one bit changes at a time, the Gray code is called as a *unit distance code* and is also known as a *cyclic code*. The Gray code cannot be used for arithmetic operation.

Table 2.5.2: Four-bit Gray code.

Decimal	Binary	Gray code	Decimal	Binary	Gray code
0	0000	0000	8	1000	1100
1	0001	0001	9	1001	1101
2	0010	0011	10	1010	1111
3	0011	0010	11	1011	1110
4	0100	0110	12	1100	1010
5	0101	0111	13	1101	1011
6	0110	0101	14	1110	1001
7	0111	0100	15	1111	1000

The advantage of using a Gray code is that transient errors can be avoided. During the transition process from one code number to a new one, the switching speed of each bit may be different from the others. For instance, the transition process from binary code 0111 to 1000 can experience the following transient states: 0111→0011→1010→1000. In fact, 0011 and 1010 are transient states that are not suppose to appear and are called as the transient errors. These transient errors should be avoided in the digital systems. Table 2.5.2 lists four-bit Gray code, four-bit binary code, and its corresponding decimal numbers.

2.5.3 Alphanumeric codes

A binary digit or bit can represent only two symbols as it has only two states "0" or "1." However, this is not enough to convey information because you need not only numbers but also letters and other symbols for communication. These symbols are required to represent 26 alphabets with capital and small letters, numbers from 0 to 9, punctuation marks, and other symbols. The *alphanumeric codes* are the codes that represent numbers and alphabetic characters. Generally such codes also represent other characters such as symbols and various instructions necessary for conveying information. The commonly used alphanumeric codes include ASCII, International Standard Organization code, Chinese national standard code, and so on.

The *ASCII code* is the most common alphanumeric code, which is universally adopted in computers and electronic devices. Most computer keyboards are standardized with ASCII. When a letter, a number, a symbol, or a command is inputted via a keyboard, its corresponding ASCII code will be recognized by the computer. The ASCII code has 128 characters and symbols represented by a seven-bit binary code. The ASCII codes for some typical symbols are listed in Table 2.5.3. The first column of Table 2.5.3 lists controlling symbols for transferring information between the computer and its peripheral equipment, and cannot be printed out or displayed. Other symbols including digits, letters, and common symbols can be printed out and displayed.

Table 2.5.3: ASCII code for some symbols.

Name	ASCII	Symbol	ASCII	Symbol	ASCII	Symbol	ASCII
NUL	0000000	!	0100001	0	0110000	A	1000001
SOH	0000001	"	0100010	1	0110001	B	1000010
STX	0000010	#	0100011	2	0110010	C	1000011
ETX	0000011	$	0100100	3	0110011	D	1000100
EOT	0000100	%	0100101	4	0110100	E	1000101
ENQ	0000101	&	0100110	5	0110101	F	1000110
ACK	0000110	,	0100111	6	0110110	G	1000111
BEL	0000111	(	0101000	7	0110111	H	1001000
BS	0001000	)	0101001	8	0111000	I	1001001
HT	0001001	*	0101010	9	0111001	J	1001010

2.6 Summary

1. A digital circuit and digital system belong to a two-valued system. A binary number system is the most commonly used number system, with the base of 2 and the weight of 2^i.
2. The hexadecimal number system is used in digital systems and computers as an efficient way of representing binary quantities. It has 16 numbers and characters, 0–9 followed by A–F, with the base of 16 and the weight of 16^i.
3. The octal number system is also an efficient way of representing binary quantities. It has eight numbers, 0–7, with the base of 8 and the weight of 8^i.
4. A decimal whole number can be converted to binary by using the repeated division-by-2 method and a decimal fraction can be converted to binary by using the repeated multiplication-by-2 method.
5. A decimal whole number can be converted to octal and hexadecimal by using the repeated division-by-base method and a decimal fraction can be converted to octal and hexadecimal by using the repeated multiplication-by-base method.
6. Hexadecimal-to-binary conversion is completed by simply replacing each hexadecimal digit with the corresponding four-bit binary number. The process is reversed for binary-to-hexadecimal conversion.
7. For the conversion between octal and binary, each octal digit corresponds to three bits.
8. Arithmetic operations of binary numbers include addition, subtraction, multiplication, and division.
9. The basic rules for binary addition are as follows:

$$
\begin{array}{cccc}
0 & 0 & 1 & 1 \\
+\,0 & +\,1 & +\,0 & +\,1 \\
\hline
0 & 1 & 1 & 10
\end{array}
$$

10. The basic rules for binary subtraction are as follows:

$$
\begin{array}{cccc}
0 & 0 & 1 & 10 \\
-\,0 & -\,1 & -\,0 & -\,1 \\
\hline
0 & 1 & 1 & 01
\end{array}
$$

11. The multiplication of two binary numbers can be realized by the combination of shifting and addition operations and the division can be realized by the combination of shifting and subtraction operations.

12. For signed binary numbers, three forms of representation, including sign-magnitude representation, one's complement representation, and two's complement representation, are generally used. The left-most position is a sign bit (0 denotes positive and 1 denotes negative).

13. The one's complement of a binary number can be directly obtained by changing each 0 to a 1 and each 1 to a 0 bit by bit.

14. The two's complement of a binary number can be derived by adding a 1 to its one's complement.

15. In digital system, binary subtraction is usually accomplished with addition by using one's and two's complement method.

16. The BCD code for a decimal number is formed by converting each digit of the decimal number to its equivalent four-bit binary number. The most commonly used BCD code is the 8421 BCD code that is a weighted code.

17. The Gray code, belonging to a nonweighted code, adopts a sequence of bit patterns in which only one bit changes between successive patterns in the sequence, which can efficiently avoid transient errors.

18. An alphanumeric code is the one that uses groups of bits to represent all of the various characters and functions that are part of a typical computer's keyboard. The ASCII code is the most widely used alphanumeric code.

Key terms

Alphanumeric code: Codes that represent numbers, letters, and alphabetic characters.

ASCII: American Standard Code for Information Interchange.

BCD: Binary-coded decimal; a digital code in which each decimal digit, 0–9, is represented by a group of four-bit binary number.

Gray code: A nonweighted code in which only one bit will change each time the decimal number is incremented; also called as a unit distance code or a cyclic code.

Hexadecimal: A number system with a base of 16.

Octal: A number system with a base of 8.

Self-test

2.1 The equivalent decimal number of binary number 10111011 is _____.
(a) 107   (b) 187   (c) 287   (d) 256

2.2 The equivalent binary number of decimal number 181 is _____.
(a) 11000101   (b) 10110110   (c) 10110101   (d) 11000110

2.3 The sum of 110101+011001 is _____.
(a) 1001110   (b) 1001101   (c) 1010010   (d) 1001100

2.4 The difference of 110101−011001 is _____.
(a) 011000   (b) 011001   (c) 010111   (d) 011100

2.5 The one's complement of a signed number 10110010 is _____.
(a) 10110011   (b) 11001101   (c) 01001101   (d) 01001110.

2.6 The two's complement of a signed number 11010111 is _____.
(a) 00101000   (b) 10101001   (c) 11011000   (d) 00101100

2.7 The 8421 BCD code of decimal number 437 is _____.
(a) 010000110111   (b) 110001110011   (c) 111011010   (d) 010000111111

2.8 The two's complement of a decimal number +118 is _____(assuming there are eight bits in the corresponding binary number)
(a) 00001010   (b) 00001001   (c) 01110110   (d) 10001010

2.9 The two's complement of a decimal number −34 is _____(assuming there are eight bits in the corresponding binary number)
(a) 10100010   (b) 11011101   (c) 00100010   (d) 11011110

2.10 Which one in the following numbers is not an excess-3 code?
(a) 0000   (b) 1000   (c) 0111   (d) 1010

2.11 The equivalent binary number of $(F7A9)_{16}$ is _____.
(a) 1111111101011001   (b) 1110111110101001
(c) 1111011110101001   (d) 1110011010101001

2.12 Which one in the following codes is the 5421 BCD code?
(a) 0000, 0001, 0010, 0011, 0100, 1000, 1001, 1010, 1011, 1100
(b) 0000, 0001, 0010, 0011, 0100, 0101, 0110, 0111, 1000, 1001
(c) 0000, 0001, 0010, 0011, 0100, 0101, 0110, 0111, 1110, 1111
(d) 0011, 0100, 0101, 0110, 0111, 1000, 1001, 1010, 1011, 1100

Problems

2.1 What are the weights of digit 3 in the following decimal numbers?
(a) 325   (b) 4513   (c) 32658   (d) 236

2.2 Expand the following binary numbers in the form of sum of weights.
(a) 10010   (b) 110   (c) 1011001   (d) 11010100

2.3 Convert the following binary numbers into decimal numbers.
(a) 100   (b) 1011   (c) 0.1001   (d) 101101.011

2.4 Convert the following decimal numbers into binary numbers.
(a) 28   (b) 422   (c) 0.32   (d) 0.246

2.5 Convert the decimal numbers shown in Problem 2.1 into octal and hexadecimal numbers.

2.6 Convert each group of the following decimal numbers to binary and finish the arithmetic operation by using binary numbers.
(a) 21+18   (b) 54−23   (c) 32×11   (d) 18÷3

2.7 Obtain the sign-magnitude, the one's complement, and the two's complement representations of the following decimal numbers.
(a) +43   (b) −126   (c) +10   (d) −38

2.8 What are two ways of represent zero in one's complement form?

2.9 How is zero represented in two's complement form?

2.10 Implement the following operations using two's complement form.
(a) 00100011−00010010   (b) 00001100−00100000
(c) 01111100−01000011   (d) 00010000−00100000

2.11 Convert the following 5421 BCD codes into their equivalent decimal numbers, binary numbers, 8421 BCD codes, and excess-3 codes.
(a) 1001 0011   (b) 1100 1010 0001
(c) 0011 1000 1001   (d) 1011 0010.0100

2.12 Convert the following decimal numbers into their equivalent 8421 BCD codes and excess-3 codes.
(a) 76   (b) 175   (c) 2446   (d) 372

3 Boolean algebra and logic simplification

3.1 Introduction

Digital system belongs to a two-valued algebraic system which was invented by an Irish logician and mathematician, George Boole, in 1854 [18]. Now a two-valued algebraic system is usually called Boolean algebra or logical algebra. Boolean algebra is to formulate logic statements with symbols so that problems can be written and solved in a manner similar to ordinary algebra. Shannon was the first person to apply Boole's work to analyze and design logic circuit in 1938. Today, Boolean algebra has already become a convenient and systematic way of analyzing and designing digital circuits. Boolean algebra involves a large amount of content. In this chapter, we only introduce the limited content in Boolean algebra required for digital circuits. This chapter covers the laws, the rules, and the theorems of Boolean algebra for the requirement of digital circuits. You will learn how to simplify logic expressions using the methods of Boolean algebra and Karnaugh map.

The objectives of this chapter are to
– Apply basic laws, rules, and theorems of Boolean algebra
– Define AND, OR, NOT operation
– Describe logic function of three basic logic gates including inverter, AND gate, OR gate
– Explain the combinational operation NAND, NOR, XOR, or NXOR
– Explain standard sum-of-product expression and standard product-of-sum expression
– Simplify Boolean expressions using Boolean algebra
– Apply Karnaugh map to simplify logic expression

3.2 Boolean operations and logic gates

Boolean algebra is a mathematics tool for digital circuit and system. This section mainly introduces basic knowledge of Boolean algebra, several commonly used logical operations and logic gates. Logic symbols used to represent the logic gates are in accordance with the ANSI/IEEE Standard 91-1984 in the whole book [19].

The objectives of this section are to
– Define variable, complement, and literal
– Describe the operation of inverter, AND gate, and OR gate
– Describe the operation of NAND gate and NOR gate
– Describe the operation of exclusive-OR and exclusive-NOR gate
– Recognize the distinctive shape logic gate symbols and the rectangular outline logic symbols of the ANSI/IEEE Standard 91-1984

https://doi.org/10.1515/9783110614916-003

- Apply commutative laws, associative laws and distributive laws
- Apply rules of Boolean algebra
- Explain DeMorgan's theorems

3.2.1 Boolean variable and Boolean constant

Boolean algebra uses variables and operators to describe a logic circuit. *Variable,* *complement,* and *literal* are terms used in Boolean algebra. A *variable* in Boolean algebra is a symbol used to represent a logical quantity. Similar to ordinary algebra, it can usually be expressed by an italic uppercase letter, for example, A, B or C. Any single variable can have one of the two possible values, 0 or 1. These two values represent two possible or opposite conditions, for example, true or false of one event, off or on of a switch, high or low of voltage level, etc. Boolean algebra has only two Boolean constants, 0 and 1. A *complement* is the inverse of a variable and is indicated by a bar over the variable (overbar). For example, the complement of the variable A is $\bar{A}$. If $A = 0$, then $\bar{A} = 1$; if $A = 1$, then $\bar{A} = 0$. The complement of the variable A is read as "not A" or "A bar". Sometimes other symbols rather than an overbar are used to denote the complement of a variable; for example, B' indicates the complement of B. In this book, only the overbar is used. A *literal* refers to a variable or the complement of a variable.

3.2.2 Basic logic operations

Basically, logic is the realm of human reasoning that tells you a certain proposition is true if certain conditions are true [20]. Propositions can be classified as true or false. Several propositions, when combined, form propositional or logic functions. For example, the propositional statement "the light is on" will be true if both "the bulb is not burned out" and "the switch is on" are true. Therefore, this logical statement can be made: *the light is on only if the bulb is not burned out and the switch is on.* In this example, the first statement is true only if the last two statements are true. The first statement ("the light is on") is the basic proposition, and the other two statements are the conditions on which the proposition depends. The conditions correspond to the input variables and the proposition is the output variable. Since such functions are true/false or yes/no statements, digital circuits with their two-state characteristics are applicable.

The term *logic* is applied to digital circuits used to implement logic functions. Several kinds of digital logic circuits are the basic elements which form the building blocks for complex digital systems. We will introduce these elements and discuss their functions in the following sections.

1. AND

Figure 3.2.1 (a) shows a switch circuit. Variables A and B represent the states of the switches and variable F describes the states of the light. The function of this circuit is that *the light is on only if switches A and B are both on.* The statement ("the light is on") is a basic proposition, and the states of the two switches are the conditions on which the proposition depends. "The light is on" will be true only if "both of the two switches are on" is true. That is to say, the proposition is true only if all conditions are true. This kind of logic function is called *AND operation*. If "the switch is on" is 1, then "the switch is off" is 0. If "the light is on" is 1, then "the light is off" is 0. The AND operation produces a HIGH (1) output only when all inputs are HIGH (1) and a LOW (0) when any or all inputs are LOW (0). A circuit that performs an AND operation is called an *AND gate*.

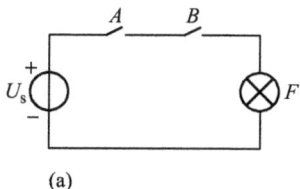

Inputs		Output
A	B	F
0	0	0
0	1	0
1	0	0
1	1	1

(a)   (b)

Figure 3.2.1: A switch circuit: (a) circuit diagram; (b) truth table.

The logical operations of an AND gate can be expressed with *a truth table* that lists all input combinations with the corresponding output, as illustrated in Figure 3.2.1 (b) for a 2-input AND gate. The inputs in the truth table can be expanded. **For an AND gate, regardless of the number of inputs, the output is HIGH only when all inputs are HIGH.**

Standard logic symbols for a 2-input AND gate are shown in Figure 3.2.2. Figure 3.2.2 (a) shows the distinctive shape symbol and Figure 3.2.2 (b) shows the rectangular outline symbol [19]. The inputs are on the left and the output are on the right in the standard logic symbols. In this textbook, distinctive shape symbols are generally used. However, the rectangular outline symbols are often found in many industry publications, thus you should be familiar with them as well.

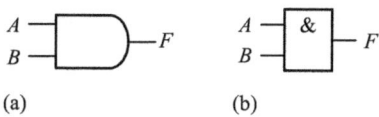

(a)   (b)

Figure 3.2.2: Standard logic symbol for 2-input AND gate: (a) distinctive shape; (b) rectangular outline with the AND (&) qualifying symbol.

The operation of a 2-input AND gate can be described by a logic expression too. If the input variable are A and B, and the output variable is F, then its Boolean expression is

$$F = AB$$

To extend the AND expression to more than two input variables, simply add a new letter representing each input variable. For example, the logic function of a 3-input AND gate can be expressed by $F = ABC$ where C represents the third input variable.

In a majority of applications, the inputs to a gate are not stationary voltage levels but voltage waveforms that change frequently between HIGH and LOW logic level. *Timing diagram* is basically a graph that accurately displays the relationship of two or more waveforms with respect to each other on a time basis. Now let's look at the operation of AND gates with pulse waveform inputs in Figure 3.2.3, keeping in mind that an AND gate obeys the truth table operation regardless of whether its inputs are constant levels or levels that change back and forth. The output level can be determined by checking the inputs with respect to each other at any given time. In Figure 3.2.3, two inputs are both HIGH (1) during the time interval t_1, and thus the output is HIGH (1). During the time interval t_2, input A is LOW (0) and input B is HIGH (1), so the output is LOW (0). During the time interval t_3, input A is HIGH (1) and input B is LOW (0), and thus the output is LOW (0). Finally, during the time interval t_4, the inputs are both LOW (0), so the output is LOW (0).

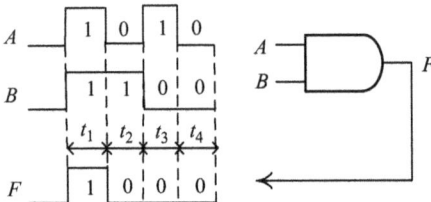

Figure 3.2.3: Timing diagram of 2-input AND gate.

2. OR

OR operation produces a HIGH output when any input is HIGH and a LOW output only when all inputs are LOW. A circuit that performs the OR operation is called an *OR gate.*

The logical symbol and the truth table for 2-input OR gate are illustrated in Figure 3.2.4. The truth table can be expanded to any number of inputs. ***For an OR gate, regardless of the number of inputs, the output is LOW only when all inputs are LOW.***

The operation of 2-input OR gate can be described by a logic expression. If the two input variables are A and B, and the output variable is F, then the Boolean expression of the 2-input OR gate is

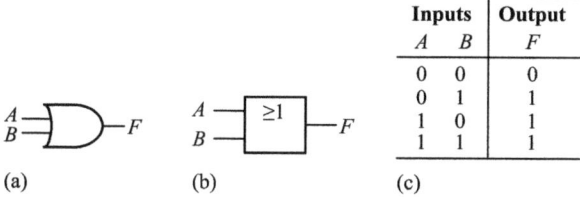

Inputs		Output
A	B	F
0	0	0
0	1	1
1	0	1
1	1	1

(a) (b) (c)

Figure 3.2.4: Standard logic symbols and truth table for 2-input OR gate: (a) distinctive shape; (b) rectangular outline with the OR (≥1) qualifying symbol; (c) truth table.

$$F = A + B$$

To extend the OR expression to more than two input variables, simply add a new letter for each input variable.

A 2-input OR gate can be used to construct a simple intrusion detection and alarm system, as shown in Figure 3.2.5 [3]. This system can be used for a room with a window and a door. The sensors A and B are magnetic switches that produce a HIGH output when opened and a LOW output when closed. As long as the window and the door are closed, all switches are closed and all the inputs of the OR gate are LOW. Thus the output is in a LOW state. When a window or a door is opened, a HIGH is produced on input and the output of the OR gate turns to a HIGH state. An alarm circuit is activated to produce the warning of the intrusion.

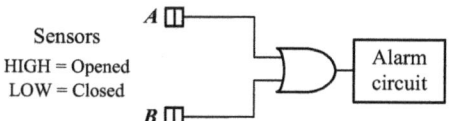

Figure 3.2.5: Application of a 2-input OR gate as an intrusion detection and alarm system.

3. NOT

NOT operation changes one logic level to the opposite logic level. That is to say, **the output is a LOW (0) when the input is a HIGH (1) and the output is a HIGH (1) when the input is a LOW (0).** The NOT operation is implemented by a logic circuit known as *an inverter*. The logical symbol and the truth table of an inverter are illustrated in Figure 3.2.6.

The operation of an inverter can be described by a logic expression. If the input variable is called A and the output variable is called F, then

$$F = \bar{A}$$

This expression shows that the output is the complement of the input. if $A = 0$, then $X = 1$; if $A = 1$, then $X = 0$.

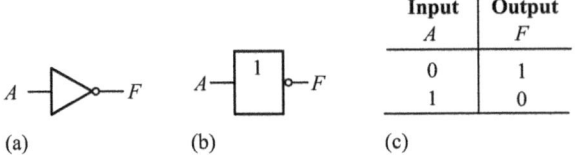

(a) (b) (c)

Figure 3.2.6: Standard logic symbol and the truth table of 2-input OR gate: (a) distinctive shape; (b) rectangular outline with the NOT (1) qualifying symbol; (c) truth table.

Figure 3.2.7 shows an application of inverters used to obtain the 1's complement of an 8-bit binary number. The bits of the binary number are applied to inputs of the inverter and the 1's complement of the number appears on the outputs.

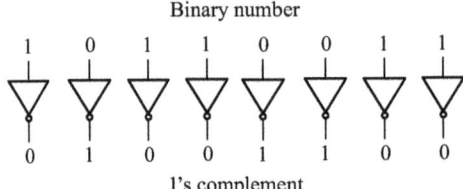

Figure 3.2.7: Inverters used to obtain the 1's complement of an 8-bit binary number.

3.2.3 Combinational logic operations

With AND, OR, and NOT operations, some combinational logic operations can be implemented. Common combinational logic operations include NAND, NOR, XOR, and NXOR.

1. NAND

The term NAND is the contraction of NOT-AND and implies an AND function with a complemented (inverted) output. The standard logic symbol and truth table of a 2-input NAND gate are shown in Figure 3.2.8.

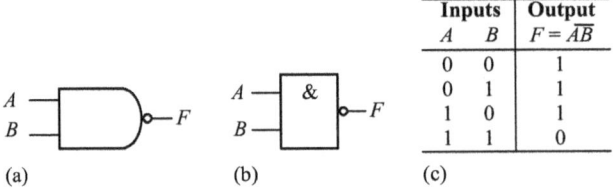

(a) (b) (c)

Figure 3.2.8: Standard logic symbol and truth table of a 2-input NAND gate: (a) distinctive shape; (b) rectangular outline with the NAND qualifying symbol (&); (c) truth table.

The NAND gate produces a LOW output only when all the inputs are HIGH. When any input is LOW, the output will be HIGH. The Boolean expression of a NAND gate is

$$F = \overline{A \cdot B}$$

Usually AND operator "·" can be omitted.

2. NOR

The term NOR is the contraction of NOT-OR and implies an OR function with a complemented (inverted) output. The standard logic symbol and truth table of a 2-input NAND gate are shown in Figure 3.2.9.

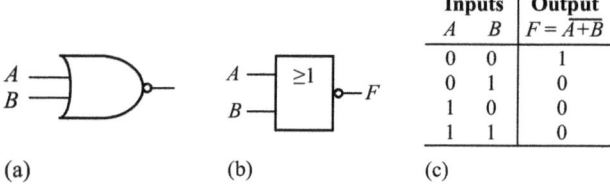

Inputs		Output
A	B	$F = \overline{A+B}$
0	0	1
0	1	0
1	0	0
1	1	0

(a)   (b)   (c)

Figure 3.2.9: Standard logic symbol and truth table of a 2-input NOR gate: (a) distinctive shape; (b) rectangular outline with the NOR qualifying symbol (≥1); (c) truth table.

The NOR gate produces a LOW output when any input is HIGH; and it produces a HIGH output only when all inputs are LOW. The Boolean expression of a NOR gate operation is

$$F = \overline{A+B}$$

The NOR gate, like the NAND gate, is a commonly used logic gate because it can be used as a universal gate; that is, NOR gates can be used to perform the AND, OR and inverter opetations.

3. Exclusive-OR (XOR) and Exclusive-NOR (XNOR)

The Boolean expression of an XOR gate is

$$F = A \oplus B = \bar{A}B + A\bar{B}$$

where " ⊕ " is the operator of an exclusive-OR operation. It can be seen from the expression that an XOR operation is formed by a combination of other operations. However, because of its fundamental importance in many applications, XOR gates are often treated as logic gates with their own unique symbols, as shown in Figure 3.2.10.

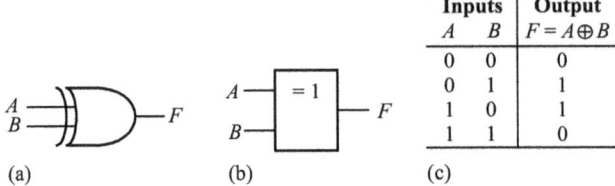

Figure 3.2.10: Standard logic symbol and truth table of an exclusive-OR gate: (a) distinctive shape; (b) rectangular outline with the XOR qualifying symbol; (c) truth table.

The output of an exclusive-OR gate is HIGH only when the two inputs are at opposite logic levels.

The Boolean expression of an XNOR gate is

$$F = A \odot B = \bar{A}\bar{B} + AB = \overline{A \oplus B}$$

where "$\odot$" is the operator of an XNOR operation. It can be seen from the expression that an XNOR operation is the complement of an exclusive-OR. However, due to its fundamental importance in many applications, an exclusive-NOR gate has its own unique symbol, as shown in Figure 3.2.11.

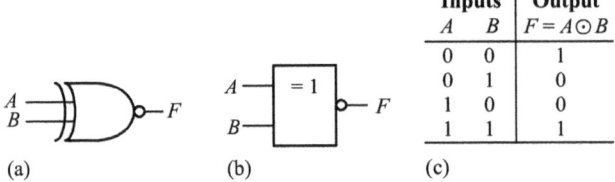

Figure 3.2.11: Standard logic symbol and truth table of an exclusive-NOR gate: (a) distinctive shape; (b) rectangular outline with the XNOR qualifying symbol (=1); (c) truth table.

The output of an XNOR gate is HIGH only when the two inputs are at the same logic level.

3.3 Laws, rules, and theorems of Boolean algebra

In Boolean algebra, there are certain well-developed laws, rules, and theorems that must be followed in order to properly apply Boolean algebra. This section only introduces the most important Boolean algebra laws, rules, and theorems for analyzing and designing digital circuits.

The objectives of this section are to
- Apply commutative laws, associative laws, and distributive laws
- Apply basic rules of Boolean algebra
- Apply DeMorgan's theorems

3.3.1 Boolean addition and Boolean multiplication

1. Boolean addition
Boolean addition is equivalent to the OR operation and the basic rules are illustrated with their relation to the OR gate as shown in Figure 3.3.1.

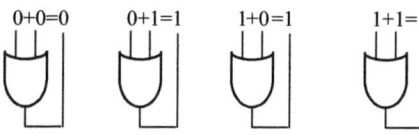

Figure 3.3.1: Illustration of the relation between Boolean addition and OR gate.

In Boolean algebra, a *sum term* is a sum of literals. In logic circuit, a sum term could be produced by an OR operation without AND operation involved. For example, $A+C$, $A+\bar{B}$ and $\bar{A}+\bar{B}+\bar{C}$ are sum terms.

2. Boolean multiplication
Boolean multiplication is equivalent to AND operation and the basic rules are illustrated with their relation to AND gate as shown in Figure 3.3.2.

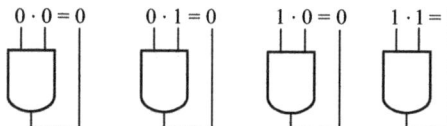

Figure 3.3.2: Illustration of the relation between Boolean multiplication and AND gate.

In Boolean algebra, a *product term* is a product of literals. In logic circuit, a product term could be produced by an AND operation without OR operation involved. For example, AC, $A\bar{B}$, $\bar{A}\bar{B}\bar{C}$, and $\bar{A}\bar{B}CD$ are product terms.

3.3.2 Laws of Boolean algebra

Similar to ordinary algebra, Boolean algebra has three basic laws including commutative laws, associative laws, and distributive laws. Each of the laws is illustrated with two or three variables, but the number of variables can be extended.

1. Commutative laws

The commutative laws for the multiplication and the addition of two variables are expressed as below:

$$AB = BA$$

$$A + B = B + A$$

2. Associative laws

The associative laws for the multiplication and the addition of three variables are written as follows:

$$(AB)C = A(BC)$$

$$(A + B) + C = A + (B + C)$$

3. Distributive laws

The distributive laws for three variables are written as follows:

$$A(B + C) = AB + AC$$

$$A + BC = (A + B)(A + C)$$

3.3.3 Rules of Boolean algebra

Table 3.3.1 lists basic rules which are useful in the manipulation and simplification of Boolean expressions.

Table 3.3.1: Basic rules of Boolean algebra.

1. $A \cdot 0 = 0$	9. $\bar{\bar{A}} = A$
2. $A \cdot 1 = A$	10. $A + AB = A$
3. $A + 0 = A$	11. $A(A+B) = A$
4. $A + 1 = 1$	12. $A + \bar{A}B = A + B$
5. $A \cdot A = A$	13. $A(\bar{A} + B) = AB$
6. $A \cdot \bar{A} = 0$	14. $AB + \bar{A}C + BC = AB + \bar{A}C$
7. $A + A = A$	15. $(A + B)(\bar{A} + C)(B + C) = (A + B)(\bar{A} + C)$
8. $A + \bar{A} = 1$	

Rules 1 through 9 can be verified with truth table. If the values on both sides of the equation are equal for all possible variable combinations, then the rule is true and otherwise the rule is false.

For example, rule 1: $A \cdot 0 = 0$. This rule can be proved as follows:

If $A = 0$, $A \cdot 0 = 0$;

If $A = 1$, $A \cdot 0 = 0$.

So $A \cdot 0 = 0$.

Rules 10 through 15 can be proved with the truth table or by applying the simple rules and the laws previously discussed.

For example, rule 12: $A + \bar{A}B = A + B$.

This rule can be proved as follows:

$$
\begin{aligned}
A + \bar{A}B &= (A + AB) + \bar{A}B \quad &&\text{Rule 10: } A = A + AB \\
&= A + (A + \bar{A})B \quad &&\text{Associate laws and distributive laws} \\
&= A + 1 \cdot B \quad &&\text{Rule 8: } A + \bar{A} = 1 \\
&= A + B \quad &&\text{Rule 2: drop the 1}
\end{aligned}
$$

The proof with the truth table is shown in Table 3.3.2. It can be seen from Table 3.3.2 that the values on both sides of the equation are equal for all possible combinations of variables A and B. So the equation of $A + \bar{A}B = A + B$ is true.

Table 3.3.2: Truth table.

A	B	$\bar{A}B$	$A + \bar{A}B$	$A+B$
0	0	0	0	0
0	1	1	1	1
1	0	0	1	1
1	1	0	1	1

Let's check rule 14: $AB + \bar{A}C + BC = AB + \bar{A}C$.

This rule can be proved as follows:

$$
\begin{aligned}
AB + \bar{A}C + BC &= AB + \bar{A}C + (A + \bar{A})BC \quad &&\text{Rule 8: times } A + \bar{A} = 1 \\
&= AB + \bar{A}C + ABC + \bar{A}BC \quad &&\text{Distributive law} \\
&= (AB + ABC) + (\bar{A}C + \bar{A}BC) \quad &&\text{Associate law} \\
&= AB(1 + C) + \bar{A}C(1 + B) \quad &&\text{Factoring and rule 4: } A+1 = 1 \\
&= AB + \bar{A}C \quad &&\text{Rule 2: } A \cdot 1 = A
\end{aligned}
$$

3.3.4 DeMorgan's theorems

DeMorgan's first theorem is stated as follows [3]:

The complement of a product of variables is equal to the sum of the complements of the individual variables.

Stated in another way:

The complement of two or more ANDed variables is equivalent to the OR of the complements of the individual variables.

For the case of two variables, the formula of this theorem can be expressed as

$$\overline{AB} = \bar{A} + \bar{B} \qquad (3.3.1)$$

The DeMorgan's second theorem is stated as follows:

The complement of a sum of variables is equal to the product of the complements of the variables.

Stated in another way:

The complement of two or more ORed variables is equivalent to the AND of the complements of the individual variables.

For the case of two variables, the formula of this theorem can be expressed as

$$\overline{A + B} = \bar{A}\bar{B} \qquad (3.3.2)$$

The proofs with the truth table are shown in Table 3.3.3. and Table 3.3.4, respectively.

Table 3.3.3: Truth table.

Inputs		Outputs	
A	*B*	$\overline{AB}$	$\bar{A} + \bar{B}$
0	0	1	1
0	1	1	1
1	0	1	1
1	1	0	0

Table 3.3.4: Truth table.

Inputs		Outputs	
A	*B*	$\overline{A + B}$	$\bar{A} \cdot \bar{B}$
0	0	1	1
0	1	0	0
1	0	0	0
1	1	0	0

Each variable in the DeMorgan's theorems as stated in eqs. (3.3.1) and (3.3.2) can also be represented by a combination of other variables. For example, *B* can be equal to the term *AC+D*. Thus, if you apply DeMorgan's first theorem in eq. (3.3.1), the following result can be obtained:

$$\overline{A(AC + D)} = \bar{A} + \overline{AC + D} \qquad (3.3.3)$$

Notice that the second term $\overline{AC + D}$ can be decomposed by applying DeMorgan's second theorem as follows:

$$\overline{AC + D} = \overline{AC} \cdot \bar{D} \tag{3.3.4}$$

For $\overline{AC}$, you can apply the DeMorgan's first theorem again, so the final result is

$$\overline{A(AC + D)} = \bar{A} + (\bar{A} + \bar{C})\bar{D} = \bar{A} + \bar{A}\bar{D} + \bar{C}\bar{D} \tag{3.3.5}$$

Although this result can be simplified further by the use of Boolean rules and laws, the DeMorgan's theorems can not be applied further.

3.4 Standard forms of Boolean expression

Boolean expression provides an efficient way to express the operation of a logic circuit formed by a combination of logic gates. Boolean expression has various expression forms. Regardless of their forms, all Boolean expressions can be converted into two standard forms. This section introduces the representation of Boolean expression first. Then standard forms of Boolean expression are introduced.

The objectives of this section are to
– Construct a truth table
– Explain a sum-of-products expression and a product-of-sums expression
– Define the standard sum-of-products term and the standard product-of-sums term
– Define the standard sum-of-products expression and the standard product-of-sums expression
– Convert any sum-of-products expression into a standard form
– Convert any product-of-sums expression into a standard form

3.4.1 Representation of Boolean expression

Boolean expression can be constructed by a finite number of two-valued variables connected by the basic operations, which can be expressed as follows:

$$F = f(A, B, C)$$

where A, B, C are input variables and F is an output variable, and f is a function relation which consists of a group of basic logic operators. Input variable is also called logic variable and output variable is called logic function. If n-input variables are involved, we say that the Boolean expression is a Boolean function of n variables. Generally, a logic circuit with n-input can be described by a Boolean function of n variables.

Boolean expression has various forms [21]. The commonly used forms are the sum-of-products and the product-of-sums. For example,

$$F = AB + \bar{A}C \qquad (3.4.1)$$

Two product terms AB and $\bar{A}C$ are summed together by Boolean addition and the resulting expression is a sum-of-products (SOP), which is also called as AND-OR form.

The Boolean expression in eq. (3.4.1) can be also converted into a product-of-sums (POS) as follows:

$$F = (A + C)(\bar{A} + B) \qquad (3.4.2)$$

Two sum terms $(A+C)$ and $(\bar{A} + B)$ are multiplied together and the resulting expression is a product-of-sums (POS), which is also called as OR-AND form. Other forms, such as NAND-NAND, NOR-NOR and AND-OR-NOT forms, are equivalent to the Boolean expression in eq. (3.4.1) as follows:

$$F = \overline{\overline{AB} \bullet \overline{\bar{A}C}} \qquad \text{NAND-NAND}$$

$$F = \overline{\overline{A + C} + \overline{\bar{A} + B}} \qquad \text{NOR-NOR}$$

$$F = \overline{\bar{A} \bullet \bar{C} + A \bullet \bar{B}} \qquad \text{AND-OR-NOT}$$

You can prove the above five Boolean expressions representing the same logic function. That is to say, they have the same truth table. Different forms of Boolean expressions can be converted into each other. For example, you can convert SOP into NAND-NAND form by using the laws, the rules, and the theorems of Boolean algebra as bellow.

$$F = AB + \bar{A}C$$

$$= \overline{\overline{AB + \bar{A}C}} \qquad \text{Rule 9: } A = \bar{\bar{A}}$$

$$= \overline{\overline{AB} \cdot \overline{\bar{A}C}} \qquad \text{DeMorgan's theorems: } \overline{A + B} = \bar{A}\bar{B}$$

3.4.2 Minterms and maxterms

Minterms, called *standard product terms*, are the product terms which contain all of n variables in the domain [22]. Each variable only appears in either true or complementary form.

For example, if the domain is made up of two variables A and B, then the number of minterms is $2^2 = 4$. The corresponding minterms are $\bar{A}\bar{B}$, $\bar{A}B$, $A\bar{B}$ and AB. For the domain with n-variables, the number of minterms is 2^n.

A minterm is equal to 1 for only one combination of variable values. For example, the minterm $\bar{A}\bar{B}$ is equal to 1 only when $A = 0$ and $B = 0$, and is 0 for the other combinations of variable values. In this case, the minterm $\bar{A}\bar{B}$ has a binary value of 00 (decimal value of 0). This minterm can be represented by m_0. Generally, the minterm can be symbolized by m_i. The subscript i is the decimal value which is equivalent to the binary value of the minterm. For example, the minterm $\bar{A}BC$ can be represented by m_3.

Maxterms, called *standard sum terms*, are the sum terms which contain all of n variables in the domain. Each variable only appears in either true or complement form.

For example, if the domain is made up of two variables A and B, then the number of maxterms is $2^2 = 4$. The corresponding maxterms are $\bar{A} + \bar{B}$, $\bar{A} + B$, $A + \bar{B}$, and $A + B$. For the domain with n variables, the number of maxterms is 2^n.

A maxterm is equal to 0 for only one combination of variable values. For example, the maxterm $\bar{A} + \bar{B} + C + D$ is equal to 0 only when $A = 1$, $B = 1$, $C = 0$, and $D = 0$, and is 1 for all other combinations of the variable values. In this case, the maxterm $\bar{A} + \bar{B} + C + D$ has a binary value of 1100 (decimal value of 12). This maxterm can be represented by M_{12}. Generally, the maxterm can be symbolized by M_i. The subscript i is the decimal value which is equivalent to the binary value of the maxterm. For example, the maxterm $\bar{A} + B + C$ in the domain of 3 variables can be represented by M_4.

Table 3.4.1 lists representation of all minterms and maxterms of three variables A, B, and C.

Table 3.4.1: Representation of minterms and maxterms.

Decimal	A	B	C	Minterms	Maxterms
0	0	0	0	$m_0 = \bar{A}\bar{B}\bar{C}$	$M_0 = A + B + C$
1	0	0	1	$m_1 = \bar{A}\bar{B}C$	$M_1 = A + B + \bar{C}$
2	0	1	0	$m_2 = \bar{A}B\bar{C}$	$M_2 = A + \bar{B} + C$
3	0	1	1	$m_3 = \bar{A}BC$	$M_3 = A + \bar{B} + \bar{C}$
4	1	0	0	$m_4 = A\bar{B}\bar{C}$	$M_4 = \bar{A} + B + C$
5	1	0	1	$m_5 = A\bar{B}C$	$M_5 = \bar{A} + B + \bar{C}$
6	1	1	0	$m_6 = AB\bar{C}$	$M_6 = \bar{A} + \bar{B} + C$
7	1	1	1	$m_7 = ABC$	$M_7 = \bar{A} + \bar{B} + \bar{C}$

The important characteristics of minterms and maxterms are described as follows:

(1) For a given combination of n-variable values, only one of 2^n minterms is equal to 1 while only one of 2^n maxterms is equal to 0.

(2) The sum of all minterms is equal to 1; the product of all maxterms is equal to 0.

(3) The product of any two minterms is equal to 0; the sum of any two maxterms is equal to 1.

(4) A minterm is equal to the complement of its corresponding maxterm and a maxterm is equal to the complement of its corresponding minterm. For example,

for three variables A, B, and C, the relationships between minterms and maxterms are described as below:

$$\overline{m_0} = \overline{\bar{A}\bar{B}\bar{C}} = A + B + C = M_0 \qquad \overline{M_0} = \overline{A + B + C} = \bar{A}\bar{B}\bar{C} = m_0$$
$$\overline{m_1} = \overline{\bar{A}\bar{B}C} = A + B + \bar{C} = M_1 \qquad \overline{M_1} = \overline{A + B + \bar{C}} = \bar{A}\bar{B}C = m_1$$
$$\ldots \qquad\qquad\qquad\qquad \ldots$$
$$\overline{m_7} = \overline{ABC} = \bar{A} + \bar{B} + \bar{C} = M_7 \qquad \overline{M_7} = \overline{\bar{A} + \bar{B} + \bar{C}} = ABC = m_7$$

3.4.3 Standard SOP form

A standard SOP expression, also called minterm expression, is an SOP expression in which each product term is a standard product term or minterm. Standard SOP expressions are important in constructing truth tables and Karnaugh map simplification.

Any nonstandard SOP expression can be converted to the standard form using Boolean algebra. By using rule 8 ($A + \bar{A} = 1$) of Boolean algebra, each nonstandard product term in SOP expression can be converted into the standard product term. The detailed steps are described as follows:

Step 1 Multiply each nonstandard product term by a term made up of the sum of a missing variable and its complement ($A + \bar{A} = 1$). As you know, you can multiply anything by 1 without changing its value. This results in two product terms.

Step 2 Repeat step 1 until all resulting product terms are converted into the standard form. This makes the number of product terms doubled for each missing variable.

Example 3.1 Convert the following expression into the standard SOP form.

$$F = A + \bar{B}C$$

Solution

The domain of this SOP expression is made up of A, B, and C.

In the first term, A, the other two variables B and C are absent. So the first term is multiplied by $B + \bar{B}$ and $C + \bar{C}$ as below:

$$A = A(B + \bar{B})(C + \bar{C}) = ABC + AB\bar{C} + A\bar{B}C + A\bar{B}\bar{C}$$

In the second term, BC, the variable A is absent. So the second term is multiplied by $A + \bar{A}$ as below:

$$\bar{B}C = \bar{B}C(A + \bar{A}) + A\bar{B}C + \bar{A}\bar{B}C$$

The standard SOP form of the original expression is as follows:

$$F = A + \bar{B}C = ABC + AB\bar{C} + A\bar{B}C + A\bar{B}\bar{C} + \bar{A}\bar{B}C$$
$$= m_1 + m_4 + m_5 + m_6 + m_7 = \sum m(1, 4, 5, 6, 7)$$

where two same standard product terms $A\bar{B}C$ can be merged by using rule 7: $A+A = A$.

3.4.4 Standard POS form

A standard POS expression, referred to as a maxterm expression, is a POS expression in which each sum term is a standard sum term or maxterm.

Any nonstandard POS expression can be converted to the standard form using Boolean algebra. By using rule 6 ($A\bar{A} = 0$) and the distributive law, each nonstandard sum term in the POS expression can be converted into the standard sum term. The detailed steps are described as follows:

Step 1 Add a term made up of the product of a missing variable and its complement ($A\bar{A}$) to each nonstandard sum term. As you know, you can add anything by 0 without changing its value.

Step 2 Apply the distributive law: $A+BC = (A+B)(A+C)$ and convert one sum term into two sum terms.

Step 3 Repeat step 1 until all resulting sum terms are converted into standard form. This makes the number of sum terms doubled for each missing variable.

Example 3.2 Convert the following expression into the standard POS form.

$$F = (A + \bar{B} + C)(A + \bar{C} + D)(A + \bar{B} + C + \bar{D})$$

Solution

The domain of this POS expression is made up of A, B, C, D.

One variable D is missing in the first term. So add $D\bar{D}$ and apply the distributive law as below:

$$A + \bar{B} + C = A + \bar{B} + C + D\bar{D} = (A + \bar{B} + C + D)(A + \bar{B} + C + \bar{D}) = M_4 M_5$$

One variable B is missing in the second term. So add $B\bar{B}$ and apply the distributive law as below:

$$A + \bar{C} + D = A + \bar{C} + D + B\bar{B} = (A + B + \bar{C} + D)(A + \bar{B} + \bar{C} + D) = M_2 M_6$$

The third term, $A + \bar{B} + C + \bar{D}$ (M_5), is already a standard form.
So the standard POS form of the original expression is as follows:

$$F = (A + \bar{B} + C)(A + \bar{C} + D)(A + \bar{B} + C + \bar{D}) = M_2 M_4 M_5 M_6 = \prod M(2, 4, 5, 6)$$

3.4.5 Boolean expression and truth table

All standard Boolean expressions can be easily converted into a truth table format. The truth table is a concise way of presenting the logic operation of a circuit.

1. Converting SOP expression to truth table format

A truth table is simply a list of all possible combinations of input variable values and their corresponding output values. For a given SOP expression, you can determine the truth table format in terms of the expression's domain. If an expression has a domain of 2 variables, there are four different combinations of those variables ($2^2 = 4$). If an expression has a domain of n variables, there are 2^n different combinations of those variables. So you can construct a truth table by listing all possible combinations of binary values of the input variables in the expression first. Then, you evaluate the output variable by substituting each binary value of the input variables combination into the SOP expression and the corresponding result is placed in the output column (F). Alternatively, you can convert the SOP expression into its standard form. If each binary value makes the standard SOP expression equal to 1, a 1 is placed in the corresponding output column. Otherwise, a 0 is placed for all the remaining binary values.

Example 3.3 Derive a truth table for the SOP expression $F = A + \bar{B}C$

Solution

Step 1 There are three variables in the domain, so there are eight possible combinations of binary values of the variables as listed in the left input columns of Table 3.4.2.

Table 3.4.2: Truth table.

Input			Output
A	**B**	**C**	**F**
0	0	0	0
0	0	1	1
0	1	0	0
0	1	1	0
1	0	0	1
1	0	1	1
1	1	0	1
1	1	1	1

Step 2 Substitute each binary value into the SOP expression: $F = A + \bar{B}C$. When $ABC = 000$, $F = 0$; when $ABC = 001$, $F = 1$; when $ABC = 010$, $F = 0$; when $ABC = 010$, $F = 0$; when $ABC = 011$, $F = 0$; and so on. The resulting output values are listed in the output column of Table 3.4.2.

Alternatively, you can convert the given SOP expression into the standard form. From example 3.1, you already convert the expression $F = A + \bar{B}C$ into a standard form and the resulting expression is

$$F = A + \bar{B}C = ABC + AB\bar{C} + A\bar{B}C + A\bar{B}\bar{C} + \bar{A}\bar{B}C$$
$$= m_1 + m_4 + m_5 + m_6 + m_7 = \sum m(1, 4, 5, 6, 7)$$

Therefore, the binary values that make the standard SOP expression equal to 1 are $\bar{A}\bar{B}C$ (m_1): 001, $A\bar{B}\bar{C}$ (m_4):100, $A\bar{B}C$ (m_5):101, $AB\bar{C}$ (m_6):110, ABC (m_7):111. For each of these binary values, a 1 is placed in the output column as shown in Table 3.4.2. For every remaining binary combination, a 0 is placed in the output column. In other words, one row of the truth table corresponds to a minterm. If a minterm exists in the SOP expression, the corresponding output is a 1. If a minterm does not exist in the SOP expression, the corresponding output is a 0.

2. Converting POS expression to truth table format

For a given POS expression, you can determine the truth table format in terms of the expression's domain just as you have done for the SOP expression. Then, you evaluate the output variable by substituting each binary value of input variable combinations into POS expression and the corresponding results are placed in the output column (F). Alternatively, you can convert the POS expression into the standard form. If each binary value makes the standard POS expression equal to 0, a 0 is placed in the corresponding output column. Otherwise, a 1 is placed for all the remaining binary values.

Example 3.4 Derive a truth table for the POS expression as below:

$$F = (A + B + C)(A + \bar{C})(A + \bar{B})$$

Solution

Step 1 There are three variables in the domain, so eight possible combinations of binary values of 3-input variables exist as listed in the left three columns of Table 3.4.3.

Table 3.4.3: Truth table.

Input			Output
A	B	C	F
0	0	0	0
0	0	1	0
0	1	0	0
0	1	1	0
1	0	0	1
1	0	1	1
1	1	0	1
1	1	1	1

Step 2 Substitute each binary value into the POS expression. When $ABC = 000$, $F = 0$; when $ABC = 001$, $F = 0$; when $ABC = 010$, $F = 0$; when $ABC = 011$, $F = 0$, and so on. The resulting output values are listed in the output column of Table 3.4.3.

Alternatively, you can convert the POS expression into the standard form.
The first term is already a standard sum term.
The second term has a missing variable B, so add $B\bar{B}$ and apply the distributive law as below:

$$A + \bar{C} + B\bar{B} = (A + B + \bar{C})(A + \bar{B} + \bar{C})$$

The third term has a missing variable C, so add $C\bar{C}$ and apply the distributive law as below:

$$A + \bar{B} = A + \bar{B} + C\bar{C} = (A + \bar{B} + C)(A + \bar{B} + \bar{C})$$

So the standard POS form of the original expression is as follows:

$$F = (A + B + C)(A + B + \bar{C})(A + \bar{B} + \bar{C})(A + \bar{B} + C) = \prod M(0, 1, 2, 3)$$

The binary values that make the standard sum terms (maxterms) in expression equal to 0 are $A + B + C$ (M_0):000, $A + B + \bar{C}$ (M_1):001, $A + \bar{B} + C$ (M_2):010, $A + \bar{B} + \bar{C}$ (M_3):011. For each of these binary values, a 0 is placed in the output column as shown in Table 3.4.3. For each of the remaining binary combinations, a 1 is placed in the output column.

3. Deriving standard expressions from a truth table

To derive the standard SOP expression from a truth table, you should pick up the binary values of the input variables that make the output a 1. For each binary value, write a product term by replacing each 1 with the corresponding variable and each 0 with the corresponding variable complement. Finally, add all product terms which make the output a 1 and derive the standard SOP expression for the output.

Similarly, to derive the standard POS expression from a truth table, you should pick up the binary values of the input variables that make the output a 0. For each binary value, write a sum term by replacing each 0 with the corresponding variable and each 1 with the corresponding variable complement. Finally, multiply all sum terms which make the output a 0 and derive the standard POS expression for the output.

Example 3.5 Derive the standard SOP expression and the standard POS expression according to the truth table shown in Table 3.4.4.

Table 3.4.4: Truth table.

Input			Output
A	B	C	F
0	0	0	0
0	0	1	0
0	1	0	1
0	1	1	1
1	0	0	0
1	0	1	1
1	1	0	1
1	1	1	1

Solution

There are five 1s in the output column and the corresponding binary values of the input variables are 010, 011, 101, 110, 111. These binary values are converted to the product terms as follows:

$010 \rightarrow \bar{A}B\bar{C}$ (m_2); $011 \rightarrow \bar{A}BC$ (m_3); $101 \rightarrow A\bar{B}C$ (m_5); $110 \rightarrow AB\bar{C}$ (m_6); $111 \rightarrow ABC$ (m_7)

The resulting standard SOP expression for the output F is

$$F = \bar{A}B\bar{C} + \bar{A}BC + A\bar{B}C + AB\bar{C} + ABC = \sum m(2,3,5,6,7)$$

For POS expression, there are three 0s in the output column and the corresponding binary values of the input variables are 000, 001, 100. These binary values are converted to the sum terms as follows:

$000 \rightarrow A + B + C$ (M_0); $001 \rightarrow A + B + \bar{C}$ (M_1); $100 \rightarrow \bar{A} + B + C$ (M_4)

The resulting standard POS expression for the output F is

$$F = (A + B + C)(A + B + \bar{C})(\bar{A} + B + C) = \prod M(0,1,4)$$

Comparing the standard SOP expression with the standard POS expression, you can find that the labels in minterm expression and maxterm expression are complemented for the same truth table. That is, the labels occurring in minterm expression will not appear in maxterm expression.

3.5 Simplification using Boolean algebra

In the previous section, we introduced that there are various forms of one Boolean expression to describe a same logic function with a unique truth table. In order to implement the Boolean expression with a simple logic circuit, you have to reduce a particular expression to its simplest form or change its form into a more convenient one implemented most efficiently. This section introduces the simplification method of the Boolean expression by using Boolean algebra. In this section, you will learn how to apply the laws, rules, and theorems of Boolean algebra to simplify general expressions.

A Boolean expression can be implemented with a logic circuit by using AND gates to implement product terms and OR gates to implement sum terms. In order to implement the Boolean expression with a simple logic circuit, you have to reduce a particular expression to its **simplest form containing the fewest terms with the fewest possible variables per term.**

Simplification using Boolean algebra is to use the laws, the rules, and the theorems of Boolean algebra to convert Boolean expression into a simplified form. There are no fixed steps for obtaining a simplified expression using Boolean algebra. Here, the commonly used methods are summarized as follows:

1. Merging the product terms

Two product terms can be merged into one product term by using $A + \bar{A} = 1$ (rule 8) and thus one or more product terms can be eliminated.

Example 3.6 Simplify the Boolean expression $F = A\bar{B} + AB + ABC + \bar{A}BC$

Solution

$$F = A\bar{B} + AB + ABC + \bar{A}BC$$

$$= A(\bar{B} + B) = (A + \bar{A})BC \quad \text{Factoring } A \text{ out of the first and second terms and } BC$$

$$\text{out of the third and fourth terms}$$

$$= A + BC \qquad\qquad \text{Applying } A + \bar{A} = 1 \text{ and drop the 1}$$

2. Absorbing the product terms

One or more product terms can be absorbed by using $A + AB = A$ (rule 10) and thus the product term, AB, can be absorbed.

Example 3.7 Simplify the Boolean expression $F = A(\overline{B + C} + AC)BD + AD$

Solution

$$F = A(\overline{B + C} + AC)BD + AD \qquad \text{Associative law}$$

$$= (\overline{B + C} + AC)B \cdot AD + AD \qquad \text{Rule 10: } A + AB = A$$

$$= AD$$

3. Eliminating the product terms

By using $AB + \bar{A}C + BC = AB + \bar{A}C$ (rule 14), the product term, BC, can be eliminated.

Example 3.8 Simplify the Boolean expression $F = AB + \bar{A}C + BCD$

Solution

$$F = AB + \bar{A}C + BCD$$

$$= (AB + \bar{A}C + BC) + BCD \qquad \text{Rule 14 : } AB + \bar{A}C = AB + \bar{A}C + BC$$

$$= AB + \bar{A}C + (BC + BCD) \qquad \text{Associative law}$$

$$= AB + \bar{A}C + BC \qquad\qquad \text{Factoring } BC \text{ and drop the 1}$$

$$= AB + \bar{A}C \qquad\qquad\qquad \text{Rule 14 : } AB + \bar{A}C = AB + \bar{A}C + BC$$

4. Removing variable or variable complement

By using $A + \bar{A}B = A + B$ (rule 12), $\bar{A}$ can be removed from the term $\bar{A}B$.

Example 3.9 Simplify the Boolean expression $F = A(\bar{B} + \bar{C}) + BC$

Solution

$$F = A(\bar{B} + \bar{C}) + BC$$
$$= A \cdot \overline{BC} + BC \qquad \text{Using DeMorgan's theorem}$$
$$= A + BC \qquad\quad \text{Rule 12}: A + \bar{A}B = A + B, \text{ removing } \overline{BC}$$

5. Adding the terms

By using $A + \bar{A} = 1$ (rule 8), a product term is multiplied by $(A + \bar{A})$ and thus decomposed into two product terms which can be merged with other product terms to obtain a simplified Boolean expression.

Example 3.10 Simplify the Boolean expression $F = A\bar{B} + B\bar{C} + \bar{B}C + \bar{A}B$

Solution

$$F = A\bar{B} + B\bar{C} + \bar{B}C + \bar{A}B$$

$$= A\bar{B}(C + \bar{C}) + B\bar{C}(A + \bar{A}) + \bar{B}C + \bar{A}B \qquad \text{Rule 8}: 1 = A + \bar{A}$$

$$= A\bar{B}C + A\bar{B}\bar{C} + AB\bar{C} + \bar{A}B\bar{C} + \bar{B}C + \bar{A}B \qquad \text{Distributive law}$$

$$= (A\bar{B}C + \bar{B}C) + (AB\bar{C} + A\bar{B}\bar{C}) + (\bar{A}B\bar{C} + \bar{A}B) \qquad \text{Associative law}$$

$$= \bar{B}C + A\bar{C} + \bar{A}B \qquad \text{Factoring and drop the 1}$$

The simplification using Boolean algebra depends on your experience. It requires that you are familiar with the laws, the rules and the theorems of Boolean algebra and obtain skill by doing more exercises.

3.6 Karnaugh map simplification

In the above section, you learnt how to simplify a Boolean expression by using Boolean algebra. As you have seen, the effectiveness of algebraic simplification depends on your familiarity with all the laws, the rules and the theorems of Boolean algebra and on your ability to apply them. The Karnaugh map was proposed by M Karnaugh in 1953 [23]. It offers a systematic method of simplifying Boolean expression. If it is used properly, the simplest Boolean expression can be more easily produced than using Boolean algebra. This section introduces Karnaugh map with the variables from two to five first. Then Karnaugh map simplification is introduced for obtaining a minimum SOP expression and a minimum POS expression. Finally, we introduce the treatment of terms that don't care in Karnaugh map simplification. Basically, the Karnaugh map, on the other hand, provides a "cookbook" method for simplification.

The objectives of this section are to

- Construct a Karnaugh map for three or four variables
- Define cell adjacency
- Map a Boolean expression on a Karnaugh map
- Map a truth table into Karnaugh map
- Combine the 1s or the 0s on the map into maximum groups

- Determine the minimum SOP expression and the minimum POS expression
- Use "don't care" conditions on a Karnaugh map

3.6.1 The Karnaugh map

Similar to a truth table, a Karnaugh map is organized as an array of cells in which each cell represents a binary value of the input variables and corresponds to a minterm. Each cell corresponds to a row in a truth table. The cells are arranged in a way that the adjacent cells only differ from one literal. This offers an opportunity of converting the simplification of a given expression into a matter of properly grouping the cells. The number of cells in Karnaugh map (K-map) is equal to the total number of possible input variable combinations as is the number of rows in a truth table. Therefore, the number of cells is $2^2(=4)$ for two variables, $2^3(=8)$ for three variables, $2^4(=16)$ for four variables, $2^5(=32)$ for five variables, as shown in Figure 3.6.1. Here, we focus on discussing only 3-variable and 4-variable situations to illustrate the principles.

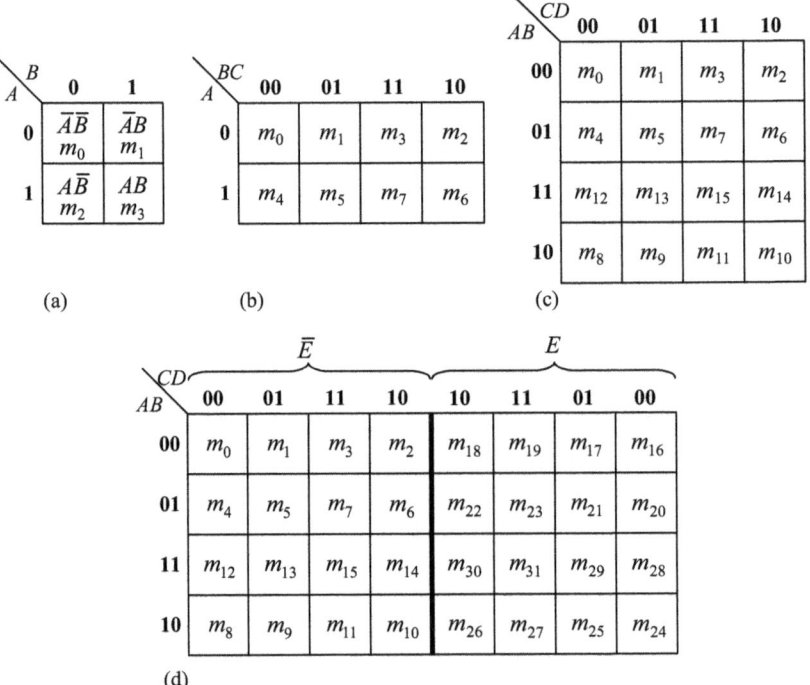

Figure 3.6.1: 2-variable (a), 3-variable (b), 4-variable (c) and 5-variable (d) Karnaugh maps.

The 3-variable Karnaugh map has an array of eight cells shown in Figure 3.6.1(b). A, B, and C are used to represent variables. Notice that other letters can also be used to represent variables. Binary values of A are along the left side and the values of B and C are across the top. The value of a given cell is determined by the binary values of A on the left in the row combined with the values of B and C on the top in the same column. The values of B and C are arranged in the sequence of Gray code to guarantee the cell adjacency. Each cell represents a minterm or a standard product term in Karnaugh map.

Similarly, the 4-variable Kaunaugh map has an array of sixteen cells shown in Figure 3.6.1(c). Binary values of A and B are along the left side and the values of C and D are across the top. The value of a given cell is determined by the binary values of A and B on the left in the same row combined with the values of C and D on the top in the same column. The binary values of A and B are arranged by the sequence of Gray code to ensure the cell adjacency. So do the binary values of C and D. Each cell represents a minterm or a standard product term in Karnaugh map.

The cells in a Karnaugh map have the logic adjacency. *Adjacency* is defined by a single-variable change. The sequence of Gray code guarantees that adjacent cells that differ by only one literal. That is to say, cells that differ by only one variable are adjacent. For instance, in the 3-variable map, the 000 cell is adjacent to the 001 cell, the 010 cell, and the 100 cell. Cells with values that differ by more than one variable are not adjacent. For example, the 000 cell is not adjacent to the 011 cell, the 101 cell, the 110 cell, or the 111 cell.

Physically, each cell is adjacent to the cells that are next to it on any of its four sides. Also, the cells in the top row are adjacent to the corresponding cells in the bottom row and the cells in the leftmost column are adjacent to the corresponding cells in the rightmost column. This is called *"wrap around"* adjacency since you can think of the map as wrapping around from top to bottom to form a cylinder or from left to right to form a cylinder. Figure 3.6.2 illustrates the cell adjacency with a 4-variable map. Moreover, the same rules for adjacency apply to Karnaugh maps with any number of cells.

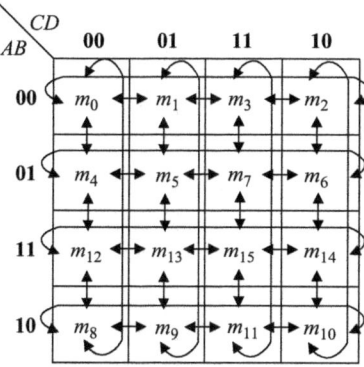

Figure 3.6.2: Cells adjacency in a Karnaugh map. Arrows indicate the adjacency between the cells.

3.6.2 Karnaugh map SOP minimization

Minimization refers to the process which results in an expression containing the fewest possible terms with the fewest possible variables [24].

Cells adjacency in Karnaugh map can be used to simplify Boolean expressions. Generally, Karnaugh map SOP minimization includes three main steps. The first step is to map an SOP expression on a Karnaugh map; the second step is to group 1s; the third step is to determine the minimized SOP expression from the map.

A minimized SOP expression contains the fewest possible product terms with the fewest possible variables per term. It can be implemented with fewer logic gates than a standard expression.

This part mainly focuses on discussing how to get a minimized SOP expression by using Karnaugh map.

1. Mapping an SOP expression

Usually, an SOP expression could be a standard SOP expression, or a nonstandard SOP expression. Sometimes an SOP expression is expressed by a truth table.

(1) Mapping a standard SOP expression

For a standard SOP expression, if a minterm is in the expression, then a 1 is placed in its corresponding cell on Karnaugh map; if a minterm is not in the expression, then a 0 is placed in its corresponding cell. When mapping a standard SOP expression is completed, the number of 1s on the Karnaugh map is equal to the number of minterms in the standard SOP expression. Usually, when mapping a standard SOP expression, the 0s can be omitted on the Karnaugh map.

Example 3.11 Map the following standard SOP expression on a Karnaugh map:

$$F = \bar{A}\bar{B}\bar{C} + \bar{A}B\bar{C} + \bar{A}\bar{B}C + ABC$$

Solution

The given SOP expression has a domain of three variables A, B, and C. You can map this standard SOP expression on a 3-variable Karnaugh map. A 1 is placed on the corresponding cell on Karnaugh map for each minterm in the expression, as shown in Figure 3.6.3.

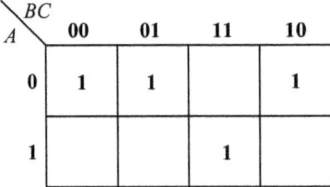

A \ BC	00	01	11	10
0	1	1		1
1			1	

Figure 3.6.3: Karnaugh map of example 3.11.

---

(2) Mapping a nonstandard SOP expression

A Boolean expression must be in the standard form before using the Karnaugh map. If an expression is not in the standard form, you can convert it to the standard form by the method introduced in section 3.4. Alternatively, you can directly map a nonstandard SOP expression on a Karnaugh map. Example 3.12 shown below illustrates the principle of directly mapping a nonstandard SOP expression on a Karnaugh map.

Example 3.12 Map the following SOP expression on a Karnaugh map:

$$F(A, B, C) = B\bar{C}$$

Solution

One method is to convert this nonstandard SOP expression into a standard SOP expression firstly.

$$F(A, B, C) = B\bar{C} = (\bar{A} + A)B\bar{C} = \bar{A}B\bar{C} + AB\bar{C} = m_2 + m_6$$

Since m_2 and m_6 are in the standard SOP expression, their corresponding cells are placed 1s and the other cells are placed 0s on Karnaugh map. The mapping result is shown in Figure 3.6.4.

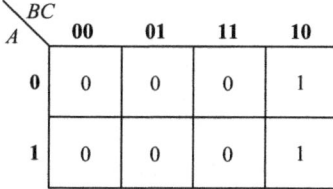

A \ BC	00	01	11	10
0	0	0	0	1
1	0	0	0	1

Figure 3.6.4: Karnaugh map of example 3.12.

Another method is to directly map the product term, $B\bar{C}$, on a Karnaugh map.

Product term $B\bar{C}$ is not a minterm due to the absence of a variable A. You can multiply it by $\bar{A} + A$ and convert it into two minterms, m_2 and m_6. The binary value of $B\bar{C}$ is 10 which corresponds to the column of 10 on a Karnaugh map. m_2 is located at the cross cell between the row of $A = 0$ and the column of $BC = 10$; m_6 is located at the cross cell between the row of $A = 1$ and the column of $BC = 10$. This means that the product term, $B\bar{C}$, can be mapped directly by placing 1s on the cross cells between the column of $BC = 10$ and the row of A = 0 and A = 1, which actually places 1s on the two cells corresponding to the column of $BC = 10$. Therefore, you can directly map the product term on a Karnaugh map.

Example 3.13 Map the following SOP expression on a Karnaugh map:

$$F(A, B, C, D) = AB\bar{D} + \bar{A}C$$

Solution

The given SOP expression has a domain of four variables A, B, C, and D. You can directly map the product term in this SOP expression on a 4-variable Karnaugh map. The mapping process is shown in Figure 3.6.5

The product term $AB\bar{D}$ is mapped by placing 1s in the two cross cells between the row of $AB = 11$ and the column of $D = 0$;

The product term $\bar{A}C$ is mapped by placing 1s in four cross cells between the row of $A = 0$ and the column of $C = 1$.

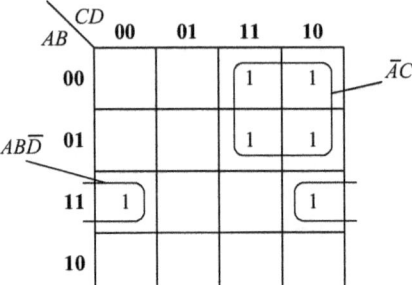

Figure 3.6.5: Karnaugh map of example 3.12.

(3) Mapping directly from a truth table

From the above introduction, you have known how to map an SOP expression. If the logic function is expressed by a truth table, how do you directly get a Karnaugh map from the truth table?

A truth table lists the output of a Boolean expression for all possible input variable combinations. Recall that each cell on a Karnaugh map corresponds to a row in a truth table which corresponds to a minterm. If the output is 1 for input variable combinations, place a 1 into the cells corresponding to the values of the input variable combinations; if the output is 0 for the input variable combinations, place a 0 into the cells corresponding to the values of the input variable combination.

For example, a logic function is expressed by a truth table, as shown in Figure 3.6.6 (a) and a mapping process from the truth table to a Karnaugh map is illustrated in Figure 3.6.6 (b).

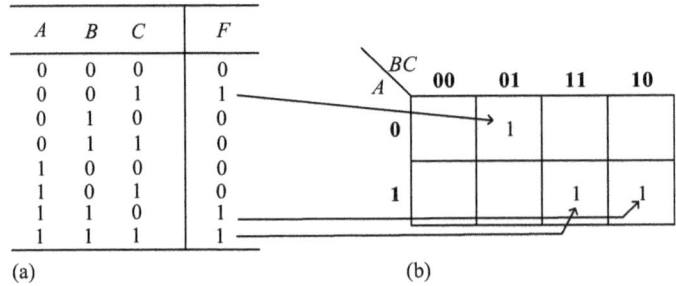

(a)   (b)

Figure 3.6.6: Mapping from a truth table (a) to a Karnaugh map (b).

Till now, you can see that a logic function can be represented by different ways such as the Boolean expression, the truth table, and the Karnaugh map.

2. Group the 1s

After an SOP expression has been mapped, the following step is to group the 1s. The goal is to maximize the size of the group and minimize the number of the groups. You can group 1s on a Karnaugh map by enclosing those adjacent cells containing 1s. The rules of group 1s are as follows:

Rule 1 A group must contain 2^i (i = 0,1,2,…) cells enclosed a rectangular or square frame. Remember that each cell in a group must be adjacent to one or more cells in the same group, but all cells in a group do not have to be adjacent to each other. In the case of a 4-variable map, a group can enclose 1, 2, 4, 8, and 16 cells.

Rule 2 The number of the groups should be minimized as possible and each group includes the largest possible number of 1s in accordance with rule 1.

Rule 3 Each 1 on the map must be included in at least one group. The 1s already in a group can be included in another group. But the overlapping groups must include non-common 1s.

Example 3.14 Group 1s in each of the Karnaugh map in Figure 3.6.7.

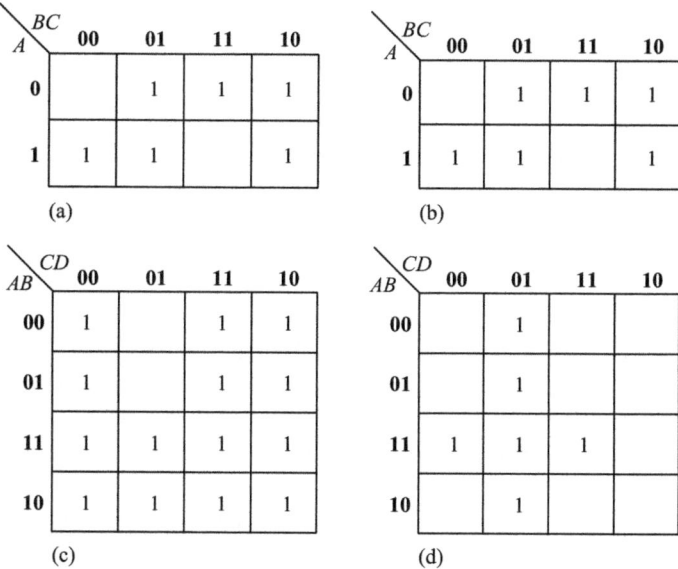

Figure 3.6.7: Karnaugh maps.

Solution

According to the rules of grouping 1s, the result of grouping 1s is shown in Figure 3.6.8.

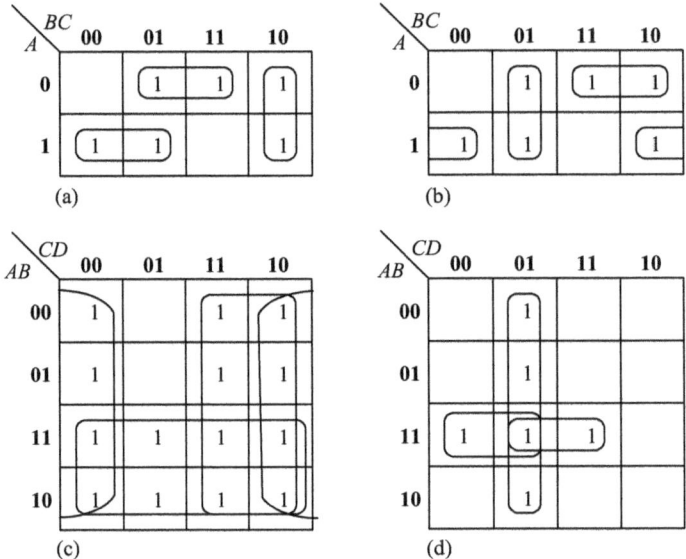

Figure 3.6.8: Grouping result for the Karnaugh maps in Figure 3.6.7.

For each Karnaugh map, the number of the groups is minimized and the size of the groups is maximized.

Notice that Figures 3.6.8 (a) and (b) have the same Karnaugh map but different grouping results. These mean that there may be more than one way to group the 1s for a Karnaugh map.

3. Determining the minimum SOP expression from the Karnaugh map

When all the 1s representing the minterms in an expression are properly mapped and grouped, the process of determining the resulting minimum SOP expression begins. The following rules are applied to find the minimum product terms and thus the minimum SOP expression:

(1) Each group of cells containing 1s produces one product term. This product term is composed of variables that have the same binary value in the group. The binary value "1" is represented by the variable and "0" is denoted by the variable complement. Variables with different binary values are eliminated.

(2) When all the minimum product terms are determined from the Karnaugh map, they are summed to form the minimum SOP expression.

Example 3.15 Use a Karnaugh map to minimize the following standard SOP expression:

$$F(A, B, C) = A\bar{C} + \bar{A}C + B\bar{C} + \bar{B}C$$

Solution

The given SOP expression has a domain of three variables A, B, and C. You can map this SOP expression on a 3-variable Karnaugh map.

The product term $A\bar{C}$ is mapped by placing 1s in the two cross cells between the row of $A = 1$ and the column of $C = 0$;

The product term $\bar{A}C$ is mapped by placing 1s in two cross cells between the row of $A = 0$ and the column of $C = 1$;

The product term $B\bar{C}$ is mapped by placing 1s in the two cross cells in the column of $BC = 10$;

The product term $\bar{B}C$ is mapped by placing 1s in the two cross cells in the column of $BC = 01$.

After mapping the SOP expression in a Karnaugh map, you will turn to next step of grouping 1s. The cells are grouped as shown in Figure 3.6.9.

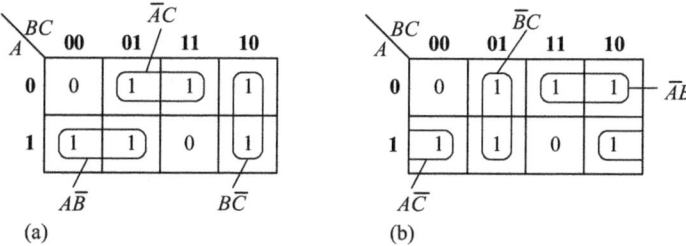

(a) (b)

Figure 3.6.9: Illustration of Karnaugh map simplification process.

Figure 3.6.9 shows two kinds of grouping method. Totally, there are at least three groups and each group of cells containing 1s creates one product term by eliminating the variables with different binary value and leaving the variables with the same binary value.

Add the three product terms and determine the resulting minimum SOP expression as follows: The minimum SOP expression in Figure 3.6.9(a) is

$$F(A, B, C) = A\bar{B} + \bar{A}C + B\bar{C}$$

The minimum SOP expression in Figure 3.6.9(b) is

$$F(A, B, C) = A\bar{C} + \bar{B}C + \bar{A}B$$

The above two SOP expressions are both minimized SOP expressions.

This illustrates that it is possible for you to get more than one result through Karnaugh map simplification.

Example 3.16 Use a Karnaugh map to minimize the following standard SOP expression:

$$F(A, B, C, D) = \bar{A}B\bar{C}D + A\bar{B}\bar{C}D + \bar{A}\bar{B}\bar{C}D + AB\bar{C}D + AB\bar{C}\bar{D} + ABCD$$

Solution

The given SOP expression has a domain of four variables A, B, C, and D. You can map this SOP expression on a 4-variable Karnaugh map.

The simplification step is illustrated in Figure 3.6.10.

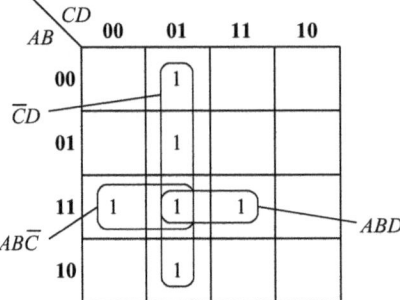

Figure 3.6.10: Illustration of simplification process of example 3.16.

The resulting minimum SOP expression is

$$F(A, B, C, D) = \bar{C}D + AB\bar{C} + ABD$$

Example 3.17 Use a Karnaugh map to minimize the following SOP expression:

$$F(A, B, C, D) = \bar{A}\bar{B}\bar{C} + A\bar{C}\bar{D} + A\bar{B} + ABC\bar{D} + \bar{A}\bar{B}C$$

Solution

Map the SOP expression on a 4-variable Karnaugh map. Notice that both groups exhibit "wrap around" adjacency. Figure 3.6.11 shows the simplification process. The group of eight cells is formed because the top and bottom cells are adjacent; the group of four cells is form to pick up the remaining two 1s because the cells in the outer columns are adjacent.

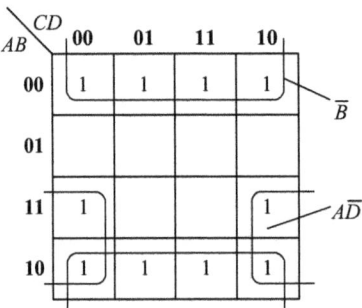

Figure 3.6.11: Illustration of simplification process of example 3.17.

The resulting minimum SOP expression is

$$F(A, B, C, D) = \bar{B} + A\bar{D}$$

Example 3.18 Use a Karnaugh map to minimize the following SOP expression:

$$F(A, B, C, D) = \bar{A}\bar{D} + A\bar{B}\bar{D} + \bar{A}\bar{C}D + \bar{A}CD$$

Solution

Map the SOP expression on a 4-variable Karnaugh map. Note that the four cells in the four corners are also adjacent and they can form a group in Figure 3.6.12.

The resulting minimum SOP expression is

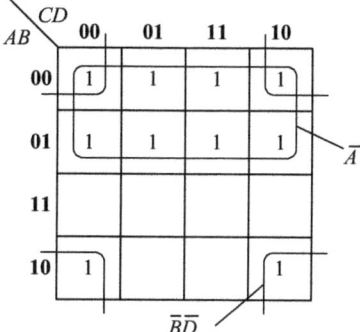

Figure 3.6.12: Illustration of simplification process of example 3.18.

$$F(A, B, C, D) = \bar{A} + \bar{B}\bar{D}$$

3.6.3 Karnaugh map POS minimization

A minimized POS expression contains the fewest possible sum terms with the fewest possible variables per term. The approach is the same as the simplification of SOP expression except that 0s representing the standard sum terms in the POS expression are placed on the Karnaugh map instead of 1s. 1s can be left off.

Generally, Karnaugh map POS minimization also includes three main steps. The first step is to map a POS expression on a Karnaugh map; the second step is to group 0s; the third step is to determine the minimized POS expression from the map [3].

Example 3.19 Use a Karnaugh map to minimize the following POS expression:

$$F = (A + B + C)(A + B + \bar{C})(A + \bar{B} + C)(A + \bar{B} + \bar{C})(\bar{A} + \bar{B} + C)$$

Solution

The given POS expression has a domain of three variables A, B, and C. You can map this POS expression on a 3-variable Karnaugh map.

Since all sum terms are standard sum terms which can be mapped directly in the corresponding cells.

$$A + B + C \rightarrow 000;$$
$$A + B + \bar{C} \rightarrow 001;$$
$$A + \bar{B} + C \rightarrow 010;$$
$$A + \bar{B} + \bar{C} \rightarrow 011;$$
$$\bar{A} + \bar{B} + C \rightarrow 110;$$

The mapping result is shown in Figure 3.6.13.

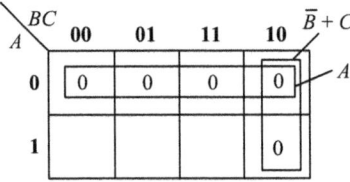

Figure 3.6.13: Illustration of simplification process of example 3.19.

Totally, there are five cells placed 0s, and the others are placed 1s. 1s can be left off.

Next step is to group 0s. One group of four cells is formed since the cells are adjacent. Another group of two cells is formed to pick up the remaining one 0.

The final step is to determine the minimum sum terms for each group and multiply them together. Thus the resulting minimum POS expression is

$$F(A, B, C) = A(\bar{B} + C)$$

3.6.4 Karnaugh map simplification with don't care terms

Sometimes a situation arises in which some input variable combinations are not allowed, or their corresponding output is not specified.

For example, there are six invalid combinations from 1010 to 1111 unused in BCD code. Since these unallowed states will never occur in an application involving BCD code, they can be treated as *"don't care"* terms with respect to their effect on the output. That is, for these "don't care" terms either a 1 or a 0 may be assigned to the output; it really does not matter since they will never occur.

Generally, the "don't care" terms are denoted by the letter x, or d, or Φ in the Boolean expression and a Karnaugh map. It can be used to help simplifying Boolean expression further in Karnaugh map. They could be chosen to be either '1' or '0', depending on whether they are beneficial to obtain a simpler expression.

Usually, when grouping the 1s, they can be treated as the 1s to make a larger group or as 0s if they cannot be used. The larger a group is, the simpler the resulting term will be.

Example 3.20 Minimize the following SOP expression with don't care terms:

$$F(A, B, C, D) = \sum m(0, 2, 3, 4, 6, 8, 10) + \sum d(11, 12, 14, 15)$$

where $\sum m(\cdot)$ lists the minterms and $\sum d(\cdot)$ represents "don't care" terms in an SOP expression. Another representation of don't care term can be expressed with a single equation. For example, $\sum d(11, 12, 14, 15) = 0$

Solution

Map a given SOP expression on a 4-variable Karnaugh map where the cells corresponding to the minterms in the expression are placed 1s, and the cells corresponding to the "don't care" terms in the expression are placed Xs, as shown in Figure 3.6.14.

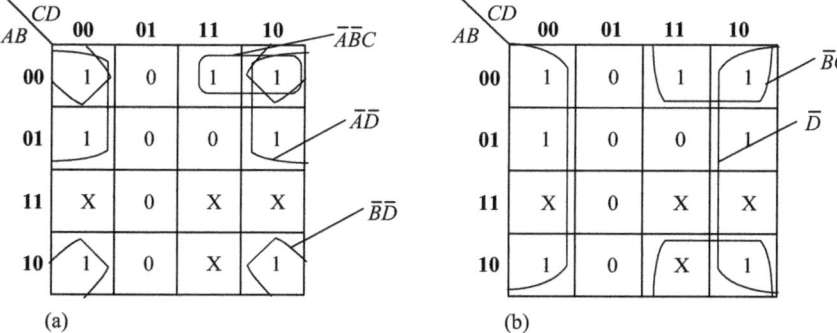

(a)   (b)

Figure 3.6.14: Comparison of "don't care" conditions unused (a) with used (b) to simplify an expression.

If the "don't care" terms are not used as 1s in Figure 3.6.14(a), the resulting expression is

$$F = \bar{A}\bar{D} + \bar{B}\bar{D} + \bar{A}\bar{B}C$$

If some "don't care" terms are used as 1s to get the simplest expression in Figure 3.6.14(b), the resulting expression is

$$F = \bar{D} + \bar{B}C$$

So you can see the advantages of using "don't care" terms to get the simplest expression.

3.7 Summary

1. Boolean algebra, also called logic algebra, is a mathematical tool used in the analysis and design of digital circuits.
2. The basic Boolean operations are the OR, AND, and NOT operations.
3. The commonly used combinational operations are NAND, NOR, XOR, and XNOR operations.

4. An AND gate produces a HIGH output only when all inputs are HIGH.
5. An OR gate outputs a HIGH when any input is HIGH.
6. An INVERTER output is the complement of its input.
7. A NOR gate is the same as an OR gate with its output connected to an INVERTER.
8. A NAND gate is the same as an AND gate with its output connected to an INVERTER.
9. The exclusive-OR (XOR) gate outputs a HIGH when its inputs are not the same.
10. The exclusive-NOR (XNOR) gate outputs a LOW when its inputs are not the same.
11. Boolean basic laws, rules and theorems can be used to simplify the expression of a logic circuit and can lead to a simpler way of implementing the circuit.
12. Two standard forms for logic expressions are standard sum-of-products form and standard product-of-sums form.
13. Karnaugh map is a graphical method for representing a circuit's truth table and generating a simplified expression for the circuit output.
14. Karnaugh map offers a systematic method for simplifying Boolean expression. Generally, Karnaugh map SOP minimization includes three main steps: mapping a SOP expression on a Karnaugh map, grouping 1s, deriving the minimized SOP expression.
15. Karnaugh map POS minimization also includes three main steps: mapping a POS expression on a Karnaugh map, grouping 0s, determining the minimized POS expression.
16. For Karnaugh map simplification, "don't care" terms can be chosen to be either '1' or '0', depending on which gives a simpler expression.

Key terms

Variable: A symbol used to represent a logical quantity that can have a value of 1 or 0, usually designated by an italic letter.

Complement: The inverse of a variable, usually indicated by a bar over the variable (overbar).

Literal: A variable or the complement of a variable.

Boolean expression: An arrangement of variables and logical operators used to express the operation of a logic circuit.

Truth table: A table listing all possible combinations of input variables with the corresponding outputs.

Timing diagram: A graph displaying the relationship of two or more waveforms with respect to each other on a time basis.

Product term: The Boolean product of two or more literals equivalent to an AND operation.

Sum term: The Boolean sum of two or more literals equivalent to an OR operation.

Product-of-sums (POS): A form of Boolean expression that is basically the ANDing of ORed terms.

Sum-of-product (SOP): A form of Boolean expression that is basically the ORing of ANDed terms.

Minterm: A product term containing all of n variables in the domain, in which each variable only appears in either true or complement form. Also called a standard product term.

Maxterm: A sum term containing all of n variables in the domain, in which each variable only appears in either true or complement form. Also called a standard sum term.

A standard POS expression: A POS expression in which each product term is a standard sum term or maxterm. Also called a maxterm expression.

A standard SOP expression: An SOP expression in which each product term is a standard product term or minterm. Also called a minterm expression.

Cell: An area on a Karnaugh map that represents a unique combination of variables in product form.

Karnaugh map: An arrangement of cells in which each cell represents a binary value of the input variables corresponding to a minterm.

"Don't care" term: A combination of input variables that is not allowed, or its corresponding output is not specified.

Adjacency: Characteristic of cells in a Karnaugh map in which there is a single-variable change from one cell to another cell next to it on any of its four sides.

Minimization: Process that results in an expression containing the fewest possible terms with the fewest possible variables.

Self-test

3.1 The complement of a variable is always _____.
   (a) 0   (b) 1   (c) the inverse of the variable   (d) equal to the variable

3.2 Literal refers to _____.
   (a) a variable   (b) a sum term
   (c) a product term   (d) a variable or the complement of variable

3.3 Which one of the following is a minterm of four variables?
   (a) $ABC\bar{C}$ (b) ABC (c) $\bar{A}B\bar{C}D$ (d) $\bar{A}B\bar{C}\bar{B}$

3.4 Which one of the following is a maxterm of three variables?
   (a) $A + B$ (b) $A + B + \bar{C} + D$
   (c) $\bar{A} + B + \bar{C}$ (d) $\bar{A} + B$

3.5 The Boolean expression $A + B + \bar{C}$ is _____.
   (a) a sum term   (b) a literal term
   (c) a product term   (d) a complemented term

3.6 The domain of the Boolean expression $A + \bar{A}BC + B\bar{C} + \bar{A}B\bar{C}D$ is
   (a) A and D (b) A only   (c) A, B, C, and D   (d) none of these

3.7 Three basic logic operations in logic algebra are _____.
   (a) AND, OR, and NOT   (b) NAND, NOR, and AND-OR-NOT
   (c) NAND, OR, and AND-OR   (d) NOR, AND-OR, and AND-OR-NOT

3.8 According to DeMorgan's theorems, which one of the following equalities is correct?
   (a) $\overline{A + B + C} = \bar{A} + \bar{B} + \bar{C}$ (b) $\overline{ABC} = \bar{A}\bar{B}\bar{C}$
   (c) $\overline{ABC} = \bar{A} + \bar{B} + \bar{C}$ (d) none of these

3.9 Which one of the following expressions is an SOP expression?
   (a) $(\bar{A} + \bar{B} + \bar{C})(A + B + C)$ (b) $A + \bar{B}(C + \bar{D})$
   (c) $\bar{A}\bar{B} + A\bar{C}D$ (d) both (b) and (c)

3.10 Which one of the following expressions is a POS expression?
   (a) $AB + ACD$ (b) $A + \bar{B}(C + \bar{D})$
   (c) $(\bar{A} + \bar{B} + \bar{C})(A + B + C)$ (d) both (b) and (c)

3.11 Which one of the following Boolean expressions is a standard SOP expression?
   (a) $AB + ACD + BC$ (b) $AB + A\bar{B} + \bar{A}\bar{B}$
   (c) $\bar{A}\bar{B}\bar{C} + ABC$ (d) both (b) and (c)

3.12 Which one of the following Boolean expressions is a standard POS expression?
   (a) $AB + ACD + BC$ (b) $(A + B)(A + \bar{B}) + \overline{AB}$
   (c) $(\bar{A} + \bar{B} + \bar{C})(A + B + C)$ (d) both (b) and (c)

3.13 Which one of the following combinations can make the value of Boolean expression $F = AB + CD$ a 1?
   (a) $A = 0$, $BC = 0$, $D = 0$ (b) $A = 0$, $BD = 0$, $C = 0$
   (c) $AB = 1$, $C = 0$, $D = 0$ (d) $AC = 1$, $B=0$, $D = 0$

3.14 The result of $1\oplus1\oplus1\oplus1 \cdots$ (2003 1s) and $1\odot1\odot1\odot1 \cdots$ (2003 1s) is
   (a) 0,0   (b) 1,0   (c) 0,1   (d) 1,1

3.15 Which one of the following Karnaugh map simplifications in Figure T3.1 is not the minimum SOP expression?

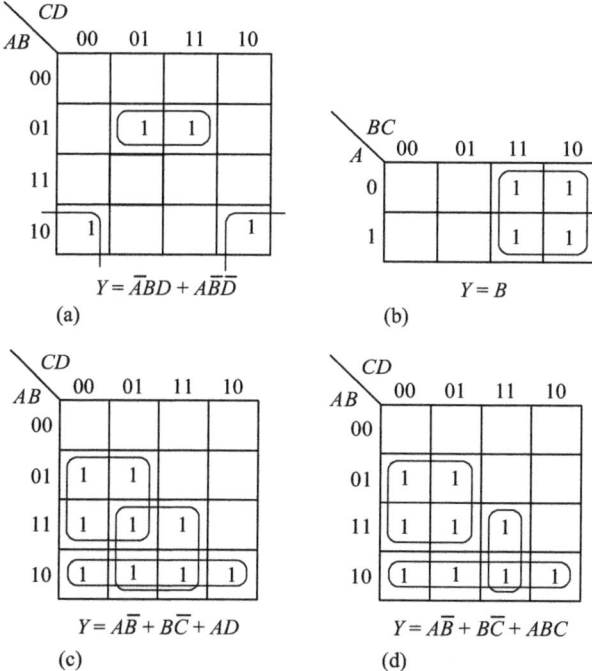

$$Y = \bar{A}BD + A\bar{B}\bar{D}$$

(a)

$$Y = B$$

(b)

$$Y = A\bar{B} + B\bar{C} + AD$$

(c)

$$Y = A\bar{B} + B\bar{C} + ABC$$

(d)

Figure T3.1

Problems

3.1 Write a Boolean expression that is a 1 only if all of its variables (A, B, C, and D) are 1s.

3.2 Write a Boolean expression that is a 1 when one or more variables (A, B, C, and D) are 0s.

3.3 Write a Boolean expression that is a 0 when one or more variables (A, B, C, and D) are 1s.

3.4 Find the values of the variables that make each product term 1 and each sum term 0.
(a) $\bar{A}\bar{B}$
(b) $A + \bar{B} + C + \bar{D}$
(c) $\bar{A}\bar{B}C$
(d) $A + \bar{B} + C$

3.5   Construct a truth table for each of the following Boolean expressions.
(1) $F = AB + \bar{A}C + BC$
(2) $F = (A + B)(\bar{A} + C)(B + C)$
(3) $F = ABCD + A\bar{B}\bar{C} + \bar{A}B$

3.6   Write Boolean expression and truth table for the logic circuits in Figure P3.1.

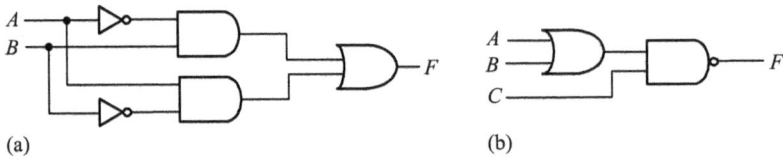

(a)   (b)

Figure P3.1

3.7   Prove the following equalities by using Boolean algebra.
(1) $A + BC = (A + B)(A + C)$
(2) $BC + AD = (A + B)(B + D)(A + C)(C + D)$
(3) $\bar{A}\bar{B}C + \bar{A}B\bar{C} + A\bar{B}\bar{C} + ABC = A \oplus B \oplus C$
(4) $(AB + C)B = AB\bar{C} + \bar{A}BC + ABC$

3.8   Using Boolean algebra, simply the following expressions as much as possible.
(1) $F = AB + AC + \bar{A}B + B\bar{C}$
(2) $F = (AB + \bar{A}B + A\bar{B})(\bar{A}\bar{B} + CD)$
(3) $F = ABCD + \bar{A}BC\bar{D} + \overline{BC\bar{D}}$
(4) $F = ABC + \overline{A + B + \bar{C}} + \bar{A}\bar{B}CD$
(5) $F = BD + B(D + E) + \bar{D}(A + D)$
(6) $F = A + A\bar{B}\bar{C} + \bar{A}CD + \bar{C}E + \bar{D}E$

3.9   Using Boolean algebra, simplify the following expressions as much as possible.
(1) $F = A\bar{B}C + CD + B\bar{D} + \bar{C}$
(2) $F = A + \bar{B} + \overline{CD} + \overline{AD} \cdot \bar{B}$
(3) $F = (\overline{A \oplus B})(B \oplus \bar{C})$
(4) $F = A\bar{B} + \bar{A}C + \bar{B}C$

3.10   Convert the following Boolean expressions into the standard SOP form.
(1) $F = A\bar{B}C + \bar{A}\bar{B} + AB\bar{C}D$
(2) $F = AB + A\bar{B}C$

3.11   Convert the following Boolean expressions into the standard POS form.
(1) $F = (A + \bar{B} + C)(\bar{B} + C + \bar{D})(A + \bar{B} + \bar{C} + D)$
(2) $F = A(A + \bar{B})(\bar{A} + C)$

3.12 Write the standard SOP expression and the standard POS expression from the truth table as shown in Table P3.1

Table P3.1: Truth table.

Inputs			Output
A	B	C	F
0	0	0	0
0	0	1	0
0	1	0	0
0	1	1	1
1	0	0	0
1	0	1	1
1	1	0	1
1	1	1	1

3.13 Convert the following minterm expressions into maxterm expression.
(1) $F(A, B, C) = \sum m(0, 1, 2, 5, 7)$
(2) $F(A, B, C, D) = \sum m(0, 2, 3, 5, 7, 8, 9, 10, 13)$

3.14 Convert the following maxterm expressions into minterm expression.
(1) $F(A, B, C) = \prod M(3, 6, 7)$
(2) $F(A, B, C, D) = \prod M(6, 7, 9, 13, 15)$

3.15 Use Karnaugh map to simplify the following minterm expressions into minimum SOP expression and minimum POS expression, respectively.
(1) $F(A, B, C) = \sum m(0, 1, 2, 5, 6)$
(2) $(A, B, C, D) = \sum m(0, 1, 2, 3, 4, 6, 7, 8, 9, 10, 11, 14)$
(3) $F(A, B, C, D) = \sum m(0, 1, 4, 6, 8, 9, 10, 12, 13, 14, 15)$
(4) $F(A, B, C, D) = \sum m(0, 1, 2, 3, 4, 5, 8, 10, 11, 12)$

3.16 Use Karnaugh map to simplify the following SOP expressions into minimum SOP expression and minimum POS expression, respectively.
(1) $F = \bar{B}\bar{C} + AB + \bar{A}B\bar{C}$
(2) $F = \bar{A}\bar{B} + BC + B\bar{C}$
(3) $F = A\bar{C} + \bar{A}C + \bar{B}C + B\bar{C}$
(4) $F = ABC + ABD + A\bar{C}D + \bar{C}D + A\bar{B}C + \bar{A}C\bar{D}$
(5) $F = A\bar{B}\bar{C} + AC + \bar{A}\bar{B}D$
(6) $F = AB + \bar{C}\bar{D} + \bar{A}\bar{B}C + AD + A\bar{B}C$
(7) $F = \bar{A}\bar{C} + \bar{A}\bar{B} + \bar{B}\bar{C}\bar{D} + BD + A\bar{B}\bar{D} + \bar{A}BC\bar{D}$

3.17 Use Karnaugh map to simplify the following expressions into minimum SOP expression and minimum POS expression, respectively.

(1) $F = A(B + \bar{C})(A + D)(B + C + \bar{D})(\bar{A} + \bar{B} + \bar{C} + D)$

(2) $F(A, B, C, D) = \prod M(1, 2, 7, 9, 10)$

(3) $F(A, B, C, D) = \prod M(0, 2, 4, 6, 9, 11, 13)$

3.18 Use Karnaugh map to simplify the following expressions containing "don't care" term into the minimum SOP expression.

(1) $F(A, B, C, D) = \sum m(3, 6, 8, 9, 11, 12) + \sum d(0, 1, 2, 13, 14, 15)$

(2) $F(A, B, C, D) = \sum m(0, 2, 3, 4, 5, 6, 11, 12) + \sum d(8, 9, 10, 13, 14, 15)$

(3) $F(A, B, C, D) = \sum m(0, 1, 2, 3, 6, 8) + \sum d(10, 11, 12, 13, 14, 15)$

(4) $\begin{cases} F_3 = \overline{A + C + D} + \bar{A}\bar{B}C\bar{D} + A\bar{B}\bar{C}D \\ AB + AC = 0 \end{cases}$

3.19 For each group of the following Boolean expressions, use a Karnaugh map to implement the logic operations:

(a) $F \cdot G$;

(b) $F + G$;

(c) $F \oplus G$

(1) $\begin{cases} F = AB\bar{C} + CD(A + \bar{B}) + \bar{B}C\bar{D} \\ G = (AB + C\bar{D})AB\bar{C} + \bar{A}\bar{B}\bar{D} \end{cases}$

(2) $\begin{cases} F(A, B, C, D) = \sum m(2, 4, 6, 9, 13, 14) + \sum d(0, 1, 3, 8, 11, 15) \\ G(A, B, C, D) = \sum m(4, 5, 7, 9, 12, 13, 14) + \sum d(1, 3, 8, 10) \end{cases}$

4 Combinational logic circuits

4.1 Introduction

In Chapters 2 and 3, fundamental mathematical tool and basic logic gates were introduced to analyze and design digital circuits. Basic logic gates can be combined to form various types of logic circuits with different functions: comparison, encoding, decoding, counting, storage, and so on. Generally, logic circuits are divided into two categories: combinational logic circuits and sequential logic circuits . When logic gates are connected together to form a specified output for certain specified combination of input variables and no storage involved, the resulting circuit is in the category of combinational logic circuits. In a combinational circuit, outputs solely depend on current inputs. While in a sequential logic circuit, outputs depend not only on current inputs but also on previous inputs. This chapter introduces the analysis and design of logic circuits with logic gates. Races and hazards in a combinational logic circuit are also discussed.

With the rapid development of integrated circuit technology, methods of constructing digital circuits also evolve. Currently, there are two methods to implement a more complicated digital circuit. One is to construct a specified logic circuit with the universal integrated chips that mainly involve medium-scale integration (MSI) and large-scale integration (LSI) chips. The other is to implement a specified integrated logic circuit with programmable logic device (PLD) by electronic design automation software.

This chapter also introduces several types of MSI combinational logic circuits, including adders, decoders, encoders, multiplexers, and their application for designing a more complicated logic circuit. The commonly used hardware description language (HDL), such as Verilog HDL, and typical Verilog HDL descriptions of commonly used MSI are also introduced.

The objectives of this chapter are to
- Define a combinational circuit
- Analyze a logic function of a combinational logic circuit constructed by logic gates
- Design a combinational circuit by using logic gates to satisfy practical requirements
- Explain races and hazards in a combinational logic circuit
- Describe the functions of several types of MSI combinational logic circuit
- Apply these MSI combinational logic circuits to design a specified logic circuit

4.2 Analysis and design of combinational logic circuits

Combinational logic circuits are composed of logic gates. The obvious characteristics of combinational circuits are that neither storage nor a signal feedback path from the

https://doi.org/10.1515/9783110614916-004

output to the input is involved. That is to say, the output level of a combinational logic circuit is only determined by the combination of current input levels and has nothing to do with the previous state of circuit [25]. This section introduces the analysis and design procedure of combinational logic circuits.

The structure of combinational circuits is shown in Figure 4.2.1, which has multiple input variables $(X_1, X_2, \ldots, X_n)$ and output variables $(P_1, P_2, \ldots, P_m)$, and signals can be transmitted only forwardly from inputs to outputs.

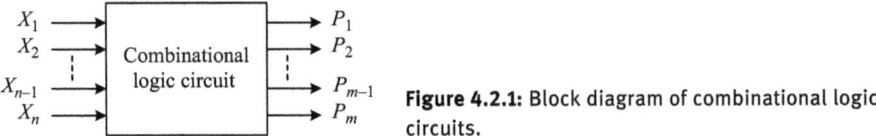

Figure 4.2.1: Block diagram of combinational logic circuits.

The logic relationship between each output P_i $(i = 1,2,\ldots,m)$ and n inputs can be described by the following logic function:

$$\begin{cases} P_1 = f_1(X_1, X_2, \ldots, X_{n-1}, X_n) \\ P_2 = f_2(X_1, X_2, \ldots, X_{n-1}, X_n) \\ \vdots \\ P_m = f_m(X_1, X_2, \ldots, X_{n-1}, X_n) \end{cases} \tag{4.2.1}$$

The objectives of this section are to
- Define combinational logic circuits
- Explain the analysis procedure of a combinational logic circuit
- Analyze the function of a combinational logic circuit by using logic gates
- Explain the design procedure of a combinational logic circuit
- Design a specified combinational logic circuit by using logic gates

4.2.1 Analysis of combinational logic circuits

The analysis of combinational logic circuits is to describe a given circuit with the logic expressions and find the relationship between output variables and input variables, and thus deduce the implementing functionality. The detailed analysis procedure is shown in Figure 4.2.2.

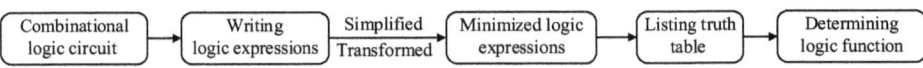

Figure 4.2.2: Analysis procedure of a combinational logic circuit.

First, determine the inputs and outputs of a given combinational logic circuit. Second, write the logic expressions from inputs to outputs stage by stage according to a logic circuit diagram and obtain the final logic expressions describing the logic relationship between the input variables and output variables. Simplify or transform the logic expressions appropriately. If the structure of the circuit is complex, the logic expressions should be simplified with the Boolean algebra or the Karnaugh map so that the logic relationship between inputs and outputs becomes clear. Third, list the corresponding truth table from the resulting logic expression. Finally, deduce the logic function of a given combinational logic circuit by analyzing the relationship between outputs and inputs from the truth table.

Example 4.1 Analyze the logic function of the circuit shown in Figure 4.2.3.

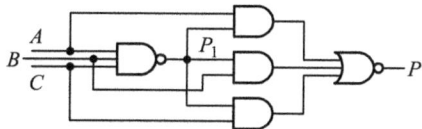

Figure 4.2.3: Logic diagram of Example 4.1.

Solution

(1) This circuit is a combinational logic circuit with three input variables A, B, and C; an intermediate output variable P_1; and a final output variable P.
(2) Write the logic expression from inputs to outputs stage by stage as follows:

$$P_1 = \overline{ABC} \tag{4.2.2}$$

$$P = \overline{AP_1 + BP_1 + CP_1} \tag{4.2.3}$$

Substitute eq. (4.2.2) into eq. (4.2.3) and then transform the logic expression into the standard SOP form:

$$P = \overline{A\,\overline{ABC} + B\,\overline{ABC} + C\,\overline{ABC}}$$

$$= \overline{\overline{ABC}(A+B+C)}$$

$$= ABC + \overline{A+B+C} \tag{4.2.4}$$

$$= ABC + \overline{A}\,\overline{B}\,\overline{C} = m_7 + m_0 = \sum m(0,7)$$

(3) List the truth table from the final logic expression in eq. (4.2.4). There are three input variables and one output variable; thus, there are $2^3 = 8$ possible combinations of input variables. As only m_0 and m_7 appear in the logic expression, only two combinations of input variables, 000 and 111, make the corresponding output yield a 1. The resulting truth table is listed in Table 4.2.1.

(4) Deduce the logic function of the given combinational logic circuit from the truth table.
It can be found that the output is a HIGH (1) only when all the variables A, B, and C have the same value and otherwise the output is a LOW (0). Therefore, the logic circuit in Figure 4.2.3 is used to judge whether three inputs are the same or not and thus it is called "consistent judgment circuit."

Table 4.2.1: Truth table.

A	B	C	P
0	0	0	1
0	0	1	0
0	1	0	0
0	1	1	0
1	0	0	0
1	0	1	0
1	1	0	0
1	1	1	1

The consistent judgment circuit can be used in detecting the working state of equipment with high reliability. Several devices are switched on at the same time; only one device is actually working while all others are standby. As long as the working device breaks down, the coincidence judgment circuit will send a trigger signal to cut off the faulty device and put one of the standby devices into the working state.

Example 4.2 Analyze the logic function of the circuit shown in Figure 4.2.4.

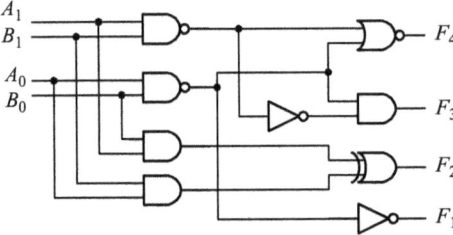

Figure 4.2.4: Logic diagram.

Solution
(1) This circuit is a combinational logic circuit with four input variables A_1, A_0, B_1, and B_0 and four output variables F_4, F_3, F_2, and F_1.
(2) Write the logic expression from inputs to outputs and transform them into simpler forms:

$$\begin{cases} F_4 = \overline{\overline{A_1 B_1} + \overline{A_0 B_0}} = \overline{\overline{A_1 B_1}} \cdot \overline{\overline{A_0 B_0}} = A_1 B_1 A_0 B_0 \\ F_3 = \overline{\overline{A_1 B_1} \cdot \overline{A_0 B_0}} = A_1 B_1 \cdot (\overline{A_0} + \overline{B_0}) = A_1 \overline{A_0} B_1 + A_1 B_1 \overline{B_0} \\ F_2 = (A_1 B_0) \oplus (A_0 B_1) = (A_1 B_0) \cdot \overline{(A_0 B_1)} + \overline{(A_1 B_0)} \cdot (A_0 B_1) \\ F_1 = \overline{\overline{A_0 B_0}} = A_0 B_0 \end{cases} \qquad (4.2.5)$$

(3) List the truth table from the final logic expression in eq. (4.2.5).
The truth table is obtained by substituting 16 possible combinations of the four input variables into eq. (4.2.5) and calculating the corresponding values of the output variables for each combination of input variables, as shown in Table 4.2.2.

(4) Deduce the logic function of the circuit from the truth table.
This circuit has multiple outputs, and thus we should examine the values of four input variables and four output variables at the same time to analyze the function of a circuit. It can be found from Table 4.2.2 that this circuit implements the multiplication of two-bit binary numbers, where A_1A_0 represents multiplicand, B_1B_0 represents multiplier, and $F_4F_3F_2F_1$ represents the product in the form of four-bit binary numbers.

Table 4.2.2: Truth table.

A_1	A_0	B_1	B_0	F_4	F_3	F_2	F_1	Comments
0	0	0	0	0	0	0	0	0×0
0	0	0	1	0	0	0	0	0×1
0	0	1	0	0	0	0	0	0×2
0	0	1	1	0	0	0	0	0×3
0	1	0	0	0	0	0	0	1×0
0	1	0	1	0	0	0	1	1×1
0	1	1	0	0	0	1	0	1×2
0	1	1	1	0	0	1	1	1×3
1	0	0	0	0	0	0	0	2×0
1	0	0	1	0	0	1	0	2×1
1	0	1	0	0	1	0	0	2×2
1	0	1	1	0	1	1	0	2×3
1	1	0	0	0	0	0	0	3×0
1	1	0	1	0	0	1	1	3×1
1	1	1	0	0	1	1	0	3×2
1	1	1	1	1	0	0	1	3×3

Example 4.3 Analyze the logic function of a circuit shown in Figure 4.2.5.

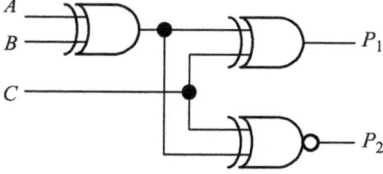

Figure 4.2.5: Logic diagram.

Solution

(1) This circuit is a combinational logic circuit with three input variables A, B, and C, and two output variables P_2 and P_1.

(2) Write the logic expression from inputs to outputs:

$$\begin{cases} P_1 = A \oplus B \oplus C = \bar{A}\bar{B}C + \bar{A}B\bar{C} + A\bar{B}\bar{C} + ABC \\ P_2 = (A \oplus B) \odot C = \bar{A}\bar{B}\bar{C} + \bar{A}BC + A\bar{B}C + AB\bar{C} \end{cases}$$

(4.2.6)

(3) List the truth table.

(4) Deduce the logic function of the circuit from the truth table.

When the number of 1s in the inputs is odd, then the output P_1 is a 1 and the output P_2 is a 0. When the number of 1s in the inputs is even, P_1 is a 0 and P_2 is a 1. This circuit is a parity generator. When odd number of 1s are applied to the inputs, a HIGH is produced on the output P_1, which acts as a flag representing an odd number of 1s are in the inputs; when even number of 1s are applied to inputs, a HIGH is produced on the output P_2, which acts as a flag representing an even number of 1s are in the inputs (Table 4.2.3).

Table 4.2.3: Truth table.

A	B	C	P_1(odd number)	P_2(even number)
0	0	0	0	1
0	0	1	1	0
0	1	0	1	0
0	1	1	0	1
1	0	0	1	0
1	0	1	0	1
1	1	0	0	1
1	1	1	1	0

4.2.2 Design of combinational logic circuits

The design of combinational logic circuits is also referred to as the synthesis of combinational logic circuits. The requirements of a logic circuit are generally described in words, or in the form of truth tables or waveforms. As logic problems involved in practical engineering applications are different, various methods can be used for different objects and requirements; even with the same design object, design methods or ideas can be different. The implementation of digital circuits can use small-scale integration (SSI), MSI, LSI, or PLD. Although the design method of a combinational circuit is different for different objects, the description of basic logic problem, design idea, and consideration of practical engineering problems have some common objectives. There are three main objectives for the design of circuits: one is to reduce the cost, which means that the number of gates or integrated chips used is as few as possible for a logic circuit. Then, the design can guarantee a high speed of performance for a circuit. Finally, the design can be simpler, which means some parts can be reused as possible. Here, we focus on the design method of a combinational logic circuit by using logic gates.

The design method for combinational logic circuits with logic gates is as follows [26]:

Step 1 Make logical abstraction in terms of the given practical problems by determining the required number of input and output variables, and then assign logical values to variables.

An event can be abstracted as a logic problem only when causes and results of the event can be clearly answered as "yes" and "no", or "true" and "false". The causes (or conditions) of the event are referred to as input variables, and the results of the event are referred to as output variables. The values of each input variable and each output variable are either "1" or "0" representing "yes" and "no", or "true" and "false".

Step 2 Derive the truth table that defines the required relationship between inputs and outputs from the given logic requirements.

It is an easy way to describe a practical logic problem with a truth table. The correctness of the truth table is the prerequisite for a successful design. A truth table should include all possible combinations of input values and their corresponding output values. Additionally, the constraints of the problem should be considered as well.

Step 3 Write the logic expression and make necessary simplification and transformation for the requirement of different devices.

When a combinational circuit is implemented with logic gates, the logic expression should be simplified and transformed into a suitable form in order to implement with the given logic gates.

Step 4 Draw a logic diagram and verify the correctness of the design.

Draw a logic diagram in terms of the final logic expression. After completing the circuit design, it is also necessary to carry out an analysis of the designed circuit to verify if the circuit meets the requirements of practical logic problem.

The above-described steps are general procedure of designing a combinational logic circuit by using logic gates. In a real engineering, a complete process of logic-circuit design also includes additional steps as follows: replace the logic symbols in the diagram with electronic components or integrated chips; then design a printed circuit board, and finally complete the soldering, assembling, and debugging of the circuit board.

Example 4.4 Design a logic circuit to control a bulb "on" and "off." As shown in Figure 4.2.6, the bulb can be switched on or off no matter when people go downstairs or upstairs.

Solution

Step 1 Make logical abstraction in terms of the given practical problems by determining the required number of input and output variables and then assign logical values to variables.

According to the causality of this event, the controlled object is a bulb, and conditions to realize the control are the states of two switches: upstairs and downstairs. The bulb has two states: "on" and

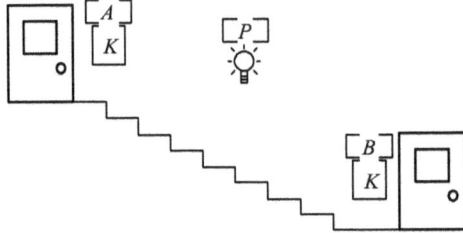

Figure 4.2.6: Illustration diagram.

"off"; and the switches have two states: "upward" and "downward." So this event can be abstracted as a logic proposition. Two switches are denoted by variables A and B, whose value equals to 1 when they are switched upward and equals to 0 when they are switched downward; the bulb is denoted by variable P, whose value equals to 1 when the bulb is on and equals to 0 when the bulb is off.

Step 2 Derive the truth table that defines the required relationship between inputs and outputs from functional description of the circuit.

Assume that initial state is $A = 0$, $B = 0$, and the bulb is on. Note that if the assumption of initial state is different, the designed circuit is different too. When the state of one switch changes, the state of the bulb changes accordingly, but when the states of the two switches change at the same time, the state of the bulb remains the same. The truth table of the circuit is shown in Table 4.2.4.

Step 3 Write the logic expression from the truth table.

Table 4.2.4: Truth table.

A	B	P
0	0	1
0	1	0
1	0	0
1	1	1

The logic expression can be directly obtained from the truth table by summing the minterms whose corresponding output equals to 1 as follows:

$$P = AB + \bar{A}\bar{B} \tag{4.2.7}$$

Equation (4.2.7) shows that the relationship between the two inputs is XNOR. It is already a minimized SOP form, and no further simplification is needed.

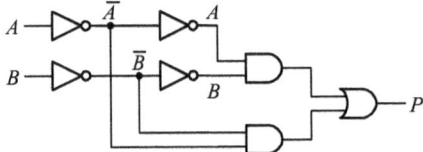

Figure 4.2.7: Logic diagram of Example 4.4.

Step 4 Draw a logic diagram from the logic expression.

Draw a logic diagram from the logic expression in eq. (4.2.7). The two stages of inverters in the logic diagram serve as input buffers. They not only provide the input variables and its complement forms, but also enhance the driving capability of the input signals (Figure 4.2.7).

Example 4.5 Design a car buzzer alarm circuit by using NAND gates. The buzzer alarms when both the windows and the doors of the car are open, or when the car key is in the ignition position and the car doors are open.

Solution

Step 1 Make logical abstraction in terms of the given practical problems by determining the required number of input and output variables and then assign logical values to the variables.

According to the causality of this event, the controlled object is a buzzer, and conditions to control the buzzer are the states of windows, the doors, and the position of key. The buzzer has two states: "alarm on" and "alarm off"; the windows and the doors have two states "opened" and "closed"; and the position of the key is "ignition" or "shut". So this event can be abstracted as a logic proposition. The windows and doors are denoted by variables W and D, whose value equals to 1 when they are opened and equals to 0 when they are closed; $K = 1$ denotes the key is in the ignition position, and $F = 1$ denotes the buzzer alarm is on.

Step 2 Write the logic expression directly and list the truth table.

In terms of logic requirements, if both $W = 1$ and $D = 1$ or both $D = 1$ and $K = 1$, then $F = 1$. So the logic expression can be directly written out as follows:

$$F = KD + WD \tag{4.2.8}$$

And the truth table of a car buzzer alarm circuit is shown in Table 4.2.5.

Table 4.2.5: Truth table.

D	W	K	F
0	0	0	0
0	0	1	0
0	1	0	0
0	1	1	0
1	0	0	0
1	0	1	1
1	1	0	1
1	1	1	1

Step 3 To implement the circuit only with NAND gates, eq. (4.2.8) should be converted into NAND-NAND form as follows:

$$F = \overline{\overline{KD + WD}} = \overline{\overline{KD} \cdot \overline{WD}} \tag{4.2.9}$$

Draw the logic circuit from eq. (4.2.9), as shown in Figure 4.2.8.

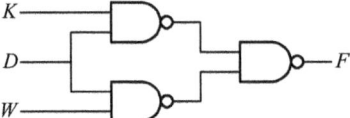

Figure 4.2.8: Logic diagram.

Example 4.6 Design a code converter to convert an excess-3 code into the corresponding 8421BCD code.

Solution

Step 1 According to the logic requirement, a code converter converts an excess-3 code into the corresponding 8421BCD code. Thus, there are four input variables represented by E_3, E_2, E_1, and E_0, and four output variables denoted by B_3, B_2, B_1, and B_0.

Step 2 List the truth table of the code converter shown in Table 4.2.6. The four input variables have 16 combinations. There are only ten valid combinations in excess-3 code, and thus the rest six combinations do not exist, which can be treated as "don't care" terms.

Table 4.2.6: Truth table.

Decimal	E_3	E_2	E_1	E_0	B_3	B_2	B_1	B_0
0	0	0	1	1	0	0	0	0
1	0	1	0	0	0	0	0	1
2	0	1	0	1	0	0	1	0
3	0	1	1	0	0	0	1	1
4	0	1	1	1	0	1	0	0
5	1	0	0	0	0	1	0	1
6	1	0	0	1	0	1	1	0
7	1	0	1	0	0	1	1	1
8	1	0	1	1	1	0	0	0
9	1	1	0	0	1	0	0	1
	0	0	0	0	×	×	×	×
	0	0	0	1	×	×	×	×
	0	0	1	0	×	×	×	×
	1	1	0	1	×	×	×	×
	1	1	1	0	×	×	×	×
	1	1	1	1	×	×	×	×

Step 3 Use the Karnaugh maps to simplify the logic functions with a consideration of the constraint condition (don't care terms), as shown in Figure 4.2.9.

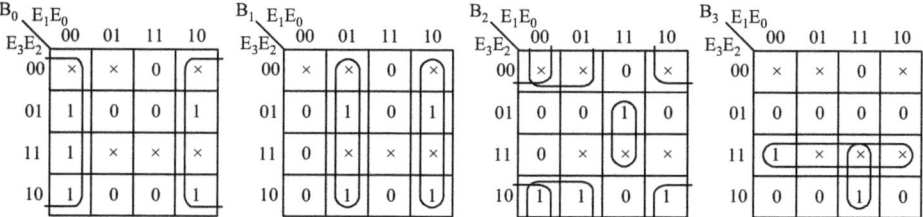

Figure 4.2.9: The Karnaugh map.

The logic expressions for four outputs are obtained from the Karnaugh map as follows:

$$\begin{cases} B_0 = \bar{E}_0 \\ B_1 = E_0\overline{E_1} + \overline{E_0}E_1 = E_0 \oplus E_1 \\ B_2 = \bar{E}_2\bar{E}_0 + \bar{E}_2\bar{E}_1 + E_2E_1E_0 = \bar{E}_2\overline{E_1E_0} + E_2E_1E_0 = E_2 \odot (E_1E_0) \\ B_3 = E_3E_2 + E_3E_1E_0 \end{cases} \qquad (4.2.10)$$

Step 4 According to the logic expressions, the logic diagram converting excess-3 code into 8421BCD code is shown in Figure 4.2.10.

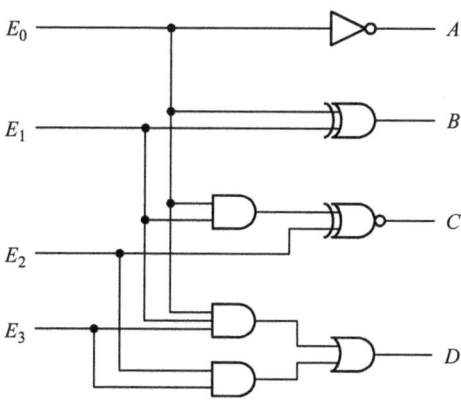

Figure 4.2.10: Logic diagram.

4.3 Adders

Adders are important not only in computers but also in many types of digital systems in which numerical data are processed. The basic function of a computer is arithmetic operations, and the adder operation is fundamental to the study of computers and digital systems. Subtraction operation can be implemented with adders. In this section, the half-adder and the full-adder are introduced first. Then you will learn how to construct a multi-bit adder and applications of adders.

The objectives of this section are to

– Explain the function of a half-adder and draw a half-adder logic diagram

- Explain the function of a full-adder and draw a full-adder logic diagram using two half-adders
- Apply full-adders to construct a multi-bit binary adder
- Explain ripple carry and look-ahead carry
- Expand the MSI adder to a more bit binary adder
- Apply the MSI adder to implement the subtraction of two binary numbers

4.3.1 Half-adders

The addition of two one-bit binary numbers is illustrated in Figure 4.3.1, where the two one-bit binary numbers are denoted by A_0 and B_0; the sum of A_0 and B_0 is denoted by S_0; and the carry is denoted by C_0. A logic circuit implementing the addition operation of two one-bit binary numbers is called a *half-adder*.

The half-adder accepts two binary digits A_0 and B_0 on its inputs and produces two binary digits on its outputs, a sum bit S_0 and a carry bit C_0.

From Figure 4.3.1, the truth table of a half-adder is listed below.

$$
\begin{array}{c}
A_0 \\
+\, B_0 \\
\hline
C_0\, S_0
\end{array}
\Rightarrow
\begin{array}{cc}
0 \\
+\ 0 \\
\hline
0\ 0
\end{array}
\quad
\begin{array}{cc}
0 \\
+\ 1 \\
\hline
0\ 1
\end{array}
\quad
\begin{array}{cc}
1 \\
+\ 0 \\
\hline
0\ 1
\end{array}
\quad
\begin{array}{cc}
1 \\
+\ 1 \\
\hline
1\ 0
\end{array}
$$

Figure 4.3.1: Addition of two one-bit binary numbers.

The logic expressions of outputs of a half-adder are deduced from Table 4.3.1 as follows:

$$
\begin{cases}
S_0 = A_0\bar{B}_0 + \bar{A}_0 B_0 = A_0 \oplus B_0 \\
C_0 = A_0 B_0
\end{cases}
\tag{4.3.1}
$$

Table 4.3.1: Truth table of a half-adder.

A_0	B_0	C_0	S_0
0	0	0	0
0	1	0	1
1	0	0	1
1	1	1	0

From eq. (4.3.1), the logic implementation required for the half-adder function can be developed. A half-adder can be implemented by an AND gate and an XOR gate, as shown in Figure 4.3.2(a). An alternative scheme is to use three AND gates and an OR gate, as shown in Figure 4.3.2(b). The logic symbol of a half-adder is shown in Figure 4.3.2(c).

4.3.2 Full-adders

A half-adder can only realize an addition of two one-bit binary numbers. However, for the addition of two multi-bit binary numbers, a half-adder is not feasible as the input carry is not considered.

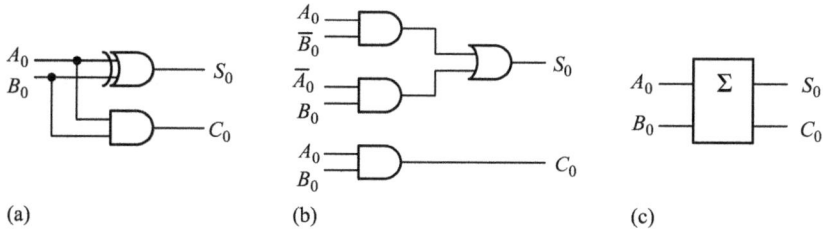

(a)   (b)   (c)

Figure 4.3.2: Logic diagrams (a) and (b), and logic symbol (c) for a half-adder.

The full-adder accepts two input bits and an input carry and generates a sum output and an output carry.

The logic symbol of a full-adder is shown in Figure 4.3.3, where A_i and B_i are input variables (operands), C_{i-1} is an input carry, S_i is the sum output, and C_i is the output carry.

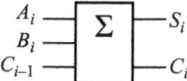 **Figure 4.3.3:** Logic symbol for a full-adder.

The truth table in Table 4.3.2 shows the operation of a full-adder.

Table 4.3.2: Truth table for a full-adder.

A_i	B_i	C_{i-1}	C_i	S_i
0	0	0	0	0
0	0	1	0	1
0	1	0	0	1
0	1	1	1	0
1	0	0	0	1
1	0	1	1	0
1	1	0	1	0
1	1	1	1	1

The logic expressions of outputs for a full-adder can be deduced for the truth table in Table 4.3.2 as follows:

$$\begin{cases} S_i = \sum m(1,2,4,7) \\ \quad = \bar{A}_i\bar{B}_iC_{i-1} + \bar{A}_iB_i\bar{C}_{i-1} + A_i\bar{B}_i\bar{C}_{i-1} + A_iB_iC_{i-1} \\ \quad = A_i \oplus B_i \oplus C_{i-1} \\ C_i = \sum m(3,5,6,7) \\ \quad = A_iB_i\bar{C}_{i-1} + A_iB_iC_{i-1} + A_i\bar{B}_iC_{i-1} + \bar{A}_iB_iC_{i-1} \\ \quad = A_iB_i + (A_i \oplus B_i)C_{i-1} \end{cases} \tag{4.3.2}$$

It can be seen from eq. (4.3.2) that a full-adder can be implemented with two half -adders and an OR gate, as shown in Figure 4.3.4.

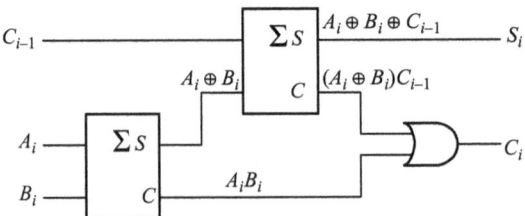

Figure 4.3.4: A full-adder implemented with two half-adders and one OR gate

Example 4.7 A logic circuit is shown in Figure 4.3.5.

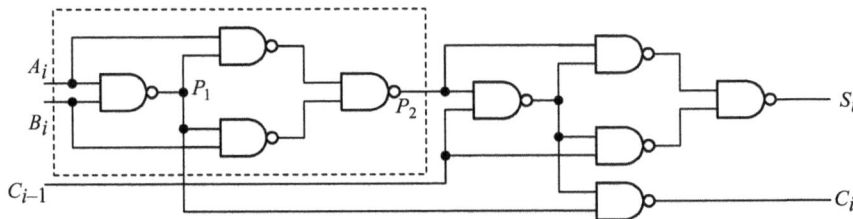

Figure 4.3.5: Logic diagram of Example 4.7.

(1) Write the logic expression of the output in the dashed box.
(2) Analyze the logic function of the circuit.
(3) Simulate the logic implementation with Multisim software.

Solution

(1) The circuit in the dashed box consists of four NAND gates, and its logic expression of the output P_2 is

$$P_2 = \overline{\overline{A_i P_1} \cdot \overline{B_i P_1}}$$

$$= \overline{\overline{A_i \overline{A_i B_i}} \cdot \overline{B_i \overline{A_i B_i}}}$$

$$= A_i \overline{A_i B_i} + B_i \overline{A_i B_i} \qquad (4.3.3)$$

$$= A_i \, \overline{B_i} + \overline{A_i} \, B_i$$

$$= A_i \oplus B_i$$

It can be seen from eq. (4.3.3) that the circuit in the dashed box implements an XOR operation.

(2) The circuit shown in Figure 4.3.5 is a combinational circuit with multiple outputs. Its logic expressions of the outputs are

$$\begin{cases} S_i = P_2 \oplus C_{i-1} = A_i \oplus B_i \oplus C_{i-1} \\ C_i = \overline{\overline{P_1} \cdot \overline{P_2 \cdot C_{i-1}}} = \overline{P_1} + P_2 \cdot C_{i-1} = A_i B_i + (A_i \oplus B_i) C_{i-1} \end{cases} \qquad (4.3.4)$$

Comparing eq. (4.3.4) with eq. (4.3.2), you can find that the circuit shown in Figure 4.3.5 is a full-adder, which can realize the addition of two input bits and an input carry.

(3) Multisim software is used to simulate the logic implementation by the following steps. The first step is to draw the logic diagram in terms of Figure 4.3.5, as shown in Figure 4.3.6; the second step is to connect the inputs of the circuit with a digital signal generator and set the output of the signal generator to generate a three-bit binary digit representing A_i, B_i, and C_{i-1}; the final step is to connect a logic analyzer for displaying the signals of A_i, B_i, C_{i-1}, S_i, and C_i.

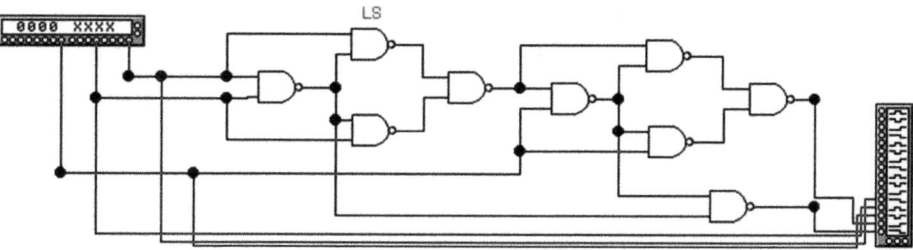

Figure 4.3.6: Simulation logic diagram.

The simulation waveform by the logic analyzer is shown in Figure 4.3.7. Through checking the logic relationship between outputs and inputs, you can find that the logic function of the circuit in Figure 4.3.5 accords with a full-adder.

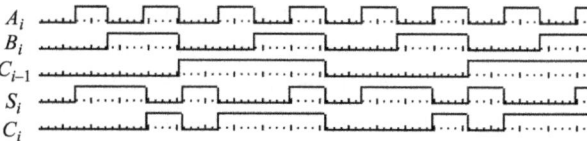

Figure 4.3.7: A simulation waveform.

4.3.3 Multi-bit binary adders

A single full-adder can only add two one-bit binary numbers and an input carry. To add binary numbers with more than one bit, additional full-adders must be used. For example, two four-bit binary numbers, $A_3A_2A_1A_0$ and $B_3B_2B_1B_0$, are added together for illustrating how to add two multi-bit binary numbers. To add two four-bit binary numbers, four full-adders are used. It begins with the addition of two least significant bits (LSB), A_0 and B_0, and thus the LSB sum bit, S_0, and carry output, C_0, are produced. The carry output, C_0, is connected to the carry input of the next higher-order full-adder and the addition operation of the adjacent higher-order bit, A_1 and B_1, starts. This addition operation continues until the addition of the most significant bit (MSB), A_3 and B_3, is accomplished. Figure 4.3.8 shows the block diagram of a four-bit binary adder using four full-adders. Notice that only an LSB full-adder can be substituted by a half-adder since the carry input of a LSB full-adder is a 0.

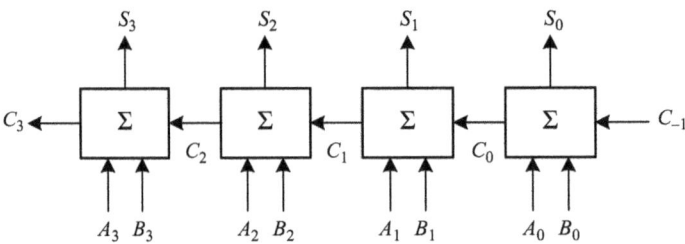

Figure 4.3.8: A four-bit binary adder using four full-adders.

When the addition of two multi-bit binary digits is implemented by using several full-adders, the carry output of each full-adder is connected to the carry input of the next higher-order stage, which is called a *ripple carry*. For a ripple carry adder , the sum and the output carry of any stage cannot be produced until the input carry occurs; as a result, a time delay will occur in the addition process.

To speed up the addition process by eliminating this ripple carry delay, a *look-ahead carry* adder is designed. For the look-ahead carry adder [27], the output carry of each stage is obtained directly from the input bits of each stage by using logic circuits rather than being generated and propagated bit by bit. Thus, the addition operation of different bits is carried out simultaneously. As a result, the time delay will be reduced. However, with the increase of bits, the complexity of the look-ahead carry adder increases.

The structure diagram of a look-ahead carry adder is shown in Figure 4.3.9, where F_4–F_1 are combinational circuits with inputs including the input carry C_0, the addends $A_3A_2A_1A_0$ and $B_3B_2B_1B_0$.

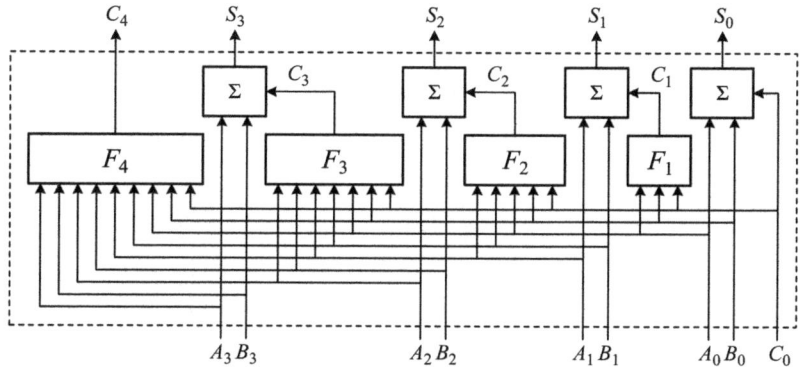

Figure 4.3.9: Structure diagram of a four-bit look-ahead carry adder.

4.3.4 MSI adders

The medium-scale integrated four-bit adder 74LS283, as a look-ahead carry full adder, can complete the addition of two four-bit binary digits. For the 74LS283, V_{CC} is pin 16 and ground is pin 8, which is a more standard configuration. Pin diagram and logic symbol for 74LS283 are shown in Figure 4.3.10, where A_3, A_2, A_1, A_0 and B_3, B_2, B_1, B_0 are inputs of two four-bit binary digits, S_3, S_2, S_1, S_0 are sum outputs, C_0 is the input carry, and C_4 is the output carry.

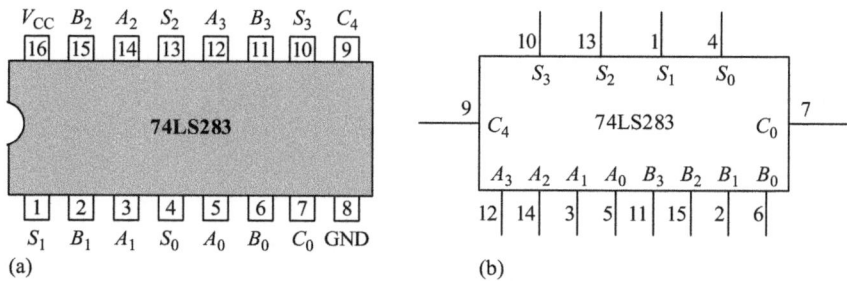

Figure 4.3.10: The MSI four-bit parallel adder 74LS283: (a) pin diagram and (b) logic symbol.

Two four-bit parallel adders can be cascaded to implement the addition of two eight-bit binary numbers, as shown in Figure 4.3.11. The carry input of the low-order adder (C_0) is connected to ground due to no carry into the LSB position, and the carry output of the lower-order adder (C_4) is sent to the carry input of the next higher-order adder. The output carry is designated C_8 that comes from the output carry of the high-order adder. The lower-order adder completes the addition of the lower or less significant four bits in the eight-bit binary numbers and the higher-order adder implements the addition of the higher or more significant four bits in the eight-bit binary numbers. Similarly, four four-bit adders can be cascaded to handle the addition of two 16-bit binary numbers.

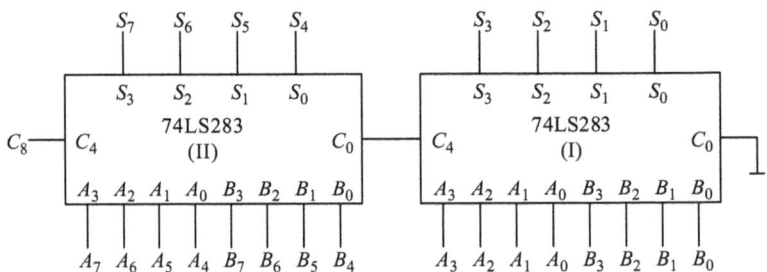

Figure 4.3.11: A cascade connection of two four-bit adders 74LS283 to form an eight-bit adder.

Example 4.8 Analyze the logic function of the circuit shown in Figure 4.3.12. The circuit consists of a four-bit adder 74LS283 and four inverters.

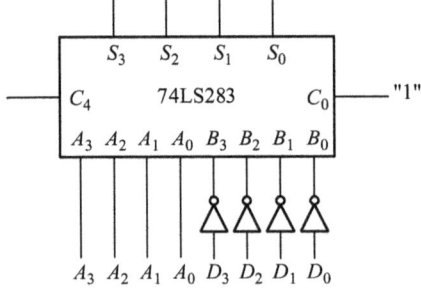

Figure 4.3.12: Logic diagram of Example 4.8.

Solution
From Figure 4.3.12, you can find that $B_3B_2B_1B_0$ is the 1's complement of $D_3D_2D_1D_0$. Since C_0 is connected to 1, the 2's complement of $D_3D_2D_1D_0$ is then generated by the addition of $B_3B_2B_1B_0$ and 1. Thus the circuit is actually the addition of $A_3A_2A_1A_0$ and the 2's complement of $D_3D_2D_1D_0$.

Assume that $A_3A_2A_1A_0 = 1100$ and $D_3D_2D_1D_0 = 0011$, and the output result of the circuit is illustrated in Figure 4.3.13(a). Since the output carry C_4 is a 1, which represents that the sum is positive, this corresponds to $12-3 = 9$ in a decimal number system.

$$
\begin{array}{cccc}
 & 1 & 1 & 0 & 0 \\
+ & 1 & 1 & 0 & 1 \\
\hline
1 & 1 & 0 & 0 & 1
\end{array}
\quad \longleftarrow \text{2's complement of 0011}
$$

(a)

$$
\begin{array}{cccc}
 & 0 & 0 & 1 & 1 \\
+ & 0 & 1 & 0 & 0 \\
\hline
0 & 0 & 1 & 1 & 1
\end{array}
\quad
\begin{array}{l}
\longleftarrow \text{2's complement of 1100} \\
\longrightarrow \text{2's complement of sum is 1001}
\end{array}
$$

(b)

Figure 4.3.13: Output of the circuit in Figure 4.3.12: (a) 12-3 and (b) 3-12.

Assume that $A_3A_2A_1A_0 = 0011$ and $D_3D_2D_1D_0 = 1100$, the output of the circuit is illustrated in Figure 4.3.13(b). The output carry C_4 is a 0, which represents that the sum is negative. The correct result should be the 2's complement of the sum, which corresponds to 3−12 = −9 in the decimal number system.

In summary, the logic function of the circuit shown in Figure 4.3.12 is the subtraction of two four-bit binary digits.

Example 4.9 A code converter consisting of a 74LS283 and NOR gates is shown in Figure 4.3.14. Assuming the input *DCBA* is a BCD8421 code, what kind of code is the output $S_3S_2S_1S_0$?

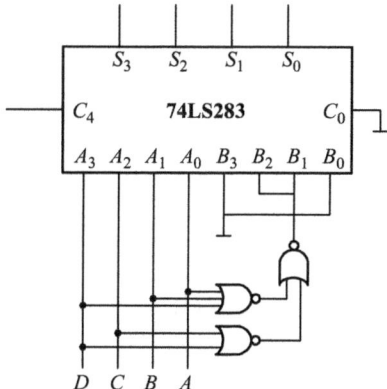

Figure 4.3.14: Logic diagram of Example 4.9.

Table 4.3.3: Function table of Example 4.9.

D	C	B	A	B_3	B_2	B_1	B_0	S_3	S_2	S_1	S_0
0	0	0	0	0	0	0	0	0	0	0	0
0	0	0	1	0	0	0	0	0	0	0	1
0	0	1	0	0	0	0	0	0	0	1	0
0	0	1	1	0	0	0	0	0	0	1	1
0	1	0	0	0	0	0	0	0	1	0	0
0	1	0	1	0	1	1	0	1	0	1	1
0	1	1	0	0	1	1	0	1	1	0	0
0	1	1	1	0	1	1	0	1	1	0	1
1	0	0	0	0	1	1	0	1	1	1	0
1	0	0	1	0	1	1	0	1	1	1	1

Solution

The logic expressions of $B_3B_2B_1B_0$ are as follows:

$$\begin{cases} B_3 = B_0 = 0 \\ B_2 = B_1 = \overline{\overline{D+C} + \overline{D+B+A}} = (D+C)(D+B+A) = D+AC+BC \end{cases}$$

According to the logic function of 74LS283, the function table of the circuit is obtained in Table 4.3.3.

It is found from Table 4.3.3 that when the input is 8421BCD code, the output $S_3S_2S_1S_0$ is 2421BCD code.

4.4 Encoders and decoders

Digital systems and computers can only process information represented by a group of binary codes, while human being is accustomed to decimal numbers and the alphabetic characters for representing various information. Therefore, it is very important for translating various information such as decimal numbers, words, and alphabetic characters into a coded form or vice versa so that computers and digit systems can recognize the information to control their peripheral devices as well as the interaction with human being. In digital systems, encoders and decoders are designed to realize the "translation" function, which are illustrated in Figure 4.4.1.

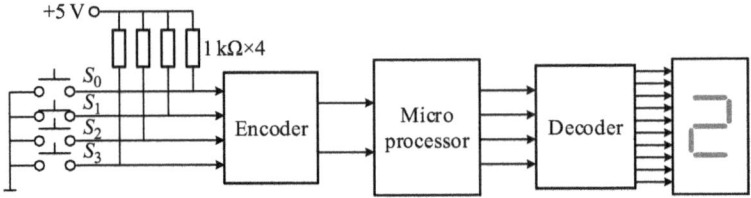

Figure 4.4.1: Application of an encoder and a decoder.

The function of the encoder in Figure 4.4.1 is to encode the four keys with binary codes so that the microprocessor can recognize which key is pressed through checking the output of the encoder. On the contrary, the decoder is to convert the coded information outputting from the microprocessor into decimal numbers or alphabetic character so that the information in decimal numbers or alphabetic characters can be directly displayed.

The process of converting from various information including words, numbers, and alphabetic characters into a coded format is called *encoding*. The encoding is performed by a logic circuit called an *encoder*, and the process of converting from a coded format to a noncoded information is called *decoding*. The decoding is performed by a logic circuit called a *decoder*. This section introduces encoders, decoders, and their applications.

The objectives of this section are to
- Define an encoder and a decoder
- Determine logic function of binary encoders and decoders

- Explain the priority feature in an encoder
- Describe the 74LS147 decimal-to-BCD priority encoder and the 74LS148 8-to-3 priority encoder
- Apply the encoder and the decoder for expansion purposes
- Describe the 74LS138 3-to-8 decoder and the 7474HC42 BCD-to-decimal decoder
- Apply the 74LS138 to design any combinational logic circuits
- Explain the principle of seven-segment display decoder
- Explain the difference between a common-cathode LED and a common-anode LED

4.4.1 Encoders

An encoder accepts an active level on one of its inputs representing specific information and converts it into a coded format output. It is a type of combinational logic circuits with multi-input and multi-output. If an encoder has m-input signals and n-bit binary code outputs, m should be less than or equal to 2^n. The commonly used encoders include binary encoders and BCD encoders.

1. Binary encoders

A *binary encoder* converts noncoded information, such as a decimal number and an alphabetic character, into a binary code. An n-bit binary code can represent 2^n events. Thus, the logic circuit encoding 2^n events into n-bit binary code are called 2^n-to-n binary encoder. A binary encoder has 2^n inputs and n outputs. Only one of the input signals of binary encoder is valid at any time and the active level can be defined by a HIGH (1) or a LOW (0). That is say, if an input is in the active state, then an n-bit binary code will be output. Notice that only one input is in the active state and other inputs are in opposite level. If multiple valid signals are inputted simultaneously, encoder cannot give the correct code.

Example 4.10 Design a 4-to-2 binary encoder to encode four keys: "up," "down," "left," and "right." Each time only one key is allowed to be pressed. When the key is pressed, the encoder can output the corresponding two-bit binary code and also offer a flag signal to inform a microprocessor to start a key processing function.

Solution

(1) Let S_0, S_1, S_2, S_3 represent the "up," "down," "left," and "right" keys, respectively. When a key is pressed, its corresponding input level is a 0; otherwise, the input is a 1. Four keys can be represented by two-bit binary code, $D_1 D_0$. The output flag signal of the encoder is denoted by INT_0. The active level of INT_0 is a HIGH, that is, $INT_0 = 1$ when a key is pressed.

(2) List the truth table from the logic requirement, as shown in Table 4.4.1. According to the requirement, only one key is pressed each time. That is say, only one input is an active level

(0) and other inputs are in opposite level (1). Therefore, if more than one zero occurs in the combination of input levels, then the combination is invalid and can be treated as "don't care" term indicated by "×."

Table 4.4.1: Truth table.

S_3	S_2	S_1	S_0	D_1	D_0	INT_0
1	1	1	1	0	0	0
1	1	1	0	0	0	1
1	1	0	1	0	1	1
1	0	1	1	1	0	1
0	1	1	1	1	1	1
0	0	0	0	×	×	×
0	0	0	1	×	×	×
0	0	1	0	×	×	×
0	0	1	1	×	×	×
0	1	0	0	×	×	×
0	1	0	1	×	×	×
0	1	1	0	×	×	×
1	0	0	0	×	×	×
1	0	0	1	×	×	×
1	0	1	0	×	×	×
1	1	0	0	×	×	×

(3) Simplify the logic expressions using the Karnaugh map. The Karnaugh map of the 4-to-2 encoder are shown in Figure 4.4.2.

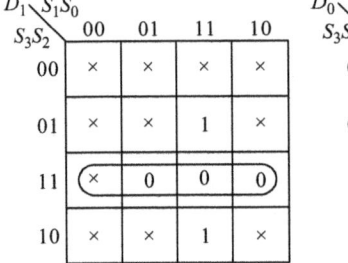

Figure 4.4.2: Karnaugh maps of Example 4.10.

Logic expressions of outputs of the 4-to-2 binary encoder are obtained as follows:

$$\begin{cases} D_1 = \overline{S_3 S_2} = \overline{S_3} + \overline{S_2} \\ D_0 = \overline{S_3 S_1} = \overline{S_3} + \overline{S_1} \\ INT_0 = \overline{S_3 S_2 S_1 S_0} = \overline{S_3} + \overline{S_2} + \overline{S_1} + \overline{S_0} \end{cases} \qquad (4.4.1)$$

It can be found from eq. (4.4.1) that if S_2 or S_3 is pressed, then $D_1 = 1$; if S_1 or S_3 is pressed, then $D_0 = 1$. No matter which key is pressed, the output flag $INT_0 = 1$.

(4)   Draw the logic diagram implemented with NAND gates, as shown in Figure 4.4.3.

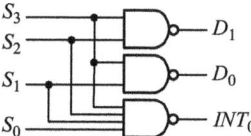

Figure 4.4.3: Logic diagram of a 4-to-2 binary encoder of Example 4.10.

2. Priority encoders

A *priority encoder* is a type of logic circuit that only produces a binary code corresponding to the active input with the higher order and will ignore any other active inputs with lower order. If there are multiple active inputs at the same time for a priority encoder, only one input with the highest-order priority is encoded.

Example 4.11. Design a priority encoder to encode four keys in Example 4.10. Assume that key S_3 has the highest-order priority and key S_0 has the lowest-order priority.

Solution

(1)   Simply the truth table according to the logic requirements.

Assume that the active input level is a LOW. S_0 has the lowest-order priority. Thus, the output $D_1D_0 = 00$ only when S_3, S_2, and S_1 are not pressed and S_0 is pressed. S_3 has the highest-order priority. Thus, no matter whether S_2, S_1, and S_0 are pressed, the output $D_1D_0 = 11$ only if S_3 is pressed, which is denoted by $S_3S_2S_1S_0 = 0\times\times\times$ in the truth table. Similarly, the simplified truth table is shown in Table 4.4.2.

Table 4.4.2: Truth table.

S_3	S_2	S_1	S_0	D_1	D_0	INT_0
1	1	1	1	0	0	0
1	1	1	0	0	0	1
1	1	0	×	0	1	1
1	0	×	×	1	0	1
0	×	×	×	1	1	1

This priority encoder performs the same 4-to-2 encoding function as in Example 4.10. But it allows multiple active inputs at the same time. All combinations of inputs have their corresponding output codes. Therefore, there is no constraint for inputs.

(2)   The Karnaugh maps are shown in Figure 4.4.4, and the minimized logic expressions of outputs for a 4-to-2 priority encoder are obtained as follows:

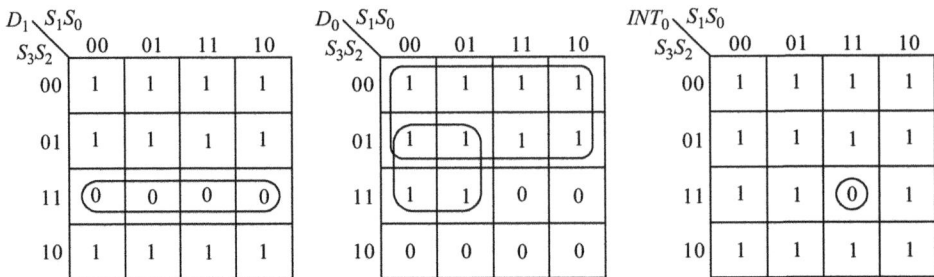

Figure 4.4.4: The Karnaugh map.

$$\left\{ \begin{aligned} D_1 &= \overline{S_3 S_2} = \overline{S_3} + \overline{S_2} \\ D_0 &= \overline{S_3} + S_2 \overline{S_1} \\ INT_0 &= \overline{S_3 S_2 S_1 S_0} = \overline{S_3} + \overline{S_2} + \overline{S_1} + \overline{S_0} \end{aligned} \right. \qquad (4.4.2)$$

(3)   A logic circuit diagram of the priority encoder is shown in Figure 4.4.5.

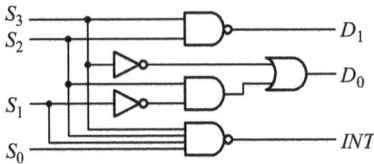

Figure 4.4.5: Logic diagram.

Comparing the priority encoder in Figure 4.4.5 with the basic binary encoder in Figure 4.4.3, you can find that the structure of a priority encoder is more complex than that of a binary encoder. However, a priority encoder is more effective for practical applications as there is no constraint for inputs.

---

3. MSI priority encoders

74LS147 and 74LS148 are commonly used MSI priority encoders. The 74LS147 is an MSI *decimal-to-BCD priority encode*r with eight active-LOW inputs (0), D_0 through D_9, representing decimal digits 1 through 9 and four active-LOW BCD outputs, $\bar{A}_3 \bar{A}_2 \bar{A}_1 \bar{A}_0$[28]. The output is zero when none of the inputs is active. Pin diagram and logic symbol are shown in Figure 4.4.6.

The 74LS148 is an MSI 8-to-3 priority binary encoder with eight active-LOW inputs, $\bar{I}_0 \sim \bar{I}_7$, and three active-LOW outputs, $\bar{Q}_C \bar{Q}_B \bar{Q}_A$. It can convert eight inputs into a three-bit binary code. Input $\bar{I}_7$ has the highest-order priority and input $\bar{I}_0$ has the lowest-order priority. $\overline{EI}$ is an active-LOW enable input to enable the device. When $\overline{EI}$=1, encoder is disabled and all outputs are at a HIGH level. It also has an enable output, $\overline{EO}$, and group signal output, $\overline{GS}$, for expansion purposes. $\overline{EO}$ is a LOW when

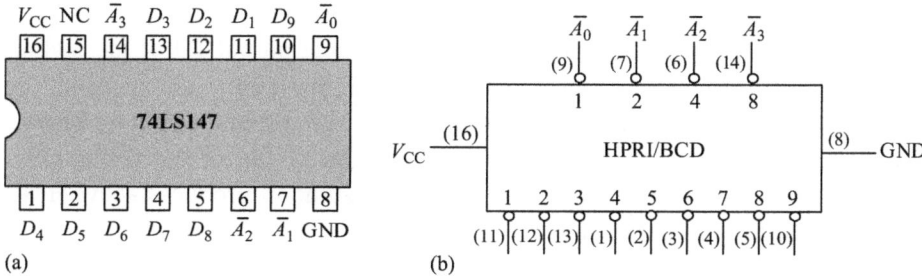

Figure 4.4.6: The 74LS147 decimal to BCD priority encoder (HPRI means highest-order input has priority): (a) pin diagram and (b) logic symbol.

$\overline{EI}$ is a LOW and none of the inputs is active. $\overline{GS}$ is a LOW when $\overline{EI}$ is a LOW and any of the inputs are active, which indicates that the encoder is encoding. The function table of the 74LS148 8-to-3 priority encoder is shown in Table 4.4.3. Figure 4.4.7 shows the logic symbol of 74LS148. The small circles represent that the active level is a LOW.

Table 4.4.3: Function table of the 74LS148 8-to-3 priority encoder.

Inputs									Outputs				
$\overline{EI}$	$\overline{I_0}$	$\overline{I_1}$	$\overline{I_2}$	$\overline{I_3}$	$\overline{I_4}$	$\overline{I_5}$	$\overline{I_6}$	$\overline{I_7}$	$\overline{GS}$	$\overline{Q_C}$	$\overline{Q_B}$	$\overline{Q_A}$	$\overline{EO}$
1	×	×	×	×	×	×	×	×	1	1	1	1	1
0	1	1	1	1	1	1	1	1	1	1	1	1	0
0	×	×	×	×	×	×	×	0	0	0	0	0	1
0	×	×	×	×	×	×	0	1	0	0	0	1	1
0	×	×	×	×	×	0	1	1	0	0	1	0	1
0	×	×	×	×	0	1	1	1	0	0	1	1	1
0	×	×	×	0	1	1	1	1	0	1	0	0	1
0	×	×	0	1	1	1	1	1	0	1	0	1	1
0	×	0	1	1	1	1	1	1	0	1	1	0	1
0	0	1	1	1	1	1	1	1	0	1	1	1	1

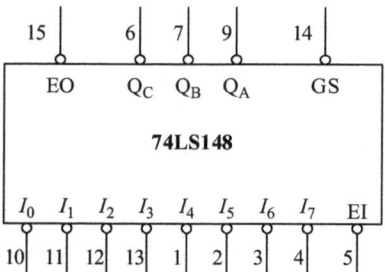

Figure 4.4.7: Logic symbol of the 74LS148 8-to-3 priority encoder.

Using $\overline{\text{EI}}$, $\overline{\text{EO}}$, and $\overline{\text{GS}}$, two 74LS148s can be cascaded to construct a 16-to-4 priority encoder, as shown in Figure 4.4.8. The enable output $\overline{\text{EO}}$ of the higher-order encoder is connected to the enable input $\overline{\text{EI}}$ of the lower-order encoder. The final outputs are obtained by the NAND gates with the corresponding binary outputs as inputs. The highest bit output is the complement of $\overline{\text{GS}}$ for the higher-order encoder by using an inverter. This particular configuration produces active-HIGH outputs for the four-bit binary number.

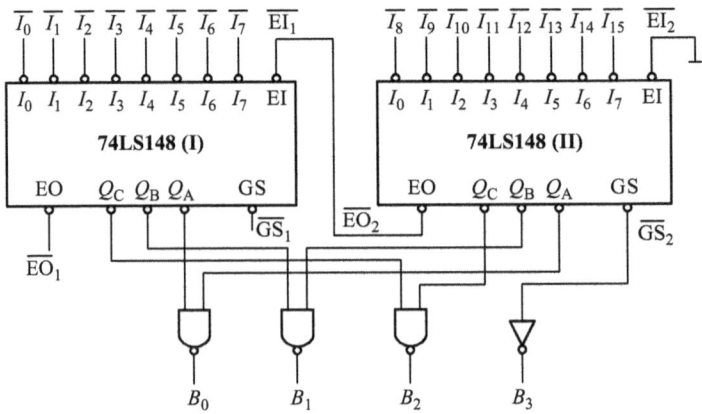

Figure 4.4.8: Expansion of 74LS148.

4.4.2 Decoders

Decoding is a reverse process of encoding. A decoder is a combinational circuit converting coded information, such as a binary number, into a noncoded form, such as a decimal form. The basic function of a decoder is to detect the presence of a specified combination of code on its inputs and to indicate the presence of that code by a specified output level. Generally, a *binary decoder* has n-input lines to handle n-bit input code and from one to 2^n output lines to represent the presence of one or more n-bit combination. Generally, there are two types of decoders: general-purpose decoders and display decoders.

General-purpose decoders include *binary decoders*, *BCD-to-decimal decoders*, and *code converters*. Binary decoders and BCD-to-decimal decoders are also called *variable decoders*. Each output of a variable decoder corresponds to a minterm of input variables. Thus variable decoders can be used as demultiplexers and the implementation of any combinational logic circuits. *Code converters* can transform one kind of code to another.

Display decoders are to convert binary codes on their inputs and provide outputs to drive display devices.

1. Variable decoders

A decoder is a multi-input and multi-output combinational logic circuit. For each group of input codes, only one output among all outputs of the decoder is in activated level and other outputs are in opposite levels. This kind of decoder is called a *variable decoder*. Assume that a variable decoder has n-bit inputs and m outputs. If m is equal to 2^n, then the variable decoder is called *binary decoder* or n-to-2^n decoder. Some commonly used binary decoders include 2-to-4 decoder, 3-to-8 decoder, and 4-to-16 decoder. If m is less than 2^n, then the variable decoder is called *partial decoder*, for example, *BCD-to-decimal decoder*. The corresponding outputs should be 2^n outputs to decode all combinations of n bits.

(1) A 2-to-4 decoder

In order to decode all possible combinations of two bits, four outputs are required ($2^2 = 4$). This type of decoder is commonly called a 2-line-to-4-line decoder or a 2-to-4 decoder since there are two inputs and four outputs. For any given code on the inputs, one of four outputs is activated. Assume that the active level is a LOW, the truth table of a 2-to-4 decoder is shown in Table 4.4.4. The logic expression of outputs can be deduced from the truth table as follows:

$$\bar{Y}_0 = \overline{\bar{S}\bar{A}\bar{B}}, \ \bar{Y}_1 = \overline{\bar{S}\bar{A}B}, \ \bar{Y}_2 = \overline{\bar{S}A\bar{B}}, \ \bar{Y}_3 = \overline{\bar{S}AB} \tag{4.4.3}$$

Table 4.4.4: Truth table.

Inputs			Outputs			
$\bar{S}$	A	B	$\bar{Y}_0$	$\bar{Y}_1$	$\bar{Y}_2$	$\bar{Y}_3$
1	×	×	1	1	1	1
0	0	0	0	1	1	1
0	0	1	1	0	1	1
0	1	0	1	1	0	1
0	1	1	1	1	1	0

where A and B are two-bit inputs, $\bar{Y}_0$–$\bar{Y}_3$ are four active-LOW outputs, and $\bar{S}$ is an enable input.

Figure 4.4.9 shows the logic diagram of a 2-to-4 decoder by using NAND gates. When $\bar{S}$ is a LOW, the decoder is enabled. For any given code on the inputs, only one of four outputs is a LOW and other outputs are in opposite level (HIGH). The logic expression of outputs can be expressed as

$$\bar{Y}_0 = \overline{\bar{A}\bar{B}} = \overline{m_0}, \ \bar{Y}_1 = \overline{\bar{A}B} = = \overline{m_1}, \ \bar{Y}_2 = \overline{A\bar{B}} = \overline{m_2}, \ \bar{Y}_3 = \overline{AB} = \overline{m_3} \tag{4.4.4}$$

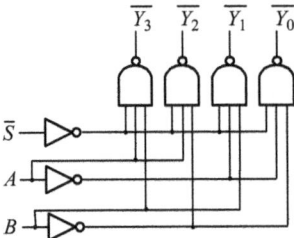

Figure 4.4.9: Logic diagram of a 2-to-4 decoder.

That is to say, each output of the 2-to-4 decoder corresponds to the complement of a minterm of the two-bit inputs.

If the active level is a HIGH, only one of four outputs is a HIGH and other outputs are in opposite level (LOW) for any given code on the inputs. When $\bar{S}$ is an active-LOW, the logic expression of outputs can be expressed as

$$Y_0 = \bar{A}\bar{B} = m_0, \ Y_1 = \bar{A}B = m_1, \ Y_2 = A\bar{B} = m_2, \ Y_3 = AB = m_3 \qquad (4.4.5)$$

In this situation, each output of the 2-to-4 decoder corresponds to a minterm of two-bit inputs, which means that four NAND gates should be replaced by four AND gates as shown in Figure 4.4.9, and a HIGH output indicates the presence of the proper binary code.

(2) An MSI 3-to-8 decoder

The 74LS138 is a good example of a MSI decoders [20]. Its logic diagram and logic symbol are shown in Figure 4.4.10. It has three inputs $A_2A_1A_0$ and eight outputs $\bar{Y}_7$–$\bar{Y}_0$, so it is called a 3-line-to-8-line decoder or a 3-to-8 decoder. An enable function (EN) is provided on this device, which is implemented with a three-input AND gate. G_1, $\bar{G}_{2A}$, and $\bar{G}_{2B}$ are three enable inputs for expansion purposes. When a LOW level is applied on $\bar{G}_{2A}$ and $\bar{G}_{2B}$, and a HIGH level on G_1, EN is a HIGH level, and thus the

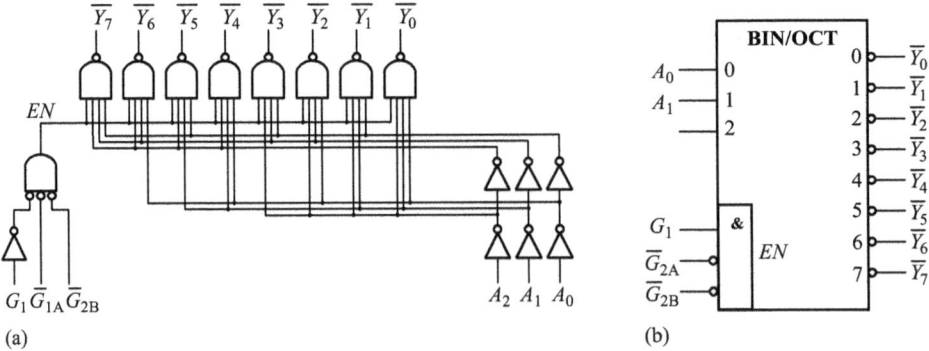

Figure 4.4.10: Logic diagram (a) and logic symbol (b) of the 74LS138 3-to-8 decoder.

decoder performs the decoding function. For any given combination of the input variables, A_2, A_1, A_0, only one of eight outputs is a LOW and other outputs are in opposite level (HIGH). When EN is a LOW, then the decoder is disabled and all decoder outputs are HIGH regardless of the states of the three inputs, A_2, A_1, and A_0.

The function table of 74LS138 is shown in Table 4.4.5.

Table 4.4.5: Function table of 74LS138.

Inputs						Outputs							
$\bar{G}_{2A}$	$\bar{G}_{2B}$	G_1	A_2	A_1	A_0	$\bar{Y}_0$	$\bar{Y}_1$	$\bar{Y}_2$	$\bar{Y}_3$	$\bar{Y}_4$	$\bar{Y}_5$	$\bar{Y}_6$	$\bar{Y}_7$
1	×	×	×	×	×	1	1	1	1	1	1	1	1
×	1	×	×	×	×	1	1	1	1	1	1	1	1
×	×	0	×	×	×	1	1	1	1	1	1	1	1
0	0	1	0	0	0	0	1	1	1	1	1	1	1
0	0	1	0	0	1	1	0	1	1	1	1	1	1
0	0	1	0	1	0	1	1	0	1	1	1	1	1
0	0	1	0	1	1	1	1	1	0	1	1	1	1
0	0	1	1	0	0	1	1	1	1	0	1	1	1
0	0	1	1	0	1	1	1	1	1	1	0	1	1
0	0	1	1	1	0	1	1	1	1	1	1	0	1
0	0	1	1	1	1	1	1	1	1	1	1	1	0

When EN = 1, the logic expressions of outputs can be derived from Table 4.4.5 as follows:

$$\bar{Y}_0 = \overline{\bar{A}_2\bar{A}_1\bar{A}_0} = \overline{m_0}, \quad \bar{Y}_1 = \overline{\bar{A}_2\bar{A}_1A_0} = \overline{m_1}, \quad \bar{Y}_2 = \overline{\bar{A}_2A_1\bar{A}_0} = \overline{m_2}, \quad \bar{Y}_3 = \overline{\bar{A}_2A_1A_0} = \overline{m_3}$$

$$\bar{Y}_4 = \overline{A_2\bar{A}_1\bar{A}_0} = \overline{m_4}, \quad \bar{Y}_5 = \overline{A_2\bar{A}_1A_0} = \overline{m_5}, \quad \bar{Y}_6 = \overline{A_2A_1\bar{A}_0} = \overline{m_6}, \quad \bar{Y}_7 = \overline{A_2A_1A_0} = \overline{m_7}$$

Each output of the 3-to-8 decoder corresponds to the complement of a minterm of three-bit input.

Figure 4.4.11 shows a simulation circuit diagram of a 74LS138. A digital signal generator is connected to inputs for providing the required four-bit natural binary codes, which are continuously applied to the enable input $\bar{E}_1$ and the three inputs, A_2, A_1, and A_0. A logic analyzer is connected to record inputs and outputs of the 74LS138.

Simulation waveforms obtained by the logic analyzer is shown in Figure 4.4.12. When $\bar{E}_1 = 1$, the decoder is prohibited and thus all outputs are at a HIGH level. When $\bar{E}_1 = 0$, the decoder performs the decoding function. The three-bit natural binary codes are continuously applied to inputs A_2, A_1, and A_0. The outputs $\bar{Y}_0$–$\bar{Y}_7$ will

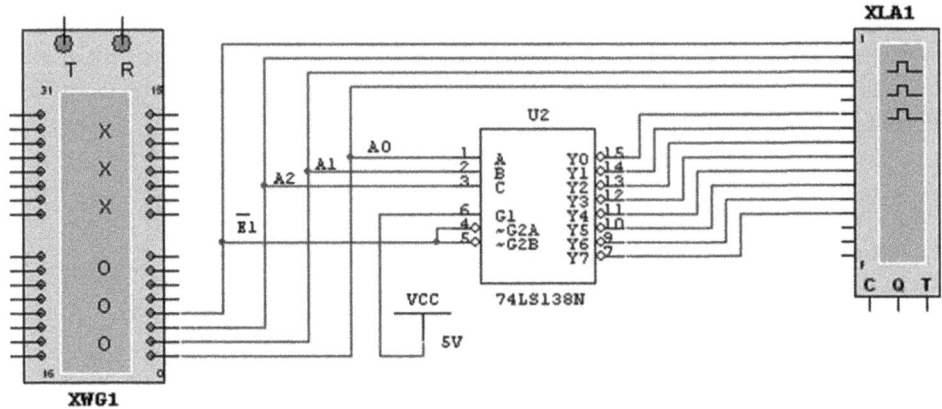

Figure 4.4.11: Simulation circuit diagram of Example 4.11.

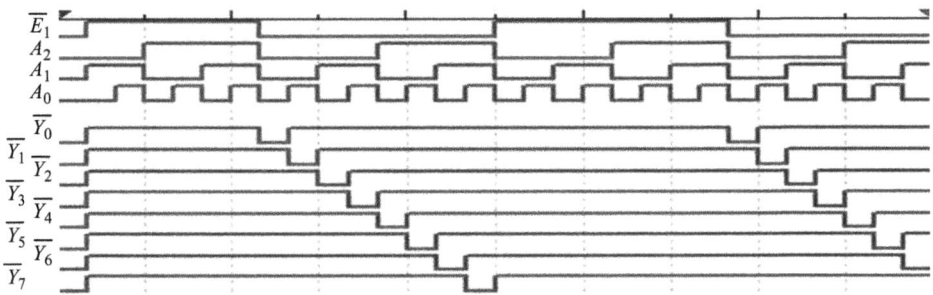

Figure 4.4.12: Simulation waveforms of Example 4.11.

produce a LOW output in sequence with an increase of the input value from 000 to 111. At any time, only one output is an active-LOW output and the other outputs are at a HIGH level.

(3) 8421BCD-to-decimal decoders

An *8421BCD-to-decimal decoder* converts each 8421BCD code into one of ten possible decimal digit indications. It is often called a 4-line-to-10-line decoder or a 1-of-10 decoder.

The 74HC42 is a commonly used MSI 8421BCD-to-decimal decoder [6]. Its functional table is given in Table 4.4.6 and the corresponding logic symbol is shown in Figure 4.4.13. For any given combination of input variables, A_3, A_2, A_1, A_0, only one of ten outputs is a LOW and other outputs are in opposite level (HIGH). Because four-bit binary codes from 1100 to 1111 do not belong to 8421BCD, they are called *pseudocodes* or *"unallowable codes."* When these six binary codes are applied to inputs, all outputs are at HIGH levels.

Table 4.4.6: Function table of 74HC42.

Inputs				Outputs									
A_3	A_2	A_1	A_0	$\overline{Y_0}$	$\overline{Y_1}$	$\overline{Y_2}$	$\overline{Y_3}$	$\overline{Y_4}$	$\overline{Y_5}$	$\overline{Y_6}$	$\overline{Y_7}$	$\overline{Y_8}$	$\overline{Y_9}$
0	0	0	0	0	1	1	1	1	1	1	1	1	1
0	0	0	1	1	0	1	1	1	1	1	1	1	1
0	0	1	0	1	1	0	1	1	1	1	1	1	1
0	0	1	1	1	1	1	0	1	1	1	1	1	1
0	1	0	0	1	1	1	1	0	1	1	1	1	1
0	1	0	1	1	1	1	1	1	0	1	1	1	1
0	1	1	0	1	1	1	1	1	1	0	1	1	1
0	1	1	1	1	1	1	1	1	1	1	0	1	1
1	0	0	0	1	1	1	1	1	1	1	1	0	1
1	0	0	1	1	1	1	1	1	1	1	1	1	0

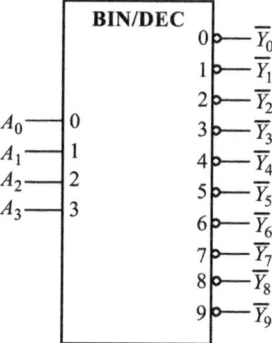

Figure 4.4.13: The logic symbol of the 74HC42 1-of-10 decoder.

(4) Applications of decoders

Decoders have many typical applications. A common application is in microprocessors for input/output (I/O) selection, as shown in Figure 4.4.14 [6]. Microprocessors must communicate with a variety of external devices called *peripherals*. The peripherals can be printers, modems, scanners, and others. A 3-to-8 decoder is connected between the microprocessor and peripherals. Its inputs are connected with the address lines of the microprocessor and its outputs are connected with the $\overline{CS}$ (chip select) of peripherals to select the peripheral as determined by the microprocessor so that the data can be sent or received from a specific peripheral. Each peripheral has a unique number, called an address, for identifying itself in the system. When the microprocessor needs to communicate with a specific peripheral, it arranges an appropriate address code for each peripheral. This binary address is decoded and the corresponding decoder output is activated to enable a particular peripheral. The parallel data lines D_0–D_7 of the peripherals are connected with the eight-bit data bus to the microprocessor so that the binary data are transferred within the micropro-cessor on a data bus.

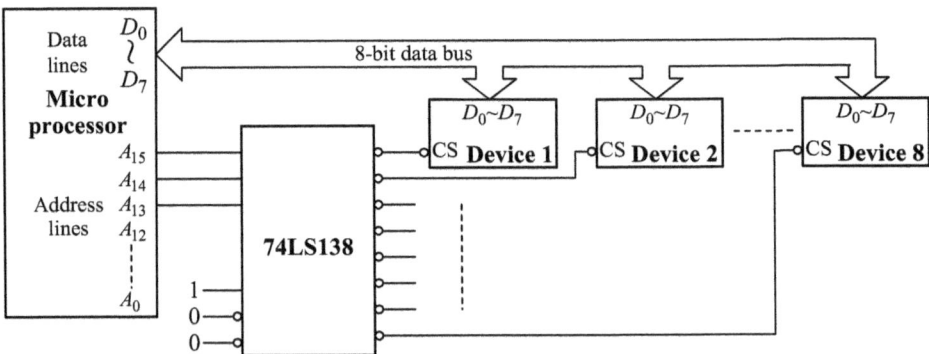

Figure 4.4.14: Example of a 3-to-8 decoder for an address decoder connected with a microprocessor and peripherals.

The 74LS138 3-to-8 decoder is arranged with three enable inputs for expansion purposes. With these enable inputs, decoders can be cascaded to form a larger-scale decoder. For instance, two 74LS138 chips can be cascaded to construct a 4-to-16 decoder, as shown in Figure 4.4.15.

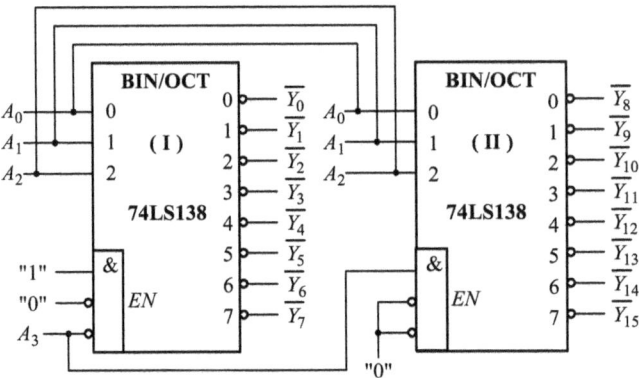

Figure 4.4.15: A 4-to-16 decoder using the cascaded 74LS38s.

Since one 74LS138 can only handle three-bit binary code for inputs that can select eight outputs, in order to implement a 4-to-16 decoder, two decoders must be used to decode four bits. The lower three-bit inputs, A_2, A_1, and A_0, of two decoders are connected together and the MSB input A_3 is connected to the enable inputs of the two decoders. When $A_3 = 0$, chip II is disabled, and its outputs are at a HIGH level; chip I is enabled, and the binary codes 0000–0111 are decoded. When $A_3 = 1$, chip I is disabled and chip II is enabled to decode the binary codes 1000–1111. Therefore, when the binary number is less than eight, the lower-order (chip I) is enabled and the higher-order decoder (chip II) is disabled, whereas when the binary number is greater than or equal to eight, the higher-order decoder is enabled and the lower-order one is disabled.

In addition, an important application of decoders is to design arbitrary combinational logic circuits. In the previous part, you learnt that each output of binary decoders corresponds to a minterm or the complement of a minterm, and any logic expression can be converted into a standard sum-of-products form. Therefore, any combinational logic circuit can be constructed with decoder and additional several logic gates.

Example 4.12 Implement a full adder with a 74LS138 and two NAND gates.

Solution
The output logic expressions of a full adder have been already deduced in Section 4.3 as follows:

$$\begin{cases} S_i = \sum m(1, 2, 4, 7) \\ C_i = \sum m(3, 5, 6, 7) \end{cases}$$

Each output of the 74LS138 is the complement of a minterm of input variables, so the output logic expression of a full adder should be transformed into the form of the complement of a minterm. Boolean algebra is applied to transform logic expression as follows:

$$\begin{cases} S_i(A_i, B_i, C_{i-1}) = \overline{\overline{m_1 + m_2 + m_4 + m_7}} = \overline{\overline{m_1} \cdot \overline{m_2} \cdot \overline{m_4} \cdot \overline{m_7}} \\ C_i(A_i, B_i, C_{i-1}) = \overline{\overline{m_3 + m_5 + m_6 + m_7}} = \overline{\overline{m_3} \cdot \overline{m_5} \cdot \overline{m_6} \cdot \overline{m_7}} \end{cases} \tag{4.4.6}$$

It can be seen from eq. (4.4.6) that a full adder can be implemented with a 74LS138 and two NAND gates, as shown in Figure 4.4.16, where A_i, B_i, C_{i-1} are applied to A_2, A_1, and A_0, respectively.

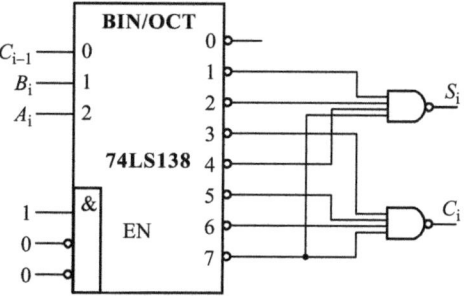

Figure 4.4.16: Logic diagram.

Example 4.13 Implement the logic expression $P(A, B, C) = AC + BC$ with a 74LS138 and an AND gate.

Solution

Since the given logic expression is a nonstandard SOP form, you should convert it into a standard SOP form by using Boolean algebra as follows:

$$P(A, B, C) = A(B + \bar{B})C + (A + \bar{A})BC = m_3 + m_5 + m_7 = \overline{m_0 + m_1 + m_2 + m_4 + m_6} = \overline{m_0} \cdot \overline{m_1} \cdot \overline{m_2} \cdot \overline{m_4} \cdot \overline{m_6}$$

The logic diagram implemented with a 74LS138 and an AND gate is shown in Figure 4.4.17.

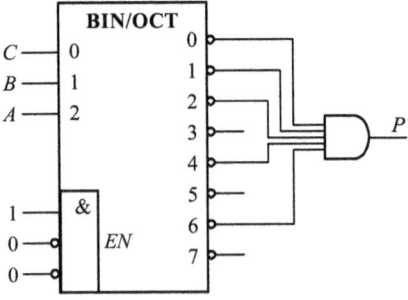

Figure 4.4.17: Logic diagram.

---

2. Display decoders
(1) Display devices and decoders

Currently, light-emitting diodes (LEDs) [29] and liquid-crystal displays (LCDs) [30] are commonly used display devices in electronic systems. Both of them can be divided into two categories: segment display and dot-matrix display. A segment display device usually consists of several segments. Numbers and alphabetic characters can be displayed by applying the required voltage to the corresponding segments, as shown in Figure 4.4.18(a). Matrix-dot display devices usually consist of a matrix of light-emitting dots, and numbers and alphabetic characters are displayed by driving different dots, as shown in Figure 4.4.18(b).

(a)

(b)

Figure 4.4.18: Illustration diagram of segment display (a) and dot-matrix display (b).

Here we only focus on the introduction of LEDs. LEDs are display devices that consist of LEDs made of semiconductor materials such as gallium arsenide and gallium arsenide phosphide. When a 10 mA forward-biased current flows through an LED, it will emit light with several different colors such as red, green, yellow, orange, and blue.

The dimensions of segment LED displays are small and are widely used for the display of instruments. The dimensions of dot-matrix LED displays are comparatively larger. They are typically used for large or very large display screens and can display complex Chinese characters and images.

LED display devices can be divided into two categories: common cathode and common anode. Seven-segment LED display is a commonly used LED display device, as shown in Figure 4.4.19. It has seven-segment LEDs with the seven control inputs from segment *a* to *g*. Decimal numbers from 0 to 9 can be displayed by driving the

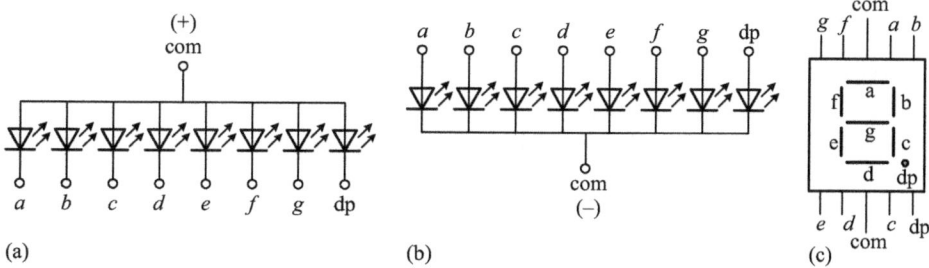

Figure 4.4.19: Seven-segment LED: (a) common anode; (b) common cathode; and (c) pin diagram.

corresponding segments and fractional point segment dp. All anodes in a common-anode seven-segment LEDs should be connected together to a positive power supply, thus this requires an active-LOW input to make the corresponding segment LED to send out light. On the contrary, all cathodes in a common-cathode seven-segment LED should be connected together to a ground, and this requires an active-HIGH input to make the corresponding segment LED to give out light. Notice that a 300–500 Ω current-limiting resistor is typically connected in series with each segment LED to protect diodes. Figure 4.4.19(c) shows the pin diagram of a commonly used seven-segment LED.

Display decoders are a type of combinational circuits to accept binary codes on its inputs and provide outputs to drive display device to produce the corresponding numbers and alphabetic characters.

Example 4.14 Design a display decoder to accept binary codes on its inputs and provide outputs to drive a common-anode seven-segment LED to display the fonts shown in Figure 4.4.20.

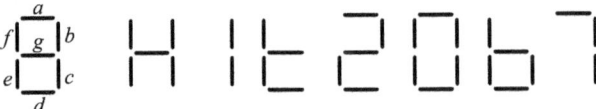

Figure 4.4.20: Fonts to be displayed.

Solution

(1) Logic abstraction

There are seven fonts to be displayed. Thus, at least three input variables should be used. Each font can be displayed by providing seven control signals to drive the seven-segment LEDs from segment a to g. The input variables are represented by A, B, and C and the output variables are denoted using the letters from a to g, which produce the control signals for seven-segment LEDs. Owing to the use of a common-anode LED, an active-HIGH output is required to drive seven-segment LED to give out light.

(2) List the truth table from logic requirement

Seven fonts will be displayed by controlling the states of the seven-segment LEDs. The three-input variables have eight combinations of input levels; seven of them correspond to the seven fonts to be displayed and the final one ($ABC = 111$) will be treated as a don't care term. The truth table is shown in Table 4.4.7.

Table 4.4.7: Truth table.

Fonts	Inputs			Outputs						
	A	B	C	a	b	c	d	e	f	g
H	0	0	0	1	0	0	1	0	0	0
I	0	0	1	1	0	0	1	1	1	1
L	0	1	0	1	1	1	0	0	0	0
2	0	1	1	0	0	1	0	0	1	0
0	1	0	0	0	0	0	0	0	0	1
6	1	0	1	1	1	0	0	0	0	0
7	1	1	0	0	0	0	1	1	1	1
	1	1	1	×	×	×	×	×	×	×

(3) Write the output expression

The Karnaugh maps are used to simplify the logic expression, as shown in Figure 4.4.21. When you design a combinational logic circuit with multiple outputs, the total cost of the design circuit should be considered, which means that the product terms in the output logic expression should be shared by different outputs as many as possible to lower the cost. For instance, $\bar{A}B\bar{C}$ exists not only in the Karnaugh map of the output a but also in that of the output b. Although the simplest form, $\bar{A}\bar{C}$, can be obtained by grouping the two 1s in the cells of m_0 and m_1, the product term, $\bar{A}B\bar{C}$, is left alone to reduce the total cost of the design circuit.

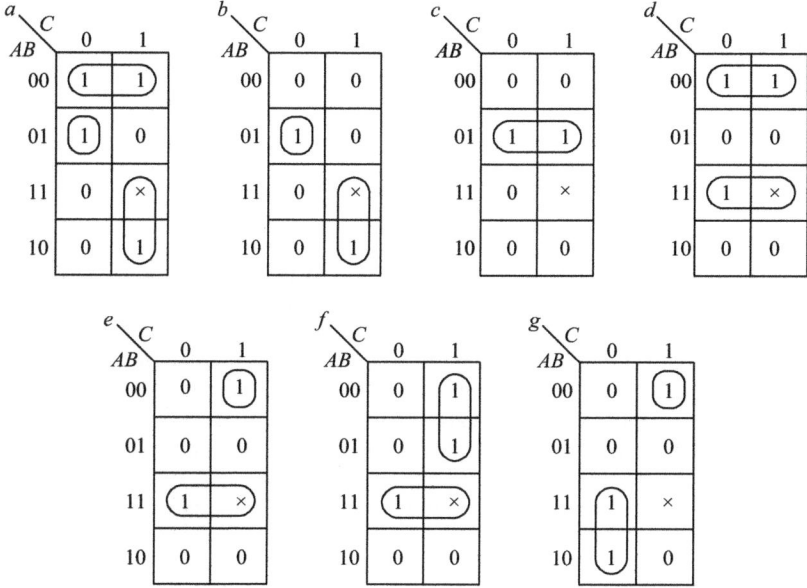

Figure 4.4.21: The Karnaugh maps.

The minimized logic expressions of the outputs are

$$\begin{cases} a = \bar{A}\bar{B} + \bar{A}B\bar{C} + AC = \overline{\overline{\bar{A}\bar{B}} \cdot \overline{\bar{A}B\bar{C}} \cdot \overline{AC}} \\ b = \bar{A}B\bar{C} + AC = \overline{\overline{\bar{A}B\bar{C}} \cdot \overline{AC}} \\ c = \bar{A}B = \overline{\overline{\bar{A}B}} \\ d = \bar{A}\bar{B} + AB = \overline{\overline{\bar{A}\bar{B}} \cdot \overline{AB}} \\ e = AB + \bar{A}B\bar{C} = \overline{\overline{AB} \cdot \overline{\bar{A}B\bar{C}}} \\ f = AB + \bar{A}C = \overline{\overline{AB} \cdot \overline{\bar{A}C}} \\ g = A\bar{C} + \bar{A}B\bar{C} = \overline{\overline{A\bar{C}} \cdot \overline{\bar{A}B\bar{C}}} \end{cases}$$

(4) Draw the logic diagram of a display decoder

Convert the logic expressions of the outputs into NAND–NAND form and implement the circuit with NAND gates, as shown in Figure 4.4.22.

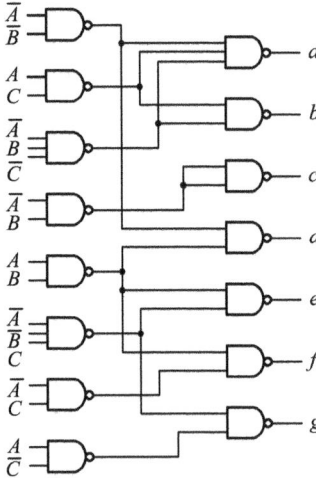

Figure 4.4.22: Logic diagram.

When a display decoder is designed and debugged, two factors should be considered. First, current-limiting resistors should be connected between the outputs of the decoder and the LED to limit the current through the LED. Second, the output current of the decoder should be strong enough to drive the LED. For a common-anode LED, the current supplied by the decoder is the sink current, whereas the current is the source current for a common-cathode LED. If the current in an LED is 8 mA, then it can be driven directly by the 74LS series logic gates whose sink current is typically 8 mA. If the current required by the LED is greater than 8 mA, then a driving circuit or a buffer should be used between the decoder and the LED.

---

(2) MSI display decoders

There are many MSI display decoders for LEDs and LCDs. Two types of integrated decoders have been designed for seven-segment LEDs to display the ten decimal digits 0–9. One is an active-HIGH for driving common-cathode LEDs, typically 74LS48; the other is an active-LOW for driving common-anode LEDs, typically 74LS47. The logic diagram of 74LS48 is shown in Figure 4.4.23 and its function table is shown in Table 4.4.8.

The logic function and pins of 74LS47 are the same as 74LS48, but 74LS47 is active-LOW for driving common-anode LEDs. The logic symbol and pin diagram of 74LS48 are shown in Figure 4.4.24(a), and the connection with a common-cathode LED is shown in Figure 4.4.24(b).

74LS47 and 74LS48 decode a four-bit input ($DCBA$) into seven outputs a–g to drive the seven-segment display. The display fonts corresponding to the four-bit inputs are shown in Figure 4.4.25. When inputs are 8421BCD code, "0–9" will be displayed. When the input is greater than 9, there are corresponding fonts too. When the input is 1111, the LED is off.

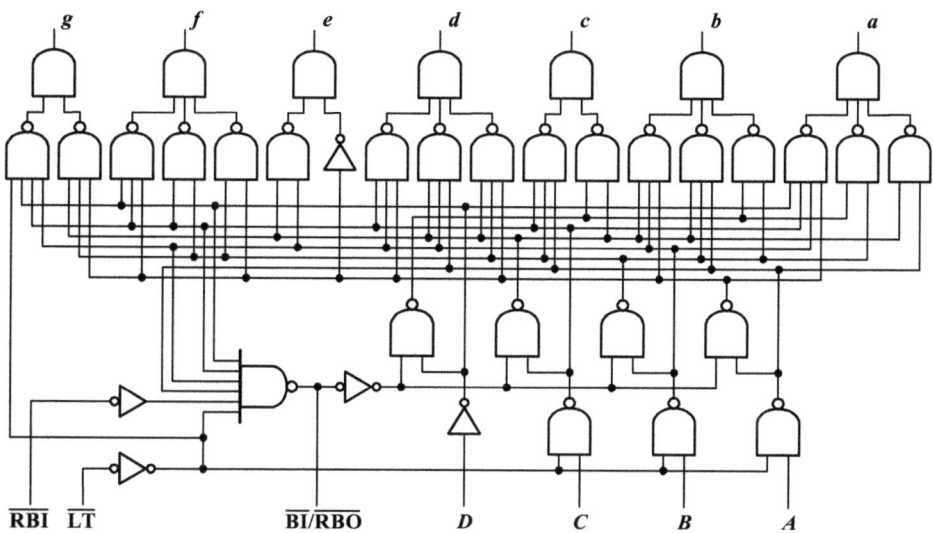

Figure 4.4.23: Logic diagram of the 74LS48 display decoder.

In addition, 74LS48 has several features as indicated by $\overline{LT}$, $\overline{RBI}$, and $\overline{BI/RBO}$ functions.

Blanking input, $\overline{BI}$, is an active-LOW. It shares a pin with ripple-blanking output $\overline{RBO}$. When a LOW is applied on $\overline{BI}$ as an input, all segment outputs are LOW and the seven-segment display is turned off.

Lamp test input, $\overline{LT}$, is an active-LOW. When $\overline{BI}$= 1 and $\overline{LT}$= 0, all segment outputs are at the HIGH level. so the seven-segment in the display is turned on. $\overline{LT}$ is used to check that no segments are burned out. The priority of $\overline{LT}$ is inferior to $\overline{BI}$.

Zero suppression in the 74LS48 is accomplished by using the ripple blanking input $\overline{RBI}$ and the ripple blanking output $\overline{RBO}$ together. $\overline{RBI}$ is an active-LOW. If $\overline{RBI}$ is a LOW and input is 0000, all outputs are at the LOW level. This causes the display to be blank and produces a LOW $\overline{RBO}$. If the input is not equal to 0000, the decoder will operate normally and the seven-segment in the display gives the corresponding fonts. Among the three control inputs, $\overline{BI}$ has the highest-order priority and $\overline{RBI}$ has the lowest-order priority. Only when $\overline{LT}$= 1, $\overline{RBI}$= 0, and $DCBA$ = 0000, $\overline{RBO}$ = 0. Ripple blanking output $\overline{RBO}$ is an active-LOW and it shares a pin with $\overline{BI}$. The $\overline{RBO}$ is often used to connect with the neighboring decoder's $\overline{RBI}$ in a cascade connection for zero suppression in the integral and fractional part, as shown in Figure 4.4.26. The $\overline{RBO}$ of each decoder is connected to the $\overline{RBI}$ of the next lower order so that all zeros to the left of the first nonzero digit are blanked. In Figure 4.4.26, the three highest-order digits are zeros in the integral part and thus are blanked. Only the digit 3 in the

Table 4.4.8: Function table of the 74LS48 display decoder.

Decimal	Inputs							Outputs						
	$\overline{LT}$	$\overline{RBI}$	D	C	B	A	$\overline{BI}/\overline{RBO}$	a	b	c	d	e	f	g
0	1	1	0	0	0	0	1 /	1	1	1	1	1	1	0
1	1	x	0	0	0	1	1 /	0	1	1	0	0	0	0
2	1	x	0	0	1	0	1 /	1	1	0	1	1	0	1
3	1	x	0	0	1	1	1 /	1	1	1	1	0	0	1
4	1	x	0	1	0	0	1 /	0	1	1	0	0	1	1
5	1	x	0	1	0	1	1 /	1	0	1	1	0	1	1
6	1	x	0	1	1	0	1 /	0	0	1	1	1	1	1
7	1	x	0	1	1	1	1 /	1	1	1	0	0	0	0
8	1	x	1	0	0	0	1 /	1	1	1	1	1	1	1
9	1	x	1	0	0	1	1 /	1	1	1	0	0	1	1
10	1	x	1	0	1	0	1 /	0	0	0	1	1	0	1
11	1	x	1	0	1	1	1 /	0	0	1	1	0	0	1
12	1	x	1	1	0	0	1 /	0	1	0	0	0	1	1
13	1	x	1	1	0	1	1 /	1	0	0	1	0	1	1
14	1	x	1	1	1	0	1 /	0	0	0	1	1	1	1
15	1	x	1	1	1	1	1 /	0	0	0	0	0	0	0
$\overline{BI}$	x	x	x	x	x	x	0 /	0	0	0	0	0	0	0
$\overline{RBI}$	1	0	0	0	0	0	/ 0	0	0	0	0	0	0	0
$\overline{LT}$	0	x	x	x	x	x	1 /	1	1	1	1	1	1	1

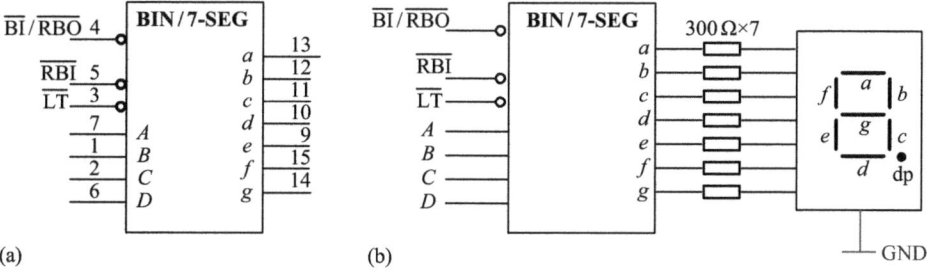

Figure 4.4.24: Logic symbol (a) of 74LS48 and circuit diagram (b) of driving a common-cathode LED.

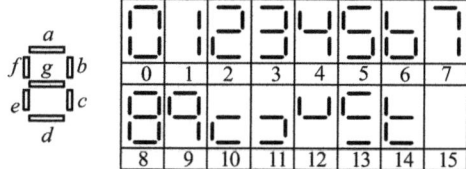

Figure 4.4.25: Displayed fonts and corresponding inputs of 74LS47 and 74LS48.

integer part is displayed. Besides, the $\overline{\text{RBO}}$ of the lowest decoder is connected to the $\overline{\text{RBI}}$ of the next higher-order decoder so that all zeros to the right of the first nonzero digit in the fractional part are blanked. Only the digit 2 in the fractional part is displayed.

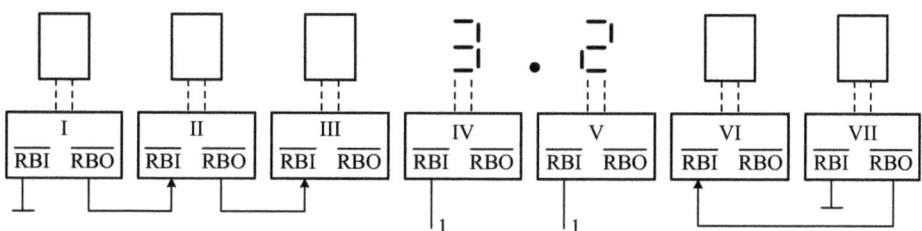

Figure 4.4.26: Cascade connection of the 74LS48 display decoders.

4.5 Multiplexers and demultiplexers

A *multiplexer* is a device that allows data from different sources to be merged on a single line through which data from different sources can be transmitted to a remote common destination by using time division multiplex. Multiplexers have multiple data input lines and a single output line; they also have data-select inputs, which select one of the data input lines to be switched to the output line. So multiplexers are also called *data selectors*. A demultiplexer basically reverses the multiplexing function. It has a single input line and multiple output lines; data-select inputs are used to take data from an input line and distribute them to one of the output lines. So demultiplexers are also known as *data distributors*. Demultiplexers can be realized by binary decoders.

The objectives of this section are to
- Explain the basic operation of a multiplexer and a demultiplexer
- Describe logic function of MSI multiplexers involving 74LS151 and 74LS153
- Describe how to use a decoder as a demultiplexer
- Apply the multiplexer to generate a logic function

4.5.1 Multiplexers

1. A 4-to-1 multiplexer

A 4-to-1 multiplexer has four data input lines, a single output line, and two data-select inputs. Two data-select inputs select data from one of four data input lines then the selected data are then transmitted to the output line. The logic symbol of a 4-to-1 multiplexer is shown in Figure 4.5.1(a). D_0, D_1, D_2, and D_3 are four data input lines and Y is an output line; A_1 and A_0 are two data-select inputs for determining which data input will be selected and sent to the output. If $A_1 = 0$ and $A_0 = 0$, the data on input D_0 appears on the data output line. If $A_1 = 0$ and $A_0 = 1$, the data on input D_1 appear on the data output line. If $A_1 = 1$ and $A_0 = 0$, the data on input D_2 appear on the data output line. If $A_1 = 1$ and $A_0 = 1$, the data on input D_3 appear on the data output line. A summary of this operation of a 4-to-1 multiplexer is given in Table 4.5.1.

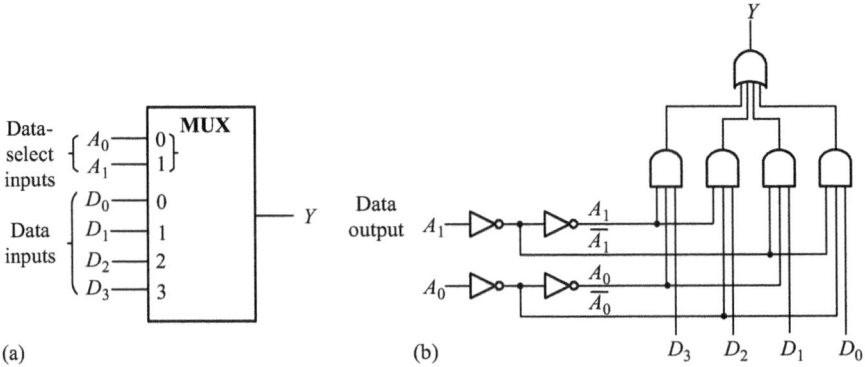

Figure 4.5.1: The 4-to-1 multiplexer: (a) logic symbol and (b) logic diagram.

The logic expression for the output of the multiplexer can be deduced from Table 4.5.1 as follows:

Table 4.5.1: Function table of a 4-to-1 multiplexer.

Select inputs		Data inputs				Output
A_1	A_0	D_3	D_2	D_1	D_0	Y
0	0	×	×	×	D_0	D_0
0	1	×	×	D_1	×	D_1
1	0	×	D_2	×	×	D_2
1	1	D_3	×	×	×	D_3

$$Y = \bar{A}_1\bar{A}_0 D_0 + \bar{A}_1 A_0 D_1 + A_1\bar{A}_0 D_2 + A_1 A_0 D_3 = \sum_{i=0}^{3} m_i D_i \qquad (4.5.1)$$

where m_i ($i = 0, 1, 2, 3$) is the minterm of data-select input variables A_1 and A_0.

From eq. (4.5.1), you can draw the logic diagram of the 4-to-1 multiplexer, as shown in Figure 4.5.2(b).

2. MSI multiplexers

Quad 2-to-1 multiplexer 74LS157, dual 4-to-1 multiplexer 74LS153, and 8-to-1 multiplexer 74LS151 are commonly used MSI multiplexers. The 74LS153 consists of two separate 4-to-1 multiplexers. Each of the two multiplexers has four data input lines, one output line, and an enable input. They share two data-select inputs. A LOW on the enable input allows the selected input data to pass through to the output. The 74LS151 is an 8-to-1 multiplexer with eight data input lines, three data-select inputs, one enable input, and two complemented outputs. The enable input is active-LOW. When the enable input is a HIGH, the multiplexer is disabled and the data input cannot pass through to the output. In this situation, the output is at a LOW level. Figure 4.5.2 shows their logic symbols and pin diagrams. The logic expression for the output of an 8-to-1 multiplexer is described as

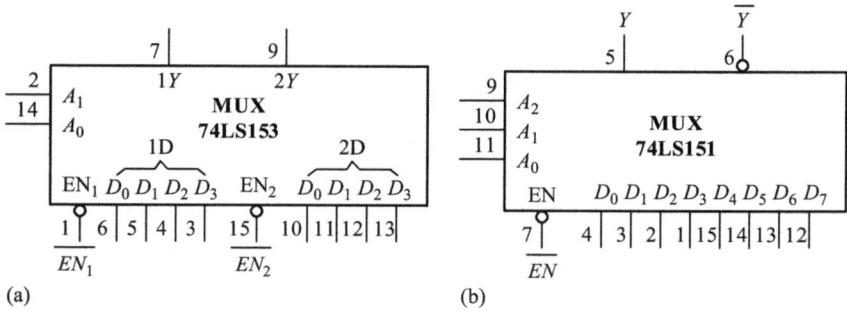

Figure 4.5.2: Logic symbol and pin diagram of 74LS153 (a) and 74LS151 (b).

$$Y = \overline{\overline{EN}} \cdot \sum_{i=0}^{7} m_i D_i \tag{4.5.2}$$

With the enable input, two 4-to-1 multiplexers can be expanded to an 8-to-1 multiplexer, as shown in Figure 4.5.3. If $A_2 A_1 A_0 = 000-011$, then $Y_2 = 0$ and an input of D_0-D_3 is selected to pass through to Y; if $A_2 A_1 A_0 = 100-111$, then $Y_1 = 0$ and an input of D_4-D_7 is selected to pass through to Y.

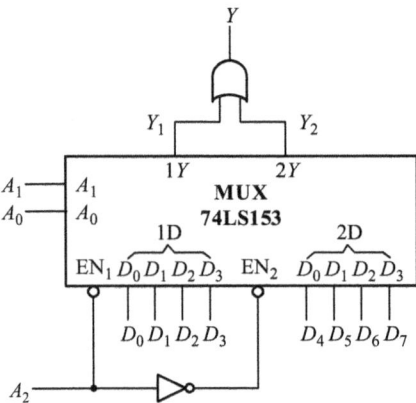

Figure 4.5.3: An 8-to-1 multiplexer with a 74LS153 two 4-to-1 multiplexers.

3. Applications of multiplexers

Multiplexers have many types of applications, in which a typical application is to implement combinational logic functions.

A 2^n-to-1 multiplexer can implement a logic function of n input variables. First, the logic function should be expressed in the standard sum-of-products expression. Then the n input variables are connected to n data-select inputs. Finally, each data input is set to a fixed logic level required. If a minterm of data-select inputs is present in the logic expression, the corresponding data input is set to a 1 and if a minterm of data-select inputs does not present in the logic expression, the corresponding data input is set to a 0.

Example 4.15 Implement the logic function $P = A\bar{C} + \bar{A}C + \bar{B}$ with an 8-to-1 multiplexer 74LS151.

Solution An 8-to-1 multiplexer can be used to implement any specified three-variable logic function. Assuming A is the MSB and C is the LSB, the logic function can be converted into a standard sum-of-products form as follows:

$$P(A, B, C) = A(\bar{B} + B)\bar{C} + \bar{A}(\bar{B} + B)C + (\bar{A} + A)\bar{B}(\bar{C} + C)$$
$$= \sum m(0, 1, 3, 4, 5, 6) \tag{4.5.3}$$

Connect the three variables A, B, and C to the three select inputs. Comparing eq. (4.5.3) with eq. (4.5.2), the value of D_i is 1 when its corresponding minterm m_i exists in eq. (4.5.3); otherwise, $D_i = 0$. So $D_0 = D_1 = D_3 = D_4 = D_5 = D_6 = 1$ and $D_2 = D_7 = 0$.

Alternatively, the Karnaugh map can be used to determine the logic level of each data input. In terms of the logic expression for the output of the multiplexer, the subscript of m_i is the same as the subscript of the data input D_i. Therefore, we can directly obtain the logic level of each data input in a Karnaugh map. The Karnaugh map of the given logic function is shown in Figure 4.5.4(a). From the Karnaugh map, the logic levels of the data inputs of the 8-to-1 multiplexer are $D_0 = D_1 = D_3 = D_4 = D_5 = D_6 = 1$, $D_2 = D_7 = 0$. So the logic diagram can be drawn in Figure 4.5.4(b).

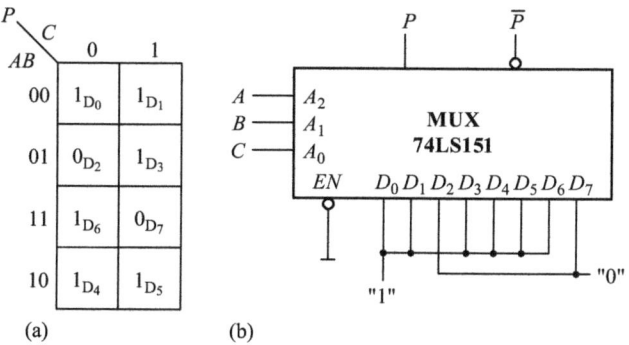

Figure 4.5.4: The Karnaugh map (a) and the logic diagram (b).

A single smaller multiplexer, for example, a 2^{n-1}-to-1 multiplexer also can be used to implement any logic function of n input variables. You can select $n - 1$ input variables as data-select inputs and the remaining variable as data inputs.

Example 4.16 Implement a four-variable logic function, $Y = A\bar{B}C + BD + \bar{A}\,\bar{C}$, with an 8-to-1 multiplexer 74LS151.

Solution
Since the given logic function has four variables and an 8-to-1 multiplexer has three data-select inputs, you can select any three variables, for instance, A, B, and C, as the select inputs of the multiplexer. Assume that A is the MSB, the given logic expression can be converted into a standard sum-of-products form, $Y(A,B,C)$, as follows:

$$Y(A, B, C) = \bar{A}\bar{B}\bar{C}\bar{D} + \bar{A}\bar{B}\bar{C}D + \bar{A}B\bar{C}\bar{D} + \bar{A}B\bar{C}D + \bar{A}BCD + A\bar{B}C\bar{D} + A\bar{B}CD + AB\bar{C}D + ABCD$$

$$= m_0\bar{D} + m_0 D + m_2\bar{D} + m_2 D + m_3 D + m_5\bar{D} + m_5 D + m_6 D + m_7 D \qquad (4.5.4)$$

$$= m_0 + m_2 + m_5 + m_3 D + m_6 D + m_7 D$$

Comparing the above equation with eq. (4.5.2), the logic level of each data input of 74LS151 can be determined as follows:

$$\begin{cases} D_0 = D_2 = D_5 = 1 \\ D_1 = D_4 = 0 \\ D_3 = D_6 = D_7 = D \end{cases} \qquad (4.5.5)$$

Notice that if different variables are chosen as the data-select inputs, the results will be different. For instance, if variables, B, C, and D are chosen as the data-select inputs, the standard sum-of-products form $Y(B,C,D)$ is as follows:

$$Y(B,C,D) = \bar{A}\bar{B}\bar{C}\bar{D} + \bar{A}\bar{B}C\bar{D} + \bar{A}B\bar{C}\bar{D} + \bar{A}BC\bar{D} + \bar{A}BCD + A\bar{B}C\bar{D} + AB\bar{C}D + AB\bar{C}\bar{D} + ABCD$$

$$= \bar{A}m_0 + \bar{A}m_1 + \bar{A}m_4 + \bar{A}m_5 + \bar{A}m_7 + Am_2 + Am_3 + Am_5 + Am_7 \qquad (4.5.6)$$

$$= m_5 + m_7 + \bar{A}m_0 + \bar{A}m_1 + \bar{A}m_4 + Am_2 + Am_3$$

If B, C, D are chosen as the data-select inputs, the corresponding logic level of each data input of 74LS151 is $D_0 = D_1 = D_4 = \bar{A}$, $D_2 = D_3 = A$, $D_5 = D_7 = 1$, $D_6 = 0$.

The logic diagram using variables A, B, and C as the data-select inputs is shown in Figure 4.5.5 (a), and the logic diagram using variables B, C, and D as the data-select inputs is shown in Figure 4.5.5(b). Obviously, the first solution is simpler than the second.

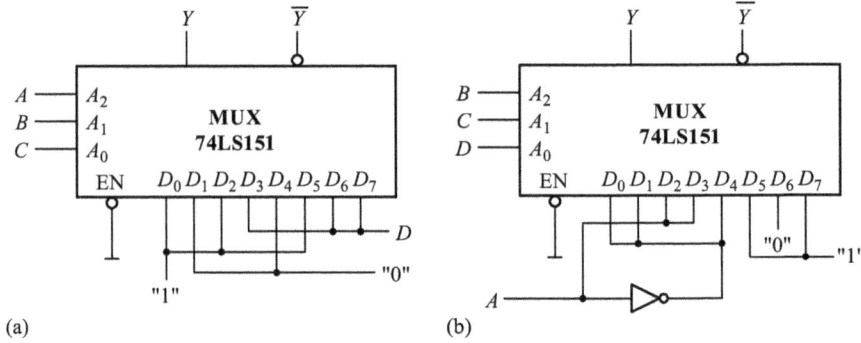

(a)　　　　　　　　　　　　　　(b)

Figure 4.5.5: Logic diagram: (a) variables A, B, and C; and (b) variables B, C, and D as data-select inputs.

If the data inputs of a multiplexer are not enough to satisfy the requirement, you can use several multiplexers for expansion. Figure 4.5.6 shows a 32-to-1 multiplexer consisting of four 8-to-1 multiplexers and one 4-to-1 multiplexer. For a 32-to-1 multiplexer, five data-select inputs $A_4A_3A_2A_1A_0$ are

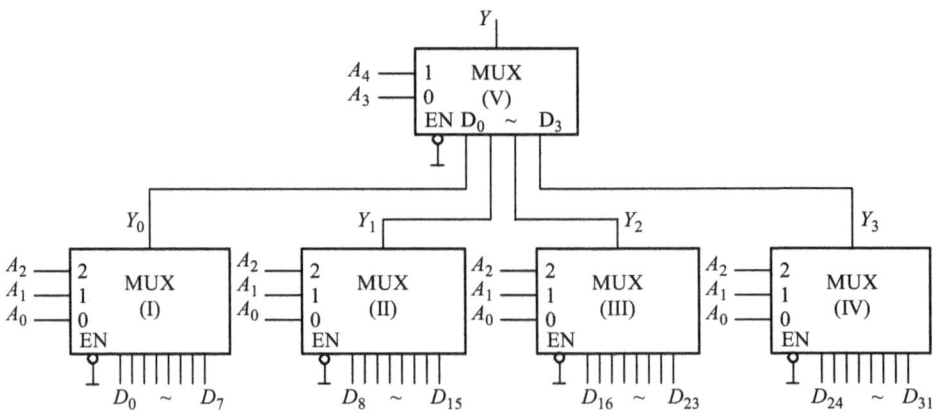

Figure 4.5.6: A 32-to-1 multiplexer.

required for selecting one of 32 data input lines to be switched to one output line. These five data-select inputs are divided into two groups. One group is the less significant three-bit inputs, $A_2A_1A_0$, as the data-select inputs of four 8-to-1 multiplexers. Each combination of the binary values of $A_2A_1A_0$ will separately select a data from four groups of data inputs D_0–D_7, D_8–D_{15}, D_{16}–D_{23}, and D_{24}–D_{31}, and send them to Y_0, Y_1, Y_2, and Y_3, respectively. Another group is the most significant two-bit inputs, A_4A_3, as the data-select inputs of a 4-to-1 multiplexer. Each combination of the binary values of A_4A_3 will select a data input coming from four outputs of four 8-to-1 multiplexers and send it to the output Y. When the data-select inputs, $A_4A_3A_2A_1A_0$, change from 00000 to 00111, one of data inputs D_0–D_7 is selected according to the binary values of $A_2A_1A_0$ and sent to the output Y. When the data-select inputs, $A_4A_3A_2A_1A_0$, change from 01000 to 01111, one of data inputs D_8–D_{15} is selected according to the binary values of $A_4A_3A_2A_1A_0$ and sent to the output Y. The rest can be deduced in the same manner.

4.5.2 Demultiplexers

A *demultiplexer* basically reverses the multiplexing function. It has a single input line and multiple output lines; data-select inputs are used to take data from an input line and distribute them to one of the output lines. So demultiplexers are also known as *data distributors*. A demultiplexer has one data input line, four output lines, and two data-select inputs. In terms of two data-select inputs, the demultiplexer distributes the data from data input line to one of four output lines. So it is called a 1-line-to-4-line demultiplexer or 1:4 demultiplexer. The logic circuit of a 1-line-to-4-line demultiplexer is shown in Figure 4.5.7. The data input line is an input of the AND gates. The two data-select lines enable only one gate at a time and the data appearing on the data input line will be selected and sent to the corresponding data-output line. It is shown in Figure 4.5.7 that a 1-line-to-4-line demultiplexer is actually a 2-to-4 decoder. The data input corresponds to the enable input and the two data-select inputs are the two-bit input code in a 2-to-4 decoder. That is to say, a demultiplexer can be implemented by a decoder.

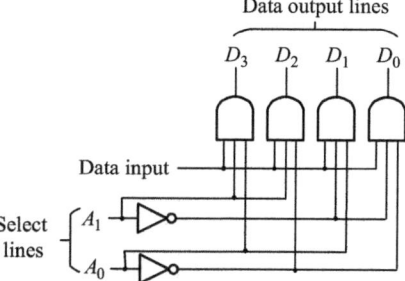

Figure 4.5.7: Logic diagram of a 1-line-to-4-line demultiplexer.

4.6 Magnitude comparators

In addition to the basic arithmetic operations such as addition, subtraction, multiplication, and division, comparison operation is also indispensable for computers and other electronic systems. A *magnitude comparator* is a logic circuit that performs the comparison of two quantities and indicates whether they are equal or not. The basic function of a comparator in digital circuit is to compare the magnitudes of two binary quantities to determine the relationship of those quantities.

The objectives of this section are to
- Describe the principle of a multi-bit binary number comparison
- Explain the logic function of a 74LS85 four-bit comparator
- Apply the 74LS85 to expand to a more-bit comparator

4.6.1 One-bit magnitude comparator

A one-bit magnitude comparator performs the comparison of two one-bit binary digits, A and B, and produces three possible output results, $Y_{A=B}$, $Y_{A>B}$, and $Y_{A<B}$. The truth table can be listed in Table 4.6.1.

Table 4.6.1: Truth table of a one-bit comparator.

A	B	$Y_{A<B}$	$Y_{A=B}$	$Y_{A>B}$
0	0	0	1	0
0	1	1	0	0
1	0	0	0	1
1	1	0	1	0

From the truth table, the corresponding logic expression for three outputs is deduced as follows:

$$\begin{cases} Y_{A<B} = \bar{A}B \\ Y_{A>B} = A\bar{B} \\ Y_{A=B} = \bar{A}\bar{B} + AB = A \odot B \end{cases} \tag{4.6.1}$$

The logic circuit of a one-bit magnitude comparator is shown in Figure 4.6.1.

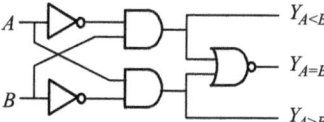

Figure 4.6.1: Logic diagram of a one-bit magnitude comparator.

4.6.2 An MSI four-bit magnitude comparator

The 74LS85 is an MSI four-bit magnitude comparator, which performs the comparison of two four-bit binary numbers, $A_3A_2A_1A_0$ and $B_3B_2B_1B_0$, and produces three possible output results, $Y_{A=B}$, $Y_{A>B}$, and $Y_{A<B}$. In addition, it has three cascading inputs $(A=B)_i$, $(A>B)_i$, and $(A<B)_i$ for cascaded expansion, as shown in Figure 4.6.2.

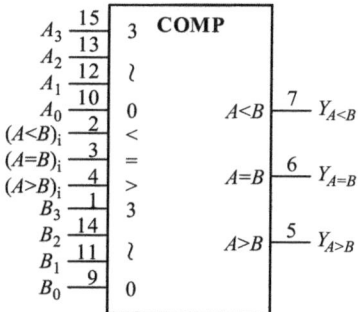

Figure 4.6.2: Logic symbol and pin diagram of the 74LS85 four-bit magnitude comparator.

To determine the relationship of two binary numbers, the general procedure is to check for an inequality in a bit position, starting with the highest order bit. When such an inequality is found, the relationship of two numbers is established, and any other inequality in lower-order bit positions must be ignored. Take the comparison of two four-bit binary numbers $A_3A_2A_1A_0$ and $B_3B_2B_1B_0$, for example, we should start with the comparison of the highest-order bit, A_3 and B_3. If $A_3>B_3$, then number $A_3 A_2 A_1 A_0$ is greater than number $B_3 B_2 B_1 B_0$. If $A_3 < B_3$, then number $A_3 A_2 A_1 A_0$ is less than number $B_3 B_2 B_1 B_0$. If $A_3=B_3$, then the next lower-order bit A_2 and B_2 should be compared. The rest can be done in the same manner. When number $A_3 A_2 A_1 A_0$ is equal to number $B_3 B_2 B_1 B_0$, the states of the cascading inputs $(A=B)_i$, $(A>B)_i$, and $(A<B)_i$ determine the final outputs of the comparator. If $(A>B)_i$ is a 1, then number $A_3 A_2 A_1 A_0$ is greater than number $B_3 B_2 B_1 B_0$; if $(A<B)_i$ is a 1, then number $A_3 A_2 A_1 A_0$ is less than number $B_3 B_2 B_1 B_0$; if $(A=B)_i$ is a 1, then number $A_3 A_2 A_1 A_0$ is equal to number $B_3 B_2 B_1 B_0$. The function table of 74LS85 is listed in Table 4.6.2.

Table 4.6.2: Function table of the 74LS85 four-bit magnitude comparator.

Number inputs				Cascading inputs			Outputs		
$A_3\,B_3$	$A_2\,B_2$	$A_1\,B_1$	$A_0\,B_0$	$(A{>}B)_i$	$(A{<}B)_i$	$(A{=}B)_i$	$Y_{A>B}$	$Y_{A<B}$	$Y_{A=B}$
$A_3{>}B_3$	×	×	×	×	×	×	1	0	0
$A_3{<}B_3$	×	×	×	×	×	×	0	1	0
$A_3{=}B_3$	$A_2{>}B_2$	×	×	×	×	×	1	0	0
$A_3{=}B_3$	$A_2{<}B_2$	×	×	×	×	×	0	1	0
$A_3{=}B_3$	$A_2{=}B_2$	$A_1{>}B_1$	×	×	×	×	1	0	0
$A_3{=}B_3$	$A_2{=}B_2$	$A_1{<}B_1$	×	×	×	×	0	1	0
$A_3{=}B_3$	$A_2{=}B_2$	$A_1{=}B_1$	$A_0{>}B_0$	×	×	×	1	0	0
$A_3{=}B_3$	$A_2{=}B_2$	$A_1{=}B_1$	$A_0{<}B_0$	×	×	×	0	1	0
$A_3{=}B_3$	$A_2{=}B_2$	$A_1{=}B_1$	$A_0 = B_0$	1	0	0	1	0	0
$A_3{=}B_3$	$A_2{=}B_2$	$A_1{=}B_1$	$A_0 = B_0$	0	1	0	0	1	0
$A_3{=}B_3$	$A_2{=}B_2$	$A_1{=}B_1$	$A_0 = B_0$	0	0	1	0	0	1
$A_3{=}B_3$	$A_2{=}B_2$	$A_1{=}B_1$	$A_0 = B_0$	0	0	0	0	0	0
$A_3{=}B_3$	$A_2{=}B_2$	$A_1{=}B_1$	$A_0 = B_0$	0	1	1	0	1	1
$A_3{=}B_3$	$A_2{=}B_2$	$A_1{=}B_1$	$A_0 = B_0$	1	0	1	1	0	1
$A_3{=}B_3$	$A_2{=}B_2$	$A_1{=}B_1$	$A_0 = B_0$	1	1	0	1	1	0
$A_3{=}B_3$	$A_2{=}B_2$	$A_1{=}B_1$	$A_0 = B_0$	1	1	1	1	1	1

The logic expressions for outputs of the 74LS85 are deduced as follows:

$$\begin{cases}
Y_{A=B} = \overline{A_3 \oplus B_3} \cdot \overline{A_2 \oplus B_2} \cdot \overline{A_1 \oplus B_1} \cdot \overline{A_0 \oplus B_0} \cdot (A = B)_i \\
Y_{A<B} = \overline{A_3}B_3 + \overline{A_3 \oplus B_3} \cdot \overline{A_2}B_2 + \overline{A_3 \oplus B_3} \cdot \overline{A_2 \oplus B_2} \cdot \overline{A_1}B_1 + \overline{A_3 \oplus B_3} \cdot \overline{A_2 \oplus B_2} \cdot \overline{A_1 \oplus B_1} \cdot \overline{A_0}B_0 \\
\quad + \overline{A_3 \oplus B_3} \cdot \overline{A_2 \oplus B_2} \cdot \overline{A_1 \oplus B_1} \cdot \overline{A_0 \oplus B_0}(A < B)_i \\
Y_{A>B} = A_3\overline{B_3} + \overline{A_3 \oplus B_3} \cdot A_2\overline{B_2} + \overline{A_3 \oplus B_3} \cdot \overline{A_2 \oplus B_2} \cdot A_1\overline{B_1} + \overline{A_3 \oplus B_3} \cdot \overline{A_2 \oplus B_2} \cdot \overline{A_1 \oplus B_1} \cdot A_0\overline{B_0} \\
\quad + \overline{A_3 \oplus B_3} \cdot \overline{A_2 \oplus B_2} \cdot \overline{A_1 \oplus B_1} \cdot \overline{A_0 \oplus B_0} \cdot (A > B)_i
\end{cases}$$

$$(4.6.2)$$

4.6.3 Applications of magnitude comparators

The 74LS85 has three cascading inputs $(A = B)_i$, $(A > B)_i$, and $(A < B)_i$. These three cascading inputs can be used to expand the comparator for any number of bits greater than four. In order to implement the comparison of two eight-bit binary numbers, two 74LS85s are required and their connection is shown in Figure 4.6.3.

To expand the comparator, the $A < B$, $A > B$, $A = B$ outputs of the lower-order comparator (I) are connected to the corresponding cascading inputs, $(A < B)_i$, $(A > B)_i$, $(A = B)_i$, of the next higher-order comparator (II). The lowest-order comparator must have a 1 on the input, $(A = B)_i$ and 0s on inputs $(A < B)_i$ and $(A > B)_i$. The final results of the comparator are presented from the outputs of the highest-order comparator.

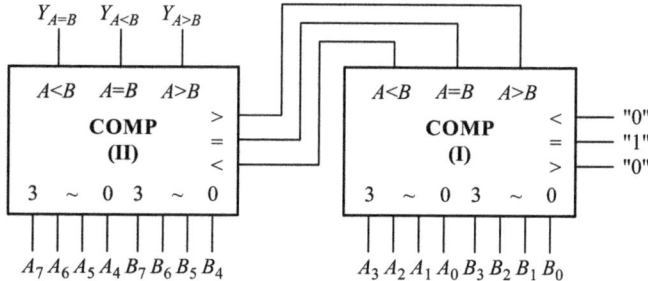

Figure 4.6.3: An eight-bit comparator using the cascading of two 74LS85s.

Another typical application of comparators is used as the control circuit of manufacture line, as shown in Figure 4.6.4 [6]. The products are delivered by the conveyor belt and the number of products can be counted by detecting the number of light pulse with a photoelectric detector. When a product passes the photoelectric detector, light signal from light source is blocked and a pulse signal is produced for the photoelectric detector. The counter counts the number of pulses representing the number of products. The counting value of the counter is compared with the preset number of products by a comparator. When the counting value of the counter is equal to the preset number, the comparator outputs a HIGH level, and informs mechanical equipment to carry out the next procedure. Meanwhile, the counter is reset and starts a new counting cycle.

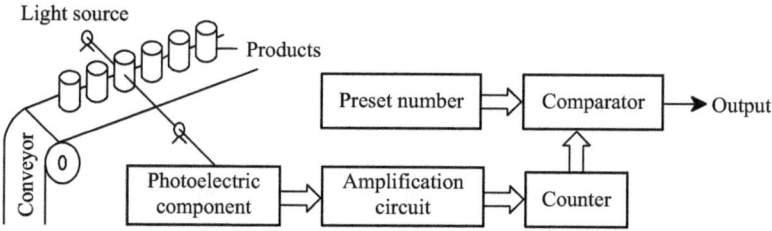

Figure 4.6.4: Block diagram of an application of magnitude comparators.

4.7 Races and hazards

The previous section discussed the analysis and design of combinational circuits in a steady state without considering the propagation delay of logic gates and the difference of the transfer time of signals. In fact, these factors would seriously affect the performance of logic circuits and even result in races and hazards. This section introduces the basic concepts of races and hazards in the combinational logic circuits and discusses the methods for eliminating hazards.

The objectives of this section are to
- Explain the static "1" and static "0" hazards
- Define races and hazards in combinational logic circuits
- Discriminate the possible hazards in combinational logic circuits
- Describe the methods of eliminating hazards

4.7.1 Basic concepts

Let us consider the combinational logic circuits in Figure 4.7.1(a). If the propagation delay of the signal is not considered, then the output $P_1 = A \cdot \bar{A} = 0$ and the output $P_2 = A + \bar{A} = 1$. However, if the propagation delay of the signal is considered, some unexpected glitches are generated in outputs, P_1 and P_2, as shown in Figure 4.7.1(b). The reason for glitches can be explained by the existence of the gate delay. In fact, every path from the input to the output is not the same length.

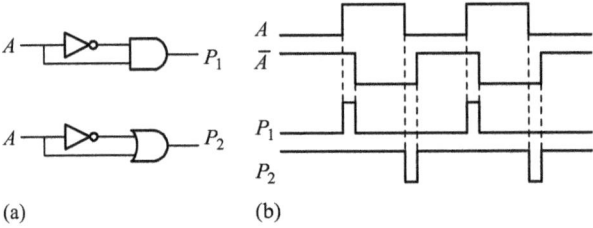

(a)   (b)

Figure 4.7.1: Combinational logic circuits (a) and its output waveform with the propagation delay (b).

The output P_1 would be constant (0) without considering the propagation delay of the inverter; however, when the propagation delay of the inverter is considered, some positive glitches occur at output P_1. This phenomenon is called the *static "0" hazard*. Similarly, output P_2 would be constant (1) without considering the propagation delay of the inverter; however, when the propagation delay of the inverter is considered, some negative glitches occur at output P_2. This phenomenon is called the *static "1" hazard*. *Races* refer to a phenomenon that several input signals of a logic gate vary at the same time, but have different transition time; or a signal is transferred to a logic gate through different paths resulting in different arrival times of signals. *Hazards* refer to a phenomenon that glitches occur due to the existence of races in a combinational logic circuit.

4.7.2 Discrimination of hazards

Hazards are caused by the following two reasons: the delay of logic gates and the race between input signals. In a complex digital system, it is very difficult to predict the accurate propagation delay of signals. To ensure the system's operation reliability, a feasible method is to find the existence of race in a combinational logic circuit and eliminate the race so that the hazards can be avoided in advance.

Boolean algebra can be used to determine whether there is a race in a combinational logic circuit. The first step is to check whether the logic expressions simultaneously have a variable and its complement. The next step is to check whether the logic expressions can be converted into the forms of $(A \cdot \bar{A})$ or $(A + \bar{A})$ for different input conditions; if yes, then a hazard may occur.

For instance, the logic expressions for the outputs of the logic circuit in Figure 4.7.2 are written as

$$\begin{cases} P_3 = \overline{A + B} + A \\ P_4 = \overline{A \cdot B} \cdot A \end{cases} \tag{4.6.3}$$

If $B = 0$, then P_3 is converted into $P_3 = A + \bar{A}$, then a static "1" hazard may occur for the circuit in Figure 4.7.2(a).

If $B = 1$, then P_4 can be changed to $P_4 = A \cdot \bar{A}$, then a static "0" hazard may occur for the circuit in Figure 4.7.2(b).

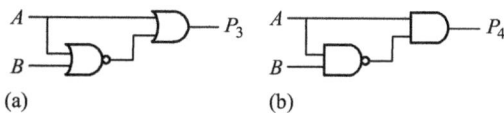

(a)   (b)

Figure 4.7.2: Logic diagram of the combinational logic circuits

---

Example 4.17 Determine whether hazards may exist in the following logic expressions:

$$\begin{cases} P_1 = AB + \bar{A}C \\ P_2 = (A + B)(\bar{A} + C) \\ P_3 = \bar{A}B + A\bar{C} + \bar{B}C \end{cases}$$

Solution
(1) When $B = C = 1$, $P_1 = A + \bar{A}$, a static "1" hazard may occur.
(2) When $B = C = 0$, $P_2 = A \cdot \bar{A}$, a static "0" hazard may occur.
(3) When $B = 1$ and $C = 0$, $P_3 = A + \bar{A}$, a static "1" hazard may occur.
     When $A = 0$ and $C = 1$, $P_3 = B + \bar{B}$, a static "1" hazard may occur.
     When $A = 1$ and $B = 0$, $P_3 = C + \bar{C}$, a static "1" hazard may occur.

Alternatively, the Karnaugh map can also be used to determine whether or not there is a possible hazard in a combinational logic circuit. The first step is to represent each product term or sum term in a logic expression by a Karnaugh map. Each term corresponds to a group. Then check whether or not the adjacent groups contain logic adjacent terms that are not grouped together. If yes, hazard may exist. For example, the Karnaugh map of a logic expression of $P_1 = AB + \bar{A}C$ is shown in Figure 4.7.3. Minterms m_3 and m_7 are two logic adjacent terms, but they are not included by other group. Thus there is a possible hazard in the logic expression of $P_1 = AB + \bar{A}C$.

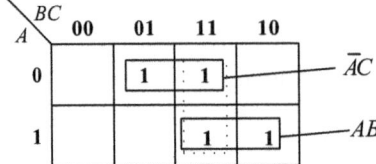

Figure 4.7.3: The Karnaugh map.

4.7.3 Methods of eliminating hazards

1. Adding a redundancy term

Take $P_1 = AB + \bar{A}C$, for example; when $B = C = 1$, a static "1" hazard exists in P_1. If a logic redundancy term, BC, is added to the logic expression, then $P_1 = AB + \bar{A}C + BC$. Thus, even if $B = C = 1$, $P_1 \equiv 1$ due to the existance of redundancy term BC. As a result, a static "1" hazard is eliminated.

Likewise, for $P_2 = (A + B)(\bar{A} + C)$, a static "0" hazard exists in P_2 when $B = C = 0$. If logic redundancy term, $B+C$, is added to the logic expression, then $P_2 = (A + B)(\bar{A} + C)$ $(B + C)$. Thus, even if $B = C = 0$, $P_2 \equiv 0$ due to the redundancy term BC. As a result, a static "0" hazard is eliminated.

A redundancy term can be easily obtained from the Karnaugh map. A group, which is denoted by the dotted group in Figure 4.7.3, is added to include the logic adjacent terms that are not grouped together. The possible hazard is eliminated. The final expression is $P_1 = AB + \bar{A}C + BC$.

2. Adding absorption capacitors

Add an absorption capacitor between the output and ground so that the glitch can be absorbed, as shown in Figure 4.7.4. Notice that the switch speed of the logic gate is reduced due to the increase in capacitance load. In practice, a suitable capacitance is chosen through the debugging process.

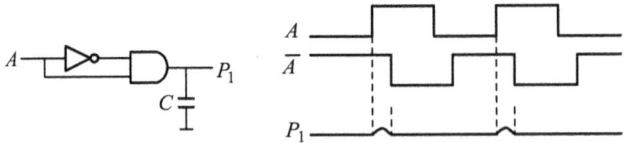

Figure 4.7.4: Adding absorption capacitors to eliminate the hazard.

As the glitch from hazards is a transient signal, it disappears quickly and usually doesn't cause severe problems for combinational logic circuits. However, if the glitch serves as an input for a sequential logic circuit, for instance, a flip-flop, it may change the operating state of the flip-flop and thus this hazard must be eliminated.

3 Adding a strobing pulse

There is an efficient method of adding a control input for a logic gate that may have a hazard. After finishing the change of input signals, a strobing pulse is applied to the control input of the logic gate and enables the logic gate. In this way, the glitches caused by the hazard can be eliminated.

4.8 Combinational logic circuits with Verilog HDL description

HDLs are a programming language used to describe the behavior or structure of a digital integrated circuit. VHDL (very high speed integrated circuit HDL) and Verilog HDL are the most popular and officially endorsed IEEE standards [10, 12, 13]. They are used to program PLD- and FPGA-based systems. This section introduces the statements of Verilog HDL through various examples so that the readers can efficiently learn and grasp simple programming methods and syntax of Verilog HDL.

The objectives of this section are to
- Explain description method of Verilog HDL
- Use Verilog HDL to describe combinational logic function

4.8.1 Description methods of Verilog HDL

Verilog HDL uses three ways to describe logic circuits, namely behavioral description, dataflow description, and structural description [31]. The *behavioral description* describes behaviors, functions, and features of a circuit without indicating what structures and logic gates should be used; this is a high-level description with strong versatility and flexibility. The *dataflow description* uses continuous assignment to describe combinational logic functions. *Structural description* describes the interconnection of entities that are primitives defined by Verilog HDL in advance. That is to say, the structural description describes logic circuits with primitives predefined by Verilog HDL.

For three description methods of Verilog HDL, behavioral description focuses on the function of a circuit; the statements are comparatively concise but may be difficult to be implemented (synthesized) with hardware. Statements with structural description are much easier to be synthesized but the statements are typically more complex. In practice, the statements of Verilog HDL are usually written with a combination of three description methods at the same time.

4.8.2 A 2-to-1 multiplexer

Many description methods can be used to describe a 2-to-1 multiplexer. Here, these examples are illustrated to describe the 2-to-1 multiplexer by using different Verilog description methods [32]. Example 4.18 illustrates the 2-to-1 multiplexer with its Verilog behavioral description; Example 4.19 shows the 2-to-1 multiplexer with its Verilog dataflow description; Example 4.21 shows the 2-to-1 multiplexer with its Verilog structural description.

Example 4.18. Design a 2-to-1 multiplexer with Verilog behavioral description.

Solution
As you know, a 2-to-1 multiplexer has two data input ports, a single output port, and one data-select input port. The data-select input select data from one of the two data inputs and the data are then transmitted to the output. The Verilog 2-to-1 multiplexer description is used as a module, as shown in Figure 4.8.1 in which a and b are two input ports, s is a select input port, and y is an output port.
    Verilog behavioral description of a 2-to-1 multiplexer is illustrated as follows.

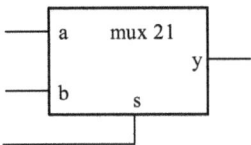

Figure 4.8.1: A 2-to-1 multiplexer module.

```
//Example 4-18
module Mux21 (a,b,s,y);     //----------------1
  input a,b;                //Declaring I/O ports
  input s;
  output y;
  assign y = (s==0)? a : b;  //----------------2
endmodule                    //----------------3
```

A *module* is the most basic unit of Verilog codes, which can specify a component or a circuit [33]. A module in Verilog HDL starts with a keyword "module" at the beginning of codes and finishes with a keyword "endmodule" at the end of codes, as shown in program lines 1 and 3 in Example 4.18. The structure of a module is as follows:

```
module <module name> (port list)
  <Declarations>
  <Module items>
Endmodule
```

Module name is the only identifier of a module. Port list specifies all ports , including input, output, and in/out (bidirectional) ports, of the module connecting with external circuits. Three types of ports, including input, output, and inout, are used in Verilog HDL, as shown in Figure 4.8.2, where the transmission direction of signals are self-explanatory [34].

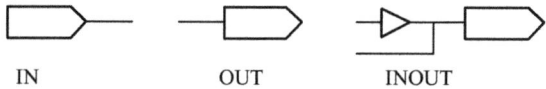

IN   OUT   INOUT

Figure 4.8.2: Three types of ports of Verilog HDL.

The contents of a module include two parts: declarations of variables and module items specifying functions, behaviors, or structures of a circuit. Declarations include the declaration of I/O ports and the declaration of internal variables used in the module. Module items define the function of the module and are the most essential part of a module.

The statement "**assign** y = (s= =0)? a: b;" in program line 2 is a continuous assignment for signal y. Continuous assignment statement can assign values to nets. The assignment of value takes place as soon as the values of variables on the right-hand side of the expression change. Continuous assignment is typically used to specify a combinational logic circuit.

The statement "s= =0?a:b" in program line 2 is a conditional statement. The conditional operator "?" is a ternary operator, whose syntax is shown as follows:

Conditional_expression ? true_expression: false_expression

First, the conditional_expression will be evaluated. If it equals to 1, then the value of the true_expression is chosen as the final result of the operation; if the conditional_expression equals to 0, then the value of the false_expression is chosen.

The logical equality operator is used in s= =0. When two expressions are equal, then the result is "1"; otherwise, the result is "0." "!=" is a logical inequality operator.

In summary, program line 2 shown in Example 4.18 assigns values to a wire-type variable y. If s is 0, then the state of y is the same as a; otherwise, the state is the same as b. The function realized in these codes is a 2-to-1 multiplexer.

Comments are generally used in Verilog HDL to explain and mark the codes to increase their readability. Programmers can add comments anywhere and these comments will be ignored by compliers and simulators. "//" is used to indicate a single-line comment. The information from "//" to the end of a line is treated as comments [34].

The simulation circuit diagram of Example 4.18 is shown in Figure 4.8.3, and the simulation waveform is shown in Figure 4.8.4.

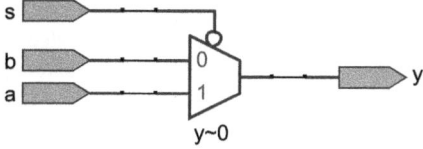

Figure 4.8.3: The simulation circuit diagram of Example 4.18.

It is found from Figure 4.8.4 that if s = 1, then y = b; if s = 0, then y = a. This verifies that the logic function of a 2-to-1 multiplexer is already implemented.

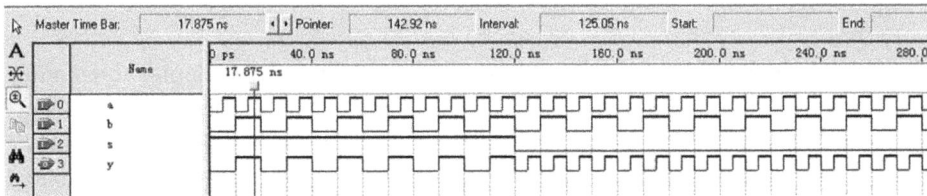

Figure 4.8.4: The simulation waveform of Example 4.18.

Example 4.19 Design a 2-to-1 multiplexer with Verilog dataflow description.

Solution
Here, two schemes of Verilog dataflow description are illustrated to design a 2-to-1 multiplexer.

Scheme 1 One Verilog dataflow description of a 2-to-1 multiplexer is illustrated as follows:

```
//Scheme 1 of Example 4-19
module Mux21 (a,b,s,y);
    input a,b;
    input s;
    output y;          //--------------1
    wire d,e;              //--------------2
    assign d = a & (~s); //--------------3
    assign e = b & s;
    assign y = d | e;
endmodule
```

Intermediate variables used in Verilog codes should be declared. The program line 2 of the above program defines variables d and e as wire type, indicating that they represent the nodes or internal connections of a circuit. By default, the output port y in program line 1 is also regarded as the wire type, representing a physical connection between logic gates. A net cannot retain data, but it can be driven through a continuous assignment statement or logic gates. In this case, a continuous driving source is required and the value of the net varies with the change of the driving source. The value of a net is high impedance (z) if no driving source is connected with it.

Bitwise operators are used in the assignment statement and operate on individual bits of operands. There are five commonly used bitwise operators in Verilog HDL, which are bitwise NOT (~), bitwise AND (&), bitwise OR (|), bitwise XOR (^), and bitwise XNOR (^~). When two digits in a bitwise operation have different bits, their LSB will be aligned automatically. 0s will be padded to the MSB of the operand with fewer bits so that it has the same length of bits with the other operand, and then the bitwise operation will be carried out.

Scheme 2. Another Verilog dataflow description of a 2-to-1 multiplexer is illustrated as follows:

```
//Example 4.8.3
module mux21 (a,b,s,y);
input a,b,s;
output y;
reg   y;
    always @( a or b or s)      //-----------------------------------1
        if (!s) y = a;          //----------------------------------2
        else y = b;             //----------------------------------3
endmodule
```

The output port y in the above program is defined as a reg-type variable. **Reg** does not necessarily represent a register. It stands for a storage cell and its latest assignment can be retained without a driving source. It can be synthesized to be a combinational circuit or a latch or a flip-flop. Note that the value of a reg-type variable can be assigned only in an always block.

The "**always @** (a or b or s)" in program line 1 is a continuous execution statement. The always block is executed continuously and cannot be interrupted unless the time control feature of Verilog utilizing symbols such as "@" is used. The part of the always block after the @ symbol, in parentheses, is called the sensitivity list. The statements inside an always block are executed by the simulator only when one or more of the signals in the sensitivity list changes their value. The sensitivity list should be as complete as possible. Otherwise, the function of the synthesized circuit may be different from the description of the module.

The if-else statement is used to indicate whether a statement is executed. Its syntax is as follows:

```
if (conditional_expression)
statement 1;
else
statement 2;
```

If the conditional_expression is evaluated to true ("1"), the statement 1 is executed; otherwise, statement 2 is executed. Consequently, the function of program lines 2 and 3 is: if s = 0, then y = a; if s = 1, then y = b. The if-else statement can only be used in always blocks.

Three logical operators are used in Verilog HDL: logic AND "&&", logic OR "||", and logic NOT "!". The value of a logic expression is either 1 (true) or 0 (false). "&&" and "||" are binary operators that involve two operands, such as (a>b)&&(b>c) and (a<b)||(b<c). The operator "!" is a unary operator and only has one operand such as! (a>b). Table 4.8.1 is the truth table of logical operations, in which the results of various logical operations are listed.

The priority of "!" is higher than "&&" and "||". The priority of logical operators is lower than the relational operators. Parentheses have the highest priority. For instance,

Table 4.8.1: Truth table of logical operations.

a	B	!a	!b	a&&b	a||b
True	True	False	False	True	True
True	False	False	True	False	True
False	True	True	False	False	True
False	False	True	True	False	False

(a>b)&&(x>y) can be written as: a>b && x>y

(a==b)||(x==y) can be written as: a==b || x==y

(!a)||(a>b) can be written as: !a || a>b

Example 4.20 Design a 2-to-1 multiplexer with the Verilog structure description.

Solution
The Verilog structure description of a 2-to-1 multiplexer is illustrated as follows.

```
//Example 4.20
    module mux21(y, a, b, s);
      input a, b, s;
      output y;
      not u1 (ns, s);//——————-1
      and u2 (sela, a, ns);
      and u3 (selb, b, s);
      or u4 (y, sela, selb);
endmodule
```

Example 4.20 invokes several bottom-layer module instances. The process of invoking the modules is called *instantiation*. After instantiation, the modules are called *instances*. The format of an instance is

```
<module_name><instance_name><port_list>;
```

Example 4.20 invokes the instances of NOT gate, AND gate, and OR gate. "not u1 (ns,s)" in Program line 1 invokes a NOT gate represented by the instance name "u1." It has two ports ns and s: ns is output and s is input. The NOT gate is predefined in Verilog HDL, and these predefined modules are often referred to as *primitives*.

Example 4.20 is a gate-level description. It provides a direct corresponding relationship between modules and real circuits. The keywords of several primitives in Verilog HDL are as follows:

not—NOT gate; **buf**— buffer; **and**—AND gate; **or**—OR gate; **nand**—NAND gate;

nor—NOR gate; **xor**—XOR gate; **xnor**—XNOR gate.

When programming with gate-level structural description, we can directly invoke these gates through instantiation. The simulation circuit diagram of Example 4.20 is shown in Figure 4.8.5, and its simulation waveform is the same as Figure 4.8.4.

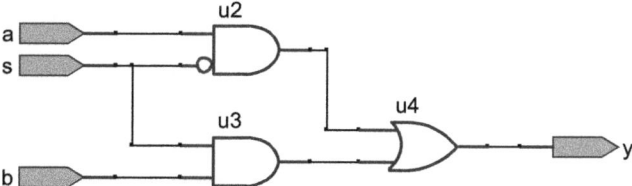

Figure 4.8.5: The simulation circuit diagram of Example 4.20.

4.8.3 4-to-1 Multiplexers

The function table of a 4-to-1 multiplexer is shown in Table 4.8.2. There are four data inputs "in0, in1, in2, in3," two select inputs "sel(1),sel(0)," and one output "out."

Table 4.8.2: Function table of a 4-to-1 multiplexer.

sel(1)	sel(0)	out
0	0	in0
0	1	in1
1	0	in2
1	1	in3

1. A 4-to-1 multiplexer implemented with a case statement

```
//Example 4.8.5
module mux4_1(out,in0,in1,in2,in3,sel);
    output out;
    input in0,in1,in2,in3;
    input[1:0] sel;              //--- -----------------------------------1
    reg out;
    always @(in0 or in1 or in2 or in3 or sel)   //----------- ---------------------2
    case(sel)
      2'b00: out=in0;
      2'b01: out=in1;
      2'b10: out=in2;
      2'b11: out=in3;
      default: out=1'bx;
    endcase
endmodule
```

(1) Vector

The "input [1:0] sel;" in program line 1 defines a vector including two variables sel[1] and sel[0]. If the bit width is not specified for a net or a reg variable in Verilog, then the variable is regarded as a scalar. If the bit width is specified, then the variable is a vector. For instance:

```
wire [7:0] bus;        //8-bit vector bus
reg [0:40] addr;       //41-bit vector register addr
```

Vectors in Verilog are treated as unsigned numbers. The bit in the leftmost position of a vector is the MSB, and the bit in the rightmost position is the LSB. For instance, the zeroth bit of addr is the MSB, whereas the seventh bit of a bus is the MSB. When the bit width of a vector is defined, the MSB and the LSB can be a negative number. For instance:

```
reg [-1:4] b;     //6-bit vector register b
```

The use of vectors is flexible, for instance:

```
bus[0]              // The 0ᵗʰ bit of bus
bus[2:0]            // The three least-significant bits of bus. Note we can't use bus
[0:2] as the use of vector should be the same as its declaration
addr[0:1]              //The two most-significant bits of addr
```

When assigning the value of a scalar to a vector, the value is assigned to the LSB of the vector. The operation of a vector is equivalent to the operation of every variable of the vector.

(2) Representation of number

Commonly used integers include decimal, hexadecimal, octal, and binary numbers. The representation of an integer in Verilog can be either a decimal number or a fixed-size number.

We can use the decimal digits 0–9 to directly represent an integer. The format of a fixed-size number representation is:

<size_in_bits> <'radix_identifier> <significant_digits (including a–f for a hex number)>

The size_in_bits can be absent. If this is the case, system will use the default size (32 bits). The size_in_bits specifies the exact size of bits, and it should be an unsigned decimal number.

Radix_identifier can be binary (b or B), decimal (d or D), hexadecimal (h or H), and octal (o or O). The single quotation mark " " must be placed in front of the radix_identifier and there is no space between them. Significant_digits are unsigned numbers or a~g for a hexadecimal number.

Any signal in Verilog can have four possible values: "x", "z", "0", and "1". "x" is used to denote an unknown logic value, which could be any of "z", "0", and "1". "z" is used to denote a high-impedance state. Both "x" and "z" can be used in binary, octal, and hexadecimal systems. For a hexadecimal number, an "x" represents a four-bit binary number and all of the four bits are "x". Likewise, an "x" in an octal number system that represents three bits of "x", and an "x" in a binary number represents a bit of "x". Same rules apply to "z".

For an unsigned number with a fixed size, when the size of the number is smaller than the specified size, 0s are usually padded to the left of the number; but if the leftmost digit is x or z, then these values are padded to the left.

To improve readability, we can use the underscore character "_" to separate long numbers. For instance, instead of writing 12′b100010001000, it can be written as 12′b1000_1000_1000 too. Note that "_" can't be placed to the leftmost digit of a number.

Examples of representation of integer are shown below:

```
' h 123F                  //Unsized hexadecimal number
' o 123                   //Unsized octal number
3 'b101                   //3-bit binary number
5 ' D 3                   //5-bit decimal number
12 ' h x                  //12-bit unknown number
16 ' o z                  //16-bit high-impedance state
16 ' b 1001_0110_1111_zzzz    //16-bit binary number
```

The following representation is incorrect:

```
123af   //'h should be added to represent a hexadecimal number
```

(3) The case statement
The case statement in Verilog is defined as

```
case (expression)
alternative1: statement1;
alternative2: statement2;
alternative3: statement3;
...
alternativej: statementj;
[default: statement;]
Endcase
```

The controlling expression and each alternative are compared bit by bit. When there are more than one matching alternatives, the statement associated with the first match is executed. When the specified alternatives do not cover all possible values of the controlling expression, the optional default clause should be included. Only one default clause can be included in a case statement. If there is neither matching alternative nor default clause, the case statement will not be executed. The size of bits of the controlling expression should be the same as the size of bits of the alternative expressions. If not, zeros (or x or z) will be padded to the left of the result which has fewer bits.

The simulation of the circuit diagram of a 4-to-1 multiplexer is shown in Figure 4.8.6, and the simulation waveform is shown in Figure 4.8.7.

It is shown in Figure 4.8.7 that when sel equals to 0, 1, 2, and 3, out equals to in0, in1, in2, and in3, respectively. The function of this circuit is a 4-to-1 multiplexer.

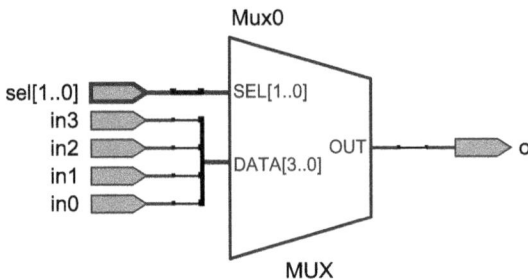

Figure 4.8.6: The simulation of the circuit diagram of a 4-to-1 multiplexer.

Figure 4.8.7: Simulation waveform of a 4-to-1 multiplexer.

2. A 4-to-1 multiplexer with the if-else statement

```
//Example 4.20
module mux4_1(out,in0,in1,in2,in3,sel);
output out;
input in0,in1,in2,in3;
input[1:0] sel;
reg out;
always @(in0 or in1 or in2 or in3 or sel)
begin
if(sel==2'b00) out=in0;
```

```
else if(sel==2'b01) out=in1;
else if(sel==2'b10) out=in2;
else if(sel==2'b11) out=in3;
else
out=1'bx;
end
endmodule
```

(1) if-else-if statement

In addition to the if-else statement, if-else-if statement can also be used as follows:

```
if (expression 1)
statement 1;
else if (expression 2)
    statement 2;
else if …
else
statement n;
```

The if-else-if statement evaluates the values of expressions in the order that they appear in the code [35]. If the value of an expression is true, then the corresponding statement is executed, and the entire if-else-if statement is terminated.

In Verilog, "else" can be absent, and an "else" in Verilog codes will automatically search the precedent codes and find the closest "if" to form an "if-else" statement. For instance:

```
if (a==0)
   if (b==0)
      c= 1;
    else
        c= 0;
```

(2) The begin-end block

The begin-end block is usually used to contain two or more statements so that they look like a single statement.

Notice that statements in a begin-end block are executed in sequential order within the block and the execution of a begin-end block will not be completed until the last statement is executed.

The syntax of the begin-end block is:

```
begin
    statement 1;
    statement 2;
    ......
    statement n;
end
```

With a begin-end block, an "else" statement can be linked with a specified "if" statement. For instance:

```
if (a==0)
begin
if (b==0)
        c=1;
end
else
        c=0;
```

(3) Default issues

Both the if-else and the case statements have default issues. The "else" in the if-else statement and the "default" in the case statement can be absent, but this will introduce latches into a combinational circuit and cause some problems. Example 4.21 shows the codes with complete "if-else" pairs.

```
//Example 4.21
module ex3reg(y, a, b, c);
input a, b, c;
output y;
reg y, rega;
always @(a or b or c)
    begin
    if(a&b)
            rega=c;
    else        //with an else as default
            rega=0;
    y=rega;
    end
endmodule
```

This example uses the always block to define a combinational circuit. As the last else exists serving as a default, reg a is synthesized to be a multiplexer, as shown in Figure 4.8.8.

The Verilog codes without the default item are as follows:

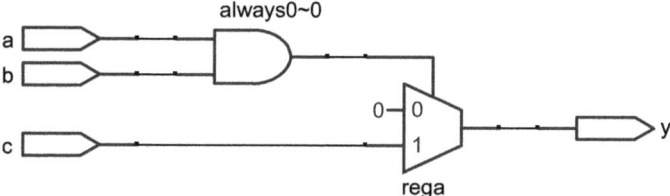

Figure 4.8.8: Synthesis result of Example 4.21.

```
//Example 4.22
module ex4reg(y, a, b, c);
input a, b, c;
output y;
reg y, rega;
 always @(a or b or c)
   begin
       if(a&b)
             rega=c;   // an "else" is absent
       y=rega;
   end
endmodule
```

An "else" is absent in Example 4.22. When a&b equals to 1, the value of c is assigned to rega. But when a&b equals to 0, rega retains its original value. During the synthesis process, rega is synthesized to be a latch, which is an extra and unnecessary part of the circuit. The synthesized circuit is shown in Figure 4.8.9.

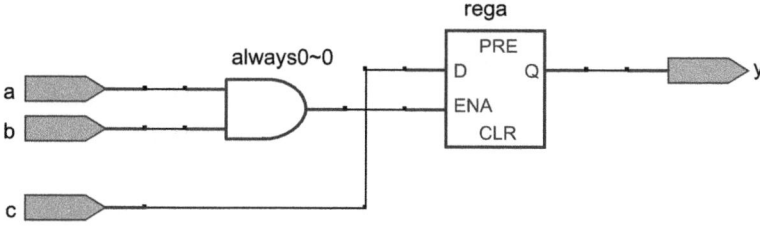

Figure 4.8.9: Simulation circuit diagram of Example 4.22.

Likewise, if the "default" is absent in the case statement, an unnecessary latch is generated too. The Verilog codes is shown in Example 4.23, and the simulation circuit diagram is shown in Figure 4.8.10.

```
//Example 4.23
module inccase(a, b, c, d, e);
input a, b, c, d;
```

```
output e;
reg e;
always @(a or b or c or d)
    case ({a,b})                    //--------------------------------------1
        2'b11: e=d ;
        2'b10: e=~c ;
    Endcase
Endmodule
```

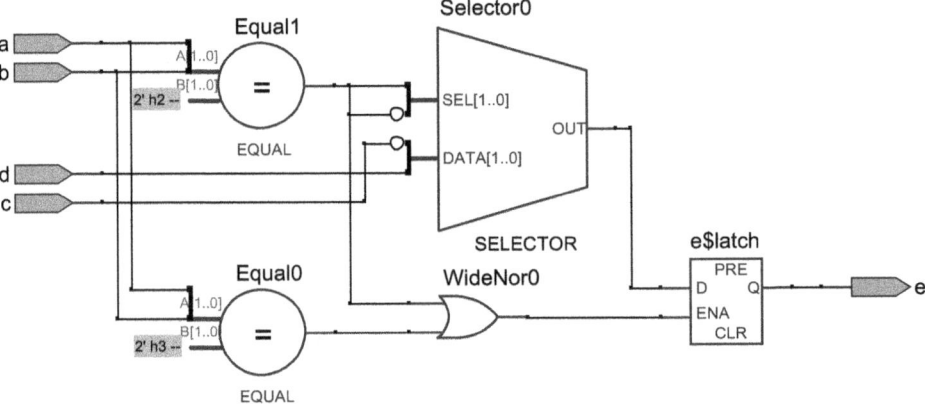

Figure 4.8.10: Simulation circuit diagram of Example 4.23.

The "{ }" in program line 1 of Example 4.23 is a concatenation operator to create a new binary number consisting of a and b.

In summary, the absence of the default item in the if statement and the case statement may produce unnecessary latches. Thus, the default item cannot be ignored in the design of combinational logic circuits with Verilog HDL.

4.8.4 Four-bit adders

The function of a four-bit adder is to realize the addition of two four-bit binary numbers [36]. The Verilog codes consist of two modules: a full-adder module and a top-layer module, as shown in Example 4.24.

```
//Example 4.24
//A full adder
module add_full(A,B,C,Carry,S);
    input A,B,C;
```

```
  output Carry,S;                //Sum and carry of the full adder
  assign S = A^B^C;
  assign Carry = (A&B)|(B&C)|(A&C);
  endmodule
//A 4-bit adder
module add_full4(A,B,C,S);
  input[3:0] A,B;
  output[3:0] S;               // Sum of the four-bit adder
  output[4:0] C;               //Carry of the four-bit adder
  assign  C[0]=0;
add_full  u1(A[0],B[0],C[0],C[1],S[0]),        //------------------1
          u2(A[1],B[1],C[1],C[2],S[1]),
          u3(A[2],B[2],C[2],C[3],S[2]),
          u4(A[3],B[3],C[3],C[4],S[3]);
endmodule
```

(1) Instance

The full adder in Example 4.24 is an instance whose name is add_full. It is a bottom-layer module of the four-bit adder. Verilog codes must have a top-layer module that cannot be instantiated but can instantiate many bottom-layer instances.

(2) Instantiation

The program line 1 in Example 4.24 means that the top-layer module add_full4 invokes the bottom-layer module add_full. This process is referred to as instantiation, and the instantiated modules are called instances. Every instance has its own name, variables, and port list.

The first format of an instantiation is:

```
<module_name>< instance_name> <port_list>;
```

add_full u1(A[0],B[0],C[0],C[1],S[0]) indicates that a bottom-layer module named add_full is invoked. The instant name is "u1" and the corresponding ports are A[0], B[0], C[0], C[1], and S[0]. In Example 4.24, the full-adder module is invoked four times. Note that the sequence of ports in the port_list of an instance should be the same as the sequence of ports in the port_list of the corresponding bottom-layer module.

The second format of an instantiation is:

```
<module_name> <instance_name> <instance_port1 (module_port 1), . instance_port (modu-
le_port2)>;
```

For instance:

```
add_full u1(.A(A[0]),.B(B[0]),.C(C[0]),.Carry(C[1]),.S(S[0])),
         u2(.A(A[1]),.B(B[1]),.C(C[1]),.Carry(C[2]),.S(S[1])),
         u3(.A(A[2]),.B(B[2]),.C(C[2]),.Carry(C[3]),.S(S[2])),
         u4(.A(A[3]),.B(B[3]),.C(C[3]),.Carry(C[4]),.S(S[3]));
```

In Verilog HDL, only one module can be defined within a module-and-endmodule block. However, we can invoke other modules in a module by multiple instantiations.

The simulation circuit diagram of Example 4.24 is shown in Figure 4.8.11, and its simulation waveform is shown in Figure 4.8.12, where we can find that the four-bit adder consists of four full adders and realizes the addition of two four-bit binary digits.

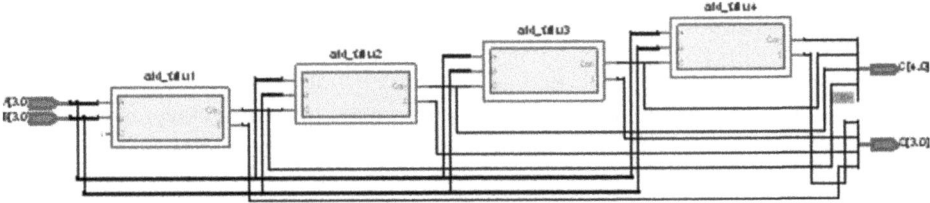

Figure 4.8.11: Simulation circuit diagram of Example 4.24.

Figure 4.8.12: Simulation waveform of Example 4.24.

4.8.5 Display decoders of seven-segment LEDs

Assuming a seven-segment LED to be driven is a common-cathode LED, Example 4.25 shows the Verilog codes of a display decoder for the LED. The circuit has four inputs and seven outputs.

```
//Example 4.25
module decode48(a,b,c,d,e,f,g,D3,D2,D1,D0);
```

```
output a,b,c,d,e,f,g;
input D3,D2,D1,D0;                   //Input a four-bit BCD code
reg a,b,c,d,e,f,g;                   //Output seven signals for the seven-segment LED
always @(D3 or D2 or D1 or D0)
begin
case({D3,D2,D1,D0}) // decoding with the case statement
    4'd0: {a,b,c,d,e,f,g}=7'b1111110;
    4'd1: {a,b,c,d,e,f,g}=7'b0110000;
    4'd2: {a,b,c,d,e,f,g}=7'b1101101;
    4'd3: {a,b,c,d,e,f,g}=7'b1111001;
    4'd4: {a,b,c,d,e,f,g}=7'b0110011;
    4'd5: {a,b,c,d,e,f,g}=7'b1011011;
    4'd6: {a,b,c,d,e,f,g}=7'b1011111;
    4'd7: {a,b,c,d,e,f,g}=7'b1110000;
    4'd8: {a,b,c,d,e,f,g}=7'b1111111;
    4'd9: {a,b,c,d,e,f,g}=7'b1111011;
    default: {a,b,c,d,e,f,g}=7'bx;
  endcase
end
endmodule
```

Concatenation operator ({}) concatenates the results of two or more expressions. The syntax of a concatenation operation is:

```
{expression 1, expression 2, …}
```

For instance:

```
{a, b[1:2], c,  4'b0010}
{4{y}}
```

The first line is equivalent to {a, b[1], b[2], c, 1′b0, 1′b0, 1′b1, 1′b0} and the second line is equivalent to {y, y, y, y}. The bit size of each operand should be certain so that the program can calculate the total size of the result of a concatenation operation.

The simulation circuit diagram of Example 4.25 is shown in Figure 4.8.13, and its simulation waveform is shown in Figure 4.8.14, where we can find that, when the inputs [D3, D2, D1, D0] are equal to 0001, the outputs [a, b, c, d, e, f, g] equal to 011000 and "1" can be displayed on the seven-segment LED.

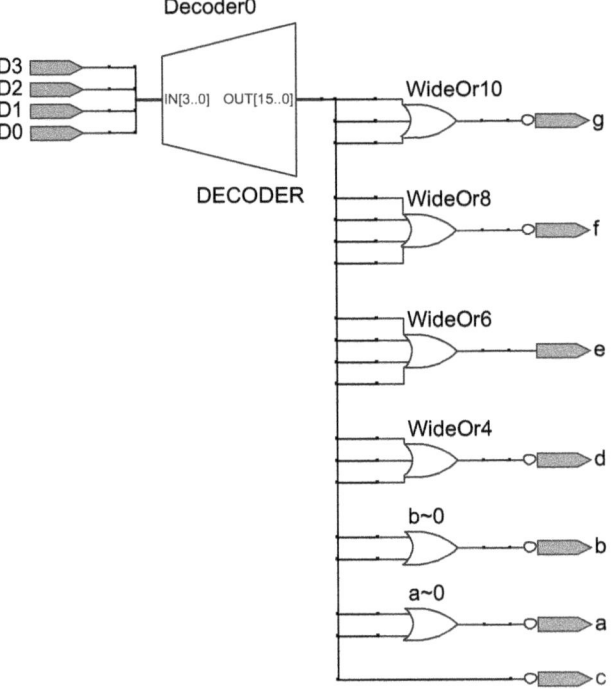

Figure 4.8.13: Simulation circuit diagram of Example 4.25.

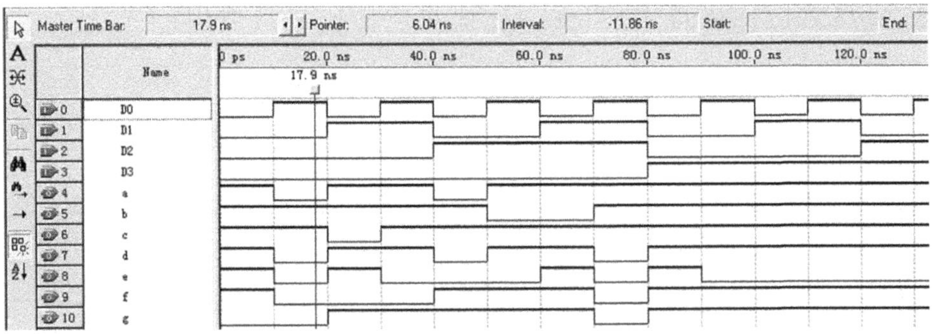

Figure 4.8.14: Simulation waveform of Example 4.25.

4.9 Summary

1. The outputs of combinational circuits depend only on the present inputs and not any storage elements are included in the circuits.
2. Logic diagram, logic expression, truth table, and the Karnaugh map are four methods of describing combinational digital circuits.

3. Analysis procedure of a combinational circuit is to write out logic expressions, simplify the expression using the Boolean algebra or the Karnaugh map, derive its truth table, and determine the implemented functions.

4. One method to design a combinatorial circuit is to construct its truth table, write and simplify the expression using the Boolean algebra or the Karnaugh map, and draw the logic diagram.

5. A full-adder performs the addition of two one-bit binary numbers and an input carry. A multibit binary adder can be constructed by cascading multiple full adders.

6. A look-ahead carry logic circuit is used to reduce excessive delays caused by carry propagation.

7. The 74LS283/HC283 MSI chip is a four-bit binary adders with a look-ahead carry logic circuit. It can be used to construct high-speed parallel adders, subtracters, and code converters.

8. A BCD adder circuit implemented by using binary adders requires special correction circuitry whenever the sum of a digit position exceeds 9 (1001).

9. A binary encoder converts 2^n noncoded information into n-bit binary code. Only one of input signals of binary encoders is valid at any time.

10. A priority encoder allows multiple active inputs at the same time, but only one input with the highest priority is encoded. The 74LS148 is an MSI 8-to-3 priority binary encoder with eight active-LOW inputs and three active-LOW outputs.

11. Binary decoders and BCD-to-decimal decoders are also called variable decoders. Each output of a variable decoder corresponds to a minterm of input variables. It can be used as demultiplexers and the implementation of any combinational logic circuits.

12. The 74LS138 is a commonly used MSI 3-to-8 decoder. It can be used for address decoders, demultiplexers, and the implementation of any combinational logic circuits.

13. Display decoders are to convert binary codes on their inputs and provide outputs to drive display devices including LEDs and LCDs.

14. The 74LS48 and 74LS47 are two types of integrated decoders for driving seven-segment LEDs. The 74LS48 is an active-HIGH for driving common-cathode LEDs while the 74LS47 is an active-LOW for driving common-anode LEDs.

15. Dual 4-to-1 multiplexer 74LS153 and 8-to-1 multiplexer 74LS151 are commonly used MSI multiplexers. They can be used for data transmission and implementing any combinational logic circuit.

16. The basic function of a comparator in a digital circuit is to compare the magnitudes of two binary quantities to determine the relationship of those quantities.

17. The 74LS85 is an MSI four-bit magnitude comparator. It is easy to expand any number of bits greater than four and to construct some special control circuit.

18. Races and hazards usually take place during the transition of the states of combinational digital circuits. Three methods of eliminating hazards is to add a redundancy term, use absorption capacitors, and apply a strobing pulse.
19. Verilog HDL is a widely used HDL. A complex logic circuit can be divided into several modules.
20. The default term should be included in the case statement and the if-else statement to avoid generating unnecessary latches.
21. There are two types of assignments in Verilog HDL: continuous assignment and procedural assignment.
22. The commonly used data types in Verilog are wire type and reg type.

Key terms

Encoder: A digital circuit that converts noncoded information, such as a decimal number and alphabetic characters into a coded form.
Priority encoder: An encoder in which only the higher-order active input is encoded and any other lower-order active inputs are ignored.
Decoder: A logic circuit that converts a coded information into a noncoded form.
Half-adder: A logic circuit that adds two one-bit binary numbers and generates a sum and an output carry.
Full-adder: A logic circuit that adds two one-bit binary numbers and an input carry to generate a sum and an output carry.
Ripple carry: A type of carry method that the carry output of each full-adder is connected to the carry input of the next higher-order stage.
Look-ahead carry: A type of carry method that the output carry of each stage is obtained directly from the input bits of each stage.
Multiplexer: A circuit that selects data from one of several inputs at a time to place it on a single output line; also called data selector.
Demultiplexer: A circuit that distributes data from one input line to one of the output lines at a time; also called data distributors.
Magnitude comparator: A circuit that performs the comparison of two quantities and indicates whether or not they are equal.
Code converter: A logic circuit that converts a type of coded information into the other type of coded form.
Race: A phenomenon that several input signals of a logic gate vary at the same time, or a signal is transferred to a logic gate through different paths resulting in different arrival time of signals.
Hazard: A phenomenon that glitches occur due to the existence of race in a logic circuit.
Glitch: An unintentional or unwanted voltage or current spike with short duration.

Self-test

4.1 The logic function implemented by the circuit shown in Figure T4.1 is.
(a) When the inputs are different, the output is 1; when the inputs are the same, the output is 0.
(b) When the inputs are different, the output is 0; when the inputs are the same, the output is 1.
(c) When the number of the inputs is odd, the output is 1; otherwise, the output is 0.
(d) When the number of the inputs is even, the output is 1; otherwise, the output is 0.

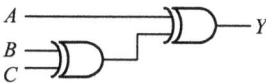

Figure T4.1

4.2 Assuming a logic gate has two inputs A and B and an output Y, their waveforms are shown in Figure T4.2. The logic function performed by this circuit is _____.
(a) AND   (b) NAND   (c) OR   (d) NOR

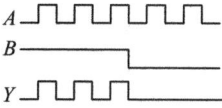

Figure T4.2

4.3 Assuming a logic gate has three inputs A, B, and C and an output Y, their waveforms are shown in Figure T4.3. The logic function performed by this circuit is _____.

(a) AND   (b) NAND   (c) OR   (d) NOR

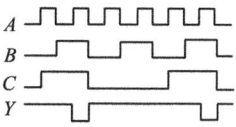

Figure T4.3

4.4 Which one of logic diagrams in Figure T4.4 corresponds to the logic expression
$Y = AB + BC$ _____.

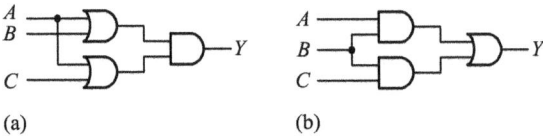

(a)   (b)

Figure T4.4

4.5 Which one of logic diagrams in Figure T4.5 corresponds to the logic function
expression $Y = A(B + C) + BC$ _____.

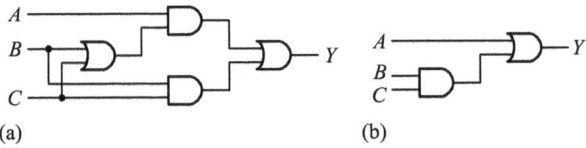

(a)   (b)

Figure T4.5

4.6 Which one of the following circuits in Figure T4.6 does not match with the
corresponding logic expression _____.

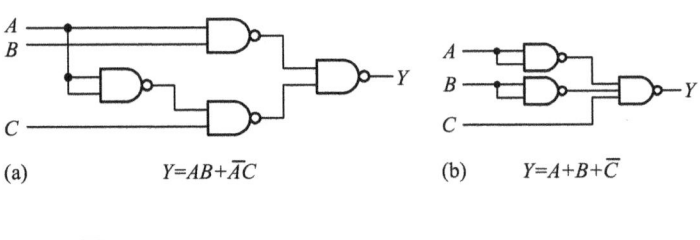

(a) $Y=AB+\overline{A}C$ (b) $Y=A+B+\overline{C}$

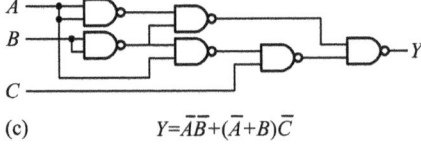

(c) $Y=\overline{\overline{A}\overline{B}}+(\overline{A}+B)\overline{C}$

Figure T4.6

4.7 The logic expression corresponding to the circuit diagram shown in Figure T4.7
is _____.

(a) $Y = A + B + \overline{AB}$
(b) $Y = AB + \overline{AB}$
(c) $Y = (A + \bar{B})(\bar{A} + B)$
(d) $Y = A\bar{B} + \bar{A}B$

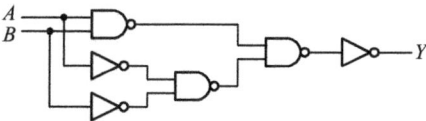

Figure T4.7

4.8 The logical function performed by the circuit shown in Figure T4.8 is _____.
(a) NAND   (b) NOR   (c) XNOR   (d) XOR

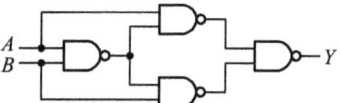

Figure T4.8

4.9 Which circuit diagram in Figure T4.9 does not perform the logic XOR function_____.

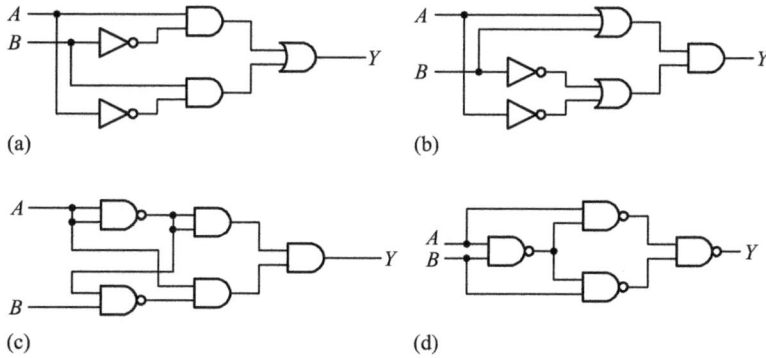

Figure T4.9

4.10 How many 74LS283 adders would be required to add two binary numbers each representing decimal numbers up through 300_{10}? _____
(a) 1   (b) 2   (c) 2   (d) 4

4.11 The binary numbers $A = 1100$ and $B = 1001$ are applied to the inputs of a 74HC85. The outputs are _____.
(a) $Y (A > B) = 1, Y (A < B) = 0, Y (A = B) = 0$
(b) $Y (A > B) = 0, Y (A < B) = 1, Y (A = B) = 0$
(c) $Y (A > B) = 0, Y (A < B) = 0, Y (A = B) = 1$
(d) $Y (A > B) = 1, Y (A < B) = 0, Y (A = B) = 1$

4.12 How many data-select lines does it take to select eight inputs?
(a) 1   (b) 2   (c) 3   (d) 4

4.13 A decoder can be used as a demultiplexer by
(a) tying all enable pins LOW.
(b) tying all data-select lines LOW.
(c) tying all data-select lines HIGH.
(d) using the input lines for data selection and an enable line for data input.

4.14 How many outputs would two 74LS148 8-line-to-3-line encoders, expanded to a 16-line-to-4-line encoder, have?
(a) 3   (b) 4   (c) 5   (d) 6

4.15 Two four-bit comparators are cascaded to form an eight-bit comparator. The cascading inputs of the most significant 4 bits should be connected to
(a) the cascading inputs of the least significant four-bit comparator.
(b) $A = B$ to a logic high, $A < B$ and $A > B$ to a logic low.
(c) the outputs from the least significant four-bit comparator.
(d) ground.

4.16 The 74LS47 is a BCD to a seven-segment decoder with ripple blanking input and output functions. The purpose of these lines is to _____.
(a) test the display to assure all segments are operational
(b) turn off the display for leading or trailing zeros
(c) turn off the display for any zero
(d) turn off the display for any nonsignificant digit

4.17 In Verilog HDL, if the else of the "if-else" statement is omitted, then the codes _____
(a) generate latches and are suitable for combinational digital circuits
(b) do not generate latches and are suitable for combinational digital circuits
(c) generate latches and are not suitable for combinational digital circuits
(d) do not generate latches and are not suitable for combinational digital circuits

Problems

4.1 Write the logic expression of the output, list the truth table, and describe the function implemented by the circuit shown in Figure P4.1.

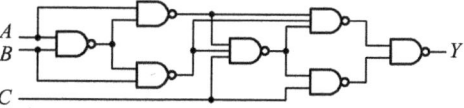

Figure P4.1

4.2 A circuit diagram is shown in Figure P4.2.
   (1) Write the logic expressions of S, C, P, and L.
   (2) What is the function of this circuit if S and C are treated as outputs?

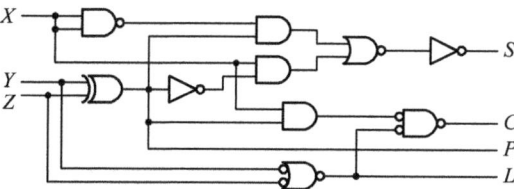

Figure P4.2

4.3 A circuit consists of a 3-to-8 decoder 74LS138 and logic gates, as shown in Figure P4.3.
   (1) Write the expressions of P_1 and P_2.
   (2) List the truth table and analyze the function performed by the circuit.

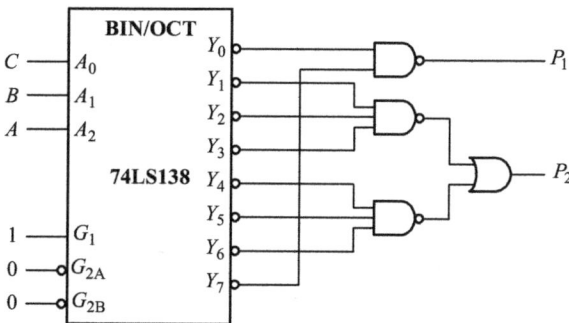

Figure P4.3

4.4 A circuit with an 8-to-1 multiplexer is shown in Figure P4.4. Write the logic expressions of Y when G_1G_0 have different combinations of logic values.

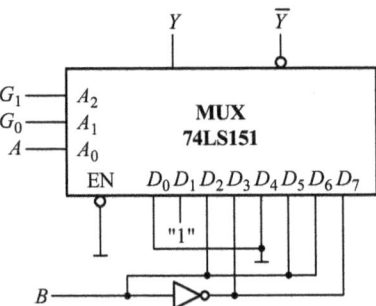

Figure P4.4

4.5 Analyze the logic function of a circuit implemented with 74LS138 and logic gates, as shown in Figure P4.5.

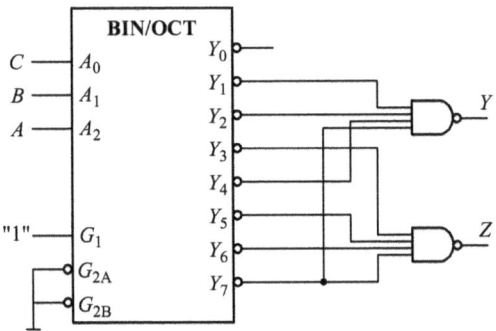

Figure P4.5

4.6 Write the logic expressions of F_1, F_2, F_3, and F_4 for a circuit shown in Figure P4.6.

4.7 Write the logic expressions of F_1 and F_2 for a circuit shown in Figure P4.7.

4.8 Implement the following logic function expressions with the fewest NAND gates.

$$\begin{cases} P_1 = \sum m(11, 12, 13, 14, 15) \\ P_2 = \sum m(3, 7, 11, 12, 13, 15) \\ P_3 = \sum m(3, 7, 12, 13, 14, 15) \end{cases}$$

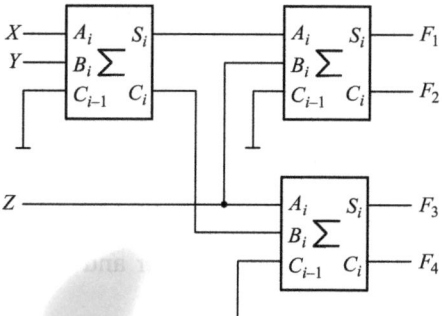

Figure P4.6

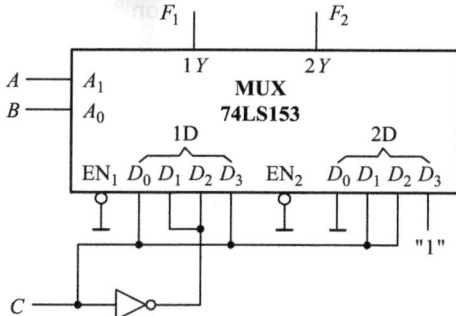

Figure P4.7

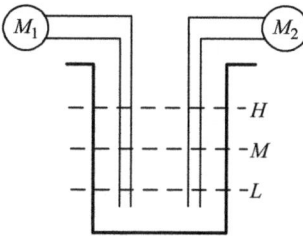

Figure P4.8

4.9 Design a logic circuit to control the operation of a big pump M_1 and a small pump M_2 in Figure P4.8. Two pumps are used to control the water level in a water tank. When the water level is above H, both pumps start to work; when the water level lies between H and M, only the big pump works; when the water level lies between M and L, only the small pump works; when the water level is below L, no pump works.

4.10 Design a full subtractor to perform a subtraction of three one-bit binary digits. Suppose A is the minuend, B is the subtrahend, J_0 is the borrow from the less significant bit, D is the difference, and J is the borrow from the more significant bit.

(1) List the truth table, and write the logic function expressions of D and J.

(2) Implement the subtractor with two-input NAND gates.

(3) Implement the subtractor with a 74LS138 3-to-8 decoder and a few logic gates.

(4) Implement the subtractor with a dual 4-to-1 multiplexer and a few logic gates.

4.11 Design a combinational digital circuit with the inputs of four-bit binary codes $B_3B_2B_1B_0$. When $B_3B_2B_1B_0$ is a 8421BCD code, the output $Y = 1$; otherwise, $Y = 0$.

4.12 Use NAND gates to design a display decoder to drive a common-anode LED. The decoder has three inputs. The LED is required to display six fonts. The six fonts can be chosen from 0 to 9 and A to Z.

4.13 Implement the following logic function expressions with a 74LS138 and a few logic gates.

$$\begin{cases} P_1(A, B) = \sum m(0, 3) \\ P_2(A, B) = \sum m(1, 2, 3) \end{cases}$$

4.14 Design a code converter to convert BCD8421 code into BCD5421 code. The converter should be implemented by a 74LS283 four-bit adder and two-input NAND gates.

4.15 Design a combinational logic circuit to realize logic functions shown in Table P4.1, where C_1 and C_0 are two inputs for selecting different functions, A and B are input variables, F is an output.

(1) List the truth table and write the logic expression of F.

(2) Implement the circuit with an 8-to-1 multiplexer and logic gates.

Table P4.1

C_1	C_0	F
0	0	$A+B$
0	1	AB
1	0	$A \oplus B$
1	1	$\overline{A \oplus B}$

4.16 A circuit diagram is shown in Figure P4.9(a).

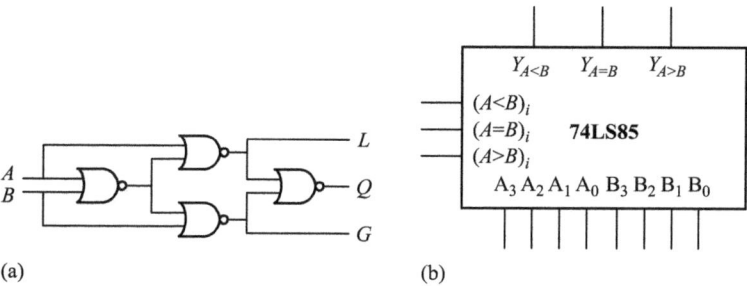

Figure P4.9

(1) Write the function expressions of L, Q, and G; list the truth table, and analyze what is the function performed by this circuit.

(2) Use the circuits shown in Figure P4.9(a) and (b) to design a five-bit magnitude comparator.

4.17 Design a logic circuit to control a room heating system. During the day time, the heating system starts to work when the room temperature is below 20 °C; in the evening, the system starts to work when the room temperature is below 17 °C.

4.18 Design a logic circuit using NAND gates to perform the agreement of passing the final driver license. There is a chief examiner, A, and two assistant examiners, B and C in a final test for driving license. A driver gets a license only when more than two (including two) examiners give a pass and the chief examiner also gives a pass.

4.19 Design a logic circuit using NAND gates to perform the judgment whether or not a student gets the certification of graduate. A student needs to take final exams of four courses to graduate. When the student passes the course, he can get 1 point for course A, 2 points for course B, 4 points for course C passes, and 5 points for course D; when the student fails in an exam, he gets 0 point for the course. The student can graduate only if the points he gets are greater than 7 points.

4.20 A circuit diagram is shown in Figure P4.10. If only one input of A, B, C, and D changes its state, are there any races and hazards phenomenon in this circuit? If yes, what are the values of the rest three inputs when there is a hazard?

4.21 Design a logic circuit to control the indicator LED for monitoring the operation of motors. Totally, there are four motors A, B, C, and D, in a factory. The indicator LED is on only when motor A and at least two of three motors B, C, and D are turned on; otherwise, the indicator LED is off.

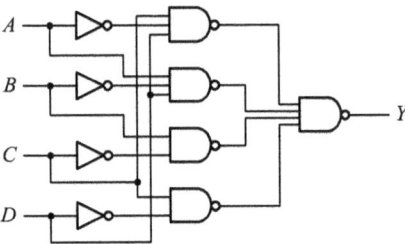

Figure P4.10

4.22 Analyze the function of the circuit of which simulation result is shown in Figure P4.11.

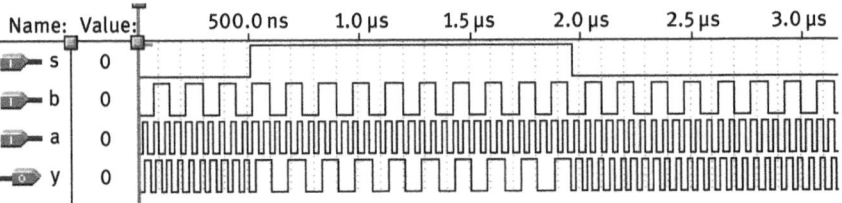

Figure P4.11

4.23 Verilog codes for a circuit are given below. Write the function table of the circuit and simulate using QuartusII software.

```verilog
module XXX_decoder(A,EN,Y);
output [7:0] Y;
input [2:0] A;
input EN;
reg[7:0] Y;
wire [3:0] temp={A,EN};
always
    case (temp)
        4'b0001 : Y=8'b00000001;
        4'b0011 : Y=8'b00000010;
        4'b0101 : Y=8'b00000100;
        4'b0111 : Y=8'b00001000;
        4'b1001 : Y=8'b00010000;
        4'b1011 : Y=8'b00100000;
        4'b1101 : Y=8'b01000000;
        4'b1111 : Y=8'b10000000;
        default : Y=8'b11111111;
    endcase
endmodule
```

4.24 Verilog codes for a circuit are shown below. Write the function table of the circuit and simulate it using QuartusII software.

```
module XXX_encoder(Y,A);
    output [2:0] A;
    input [7:0] Y;
    reg [2:0] A;
    wire [7:0] temp=Y;
    always
      case (temp)
            8'b00000001: A=3'b000;
            8'b00000010: A=3'b001;
            8'b00000100: A=3'b010;
            8'b00001000: A=3'b011;
            8'b00010000: A=3'b100;
            8'b00100000: A=3'b101;
            8'b01000000: A=3'b110;
            8'b10000000: A=3'b111;
            default      A=3'b000;
      endcase
endmodule
```

4.25 There are three radars A, B, and C. The power dissipation of A and B are the same, and the power dissipation of C is twice of the power dissipation of A. Two generators X and Y supply power to the radars. The maximum output power of generator X is the same as the power dissipation of radar A, and the maximum output power of generator Y is three times of the output power of X. Design a logic circuit to control the start and the stop of the generators X and Y ensuring the output electric power minimized. The detailed requirements are:
(1) List the truth table and indicate the logic definition of the variables.
(2) Write Verilog HDL codes for the circuit.
(3) Simulate the logic function using QuartusII software.

5 Flip-flops and related devices

Combinational circuit introduced in Chapter 4 is a type of digital circuits, whose outputs solely depend on the combination of current inputs, without storage being involved. Another kind of digital circuits is sequential logic circuit, in which outputs depend on not only the current inputs but also the previous inputs. Therefore, memory and storage are necessary parts for a sequential logic circuit. This chapter introduces bistable, monostable, and astable multivibration. Bistable multivibration has two stable states called logic 0 (or RESET) and logic 1 (or SET); it can maintain either of these states indefinitely, making it useful as memory devices. The monostable multivibration, also called the one shot, only has one stable state. A single controlled-width pulse is generated when a one shot is triggered. The astable multivibration has no stable state and is used to produce a periodic of pulse signal that can be used as a clock pulse source in a sequential logic circuit. In addition, Schmitt trigger and 555 timer are introduced for pulse generating, pulse shaping, and pulse transforming.

The objectives of this chapter are to
- Explain the operation of a basic S-R latch
- Describe the difference between a flip-flop and a latch
- Identify four kinds of edge-triggered flip-flops, and their dynamic characteristics
- Discuss the application of flip-flops
- Explain how a one shot operates
- Explain the difference of retriggerable and nonretriggerable one shots
- Explain the operation and application of Schmitt trigger
- Explain the operation of an astable devices
- Describe the function of 555 timer
- Apply a 555 timer to construct a one shot and an astable device

5.1 Latches

Latch and *flip-flop* are two categories of bistable multivibrators. Both of them have two stable states, that is, logic 0 and logic 1. Generally, a latch or a flip-flop can only store one-bit binary digit. It is the basic storage element in sequential logic. The main difference between latches and flip-flops is the method used for changing their states. This section mainly introduces the basic S-R latch and the gated D latch.

The objective of this section are to
- Explain the operation of a basic S-R latch
- Implement an S-R latch with logic gates
- Describe the logic function of basic S-R latch
- Explain the operation of a gated S-R latch and a gated D latch
- Describe the logic function of the gated D latch

https://doi.org/10.1515/9783110614916-005

5.1.1 The S-R latch

The S-R latch consists of two cross-coupled NAND gates or NOR gates.

1. The S-R latch with NAND gates

A basic S-R latch formed with two cross-coupled NAND gates is shown in Figure 5.1.1(a). It has two inputs, $\bar{R}_d$ and $\bar{S}_d$, and two outputs, Q and $\bar{Q}$. Note that the output of each gate is connected to an input of the opposite gate. This produces the regenerative feedback that is a characteristic of all latches and flip-flops. Normally, two outputs of a latch always complement each other. That is, *when Q is LOW, $\bar{Q}$ is HIGH, and when Q is HIGH, $\bar{Q}$ is LOW*. A latch has two stable states: Q is LOW, called the *RESET* state, and Q is HIGH, called the *SET* state. $\bar{R}_d$ is the RESET input, and $\bar{S}_d$ is the SET input. Ususally, both inputs are HIGH and an activated input is a LOW when a latch is triggered. Figure 5.1.1(b) shows the logic symbol for an S-R latch formed with two cross-coupled NAND gates.

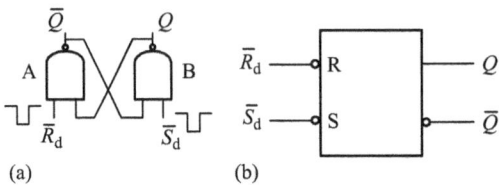

(a)   (b)

Figure 5.1.1: The basic S-R latch: (a) with NAND gates; (b) logic symbol.

    Let us begin with assuming that both inputs and output are HIGH, which is the normal latched state. Since the Q output is connected back to an input of gate A and the $\bar{R}_d$ input is HIGH, the output of gate A must be LOW. This LOW output is coupled back to an input of gate B, ensuring that its output is HIGH.

    When the Q output is HIGH, the latch is in the SET state. It will remain in this state indefinitely until a LOW is temporarily applied to the $\bar{R}_d$ input. With a LOW on the $\bar{R}_d$ input and a HIGH on $\bar{S}_d$, the output of gate A is forced HIGH. This HIGH on the $\bar{Q}$ output is coupled back to an input of gate B, and since the $\bar{S}_d$ input is HIGH, the output of gate B goes LOW. This LOW on the Q output is then coupled back to an input of gate A, ensuring that the $\bar{Q}$ output remains HIGH even when the LOW on the $\bar{R}_d$ input is removed.

    When the Q output is LOW, the latch is in the RESET state and the latch remains in the RESET state until a LOW is temporarily applied to the $\bar{S}_d$ input. With a LOW on the $\bar{S}_d$ input and a HIGH on $\bar{R}_d$, the output of gate B is forced HIGH. This HIGH on the Q output is coupled back to an input of gate A, and since the $\bar{R}_d$ input is HIGH, the output of gate A goes LOW. This LOW on the $\bar{Q}$ output is then coupled back to an input of gate B, ensuring that the Q output remains HIGH even when the LOW on the $\bar{S}_d$ input is removed.

An invalid condition occurs when LOWs are applied to both $\bar{S}_d$ and $\bar{R}_d$ at the same time. As long as the LOW levels are simutaneously held on the inputs, both the Q and $\bar{Q}$ outputs are forced HIGH, thus violating the basic complementary operation of the outputs. If the LOW are released simutaneously, both outputs will attempt to go LOW. Since there is always some difference in the propagation delay time of the gates, one of the gates will dominate in its transition to the LOW output state. This, in turn, forces the output of the slower gate to remain HIGH. In this situation, you cannot reliably predict the next state of the latch.

The logic function of the S-R latch can be summarized in a characteristic table, as listed in Table 5.1.1. Q_n and $\bar{Q}_n$ represent the *current state* or *present state*. Q_{n+1} and $\bar{Q}_{n+1}$ represent new states after each clock pulse, also called as *next state*.

Table 5.1.1: The characteristic table.

Inputs		Outputs		Comments
$\bar{R}_d$	$\bar{S}_d$	Q_{n+1}	$\bar{Q}_{n+1}$	
0	1	0	1	RESET
1	0	1	0	SET
1	1	Q_n	$\bar{Q}_n$	No change
0	0	1	1	Invalid condition

The logic function of an S-R latch can be also expressed by the characteristic equation. The characteristic equation for an active-LOW input S-R latch formed with two cross-coupled NAND gates is deduced as follows:

$$\begin{cases} Q_{n+1} = S_d + \bar{R}_d Q_n \\ \bar{R}_d + \bar{S}_d = 1 \end{cases} \tag{5.1.1}$$

Note that $\bar{R}_d + \bar{S}_d = 1$ is the constraint condition, which means $\bar{R}_d$ and $\bar{S}_d$ cannot be 0 at the same time when the latch is in the normal operation.

Example 5.1 If the $\bar{R}_d$ and $\bar{S}_d$ waveforms in Figure 5.1.2(a) are applied to the inputs of the latch in Figure 5.1.1(a), determine the waveform of the Q and $\bar{Q}$ outputs. Assume that Q is initially low.

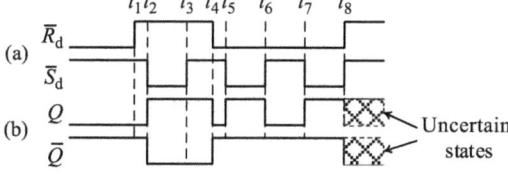

Uncertain states

Figure 5.1.2: (a) Input waveforms; (b) output waveforms.

Solution

The waveforms of the Q and $\bar{Q}$ outputs can be determined with respect to the $\bar{R}_d$ and $\bar{S}_d$ inputs, as shown in Figure 5.1.2(b). Note that the Q and $\bar{Q}$ outputs are both HIGH since both $\bar{R}_d$ and $\bar{S}_d$ are LOWs during the time interval between t_5 and t_6. At the time of t_6, only one of the $\bar{R}_d$ and $\bar{S}_d$ inputs change from LOW to HIGH; thus, the Q and $\bar{Q}$ outputs can be in a certain state. However, the situation is different at the time of t_8. During the time interval between t_7 and t_8, the Q and $\bar{Q}$ outputs are both HIGH since both $\bar{R}_d$ and $\bar{S}_d$ are LOWs. If the LOWs are released on the $\bar{R}_d$ and $\bar{S}_d$ inputs simutaneously at the time of t_8, both outputs will attempt to go LOW. Because of the small difference in the propagation delay time of the gates, one of the gates will dominate in its transition to the LOW output state and then forces the output of the slower gate to remain HIGH. Therefore, the next state of the latch is uncertain, which is indicated by the shadow.

2. The S-R latch with NOR gates

The basic S-R latch can also be formed with two cross-coupled NOR gates, as shown in Figure 5.1.3. Table 5.1.2 is a characteristic table that illustrates the operation for each of the four possible combinations of levels on the inputs. The logic function can be analyzed with the similar method as described earlier.

Table 5.1.2: The characteristic table.

Inputs		Outputs		Comments
R_d	S_d	Q_{n+1}	$\bar{Q}_{n+1}$	
0	0	Q	$\bar{Q}_n$	No change
0	1	1	0	SET
1	0	0	1	RESET
1	1	0	0	Invalid condition

According to the characteristic table, the characteristic equation for the latch shown in Figure 5.1.3 can be expressed by

$$\begin{cases} Q_{n+1} = S_d + \bar{R}_d Q_n \\ R_d S_d = 0 \end{cases} \tag{5.1.2}$$

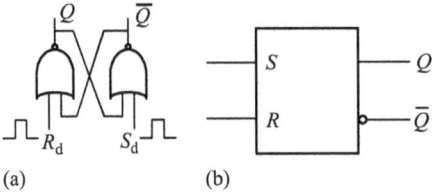

(a)   (b)

Figure 5.1.3: The basic S-R latch: (a) with NOR gates; (b) logic symbol.

Note the constraint condition is that S_d and R_d cannot be HIGH at the same time when the latch is in the normal operation.

5.1.2 The gated S-R latch

For a basic S-R latch, the outputs change as long as the input level of $\bar{R}_d$ or $\bar{S}_d$ is changed. A gated latch adds two NAND gates, gate C and gate D, and an enable control input, *EN*. Figure 5.1.4 shows the logic diagram and logic symbol for a gated S-R latch. When a LOW is applied to the *EN* input, gate C and gate D are both disabled and thus their outputs are both HIGH. The latch will maintain the current state, that is, $Q_{n+1}=Q_n$. When a HIGH is applied to the *EN* input, the *R* and *S* inputs control the state of the latch. Due to the inverter of gate C and gate D, the *R* and *S* inputs are both active-HIGH. As long as the *EN* input remains a HIGH, the output of the latch change with *R* and *S* inputs. In this circuit, the invalid state occurs when both *S* and *R* are simultaneously HIGH. The characteristic table of the gated S-R latch is the same as the basic S-R latch, as shown in Table 5.1.2. The characteristic equation can be written as

$$\begin{cases} Q_{n+1} = S + \bar{R}Q_n \\ RS = 0 \end{cases} \tag{5.1.3}$$

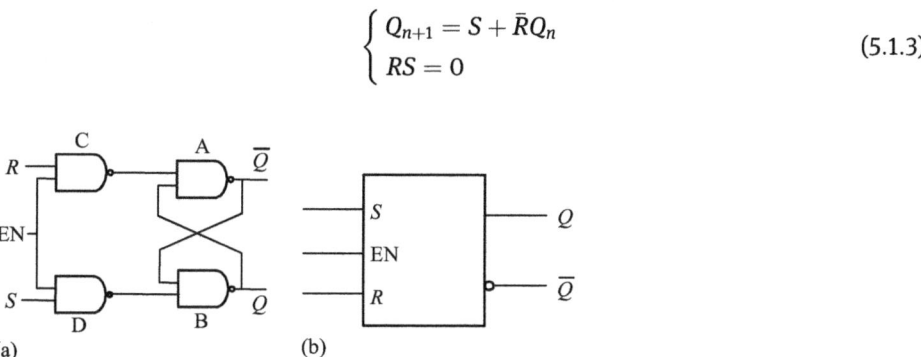

(a)   (b)

Figure 5.1.4: A gated S-R latch: (a) logic diagram; (b) logic symbol.

Note that when *EN* is HIGH, the outputs change as long as the input level of *R* or *S* is changed. If there is an interference signal affecting the level of the input signals, the output of the gated S-R latch will also be changed, as shown in Figure 5.1.5. Therefore, the gated S-R latch has weak anti-interference capability.

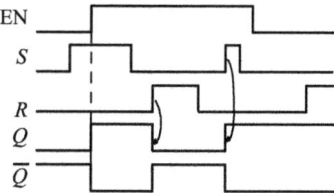

Figure 5.1.5: The output waveform of a gated S-R latch.

5.1.3 The gated D latch

The gated D latch is another kind of gated latch, which is very useful when a single data bit (0 or 1) is to be stored [37]. Different from the gated S-R latch, it has only one D input in addition to *EN* input. By adding an inverter between the *R* and *S* inputs, a gated D latch can be implemented, as shown in Figure 5.1.6.

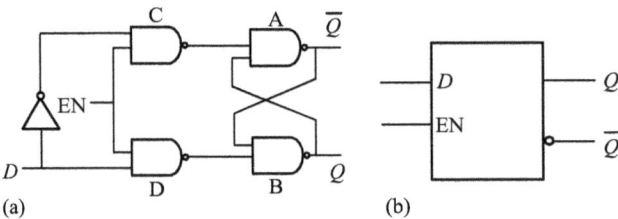

(a)   (b)

Figure 5.1.6: A gated D latch: (a) logic diagram; (b) logic symbol.

When the *D* input and the *EN* input are both HIGH, the *Q* output is HIGH and the latch will be in a SET state. When the D input is LOW and *EN* is HIGH, the *Q* output is LOW and the latch is in a RESET state. The SET state corresponds to store a 1, and the RESET state corresponds to store a 0. When the *EN* input is LOW, the latch keeps no change. That is, the output *Q* follows the input *D* when *EN* is HIGH. The characteristic equation can be expressed as

$$Q_{n+1} = D \tag{5.1.4}$$

Example 5.2 If the *D* input waveform in Figure 5.1.7(a) is applied to the input of the gated D latch in Figure 5.1.6, determine the waveform of the *Q* and $\bar{Q}$ outputs. Assume that *Q* is initially low.

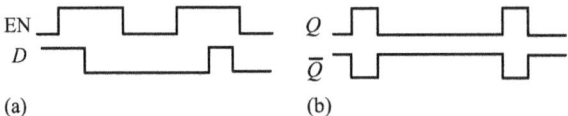

(a)   (b)

Figure 5.1.7: A Gated D latch: (a) input waveform and (b) output waveform.

Solution

For the gated D latch, the *Q* output follows the *D* input when *EN* is HIGH and the *Q* output keeps no change when *EN* is LOW. Therefore, the *Q* and $\bar{Q}$ output waveforms are shown in Figure 5.1.7(b).

5.2 Flip-flops

As mentioned in the previous section, the main difference between latches and flip-flops is the method used for changing their states. The latch is sensitive to the input level. However, the output of a flip-flop only changes at a specified point on a triggering input called the *clock pulse,* which is designated as a control input, *CP.* Generally, a flip-flop is sensitive to the positive going or negative going edge of clock. Classified by logic function, flip-flops can be categorized as S-R flip-flop, J-K flip-flop, D flip-flop, and T flip-flop. Classified by trigger method, they can be divided into master-slave flip-flops and edge-triggered flip-flops. This section introduces master-slave flip-flops and edge-triggered flip-flops.

The objectives of this section are to
- Define the master-slave flip-flop
- Explain the operation of pulse-triggered master-slave flip-flop
- Define the edge triggered flip-flop
- Explain the operation of edge-triggered S-R, D, J-K, and T flip-flop
- Explain the difference between the master-slave flip-flops and edge-triggered flip-flops
- Discuss the asynchronous inputs of a flip-flop
- Implement the transition of different kinds of flip-flops
- Recognize the dynamic characteristics of flip-flops

5.2.1 Master-slave flip-flops

Master-slave flip-flop is a kind of flip-flops with the pulse-triggered method. Data are entered into the flip-flop at the leading edge of the clock pulses, but the output does not reflect the input state until the trailing edge [38]. Two types of pulse-triggered master-slave flip-flops, S-R and J-K flip-flops, are introduced here.

1. The S-R master-slave flip-flop
The S-R master-slave flip-flop consists of two cascading gated S-R latches, as shown in Figure 5.2.1 [3]. One is the master flip-flop including four NAND gates, G_1, G_2, G_3,

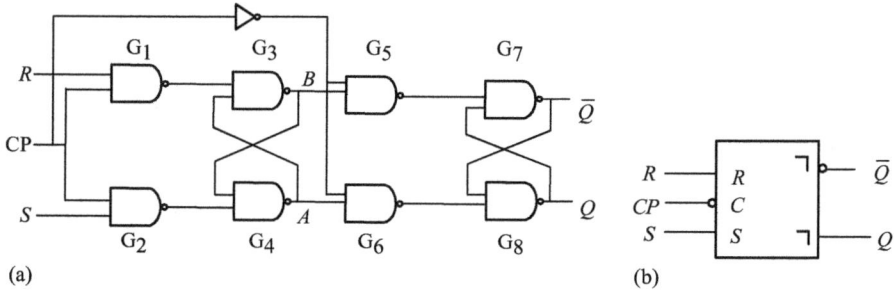

Figure 5.2.1: The S-R master-slave flip-flop: (a) logic diagram; (b) logic symbol.

and G_4. Another is the slave flip-flop involving four NAND gates, G_5, G_6, G_7, and G_8. The slave flip-flop has the same structure as the master flip-flop except that it is clocked on the inverted clock pulse and controlled by the outputs of the master flip-flop rather than the external S-R inputs.

When the positive-going edge of clock pulse arrives, the state of master flip-flop is determined by the S and R inputs. Since the outputs of the master flip-flop are connected to the inputs of the slave flip-flop, the state of the master flip-flop is sent to the slave flip-flop on the negative-going edge of clock pulse. At the negative-going edge of clock pulse, the state of the slave flip-flop then appears in the Q and $\bar{Q}$ outputs. After the negative-going edge of clock pulse, the Q and $\bar{Q}$ outputs will keep no change because the outputs of the master flip-flop keep no change when the clock pulse is a LOW. The characteristic of the S-R master-slave flip-flop is same as that of the gated S-R latch, as listed in Table 5.2.1.

Table 5.2.1: The characteristic table.

Inputs		Output	Comments
S	R	Q_{n+1}	
0	0	Q_n	No change
0	1	0	RESET
1	0	1	SET
1	1	indeterminate	Invalid condition

The characteristic equation for S-R master-salve flip-flop is the same as the gated S-R latch as follows:

$$\begin{cases} Q_{n+1} = S + \bar{R}Q_n \\ RS = 0 \end{cases} \tag{5.2.1}$$

The logic symbol for the S-R master-slave flip-flop is shown in Figure 5.2.1(b). The postponed output symbol (⌐) at the Q and $\bar{Q}$ output is the key to identify a master-slave flip-flop. This symbol indicates that the output does not reflect the S-R input until the occurrence of the clock edge (either positive-going or negative-going) following the triggering edge.

2. The J-K master-slave flip-flop

Figure 5.2.2 shows the logic diagram for the J-K master-slave flip-flop [3]. Based on the S-R master-salve flip-flop, the J-K master-slave flip-flop has a slight change. Except the J and K inputs, the Q and $\bar{Q}$ output of the slave flip-flop is connected back as an input of gate G_1 and gate G_2, respectively. The S and R inputs for S-R master-salve flip-flop become:

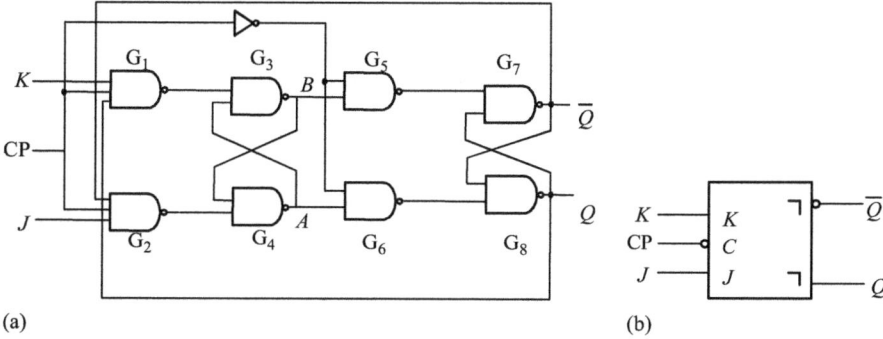

Figure 5.2.2: J-K master-slave flip-flop: (a) logic diagram and (b) logic symbol.

$$S = J\bar{Q}_n, \ R = KQ_n \tag{5.2.2}$$

Substitute eq. (5.2.2) into the characteristic eq. (5.2.1), then the characteristic equation for J-K master-salve flip-flop can be derived as follows:

$$Q_{n+1} = J\bar{Q}_n + \overline{KQ_n} \cdot Q_n = J\bar{Q}_n + \bar{K}Q_n \tag{5.2.3}$$

According to the characteristic equation, the characteristic table for J-K master-slave flip-flop is summarized in Table 5.2.2.

Table 5.2.2: The characteristic table.

Inputs		Output	Comments
J	**K**	**Q_{n+1}**	
0	0	Q_n	No change
0	1	0	Reset
1	0	1	Set
1	1	$\bar{Q}_n$	Toggle

The logic operation of the J-K master-slave flip-flop is the same as that of the S-R master-slave flip-flop in the SET, RESET, and no-change modes. The difference in operation occurs when both J and K inputs are HIGH; the flip-flop changes its state to the opposite state on each successive clock pulse. This operation mode is called *toggle* operation. Note that there is no invalid state for J-K flip-flop, which is different from the case for S-R master-slave flip-flop.

A limitation to the master-slave operation is that the *J* and *K* inputs should be kept unchanged when the clock signal is a HIGH level because the state of master latch can change during this time.

5.2.2 Edge-triggered flip-flops

An edge-triggered flip-flop changes its state either at the positive edge (rising edge) or at the negative edge (failing edge) of the clock pulse and is sensitive to its inputs only at this transition of the clock. Three types of edge-triggered flip-flops are covered in this part [3]. They are S-R, D, and J-K flip-flops. The logic symbols for all of these flip-flops are shown in Figure 5.2.3, in which each type can be either positive edge-triggered (no bubble at C input) or negative edge-triggered (bubble at C input). Especially, the ">"symbol at the C input, which is called the dynamic input indicator, is also the key to identify an edge-triggered flip-flop by its logic symbol.

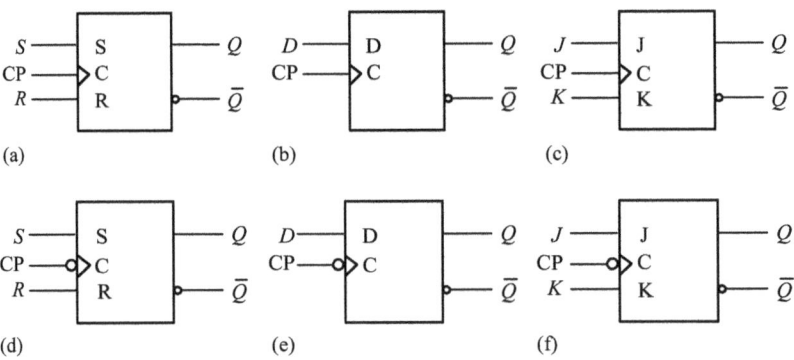

Figure 5.2.3: Positive edge-triggered S-R (a), D (b) and J-K (c) flip-flops, and negative edge-triggered S-R (d), D (e), and J-K (f) flip-flops.

1. The Edge-triggered S-R Flip-flop

Figure 5.2.4 shows a positive edge-triggered S-R flip-flop. It contains three parts: a basic NAND latch (gate G_3 and gate G_4), a pulse-steering circuit (gate G_1 and gate G_2), and a pulse transition detector (or edge detector) circuit. The pulse transition detector detects a rising (or falling) edge and produces a very short-duration (ns) spike.

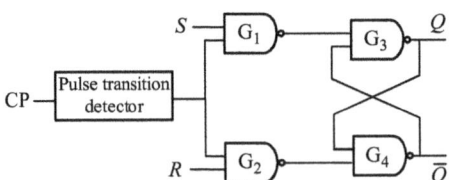

Figure 5.2.4: Positive edge-triggered S-R flip-flop.

When S is HIGH and R is LOW, the Q output goes HIGH on the triggering edge of the clock pulse, and the flip-flop is SET. When S is LOW and R is HIGH, the Q output goes LOW on the triggering edge of the clock pulse, and the flip-flop is RESET. When both S and R are LOW, the Q output keeps no change and maintains its previous state. An invalid condition exists when both S and R are HIGH. The S and R inputs of the S-R flip-flop are called *synchronous inputs* because data on these inputs are transferred to the output of flip-flop only on the triggering edge of the clock pulse. The characteristic table of this flip-flop is shown in Table 5.2.3.

Table 5.2.3: The characteristic table.

Inputs			outputs		Comments
CP	S	R	Q_{n+1}	$\bar{Q}_{n+1}$	
×	0	0	Q_n	$\bar{Q}_n$	No change
↑	0	1	0	1	RESET
↑	1	0	1	0	SET
↑	1	1	1	1	Invalid

"↑" represents the clock transition from LOW to HIGH.
"×" represents the "don't care."

The characteristic equation for the positive edge-triggered S-R flip-flop is the same as that of gated S-R latch and can be expressed as

$$\begin{cases} Q_{n+1} = S + \bar{R}Q_n \\ RS = 0 \end{cases} \tag{5.2.4}$$

Note that the flip-flop cannot change its state except on the triggering edge of a clock pulse. The S and R inputs can be changed at any time when the clock input is LOW or HIGH (except for a very short interval around the triggering transition of the clock) without affecting the output. In addition, the operation and characteristic table for a negative edge-triggered S-R flip-flop are the same as those for a positive edge-triggered flip-flop except that the falling edge of the clock pulse is the triggering edge.

Since the S-R flip-flop is not available in IC form, the most common used edge-triggered flip-flop is the D and J-K flip-flop.

2. The Edge-triggered D Flip-flop

The basic edge-triggered D flip-flop can be constructed by adding an inverter to the S-R flip-flop, as shown in Figure 5.2.5. There is only one input, the D input, in addition to the clock. If a HIGH is applied on the D input, the flip-flop will be in SET state at the arrival of positive-triggered edge of the clock pulse. Thus, a HIGH on the D input is stored by the flip-flop on the positive-going edge of the clock pulse. If a LOW is applied on the D input, the flip-flop will be in RESET state at the arrival of positive-triggered edge of the clock pulse. Thus, the LOW on the input is stored by the flip-flop on the positive-going edge of the clock pulse. In the SET state, the flip-flop is storing a 1, and

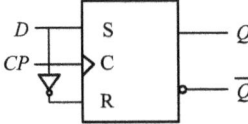

Figure 5.2.5: A positive edge-triggered D flip-flop.

in the RESET state, it is storing a 0. That is, the output Q follows the input D at active triggered-edge of clock pulse. Table 5.2.4 summarizes the logical operation of the positive edge-triggered D flip-flop. The edge-triggered D flip-flop is very useful when a single data bit (0 or 1) is to be stored.

Table 5.2.4: The characteristic table.

Inputs		Outputs		Comments
CP	D	Q_{n+1}	$\bar{Q}_{n+1}$	
↑	0	0	1	RESET (store a 0)
↑	1	1	0	SET (store a 1)

The characteristic equation for the D flip-flop can be expressed as

$$Q_{n+1} = D \tag{5.2.5}$$

Example 5.3 Given the waveform for the D input and the clock pulse in Figure 5.2.6, determine the Q output waveform if the flip-flop starts at RESET state.

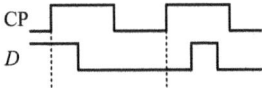

Figure 5.2.6: Input waveform for D flip-flop.

Solution

For the positive edge-trigger D flip-flop, the output Q follows the input D at the arrival of positive edge of the clock pulse. Hence, the resulting waveform of the output Q is shown in Figure 5.2.7.

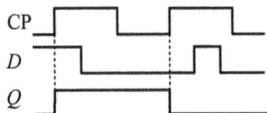

Figure 5.2.7: Output waveform for D flip-flop.

3. The Edge-triggered J-K Flip-flop

The edge-triggered J-K flip-flop is versatile and a widely used type of flip-flop. Figure 5.2.8 illustrates the basic internal logic for a positive edge-triggered J-K flip-flop. Unlike the S-R edge-triggered flip-flop, the outputs of J-K flip-flop are fed back as the input of the pulse steering NAND gates G_1 and G_2. The two inputs are labeled J and K in honor of Jack Kilby, who invented the integrated circuit.

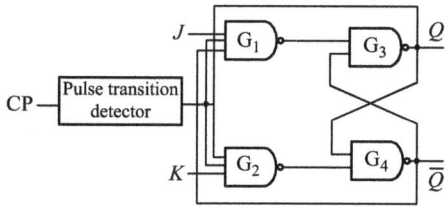

Figure 5.2.8: The logic diagram for a positive edge-triggered J-K flip-flop.

Assume that the flip-flop is RESET and that the J input is HIGH, the K input is LOW. At the positive triggering edge of clock pulse, gate G_1 is enabled. Since $\bar{Q}$ and J are both HIGH, the output of gate G_1 is LOW. This LOW level is applied on the input of gate G_3 and thus the Q output of the flip-flop changes from LOW to HIGH. The flip-flop is now SET.

If you now set J input as LOW and K input as HIGH, gate G_2 is enabled at the positive triggering edge of clock pulse. Since Q and K are both HIGH, the output of gate G_2 is LOW. This LOW level is applied on the input of gate G_4 and thus the output $\bar{Q}$ of the flip-flop changes from LOW to HIGH and Q changes from HIGH to LOW. The flip-flop is now RESET.

If you set both J input and K input LOW, the outputs of both gate G_1 and G_2 are HIGH. The HIGH are applied on the inputs of gate G_3 and G_4 and thus the flip-flop will stay in its present state at the arrival of positive triggering edge of the clock pulse.

Assume that the flip-flop is RESET. If you set both J and K inputs HIGH, the HIGH on $\bar{Q}$ enables gate G_1, and the output of G_1 becomes LOW at the arrival of positive triggering edge of the clock pulse; since this LOW is input to gate G_3, the output Q is HIGH and the flip-flop is SET. Now the HIGH on Q enable the gate G_2, the output of G_2 becomes LOW at the arrival of positive triggering edge of the clock pulse; and since this LOW is input to gate G_4, the output $\bar{Q}$ is HIGH and the flip-flop is RESET. Therefore, on each successive positive triggering edge of the clock pulse, the J-K flip-flop changes its state to the opposite state, which is a toggle operation.

Table 5.2.5 summarizes the characteristic table of the positive edge-triggered J-K flip-flop. The function of the J-K flip-flop is identical to that of the S-R flip-flop in the SET, RESET, and no-change. The difference is that the J-K flip-flop has no invalid state.

Table 5.2.5: The characteristic table.

Inputs			Outputs		comments
CP	*J*	*K*	Q_{n+1}	$\bar{Q}_{n+1}$	
↑	0	0	Q_n	$\bar{Q}_n$	No Change
↑	0	1	0	1	RESET
↑	1	0	1	0	SET
↑	1	1	$\bar{Q}_n$	Q_n	Toggle

According to the characteristic table, the characteristic equation can be expressed as

$$Q_{n+1} = J\bar{Q}_n + \bar{K}Q_n \tag{5.2.6}$$

The J-K flip-flop is a kind of versatile flip-flop. By choosing different connection, a J-K flip-flop can form other types of flip-flop.

If the *J* and *K* inputs are tying together as an input *T*, a T flip-flop is formed, as shown in Figure 5.2.9. When a HIGH is applied on the input *T*, this make *J* and *K* inputs HIGH and thus the flip-flop is in toggle operation. T flip-flop changes its state to the opposite state on each clock edge. When a LOW is applied on the input *T*, this make *J* and *K* inputs both LOW and thus the flip-flop keeps its previous state. The characteristic table for the positive edge-triggered T flip-flop is shown in Table 5.2.6.

The corresponding characteristic equation of T flip-flop can be deduced from that of J-K flip-flop as follows:

$$Q_{n+1} = T\bar{Q}_n + \bar{T}Q_n = T \oplus Q_n \tag{5.2.7}$$

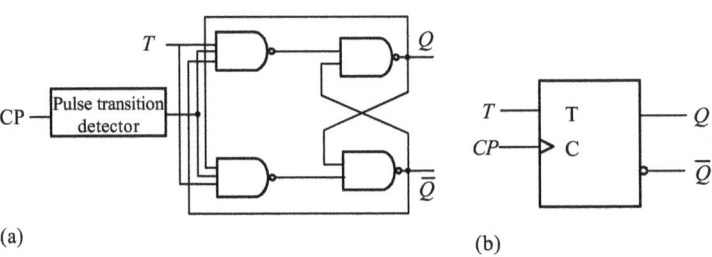

(a)   (b)

Figure 5.2.9: Logic diagram (a) and logic symbol (b) for a positive edge-triggered T flip-flop.

Table 5.2.6: The characteristic table.

Inputs		Outputs		Comments
CP	*T*	Q_{n+1}	$\bar{Q}_{n+1}$	
↑	0	Q_n	$\bar{Q}_n$	No Change
↑	1	$\bar{Q}_n$	Q_n	Toggle

A J-K flip-flop connected for toggle operation is called as T′ flip-flop or "toggle" flip-flop. T′ flip-flop changes its state to the opposite state on each clock edge, giving an output that is half the frequency of the clock pulse, as shown in Figure 5.2.10. It is useful for constructing binary counter, frequency divider, and general binary addition device.

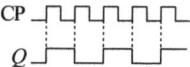

Figure 5.2.10: The output waveform of T′ flip-flop.

If $T = 1$, that is, $J = K = 1$, a T flip-flop become a T′ flip-flop. The characteristic equation of T′ flip-flop can be expressed as follows:

$$Q_{n+1} = \bar{Q}_n \tag{5.2.8}$$

5.2.3 Asynchronous inputs

For the flip-flops just discussed, the S-R, D, and J-K inputs are synchronous inputs, as data on these inputs are transferred to the flip-flop's output only on the triggered edge of the clock pulse, which means the data are transferred synchronously with the clock.

Most intergrated circuit flip-flops also have asynchronous inputs, which affect the state of the flip-flop independent of the clock. Usually, *asynchronous input*s include *preset* (PRE), also called *direct set* (S_d), and *clear* (CLR), also called as *direct reset* (R_d).

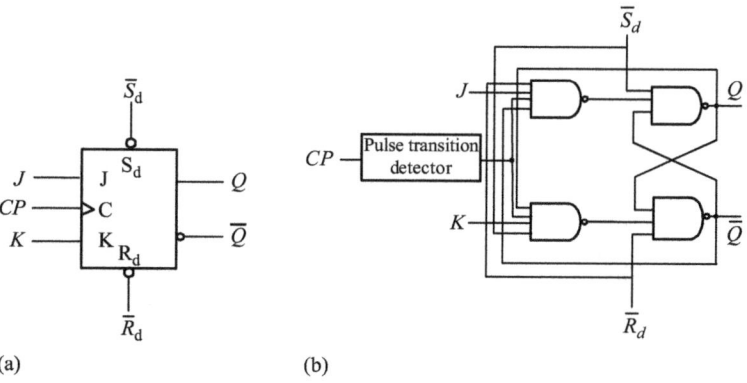

(a)   (b)

Figure 5.2.11: A J-K flip-flop with direct reset and direct set inputs: (a) logic symbol; (b) logic diagram.

An active level on the preset (or direct set) input will set the flip-flop and an active level on the clear (or direct reset) input will reset it. Figure 5.2.11 shows the logic symbol and logic diagram for a J-K flip-flop with direct set and direct reset inputs. These inputs are active-LOW, as indicated by the bubbles. Normally, when the flip-flop is in the synchronous operation, both direct set and reset inputs must be HIGH. As you can see from Figure 5.2.11(b), the inputs are connected so that they override the effect of the synchronous inputs, J, K, and the clock. With asynchronous inputs, the characteristic table for the J-K flip-flop can be revised as Table 5.2.7.

Table 5.2.7: The characteristic table.

Inputs					Outputs		Comments
cp	$\bar{R}_d$	$\bar{S}_d$	J	K	Q_{n+1}	$\bar{Q}_{n+1}$	
×	0	1	×	×	0	1	Asynchronous RESET
×	1	0	×	×	1	0	Asynchronous SET
0	1	1	×	×	Q_n	$\bar{Q}_n$	No Change
↑	1	1	0	0	Q_n	$\bar{Q}_n$	No Change
↑	1	1	0	1	0	1	Synchronous RESET
↑	1	1	1	0	1	0	Synchronous SET
↑	1	1	1	1	$\bar{Q}_n$	Q_n	Toggle

Example 5.4 For the positive edge-triggered J-K flip-flop with direct set and reset inputs in Figure 5.2.11, when J and K inputs are both HIGH, determine the Q output for the inputs shown in the timing diagram in Figure 5.2.12 if Q is initially LOW.

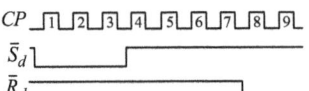

Figure 5.2.12: Input waveforms in Example 5.4.

Solution

1. During clock pulses 1, 2, and 3, the direct set ($\bar{S}_d$) is LOW, keeping the flip-flop SET regardless of the synchronous J and K inputs.
2. For clock pulses 4, 5, 6, and 7, the flip-flop is in toggle operation because J and K are both HIGH, and $\bar{S}_d$ and $\bar{R}_d$ are both HIGH.
3. For clock pulses 8 and 9, the direct reset input ($\bar{R}_d$) is LOW, keeping the flip-flop RESET regardless of the synchronous inputs.

The resulting waveform of the Q output is shown in Figure 5.2.13.

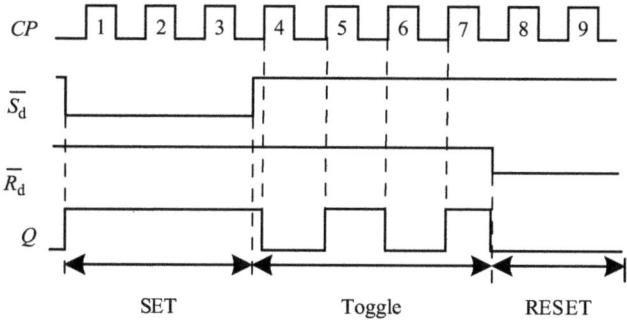

Figure 5.2.13: Output waveform in Example 5.4.

5.2.4 Flip-flop conversion

Generally, mainstream ICs of edge-triggered flip-flops are D flip-flop and J-K flip-flop. Other types of flip-flops can be implemented with D flip-flop and J-K flip-flop. The implementing method is to use characteristic equation of different types of flip-flops to determine the inputs of flip-flop.

Example 5.5 Use a D flip-flop to construct a T′ flip-flop.

Solution

The characteristic equation for D flip-flop is $Q_{n+1} = D$.
The characteristic equation for T′ flip-flop is $Q_{n+1} = \bar{Q}_n$.

Compare the characteristic equation of two type of flip-flops, when $D = \bar{Q}_n$, a T′ flip-flop can be constructed by using a D flip-flop. That is, a T′ flip-flop can be implemented by connecting the $\bar{Q}$ output of D flip-flop with the D input. The logic diagram is shown in Figure 5.2.14(a).

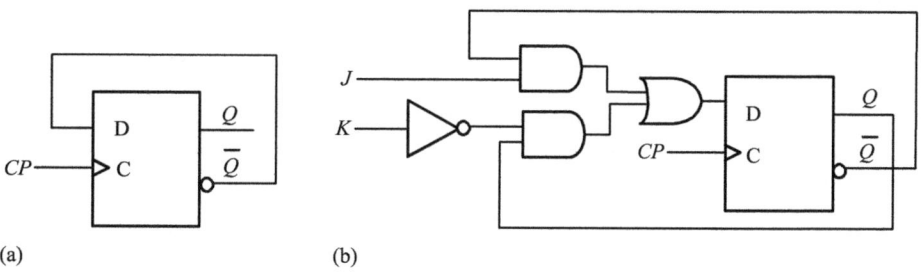

(a) (b)

Figure 5.2.14: The logic diagram.

Example 5.6 Use a D flip-flop to construct a J-K flip-flop.

Solution

The characteristic equation for D flip-flop is $Q_{n+1} = D$.
The characteristic equation for J-K flip-flop is $Q_{n+1} = J\bar{Q}_n + \bar{K}Q_n$.

Compare the characteristic equation of two types of flip-flops, when $D = J\bar{Q}_n + \bar{K}Q_n$, a J-K flip-flop can be constructed by using a D flip-flop and few logic gates. The logic diagram is shown in Figure 5.2.14(b).

5.2.5 Flip-flop operating characteristics

The operation characteristics for flip-flops includes the time for data receiving, state transition, propagation delay, or response. Generally, the specifications are applicable to all CMOS and TTL flip-flops. Here, the D flip-flop is taken as an example to discuss the operating characteristics, as shown in Figure 5.2.15.

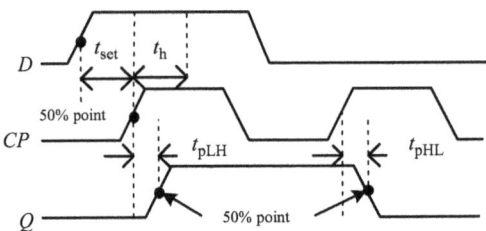

Figure 5.2.15: Operation characteristics for D flip-flop.

1. Set-up time
The set-up time (t_{set}) is the minimum interval required for the logic levels to be maintained constantly on the input prior to the triggering edge of the clock pulse in order for the levels to be reliably clocked into the flip-flop.

2. Hold time
The hold time (t_h) is the minimum interval required for the logic levels to remain on the inputs after the triggering edge of the clock pulse in order for the logic levels to be reliably clocked into the flip-flop.

3. Propagation delay times
A propagation delay time is the interval of time required after an input signal has been applied until the resulting output change to occur. t_{pLH} represents the propagation delay from the triggering edge of the clock pulse to the LOW-to-HIGH transition of the output. t_{pHL} represents the propagation delay from the triggering edge of the clock

pulse to the HIGH-to-LOW transition of the output. The average propagation delay can be calculated as

$$t_{\mathrm{pd}} = \frac{t_{\mathrm{pLH}} + t_{\mathrm{pHL}}}{2} \tag{5.2.9}$$

4. Maximum clock frequency

The maximum clock frequency (f_{CPMAX}) is the highest rate at which a flip-flop can be reliably triggered. At clock frequencies above the maximum, the flip-flop would be unable to respond quickly enough, and its operation would be impaired.

Example 5.7 Determine the maximum operating frequency for a positive edge-triggered D flip-flop with the set-up time t_{set}, and the propagation delay times $t_{\mathrm{pLH}} = t_{\mathrm{pHL}} = t_{\mathrm{pd}}$.

Solution

The operation characteristic for D flip-flop in Example 5.7 is illustrated in Figure 5.2.16.

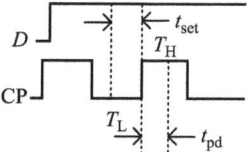

Figure 5.2.16: Operation waveforms.

In order to guarantee the logic levels propagated reliably into the flip-flop, the time interval of the LOW level of the clock pulse, T_{L}, should be greater than the set-up time t_{set}; the time interval of the HIGH level of clock pulse, T_{H}, should also be greater than the average propagation delay t_{pd}. Thus, the period of the clock pulse, T_{cp}, is at least equal to the sum of T_{L} and T_{H}. That is, $T_{\mathrm{cp}} = T_{\mathrm{L}} + T_{\mathrm{H}} \geq t_{\mathrm{set}} + t_{\mathrm{pd}}$, so $f_{\mathrm{CP}} = \frac{1}{T_{\mathrm{cp}}} \leq \frac{1}{t_{\mathrm{set}} + t_{\mathrm{pd}}}$. Therefore, the maximum operating frequency is $\frac{1}{t_{\mathrm{set}} + t_{\mathrm{pd}}}$.

5.3 One shot

One shot is also called a *monostable* multivibration, which has only one stable state and is commonly used for timing circuits in electronics system. It is normally in its stable state and will change its state from the stable state to an unstable state only when triggered. Once it is triggered, the one shot could not stay at the unstable state for a long time and will return automatically to its stable state for a predetermined length of time. The time that the device stays in its unstable state determines the pulse width of its output.

The objectives of this section are to
- Describe the basic operation of a one shot
- Explain how a nonretriggerable one shot operates
- Explain how a retriggerable one shot operates
- Discuss the function of 74LS122

5.3.1 Structure and principle

A basic S-R latch formed with two cross-coupled NAND gates has two stable states, $Q=0$ or $Q=1$, as shown in Figure 5.3.1(a). The output of each gate is directly fed back as an input of the opposite gate. If an inverter is used to replace the NAND gates G_2 in S-R latch and the $\bar{Q}$ output is fed back through a RC circuit instead of line feedback, the resulting circuit is one shot, as shown in Figure 5.3.1(b). There are only one trigger input and two complementary outputs, Q and $\bar{Q}$. The trigger pulse should be a norrow negative pulse.

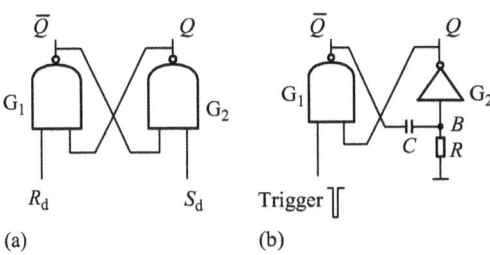

(a)   (b)

Figure 5.3.1: Comparison of S-R latch and one shot: (a) S-R latch and (b) one shot.

Normally, the trigger input is HIGH and the Q output of one shot is HIGH that is a stable state. When a negative pulse is applied to the trigger input, the output of gate G_1 goes HIGH. This transition from LOW to HIGH is coupled through the capacitor C to the input of gate G_2. The apparent HIGH on G_2 makes the Q output go LOW. The one shot will change its state from the stable state, $Q =1$, to the unstable state, $Q =0$. This LOW output goes back to the input of gate G_1 and makes the output of gate G_1 keep HIGH. Once the one shot enter into the unstable state, the capacitor immediately begins to charge through R toward the high voltage level. The rate at which it charges is determined by the RC time constant. With the charge of the capacitor, the voltage of point B (U_B) falls. When the capacitor charges to a certain level, the voltage of point B goes below a threshold voltage, and the Q output goes back HIGH since there is a LOW input to G_2. The one shot will return automatically to its stable state. The time that the one shot stays in its unstable state determines the pulse width of its output.

In summary, the Q output goes LOW in response to the trigger pulse. It remains LOW for a time set by the RC time constant. At the end of this time, it goes HIGH. A one shot produces a single pulse each time it is triggered. This operation is illustrated in Figure 5.3.2.

The time duration of the output pulse, t_w, is proportional to the time constant τ ($\tau = RC$) by $t_w = kRC$, where k is a constant of proportionality and depends on the particular device. For the one shot in Figure 5.3.1, it can be obtained by solving the

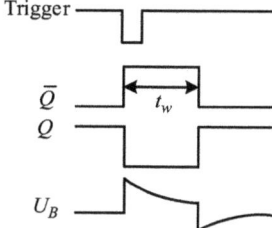

Figure 5.3.2: The operation waveform of the one shot.

voltage transient equations of point B. The initial and final voltage of point B can be expressed as follows:

$$U_B(0_+) = V_{DD} \text{ (Initial voltage)}$$

$$U_B(\infty) = 0 \text{ (Final voltage)}$$

The transient value, $U_B(t)$, at the time t can be calculated by the voltage transient equation as follows:

$$U_B(t) = U_B(\infty) + [U_B(0_+) - U_B(\infty)]e^{-\frac{t}{\tau}} = V_{DD}e^{-\frac{t}{\tau}} \tag{5.3.1}$$

Assume that the threshold voltage of logic gates is denoted by U_{th}. When $U_B(t) = U_{th} = V_{DD}e^{-\frac{t}{\tau}}$, the corresponding time is the output pulse width t_W.

$$t_W = \tau \ln \frac{V_{DD}}{U_{th}} \tag{5.3.2}$$

Especially, if $U_{th} = \frac{1}{2}V_{DD}$, then

$$t_W \approx 0.7\tau = 0.7RC \tag{5.3.3}$$

Except for the pulse width, another important parameter of the one shot is the maximum operation frequency, which can be determined by

$$f_{max} \le \frac{1}{t_W + t_{re}} \tag{5.3.4}$$

where t_{re} is the recovery time that is the discharge time of capacitor. Generally, the input equivalent resistor of logic gate can be ignored; the recovery time can be expressed by

$$t_{re} = (3\sim5)\tau_{re} \tag{5.3.5}$$

where τ_{re} is the discharge time constant of capacitor.

5.3.2 Types of one shots

There are two basic types of one shots: one is the nonretriggerable one shot and another is the retriggerable one shot.

A nonretriggerable one shot will not respond to any additional trigger pulse from the time it is triggered into its unstable state until it returns to its stable state. That is, it will neglect any triggered into its unstable state until it returns to its stable state. Figure 5.4.3 shows the nonretriggerable one shot being triggered at intervals greater than its pulse width and at intervals less than the pulse width. Note that in Figure 5.3.3(b), three additional pulses are neglected.

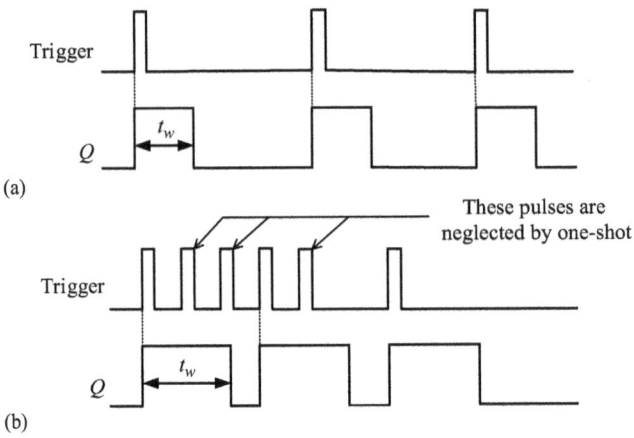

Figure 5.3.3: Nonretriggerable one shot action.

A retriggerable one shot can be triggered before it times out. As a result of retriggering, the pulse width can be extended, as illustrated in Figure 5.3.4.

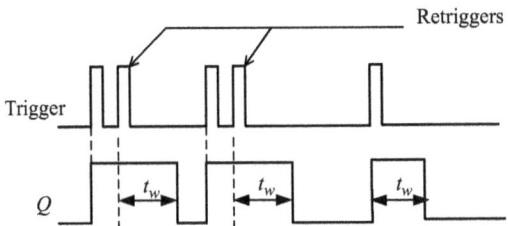

Figure 5.3.4: Retriggerable one shot action.

5.3.3 Specific integrated one shot

The commonly used one shot ICs include 54/74LS121, 54/74LS122, 54/74LS123, CD4098, CD4538, and many others [6].

For example, the 74LS122 is a retriggerable one shot, as shown in Figure 5.3.5. It has one clear input ($\bar{R}$) and four trigger inputs ($\bar{A}_1$, $\bar{A}_2$, B_1, and B_2). It can be triggered with either the negative pulse applied to the trigger inputs ($\bar{A}_1$ or $\bar{A}_2$) or the positive pulse applied to the inputs (B_1, B_2, or $\bar{R}$), as shown in the function table of 74LS122 in Table 5.3.1.

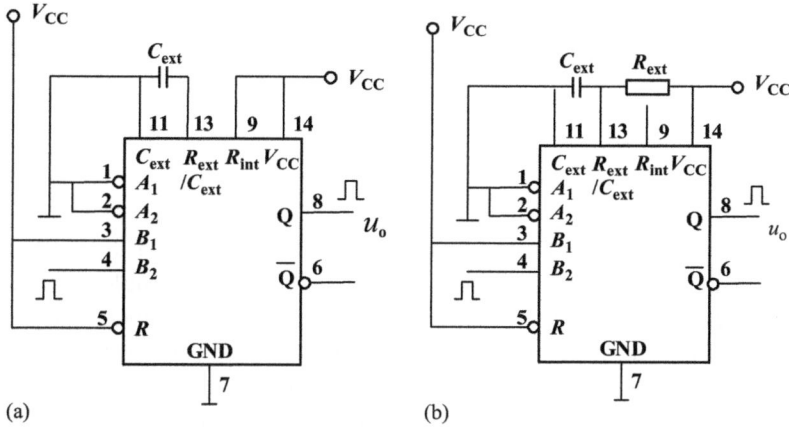

(a) (b)

Figure 5.3.5: Connection of 74LS122 for a pulse generator: (a) using internal resistor and (b) using external resistor.

The pulse width formula for the 74LS122 is

$$t_W = 0.32RC_{ext}\left(1 + \frac{0.7}{R}\right) \tag{5.3.6}$$

where the resistor R can be either the internal resistor (10 kΩ) or the external resistor (R_{ext}), C_{ext} is the external capacitor, and $0.32(1+0.7/R)$ is the constant of proportionality, k.

The connections for the 74LS122 to produce a pulse are shown in Figure 5.3.5. You can adopt either the internal resistor or the external resistor to produce the required pulse width. The value of the external resistor (R_{ext}) is from 5 to 260 kΩ and the value of the external capacitor (C_{ext}) is from 10 pF to 10 µF. Note that although the external capacitor (C_{ext}) is already connected to ground internally, an external ground also should be connected to the external capacitor in practical application to suppress the interference.

One shot can be used for timing circuit, pulse shaping, pulse broadening, and so on. For long distance transmission, it is a common for the pulse to produce a

Table 5.3.1: The function table.

$\overline{R}$	$\overline{A}_1$	$\overline{A}_2$	B_1	B_2	Q	$\overline{Q}$
0	×	×	×	×	0	1
×	1	1	×	×	0	1
×	×	×	0	×	0	1
×	×	×	×	0	0	1
1	0	×	↑	1	⊓	⊔
1	0	×	1	↑	⊓	⊔
1	×	0	↑	1	⊓	⊔
1	×	0	1	↑	⊓	⊔
1	1	↓	1	1	⊓	⊔
1	↓	↓	1	1	⊓	⊔
1	↓	1	1	1	⊓	⊔
↑	0	×	1	1	⊓	⊔
↑	×	0	1	1	⊓	⊔

degeneration due to the presence of some interference and attenuation. The one shot can be used for constructing the reshaping circuit. The degenerating pulse can be used as a trigger input of one shot and the desired pulse can be generated by controlling the time constant. Because the one shot can produce a rectangular pulse with a fixed width and amplitude, it can also be used for timing and delay circuits. In addition, by applying a narrow pulse to the trigger input of a one shot, a wide pulse can be obtained from the Q output of a one shot.

5.4 Astable multivibrator

An astable multivibrator has no stable state and thus its output change back and forth between two unstable states without the requirement of any external triggering pulse. It is often used to produce a periodic of pulse signal as a pulse generator. This section introduces astable multivibrator.

The objectives of this section are to
– Explain the basic operation of astable multivibrator
– Explain the difference between bistable, monostable, and astable multivibrator
– Describe how to construct a crystal oscillator

5.4.1 Astable multivibrator

In Section 5.3, you have learned that a simple one shot can be constructed by using an RC circuit to replace one feedback loop in S-R latch, as shown in Figure 5.4.1(a) and (b).

If both two-feedback loops are replaced by RC circuits and two NAND gates are substituted by two inverters, the resulting circuit is an astable multivibrator, as shown in Figure 5.4.1(c).

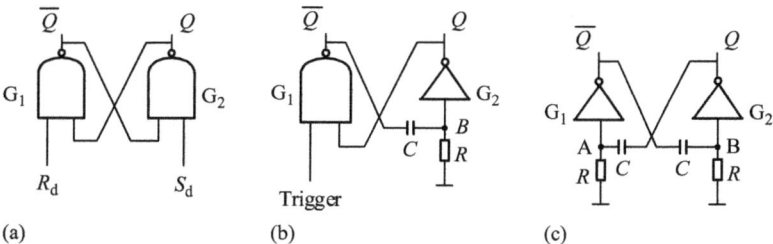

Figure 5.4.1: Comparison of S-R latch (a), one shot (b), and astable multivibrator (c).

The astable multivibrator has no stable state, but it has two unstable states: $Q = 0$ and $Q = 1$. Assume that the astable multivibrator in the first unstable state is $Q = 0$ and $\bar{Q} = 1$; the HIGH output, $\bar{Q}$, allows the capacitor to begin charging through the resistor. With the increase of voltage of the capacitor, the voltage of point B falls. As soon as the voltage of point B falls below the threshold voltage of gate G_2, the Q output changes from LOW to HIGH immediately. Through the coupling of the capacitor, the voltage of point A also becomes HIGH and thus the $\bar{Q}$ output goes LOW. The circuit enters into the second unstable state, $Q = 1$ and $\bar{Q} = 0$. Because the $\bar{Q}$ output is HIGH, it allows the capacitor to begin charging through resistor and the voltage of point A falls. As soon as the voltage of point A falls below the threshold voltage of gate G_1, the $\bar{Q}$ output changes from LOW to HIGH immediately. Through the coupling of the capacitor, the voltage of point B also becomes HIGH and thus the Q output goes LOW. As a result, the astable multivibrator changes its state from the second unstable state into the first unstable state. The state of astable multivibrator changes back and forth between two unstable states without requirement of any external triggering pulse. Figure 5.4.2 shows the voltage waveforms of the output,

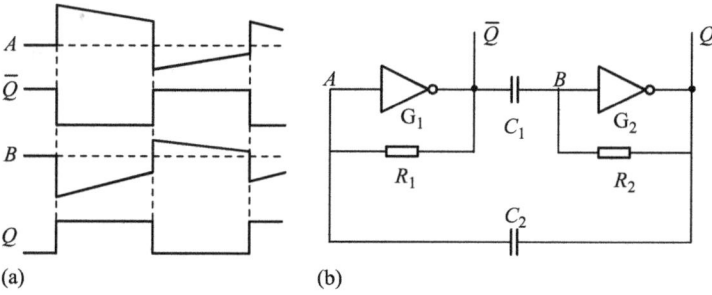

Figure 5.4.2: The astable multivibrator in Figure 5.4.1: (a) waveforms and (b) equivalent circuit.

point A and point B. The time that the astable multivibrator stays in each unstable state can be deduced by solving the voltage transient equations of point B and point A, respectively. If the circuit is symmetric and the threshold voltage of each inverter is half of the supply voltage, the time interval of each unstable state is $t_w = 0.7RC$ and the period of the output waveform is $T = 2t_w = 1.4RC$. In addition, the circuit of astable multivibrator in Figure 5.4.1(c) can be equivalent to that in Figure 5.4.2(b) by connecting the resistor to the LOW output of the corresponding inverter instead of connecting to ground.

There are also other types of astable multibrators, as shown in Figure 5.4.3. You can analyze the operation principle by yourself. Note that the key to analyze this kind of circuits including RC components is the charge and discharge process of the capacitors.

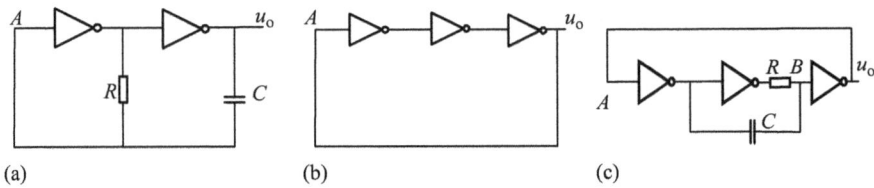

(a) (b) (c)

Figure 5.4.3: Several astable multivibrators: (a) CMOS; (b) ring loop; and (c) modified ring loop.

5.4.2 Crystal oscillator

The threshold voltage of the multivibrator that contains the resistor, capacitors, or logic circuit is sensitive to the environmental conditions, especially the temperature, so it is difficult for this kind of multivibrator to achieve a frequency stability better than 10^{-3}. For high-frequency stability, crystal oscillators are widely used because they can achieve a frequency stability higher than 10^{-9}.

Figure 5.4.4 shows the symbol, equivalent electrical circuit, and impedance characteristics of the quartz crystal. The equivalent electrical circuit contains a series RLC circuit, which represents the mechanical vibrations of the crystal, in parallel with a capacitance, C_p, which represents the electrical connections to the crystal [17].

There are two resonance frequency: series resonance frequency, f_s, and parallel resonance frequency, f_p, for quartz. They can be expressed as

$$f_s = \frac{1}{2\pi\sqrt{L_sC_s}} \tag{5.4.1}$$

$$f_p = \frac{1}{2\pi\sqrt{L_s\left(\frac{C_pC_s}{C_p + C_s}\right)}} \tag{5.4.2}$$

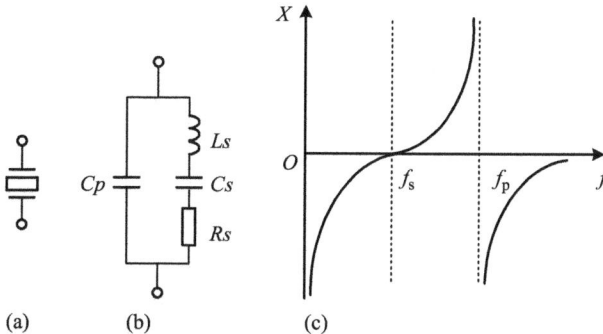

(a) (b) (c)

Figure 5.4.4: Quartz crystal's symbol (a); equivalent circuit (b); and impedance as a function of frequency (c).

A typical quartz crystal oscillator circuit is shown in Figure 5.4.5. A quartz crystal is used to replace the capacitor C_2 in Figure 5.4.2(b). The frequency of quartz crystal oscillators is determined by series resonance frequency since quartz crystal tends to operate toward their "series resonance". The role of the capacitor C_1 acts as the coupling capacitor between two inverters.

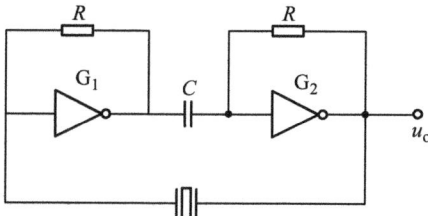

Figure 5.4.5: Quartz crystal oscillator.

5.5 Schmitt trigger

A Schmitt trigger is named a "trigger" because the output retains its value until the input changes sufficiently to trigger a change. Generally, the triggering input of flip-flop and one shot require a rapid transition input voltage (pulse signal), but Schmitt trigger can respond to a very slowly varying input voltage and belongs to level-triggered circuit. It is widely used in pulse shaping and waveform transformation. This section introduces principle, characteristics, and applications of Schmitt trigger.

The objectives of this section are to
- Describe the basic operation of a Schmitt trigger
- Explain the hysteresis characteristics of a Schmitt trigger
- Discuss the application of a Schmitt trigger

5.5.1 Hysteresis characteristics

The Schmitt trigger is a type of comparators with hysteresis characteristics, as shown in Figure 5.5.1. It has two threshold voltages: *upper threshold voltage*, U_{T+}, and *lower threshold voltage*, U_{T-}. For a inverting Schmitt trigger, when the input voltage, u_i, rises slightly above the upper threshold voltage, the output of Schmitt trigger produce a transition from HIGH to LOW; while the input voltage falls slightly below the lower threshold voltage, the output produces a transition from LOW to HIGH. For a noninverting Schmitt trigger, when the input voltage is slightly more positive than the upper threshold voltage, the output is a transition from LOW to HIGH; while the input voltage is slightly more negative than the lower threshold voltage, the output produces a transition from HIGH to LOW. Generally, the upper threshold voltage is greater than the lower threshold voltage. The difference between U_{T+} and U_{T-} is called *hysteresis voltage* denoted by ΔU_T.

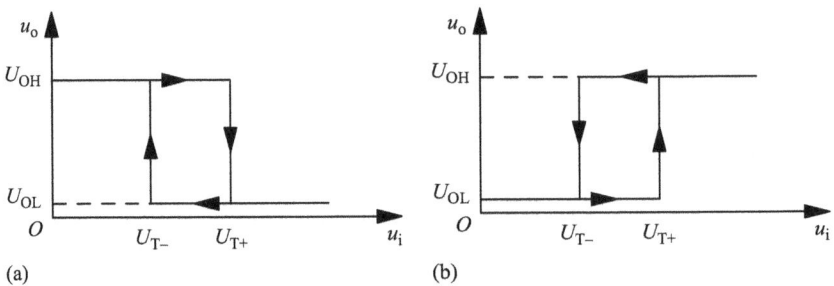

Figure 5.5.1: Hysteresis characteristics of inverting (a) and noninverting (b) Schmitt trigger.

Different from the monostable multivibrator and flip-flop triggered by the pulse edge, the Schmitt trigger can respond to a very slowly varying input voltage, which belongs to level-triggered device. Assuming that the input signal is a triangular wave, the corresponding output waveforms for the inverting Schmitt trigger and the noninverting Schmitt trigger are shown in Figure 5.5.2, respectively. For example, in the noninverting configuration, when the input is slightly above the upper threshold voltage (U_{T+}), the output produce a transition from LOW to HIGH; when the input is slightly below the lower threshold voltage (U_{T-}), the output changes from HIGH to LOW. This dual threshold action is called *hysteresis* and implies that the Schmitt trigger possesses memory and can act as a latch or flip-flop.

Logic symbols of the inverting and noninverting Schmitt trigger are shown in Figure 5.5.3. Similar to the traditional symbols for an inverter, the symbol of an inverting Schmitt trigger has an inverted hysteresis curve inside a buffer followed by a bubble; a noninverting Schmitt trigger has a noninverted hysteresis curve inside a buffer.

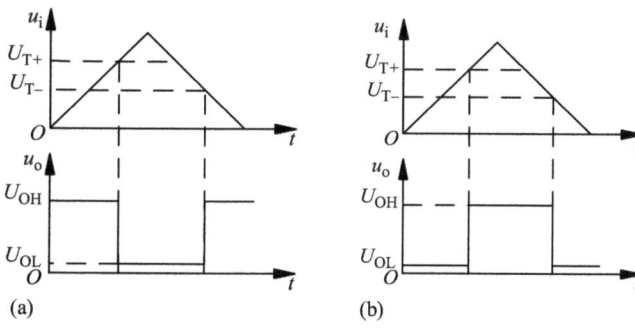

Figure 5.5.2: Input and output waveforms of inverting (a) and noninverting (b) Schmitt trigger.

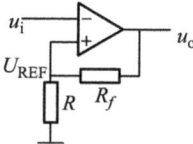

Figure 5.5.3: The logic symbol of inverting (a) and noninverting (b) Schmitt trigger.

5.5.2 Structure and principle

There are many ways to build Schmidt triggers. Figure 5.5.4 shows a circuit structure of Schmitt trigger by using an operational amplifier with a positive feedback.

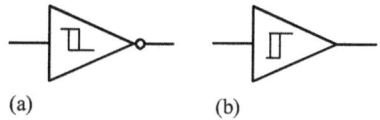

Figure 5.5.4: Schmitt trigger circuit.

Two resistors R and R_f act as a voltage divider and offer a reference voltage to the noninverting input of op-amplifier. When the Schmitt trigger is in high state, the output voltage is the positive saturation voltage, that is, $u_o = +U_{sat}$, and the reference voltage is expressed as

$$U_{REF1} = \frac{R}{R + R_f} u_o = \frac{R}{R + R_f} U_{sat} \tag{5.5.1}$$

When the input voltage u_i is less than U_{REF1}, the Schmitt trigger will stay in the high state. But when the input voltage rises slightly above U_{REF1}, the Schmitt trigger changes its state from HIGH to LOW and thus the output voltage becomes the negative saturation voltage, that is, $u_o = -U_{sat}$. As a result, the reference voltage is also changed, which can be expressed as

$$U_{\text{REF2}} = \frac{R}{R+R_f} u_o = -\frac{R}{R+R_f} U_{\text{sat}} \tag{5.5.2}$$

When the input voltage u_i is greater than U_{REF2}, the Schmitt trigger will stay in the low state. But when the input voltage falls slightly below U_{REF2}, the Schmitt trigger immediately changes its state from LOW back to HIGH. Therefore, the hysteresis characteristics are produced and the hysteresis voltage can be obtained by the difference of $U_{\text{T+}}$ and $U_{\text{T-}}$. Because $U_{\text{T+}} = U_{\text{REF1}}$ and $U_{\text{T-}} = U_{\text{REF2}}$, the hysteresis voltage can be deduced as follows:

$$\Delta U_{\text{T}} = U_{\text{T+}} - U_{\text{T-}} = \frac{2R}{R+R_f} U_{\text{sat}} \tag{5.5.3}$$

5.5.3 Applications of schmitt trigger

Schmitt triggers are widely used in the field of pulse shaping, pulse conversion, and pulse magnitude distinguishing.

1. Pulse shaping
In a digital system, the pulse signal may be distorted during transmission, for example, pulse edge deformation, oscillation on the rising or falling edge, and additional noise appears on the pulse signal, as shown in Figure 5.5.5. An inverting Schmitt trigger can be used for pulse shaping by setting a suitable value for $U_{\text{T+}}$ and $U_{\text{T-}}$.

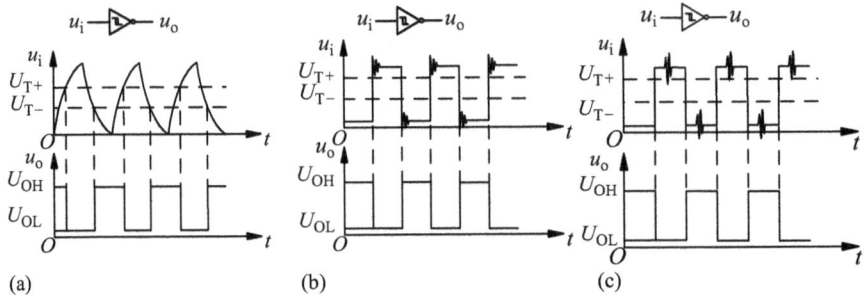

(a) (b) (c)

Figure 5.5.5: Schmitt trigger used for pulse shaping: (a) pulse edge deformation; (b) oscillation on the rising or falling edge; and (c) additional noise.

2. Waveform conversion
Schmitt triggers can be used to transform a periodic signal to a pulse signal, as shown in Figure 5.5.6. The width of pulse can be adjusted by changing the difference between $U_{\text{T+}}$ and $U_{\text{T-}}$.

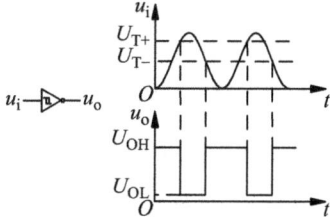

Figure 5.5.6: A Schmitt trigger used for waveform conversion.

3. Pulse magnitude distinguishing

Because the output of the Schmitt trigger depends on the magnitude of the input voltage, we can make use of this characteristic to distinguish the magnitude of pulse, as shown in Figure 5.5.7. Only when pulse magnitude is higher than the value of U_{T+}, the pulse can be recognized.

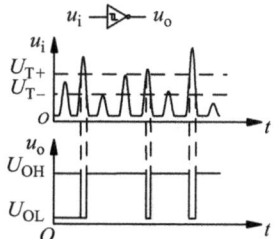

Figure 5.5.7: A Schmitt trigger used for distinguishing the magnitude of pulse.

4. Using as an astable multivibrator

A Schmitt trigger is a bistable multivibrator, and it can be used to implement the astable multivibrator, also called as the *relaxation oscillator*. This is achieved by connecting a single RC integrating circuit between the output and the input of an inverting Schmitt trigger, as shown in Figure 5.5.8. If Schmitt trigger is in the high state, the capacitor is charged and the input voltage increases. When the input voltage is slightly more positive than the upper threshold voltage, the Schmitt trigger changes its state from HIGH to LOW. When Schmitt trigger is in the LOW state, the capacitor starts to discharge and makes the input voltage fall. When input voltage is slightly more negative than the lower threshold voltage, the Schmitt trigger goes back

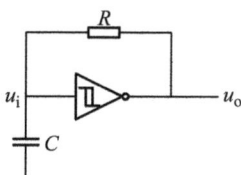

Figure 5.5.8: A multivibrator constructed with a Schmitt trigger.

to the HIGH state. The output will be a continuous pulse wave whose frequency depends on the values of R and C, and the threshold points of the Schmitt trigger. Since multiple Schmitt trigger circuits can be provided by a single integrated circuit (e.g., the 4,000 series CMOS device type No. CD40106 contains six of them), a spare section of the IC can be quickly pressed into service as a simple and reliable oscillator with only two external components.

5.6 555 timer

The 555 timer is a kind of versatile analog-to-digital hybrid integrated circuit. It can be configured as a monostable multivibtrator, an astable multivibrator, and Schmitt trigger, and is widely used in the field of pulse generation, pulse shaping, time delay, and timing circuits. This section introduces the functions and applications of 555 timer.

The objectives of this section are to
- Describe the basic operation of 555 timer
- Set up a 555 timer operate as a one shot or an astable multivibrator
- Discuss the application of 555 timer

5.6.1 Basic operation

There are two types of 555 timer ICs in the market. A type of 555 timer ICs belonging to the TTL family includes a single 555 timer named 555 and dual 555 timers named 556. Another type of IC products belonging to CMOS family contains a single 555 timer named 7555 and dual 555 timers named 7556. Irrespective of the family they belong to, the 555 timers have the same function and external pin configuration. They have several distinguished advantages when compared with other ICs. The dc supply voltage has a wide range from 5 to 15 V for TTL family and from 3 to 18 V for CMOS family; the maximum load current is offered up to 200 mA for TTL family and 4 mA for CMOS family.

Let us take the 555 timer in TTL family, for example, to introduce the functions and basic operations of the 555 timer. Figure 5.6.1 shows the pin diagram and the functional diagram of the internal components of a 555 timer. The power connections to the chip are through pins 1 (ground, GND) and 8 (+Vcc). The positive supply voltage (+Vcc) should be between 5 and 15 V. It can be seen from the functional diagram that a 555 timer consists of a few different elements: resistors, transistors, comparators, a R-S latch, and an output stage.

A voltage divider is set up by three resistors connected between Vcc and ground. Since three resistors have the same value of 5 kΩ, the voltage at the junctions between the resistors are 2/3 Vcc and 1/3 Vcc used as reference voltages for two comparators.

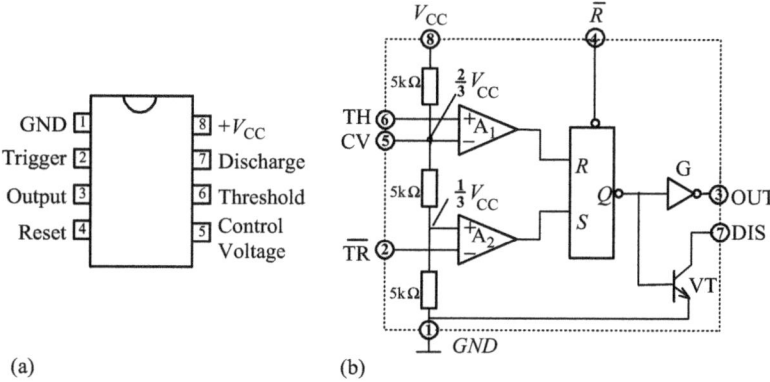

(a) (b)

Figure 5.6.1: Pin diagram (a) and functional diagram (b) of a 555 timer.

Two comparators are composed of two operational amplifiers with open loop. The comparator compares an input with a reference voltage and outputs a LOW or HIGH signal based on whether the input is a higher or lower voltage than the reference. The comparator A_2 connected to pin 2 compares the "trigger" input, $\overline{TR}$, to a reference voltage of $1/3 Vcc$ and the comparator A_1 connected to pin 6 compares the "threshold (TH)" input to a reference voltage of $2/3 Vcc$. If you want to change the value of reference voltage, the reference voltage can be externally input through pin 5 named control voltage (CV). If there is no external reference input, pin 5 is usually connected to ground through a small capacitor of 0.01 μF, which can eliminate the interference of high frequency noise.

A R-S latch is used to switch between two stable states. The latch outputs a HIGH or LOW based on the states of the two comparators. It can be reset by applying a LOW to pin 4 named the clear input or direct reset input, $\overline{R}$.

The transistor attached to pin 7 is an NPN transistor. Since pin 7 is connected to the collector of the NPN transistor VT, this type of configuration is called as *open collector*. This pin is usually connected to a capacitor and is used to discharge the capacitor each time the output pin goes low. This is the reason why pin 7 is named discharge (DIS) terminal and the transistor is called as the discharging transistor.

The Q output of the latch passes through an output buffer G to the output pin 3. Whether or not the latch outputs a HIGH or LOW is determined by the states of the two comparators and direct reset input; let us discuss the basic operation of a 555 timer.

When the direct reset input is LOW, the latch is reset directly, causing the output (pin 3) to go LOW and making the discharging transistor on.

When the direct reset input is HIGH, the output is determined by the trigger input and threshold input. If the threshold input is higher than $2/3 Vcc$ regardless of the trigger input, the output of comparator A_1 is HIGH and resets the latch, resulting in the output to go LOW and the discharge transistor turned on. If the trigger input is lower than $1/3Vcc$ and the threshold input is also lower than $2/3Vcc$, the output of comparator A_2 is HIGH and sets the R-S latch, causing the output to go HIGH and the discharging transistor turned off. If trigger input is greater than $1/3Vcc$ and the threshold input is lesser than $2/3Vcc$, the output of two comparators are both LOW and thus the latch keeps no change. The logic function of the 555 timer is shown in Table 5.6.1.

Table 5.6.1: The logic function.

$\bar{R}$	TH	$\bar{T}R$	OUT	VT
0	×	×	0	on
1	> $2/3V_{cc}$	×	0	on
1	<$2/3V_{cc}$	> $1/3V_{cc}$	No change	No change
1	<$2/3V_{cc}$	< $1/3V_{cc}$	1	off

5.6.2 One shot operation

A 555 timer can be used to construct a one shot by adding an external resistor (R) and capacitor (C), as shown in Figure 5.6.2. The capacitor 0.01μF is introduced to eliminate the effect of high-frequency noise on the trigger and threshold levels. The negative trigger pulse u_i is applied to the triggering input. The high level of the trigger pulse u_i should be greater than $2/3Vcc$, while the low level of the trigger pulse u_i should be less than $1/3Vcc$. The output signal is gotten out from pin 3.

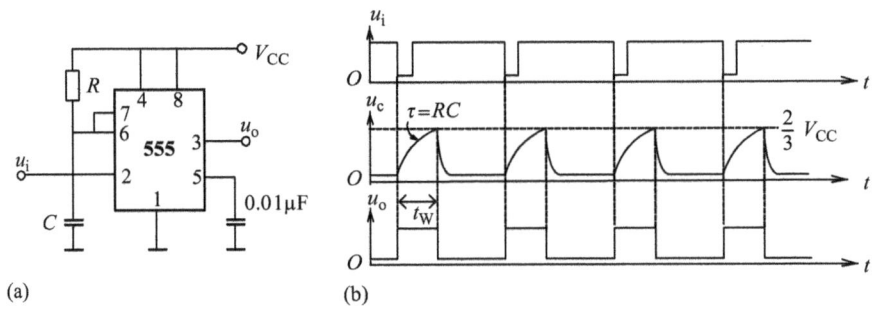

(a)   (b)

Figure 5.6.2: The 555 timer connected as a one shot: (a) circuit diagram; (b) voltage waveforms.

Before the trigger pulse is applied, the output is LOW and the discharged transistor VT is turned on. This keeps C discharged and the voltage of capacitor, u_c, is around 0.3 V. The 555 timer operates at the stable state. When the negative-going trigger pulse u_i is applied, the output goes high and VT is turned off because the trigger input is less than 1/3Vcc and the threshold input is less than 2/3Vcc. The state of the output changes from the stable state (LOW) to unstable state (HIGH). Since the discharge transistor is off, the capacitor C starts to be charged through resistor R and u_c increases. When u_c rises slightly more than 2/3Vcc, the output goes back LOW and VT is turned on immediately, allowing the capacitor C to discharge through VT and thus the output goes back to the stable state. The output pulse width, t_w, can be determined by solving the transient equation of u_c as follows:

$$u_c(t) = u_c(\infty) + [u_c(0) - u_c(\infty)]e^{-\frac{t}{\tau}} \tag{5.6.1}$$

where $u_c(0)$=0V, $u_c(\infty)$=V_{CC}, and $\tau = RC$.

When $t = t_w$, $u_c = 2/3V_{cc}$. Thus, the output pulse width, t_w, can be deduced from eq. (5.6.1) as follows:

$$t_w = 1.1RC \tag{5.6.2}$$

Generally, the value of t_w is several microseconds to several minutes, and the stability and accuracy of the timer will decrease with the increase of t_w.

5.6.3 Astable operation

A 555 timer can be also connected to operate as an astable multivibrator by adding two external resistors and one capacitor, as shown in Figure 5.6.3. Note that the threshold input *TH* is now connected to the trigger input. Initially, when the power is turned on, C is uncharged, causing both trigger input and threshold input less than the corresponding reference voltage. Hence, the output is HIGH and VT is off, allowing the capacitor to begin charging through R_A and R_B and the voltage u_c increases. When u_c reaches up to 2/3V_{cc}, the output u_o changes from HIGH to LOW and VT is on, allowing the capacitor to begin discharging through R_B and VT. As soon

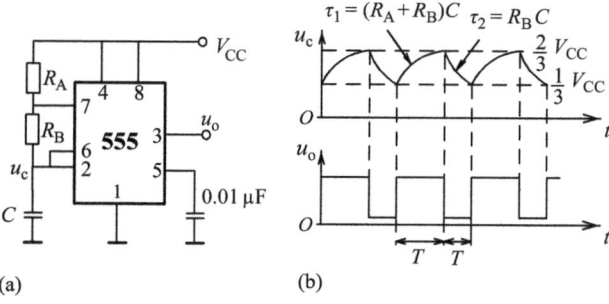

(a) (b)

Figure 5.6.3: The 555 timer connected as an astable multivibrator: (a) circuit diagram; (b) voltage waveform.

as u_c falls down to $1/3V_{cc}$, the output u_o goes HIGH and VT is off, allowing C to begin charging again through R_A and R_B and thus a new cycle begins.

The duration time, T_1, of the HIGH output corresponds to the time interval that the voltage of capacitor u_c to be charged from $1/3V_{cc}$ to $2/3V_{cc}$; the duration time, T_2, of the LOW output corresponds to the time interval that the voltage of capacitor u_c to be discharged from $2/3V_{cc}$ to $1/3V_{cc}$. They can be deduced from the transient equation of u_c. The resulting duration time T_1 and T_2 are expressed as

$$T_1 = 0.7(R_A + R_B)C \tag{5.6.3}$$

$$T_2 = 0.7R_BC \tag{5.6.4}$$

Therefore, the period of the output pulse, T, can be deduced as follows:

$$T = T_1 + T_2 = 0.7(R_A + 2R_B)C \tag{5.6.5}$$

The duty circle (D) is

$$D = \frac{T_1}{T} = \frac{T_1}{T_1 + T_2} \times 100\% = \frac{R_A + R_B}{R_A + 2R_B} \times 100\% \tag{5.6.6}$$

According to eqs. (5.6.5) and (5.6.6), the period and duty circle can be adjusted by selecting R_A and R_B. Generally, the values of the resistors and capacitor for 555 timer working as an astable multivibrator are as follows: $R_A \geq 1k\Omega$, $R_B \geq 1k\Omega$, $R_A + R_B \leq 3.3M\Omega$, and $C \geq 500pF$.

Example 5.8 The circuit is composed of two 555 timers, as shown in Figure 5.6.4. Determine the frequencies of u_{o1} and u_{o2}, and draw their waveforms.

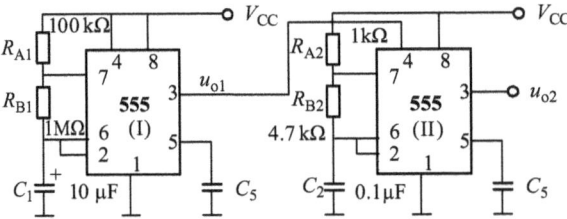

Figure 5.6.4: Circuit diagram.

Solution

Two 555 timers are connected to operate as astable operation (oscillator). The periods for timer I and timer II can be calculated by using eq. (5.6.5):

$$T_1 = 0.7(R_{A1} + 2R_{B1})C_1 = 14.7s$$

$$T_2 = 0.7(R_{A2} + 2R_{B2})C_2 = 0.728ms$$

Thus, the frequencies of u_{o1} and u_{o2} are

$$f_{o1} = \frac{1}{T_1} = \frac{1}{0.7(R_{A1} + 2R_{B1})C_1} = \frac{1}{14.7s} = 0.068\,Hz$$

$$f_{o2} = \frac{1}{T_2} = \frac{1}{0.7(R_{A2} + 2R_{B2})C_2} = \frac{1}{0.728ms} = 1.37\,kHz$$

When the output of the timer I, u_{o1}, is LOW, reset the timer II and thus the u_{o2} output is LOW. Only when the output of the timer I, u_{o1}, is HIGH, timer II operates as astable multivibration and produces the output pulse with a period of T_2. The waveforms for u_{o1} and u_{o2} are shown in Figure 5.6.5.

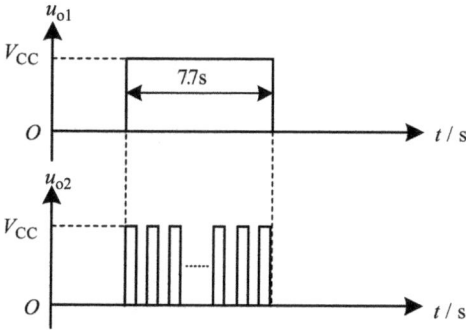

Figure 5.6.5: Waveforms of u_{o1} and u_{o2}.

5.6.4 Schmitt trigger

The Schmitt trigger can be constructed by using a 555 timer, as shown in Figure 5.6.6(a). The input signal, u_i, is sent to the trigger input $\overline{TL}$ (pin 2) and threshold input TH (pin 6) and the output, u_o, is from the output terminal (pin 3). The discharging terminal (pin 7) is connected to V_{cc2} via a resistor R.

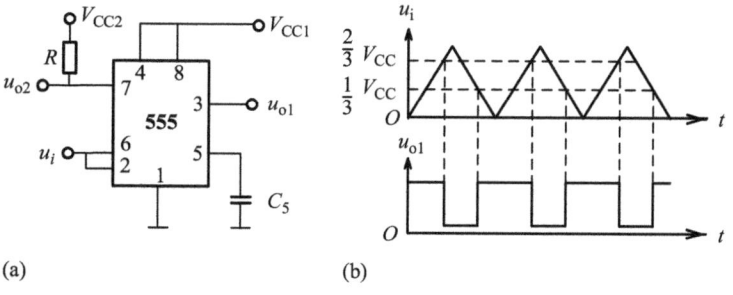

(a)   (b)

Figure 5.6.6: Schmitt trigger constructed with 555 timer (a) and its waveform (b).

Assuming that the input voltage u_i increases from 0 V, only if u_i is lower than $2/3V_{cc}$, the output u_{o1} keeps HIGH. When the input voltage u_i reaches up to $2/3V_{cc}$, the output u_{o1}

turns from HIGH to LOW and it will keep LOW until u_i falls down to $1/3V_{cc}$. When u_i is lower than $1/3V_{cc}$, the output u_{o1} goes back to HIGH. The input and output waveforms are shown in Figure 5.6.6(b). It can be concluded that the two thresholds voltage for the Schmitt trigger, V_{T+} and V_{T-} are $2/3V_{cc}$ and $1/3V_{cc}$, respectively.

5.6.5 Voltage-controlled oscillator

When the 555 timer is connected to operate as an astable multivibrator and the control voltage input (pin 5) is connected to a control voltage U_s, as shown in Figure 5.6.7, the 555 timer will operate as a voltage-controlled oscillator. The operation principle is the same as the astable operation, where the only difference is that the trigger voltage level is $U_s/2$ and the threshold voltage level is U_s. The period and duty circle of output pulse can be tuned by adjusting U_s.

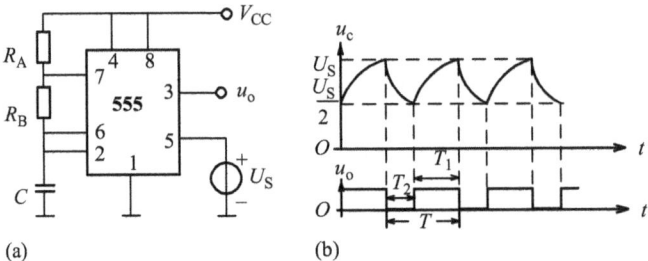

(a)   (b)

Figure 5.6.7: Voltage-controlled oscillator constructed with 555 timer (a) and its waveforms (b).

T_1 is the time duration that the capacitor is charged from $U_s/2$ to U_s, which can be expressed as

$$T_1 = (R_A + R_B)C \ln \frac{V_{CC} - U_s/2}{V_{CC} - U_s} \tag{5.6.7}$$

T_2 is the time duration that the capacitor is discharged from U_s to $U_s/2$, which can be expressed as

$$T_2 = 0.7R_B C \tag{5.6.8}$$

Therefore, the period of the output waveform can be expressed as

$$T = T_1 + T_2 = (R_A + R_B)C \ln \frac{V_{CC} - U_s/2}{V_{CC} - U_s} + 0.7R_B C \tag{5.6.9}$$

5.7 Summary

1. Multivibrators contain bistable, monostable, and astable multivibrators.
2. A bistable multivibrators is a logic circuit with a memory characteristic such that its Q outputs will go to a new state in response to an input pulse and will remain in that new state after the input pulse is terminated.
3. Latch and flip-flop are two categories of bistable multivibrators. The main difference between latch and flip-flop is the method used for changing their state. The latch is sensitive to the input level. However, a flip-flop only changes its state at a specified point of the clock pulse.
4. A NAND latch and a NOR latch are simple bistable multivibrators that respond to logic levels on their SET and RESET asynchronous inputs.
5. Master-slave flip-flop is a kind of flip-flop with the pulse-triggered method. Data are entered into the flip-flop at the leading edge of the clock pulses, but the output does not reflect the input state until the trailing edge of the clock pulses.
6. Edge-triggered flip-flop changes its state either at the positive edge (rising edge) or at the negative edge (failing edge) of the clock pulse and is sensitive to its inputs only at this transition of the clock.
7. Most clocked flip-flops have asynchronous inputs that can set or reset the flip-flops independent of the clock input.
8. Mainstream ICs are D and J-K edge-triggered flip-flops. Other types of flip-flops can be implemented with D flip-flop and J-K flip-flop by using characteristic equations of different types of flip-flops to determine the inputs.
9. The main applications of bistable multivibrators include data storage and transfer, data shifting, counting, and frequency division.
10. The operation characteristic for flip-flops includes the time for data receiving, state transition, propagation delay, or response.
11. A monostable multivibrators (one shot) is a logic circuit that can be triggered from its stable state to its unstable state in which it remains for a time interval proportional to an RC time constant.
12. Astable multivibrators have no stable state and are used to produce a period of pulse signal as a pulse generator.
13. The Schmitt trigger, which also belongs to bistable multivibrator, can respond reliably to slow-changing signals and will produce outputs with clean, sharp edges.
14. The 555 timer is a kind of versatile analog to digital hybrid integrated circuit, which can be configured as a monostable multivibtrator, an astable multivibrator, Schmitt trigger by only adding the few external resistors and capacitor.
15. Various circuits can be used to generate clock signals at a desired frequency, including Schmitt-trigger oscillators, a 555 timer, and a crystal-controlled oscillator.

Key terms

Latch: A bistable logic circuit used for storing a bit.

SET: The state of a latch or flip-flop when the output is HIGH.

RESET: The state of a latch or flip-flop when the output is LOW.

Edge-triggered flip-flop: A type of flip-flops in which input data entes into the output either at the positive edge or at the negative edge of the clock pulse.

Toggle: The action of a flip-flop in which it changes state on each clock pulse.

Preset: An asynchronous input used to set the flip-flop; also called as direct set.

Clear: An asynchronous input used to reset the flip-flop; also called as direct reset.

One shot: A monostable multivibrator.

Multivibrator: An electronic circuit used to implement a variety of simple two-state devices such as relaxation oscillators, timers, and flip-flops.

Astable: Having no stable state.

Bistable: Having two stable states.

Monostable: Having only one stable state. A monostable multivibrator produces a single pulse in response to a triggering input.

Timer: A multifunctional analog-to-digital hybrid integrated circuit, which can be configured in two different modes as either a monostable multivibtrator (one shot) or as an astable multivibrator (oscillator).

Hysteresis: A characteristic of a threshold-triggered circuit, such as the Schmitt trigger where the device turns on and off at different input levels.

Self-test

5.1 A latch has _____ stable states.
(a) one   (b) two   (c) three   (d) four

5.2 If a S-R latch has a 0 on the S input and a 1 on the R input, the latch will be _____
(a) set   (b) reset   (c) invalid   (d) clear

5.3 The invalid state of a S-R latch occurs when _____
(a) $S = 1, R = 0$ (b) $S = 0, R = 1$ (c) $S = 1, R = 1$ (d) $S = 0, R = 0$

5.4 The output of gated D latch will be SET when _____
(a) $EN = 1, D = 0$ (b) $EN = 1, D = 1$ (c) $EN = 0, D = 1$ (d) $EN = 0, D = 0$

5.5 The purpose of the clock input to a flip-flop is to _____
(a) clear the device
(b) set the device
(c) cause the output to change states
(d) cause the output to assume a state dependent on the controlling inputs

5.6 Flip-flops belongs to _____
      (a) monostable multivibrators   (b) bistable multivibrators
      (c) astable multivibrators   (d) one shot

5.7 A feature that distinguishes the J-K flip-flop from the S-R flip-flop is the _____
      (a) toggle condition   (b) preset input   (c) type of clock   (d) clear input

5.8 Fill the truth table in table T5.1 for a circuit shown in Figure T5.1.

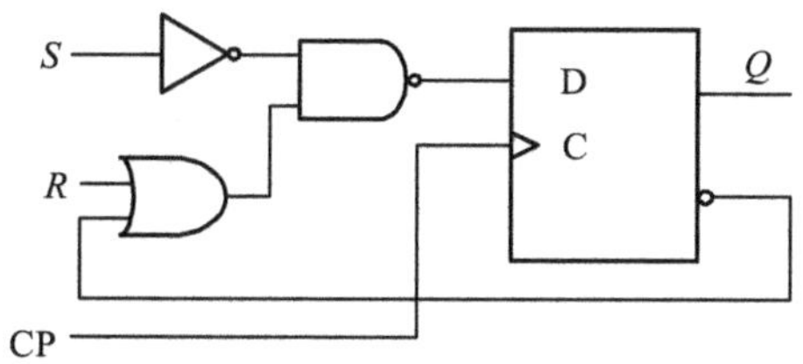

Figure T5.1

Table T5.1

R	S	Q_{n+1}

5.9 A J-K flip-flop with $J = K = 1$ has a 20 kHz clock input, the Q output is _____
      (a) a 10 kHz square wave   (b) a 20 kHz square wave
      (c) a 30 kHz square wave   (d) a 40 kHz square wave

5.10 The one shot is a device with _____ stable state
      (a) one   (b) two   (c) three   (d) four

5.11 A one shot is a type of _____
      (a) monostable multivibrator   (b) astable multivibrator   (c) timer
      (d) answers A and C   (e) answers B and C

5.12 The output pulse width of a nonretriggerable one shot depends on _____
      (a) trigger intervals   (b) supply voltage
      (c) a resistor and capacitor   (d) threshold voltage

5.13 An astable multivibrator _____
      (a) require a periodic trigger input   (b) has no stable state
      (c) produces a periodic pulse output   (d) is an oscillator
      (e) answers A, B, C, and D   (f) answers B, C, and D

5.14 A multivibrator constructed with a 555 timer is shown in Figure T5.2. The period of the output waveform is _____.

(a) $0.7(R_A+2R_B)C$ (b) $0.7(R_A+R_B)C$ (c) $(R_A+2R_B)C$ (d) $1.2(R_A+R_B)C$

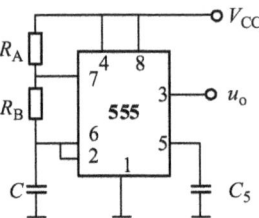

Figure T5.2

5.15 Duty circle of the output waveform for circuit shown in Figure T5.2 is _____

(a) $D = \dfrac{2R_A+R_B}{R_A+2R_B} \times 100\%$

(b) $D = \dfrac{R_A+R_B}{R_A+2R_B} \times 100\%$

(c) $D = \dfrac{R_A+R_B}{R_A+3R_B} \times 100\%$

(d) $D = \dfrac{R_A+3R_B}{R_A+2R_B} \times 100\%$

Problems

5.1 If the waveforms shown in Figure P5.1 are applied to an active-LOW input S-R latch, draw the resulting Q output waveform. Assume that Q starts LOW.

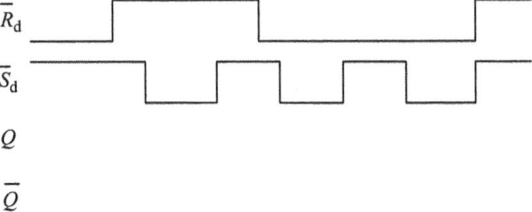

Figure P5.1

5.2 Draw the resulting Q output waveform of an S-R latch, as shown in Figure P5.2. Assume that Q is initially LOW.

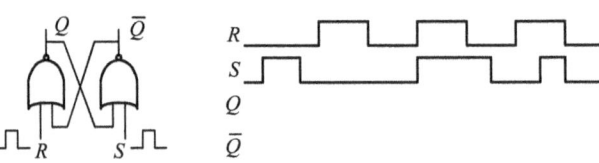

Figure P5.2

5.3 List the truth table and the characteristic equation for the circuit shown in Figure P5.3.

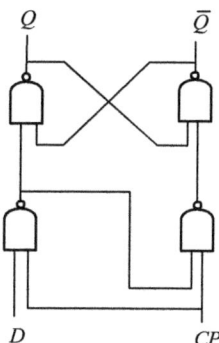

Figure P5.3

5.4 For a gated S-R latch, determine the Q output waveform in terms of the given input waveforms in Figure P5.4.

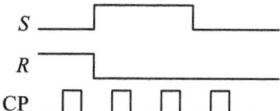

Figure P5.4

5.5 Determine the Q output waveforms of each flip-flop in relation to the clock pulse in Figure P5.5. Assume that Q starts LOW.

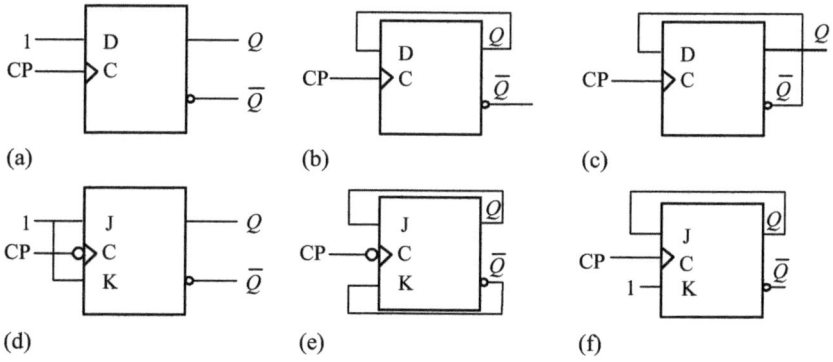

Figure P5.5

5.6 Write out the logic expressions of Q_{n+1} for each flip-flop in Figure P5.6 by using characteristic equation of the flip-flop.

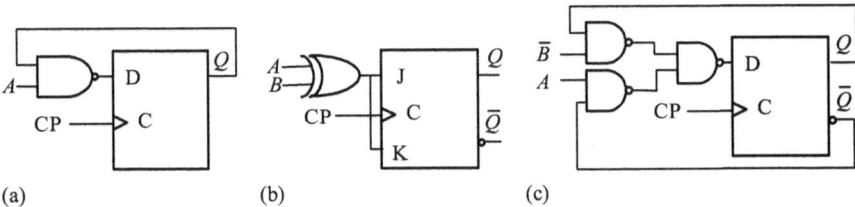

(a)   (b)   (c)

Figure P5.6

5.7 For the circuit shown in Figure P5.7, construct its state table and state diagram; draw the timing diagram when a sequential input "1011100" is applied on the input X.

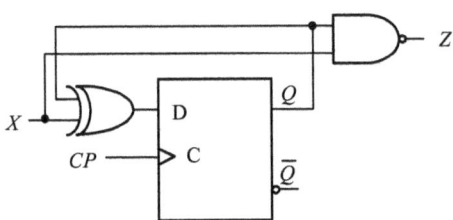

Figure P5.7

5.8 For the circuit in Figure P5.8, draw the resulting output waveform. Assume that Q_0 and Q_1 are initially LOW.

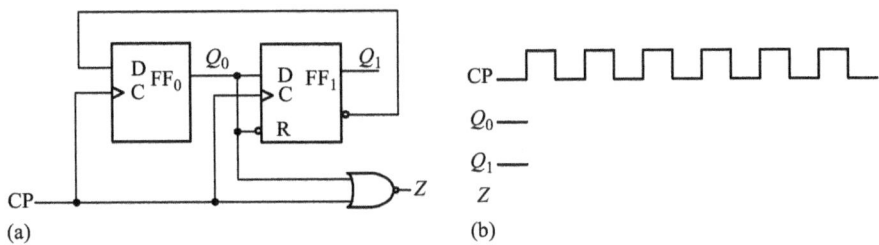

(a)   (b)

Figure P5.8

5.9 For the circuit in Figure P5.9, determine the resulting output waveform. Assume that Q_0 and Q_1 are initially LOW.

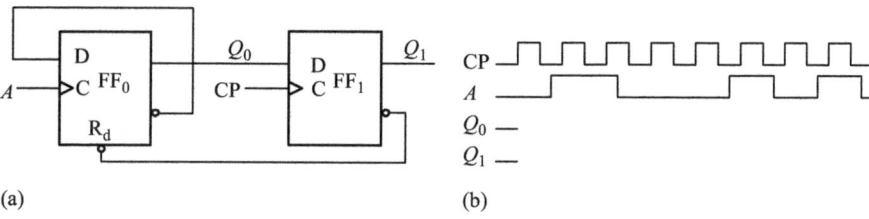

(a)   (b)

Figure P5.9

5.10 For the circuit in Figure P5.10, determine the resulting output wave form. Assume that Q_0 and Q_1 are initially LOW.

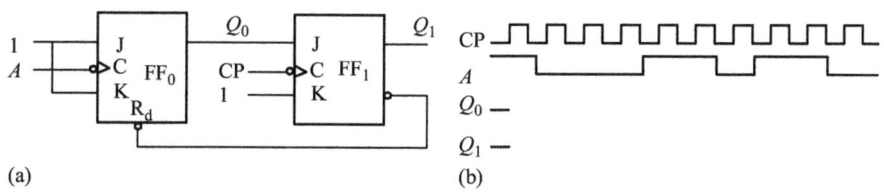

(a)   (b)

Figure P5.10

5.11 A D flip-flop is connected as shown in Figure P5.11. Determine the Q output in relation to the clock. What specific function does this device perform?

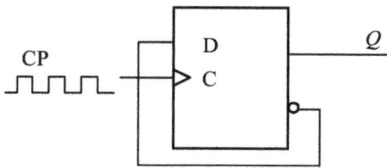

Figure P5.11

5.12 Calculate the output pulse width (t_w) of the one shot circuit (74LS122) shown in Figure P5.12 at the given the circuit parameters of $C_{ext} = 1,000$ pF, $R_{ext} = 10$ kΩ, and $V_{CC} = 5.0$ V.

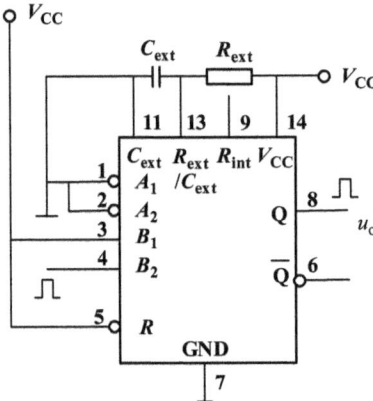

Figure P5.12

5.13 Design three circuits to implement the relation of input A and three outputs, B, C, and D, respectively. The waveform of input A and three corresponding outputs, B, C, and D are shown in Figure P5.13.

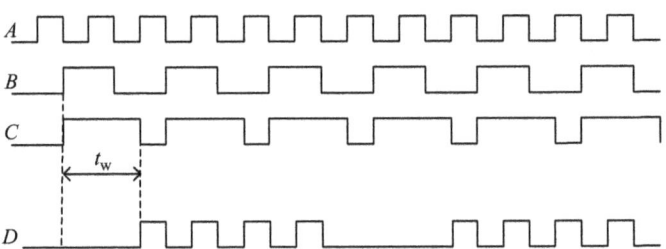

Figure P5.13

5.14 A circuit consists of a 555 timer and a J-K flip-flop, as shown in Figure P5.14. Assume that $f_{CP} = 10$ Hz, $R_1C_1 \ll T_{cp}$, $R_2 = 56$ kΩ, $C_2 = 4.7$ μF. Determine the waveforms of Q, u_i and u_o, and calculate the frequency and duty circle of u_o.

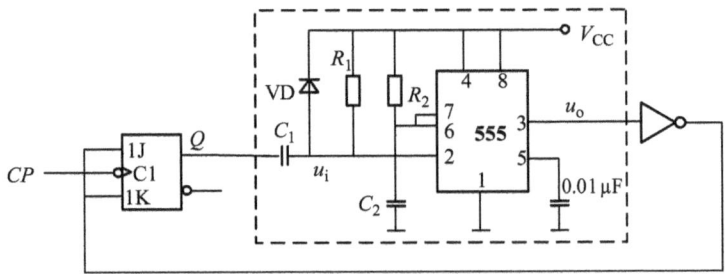

Figure P5.14

5.15 Figure P5.15(a) shows a circuit constructed by using a 555 timer with $V_{CC} = 5$ V and $U_s = 4$ V. If the input waveform in Figure P5.12(b) is applied on the u_i input, determine the u_o output waveform and explain the function of this circuit.

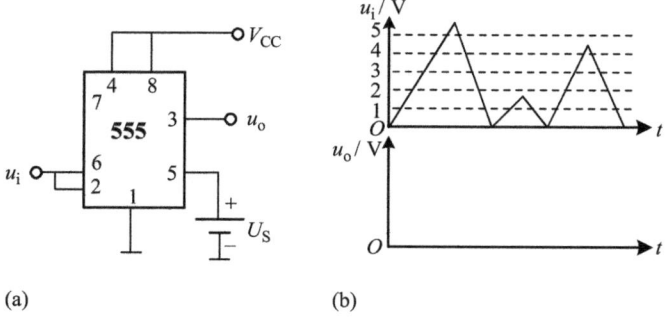

(a)   (b)

Figure P5.15

5.16 A electronic doorbell circuit constructed with 555 timer is shown in Figure P5.16. When the switch S is on, the doorbell will ring until a predetermined time is out. Determine the frequency of the doorbell and explain the function of the capacitors C_2 and C_3. In addition, modify the circuit to extend the time of doorbell ringing.

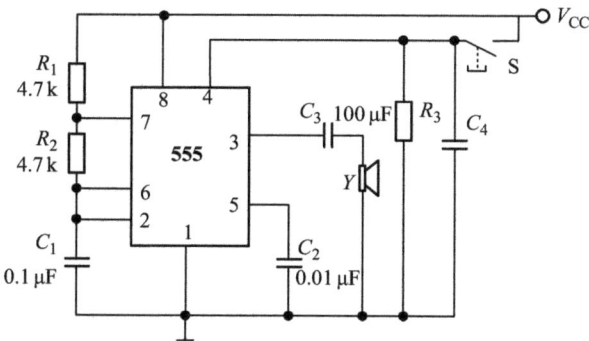

Figure P5.16

5.17 Figure P5.17 shows an anti-incursion alarm circuit by using a 555 timer. When a theft invades, the wire connected point a with point b will be cut off and output an alarm signal. Determine the function of the circuit shown in Part I and Part II, respectively. According to the values given in the Figure P5.13, calculate the frequency of the alarm signal.

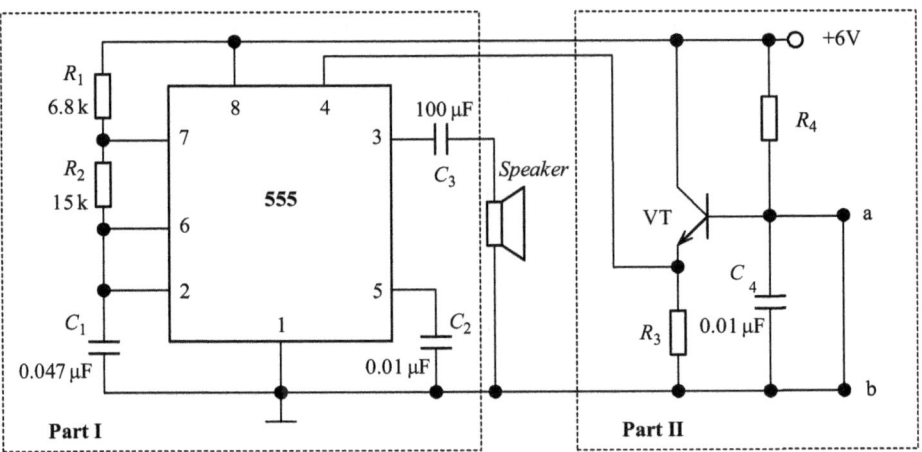

Figure P5.17

6 Sequential logic circuits

6.1 Introduction

A sequential logic circuit is a kind of digital circuits, in which the outputs depend on not only the current inputs but also the previous inputs and outputs. That is, a sequential logic circuit has state (memory) while a combinational logic circuit does not. A sequential logic circuit is a finite state machine that has a finite state. Usually, memory consists of a group of flip-flops. One flip-flop can store only one bit at a time. The state of a sequential circuit is represented by the bits stored in the flip-flops. This chapter mainly introduces analysis and design of a sequential logic circuit.

The objectives of this chapter are to
- Distinguish between sequential logic circuits and combinational logic circuits in circuit configuration and logic functions
- Explain the basic concepts of sequential logic circuit
- Analyze the logic function of any given sequential logic circuit
- Design sequential logic circuit according to the given specification
- Explain how to simplify the states of sequential circuit

6.1.1 Structure of sequential logic circuit

Figure 6.1.1 shows a block diagram of a sequential logic circuit [21]. A sequential logic circuit consists of a combinational logic and memory element. A memory element usually includes a group of flip-flops. The state of sequential logic circuit can be represented by the output states of the flip-flops. $X(x_1, x_2, ..., x_j)$ and $P(p_1, p_2, ..., p_j)$ are part of inputs and outputs of the combinational circuit, respectively. They are usually used as the interface signals with the external. $Q(q_1, q_2, ..., q_m)$ are part of inputs of the combinational circuit, which are the outputs of the flip-flops in the memory element and are used to represent the state of sequential logic circuit. $W(w_1, w_2, ..., w_k)$ are part of outputs of combinational circuit as the inputs of flip-flops in the memory element. The logic relationship between the inputs and outputs can be expressed by the following equations:

$$P(t_n) = F[X(t_n), Q(t_n)] \tag{6.1.1}$$

$$W(t_n) = G[X(t_n), Q(t_n)] \tag{6.1.2}$$

$$Q(t_{n+1}) = H[X(t_n), Q(t_n)] \tag{6.1.3}$$

where t_n and t_{n+1} represent two adjacent discrete time, and $Q(t_n)$ and $Q(t_{n+1})$ represent current state and next state of the flip-flops, which are often written as Q_n and Q_{n+1}. Equation (6.1.1) is called the *output equation*, which describes the relationship of the

https://doi.org/10.1515/9783110614916-006

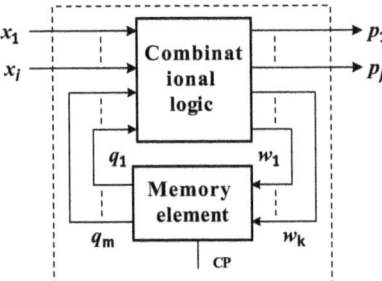

Figure 6.1.1: A structure of sequential circuits.

outputs of combinational logic with the inputs and the current states of the flip-flops. Equation (6.1.2) is called the *driving equation* or *exciting equation*, which is the logic expression for the flip-flop inputs. Equation (6.1.3) is called the *state equation* or *transition equation*, which describes the relationship of next states of the flip-flops with the inputs and current states of the flip-flops.

6.1.2 Classification of sequential circuits

Generally, a memory element consists of a group of flip-flops. According to the mode that the flip-flops are clocked, sequential circuits can be divided into *synchronous* sequential circuits (SSCs) and *asynchronous* sequential circuits. A SSC is a digital circuit in which the changes in the state of all flip-flops are synchronized by a global clock signal. This global clock signal is applied to every flip-flops and all flip-flops change their states at the same type of clock signal, the rising edge or falling edge of a clock signal. So in an ideal synchronous circuit, every change in the logic levels of all flip-flops is simultaneous. An *asynchronous circuit* is a sequential logic circuit that is not governed by a clock signal. In asynchronous circuits, there is no global clock signal for all flip-flops, thus the state of the circuit changes as soon as the inputs change. However, asynchronous circuits are more difficult to design. This is because the resulting state of an asynchronous circuit can be sensitive to the relative arrival times of inputs at gates. If transitions on two inputs arrive at almost the same time, the circuit can go into the wrong state depending on slight differences in the propagation delays of the gates. Today, most digital devices use synchronous circuits. Therefore, analysis and design of synchronous circuits are mainly introduced in the following section.

A sequential logic circuit belongs to a finite state machine. Usually a *state machine* includes a finite state in sequence. There are two basic models for the state machine. One is called *Moore model*, where the outputs only depend on current state of the sequential circuit. Another is called *Mealy model*, where the outputs depend on both current state and the present values of the combinational inputs.

6.1.3 SSC models

A sequential circuit is said to be a SSC if it satisfies the following conditions:
-   There is at least one flip-flop in every loop
-   All flip-flops have the same type of dynamic clock
-   All clock inputs of all the flip-flops are driven by the same clock signal

Any SSC can be described using structural models and behavioral models. The structural models include logic diagram, excitation equations, output equations, and the behavioral models contain state equations, output equations, state table, and state diagram. These two models can also be converted from one form to another. To analyze a SSC, you can derive one of the behavioral models from the structural models. On the contrary, you can also derive a structural model from the behavioral models to design a SSC.

6.2 Analysis of sequential logic circuits

Analysis of a sequential circuit is to derive the behavior of a given sequential circuit by looking for the rule of state transition and output change under the action of clock signal and input signals. To achieve this objective, you can derive the behavioral models from the structural models. This section introduces analysis procedure of a sequential logic circuit. *After completing this section, you should be able to analyze the behavior of any given sequential logic circuit.*

6.2.1 Analysis of synchronous sequential circuits

For SSC, all flip-flops are clocked by a global clock signal, so the analysis of SSC is simpler than that of asynchronous sequential circuit. If the logic diagram of a sequential circuit is given, the basic analysis procedure for a given SSC can be summarized as follows:

Step 1 Derive the exciting equations and output equations directly from the logic diagram of a given SSC.

Excitation equations are logic expressions for the flip-flop inputs and output equations are logic expressions for the outputs of the SSC.

Step 2 Derive the state equation by substituting the excitation equations into the characteristic equation of each flip-flop.

For a SSC, the characteristic equations of the flip-flops are the state equations of the SSC since all flip-flops are clocked by the same clock pulse (CP). In fact, the state equations of the SSC are logic expressions of next states of all flip-flop.

Step 3 Construct the state transition table from the state equations and output equations of the SSC.

The state transition table, state transition diagram, and timing diagram are three ways to express the state transition of the sequential circuit. The *state transition table* is a table showing what state a sequential circuit will move to, based on the current state and other inputs. A *state table* is essentially a truth table in which some of the inputs are the current state, and the outputs include the next state, along with other outputs. A state transition diagram is a graphical representation of state transitions, outputs, and inputs under the action of clock signal for a given sequential circuit. A *timing diagram* is a waveform representation of a set of signals in the time domain. A timing diagram can contain many signal waveforms, usually one of them being the clock. The state diagram and timing diagram can more visually display the function of sequential logic circuits.

Step 4 Draw the state diagram or timing diagram from state stable and deduce the implementing functionality of the given sequential logic circuits.

Note that not all steps in the aforementioned analysis procedure must be performed. You can determine which steps are necessary for a given sequential logic circuit. For instance, some sequential circuits have no output signals and the output equations can be omitted during the analysis process.

Let us take a look at the detailed analysis procedure through an example.

Example 6.1 Analyze the implementing functionality of a SSC, as shown in Figure 6.2.1.

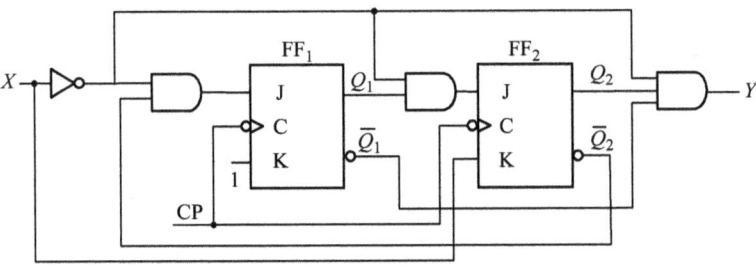

Figure 6.2.1: A logic diagram of Example 6.1.

Solution

Step 1 Derive the exciting equations and output equations directly from logic diagram in Figure 6.2.1.

There are two J-K flip-flops in the SSC. According to logic diagram, the exciting equations of the two J-K flip-flops are

$$J_1 = \overline{X} \cdot \overline{Q}_{2n}, \quad K = 1 \qquad (6.2.1)$$

$$J_2 = \bar{X}Q_{1n}, \quad K = X \tag{6.2.2}$$

The output equation can be expressed as

$$Y = \bar{X}Q_{2n}\bar{Q}_{1n} \tag{6.2.3}$$

Step 2 Derive the state equation of each flip-flop by substituting the exciting equations into the characteristic equation of each flip-flop.

The characteristic equation of J-K flip-flops is

$$Q_{n+1} = J\bar{Q}_n + \bar{K}Q_n \tag{6.2.4}$$

By substituting the excitation eqs. (6.2.1) and (6.2.2) into the characteristic eq. (6.2.4), respectively, two resulting state equations can be derived as follows:

$$Q_{1n+1} = J_1\bar{Q}_{1n} + \bar{K}_1Q_{1n} = \bar{X} \cdot \bar{Q}_{2n}\bar{Q}_{1n} \tag{6.2.5}$$

$$Q_{2n+1} = J_2\bar{Q}_{2n} + \bar{K}_2Q_{2n} = \bar{X}Q_{1n}\bar{Q}_{2n} + \bar{X}Q_{2n} = \bar{X}(Q_{1n} + Q_{2n}) \tag{6.2.6}$$

Note that these eqs. (6.2.5) and (6.2.6), which express the next state as a function of present states and inputs, are also called *transition equations*.

Step 3 Construct the state table from the state equations and output equation.

A state table is essentially a truth table in which some of the inputs are current states, and the outputs include next state, along with other outputs. For each combination of the level of input X and present state Q_{2n}, Q_{1n}, the corresponding next state Q_{2n+1}, Q_{1n+1} can be calculated by using eqs. (6.2.5) and (6.2.6), and the corresponding output Y can be also obtained from output eq. (6.2.3). The resulting value of next state and the output corresponding each input combination are filled in the corresponding row in the state table as shown in Table 6.2.1.

Table 6.2.1: A state table.

Input	Present state		Next state		Output
X	Q_{2n}	Q_{1n}	Q_{2n+1}	Q_{1n+1}	Y
0	0	0	0	1	0
0	0	1	1	0	0
0	1	0	1	0	1
0	1	1	1	0	0
1	0	0	0	0	0
1	0	1	0	0	0
1	1	0	0	0	0
1	1	1	0	0	0

Alternatively, we can simply use Q and Q^+ for representing present state and next state, respectively, so the state table can be simplified as Table 6.2.2.

Step 4 Draw a state diagram from the state table and deduce the implementing functionality of the given sequential logic circuits.

A *state diagram* is another way to illustrate the behavior of a SSC. Each state is denoted by a circle in the state diagram. The number of states in a state table or a state diagram will equal 2^m, where m is the number of flip-flops. Each combination of the flip-flop values in a circle represents a state.

Table 6.2.2: A simplified state table.

Present state		Next state				Output	
		$X=0$		$X=1$		$X=0$	$X=1$
Q_2	Q_1	Q_2^+	Q_1^+	Q_2^+	Q_1^+	Y	Y
0	0	0	1	0	0	0	0
0	1	1	0	0	0	0	0
1	0	1	0	0	0	1	0
1	1	1	0	0	0	0	0

Each arrow stands for a state transition of the sequential circuit, corresponding to a row in the state table. The number of arrows will equal to $2^m \times 2^k$, where k is the number of binary input signals. A label of the form "X/Y" is attached to each arrow corresponding to this transition. Each arrow is labeled with the "X/Y," where X denotes the input while Y denotes the output of combinational circuit, which cause the transition from present state (the source of the arrow) to next state (the destination of the arrow). In case of Example 6.1, there are two J-K flip-flops in the SSC, so the number of possible states is four. That is, Q_2Q_1 can be equal to 00, 01, 10, or 11. Each state is denoted by a circle with the value of Q_2Q_1 inside, as shown in Figure 6.2.2. There are total eight arrows since the given sequential circuit only has one input, X. Each arrow between two circles denotes a transition of the sequential circuit, corresponding to a row in the state table. For example, the first row in the state table represents one transition from present state 00 (Q_2 Q_1) to next state 01 (Q_2 Q_1) and the corresponding output Y is a 0 when X is a 0. In the state diagram, one arrow attached to 0/0 starts from state 00 and end at state 01. Similarly, you can draw the other transition in the state diagram.

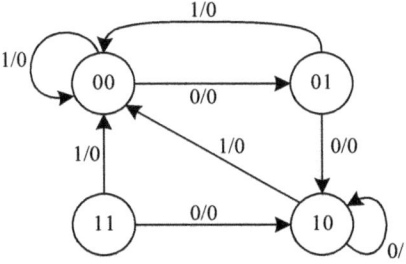

Figure 6.2.2: A state diagram.

Following these transition arrows, we can see that as long as $X = 0$, the sequential circuit start from the initial state 00 and goes through the states in the following sequence: 00, 01, 10, 10, ... After three CP, the output, Y, will be a 1. Otherwise, the output, Y, will be a 0. This means that only the Y output is a HIGH when three or more consecutive zeros are applied to the X input in sequence. Thus, the implementing functionality of the sequential logic circuits is a sequence detector that can detect three or more consecutive zeros input sequence.

Sometimes, a *timing diagram* is required to replace the state diagram.

Example 6.2 Analyze the implementing functionality of an SSC, as shown in Figure 6.2.3.

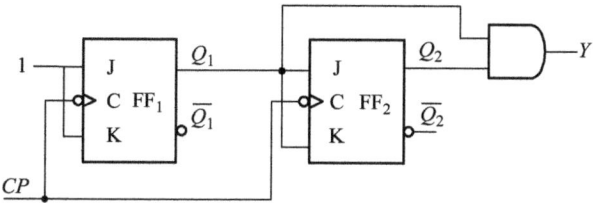

Figure 6.2.3: A sequential circuit for Example 6.2.

Solution

Step 1 Derive the exciting equations and output equations directly from the logic diagram shown in Figure 6.2.3.

There are two J-K flip-flops in the circuit; the exciting equations are

$$J_1 = K_1 = 1 \tag{6.2.7}$$

$$J_2 = K_2 = Q_1 \tag{6.2.8}$$

The output equation is

$$Y = Q_1 Q_2 \tag{6.2.9}$$

Step 2 Derive the state equation of each flip-flop by substituting the exciting equations into the characteristic equation of each flip-flop.

The characteristic equation of J-K flip-flops is

$$Q_{n+1} = J\bar{Q}_n + \bar{K}Q_n \tag{6.2.10}$$

Thus, the state equations can be expressed as

$$Q_{1n}^+ = \bar{Q}_1 \tag{6.2.11}$$

$$Q_{2n}^+ = Q_1\bar{Q}_2 + \bar{Q}_1 Q_2 = Q_1 \oplus Q_2 \tag{6.2.12}$$

Step 3 Construct a state table from the state equations and output equation.

The state table is shown as Table 6.2.3

Table 6.2.3: A state table.

Present state		Next state		Output
Q_2	Q_1	Q_2^+	Q_1^+	Y
0	0	0	1	0
0	1	1	0	0
1	0	1	1	0
1	1	0	0	1

Step 4 Draw a state diagram from the state table and deduce the implementing functionality of the given sequential logic circuits.

The state diagram is shown in Figure 6.2.4. It can be found that the sequential circuit goes through the states in the following sequence: 00, 01,10, 11,00, 01, ... When it counts to 11, the output is a HIGH. When the next clock arrives, the circuit goes back to the initial state 00, and start new count cycle again. Since this sequence is characteristic of modulus-4 counting, we can conclude that the sequential circuit in Figure 6.2.3 is a modulus-4 counter and belongs to the Moore model due to no input signal.

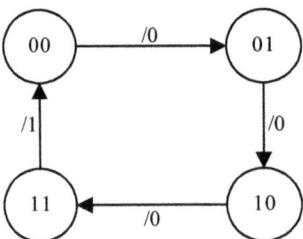

Figure 6.2.4: A state diagram.

6.2.2 Analysis of asynchronous sequential logic circuit

The main difference between asynchronous sequential logic circuits and synchronous sequential logic circuits is that all flip-flops do not share one clock signal, so the flip-flops are not triggered simultaneously. In the analysis of asynchronous sequential circuit, the most important step is to write the clock equation of each flip-flop. Only when the clock signal is active, the next state of the flip-flop can be obtained from the corresponding state equation; otherwise, the state of flip-flop will remain unchanged. The analysis procedure starts from the first flip-flop triggered by the clock signal, and then step-by-step analyzing the following flip-flops. Since all flip-flops do not share a clock signal, the state transition of asynchronous circuit has a certain delay time. The analysis procedure of an asynchronous circuit that is composed of flip-flops and logic gates is the same as that of synchronous circuit.

Let us take an example to help you understand analysis procedure of an asynchronous sequential logic circuit.

Example 6.3 Analyze the implementing functionality of an asynchronous sequential circuit, as shown in Figure 6.2.5.

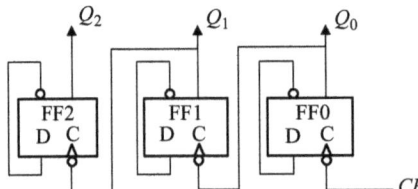

CP **Figure 6.2.5:** A logic diagram of Example 6.3.

Solution

Step 1 Derive the clock equations and exciting equations directly from the logic diagram shown in Figure 6.2.5.

The clock equations are

$$CP_0 = CP \quad CP_1 = Q_{0n} \quad CP_2 = Q_{1n} \tag{6.2.13}$$

The exciting equations for D flip-flops are

$$D_0 = \bar{Q}_{0n} \quad D_1 = \bar{Q}_{1n} \quad D_2 = \bar{Q}_{2n} \tag{6.2.14}$$

Step 2 Derive the state equation of each flip-flop by substituting the exciting equations into the characteristic equation of each flip-flop.

The characteristic equation of D flip-flops is

$$Q_{n+1} = D \tag{6.2.15}$$

The state equations can be derived as

$$Q_{0n+1} = \bar{Q}_{0n} \quad Q_{1n+1} = \bar{Q}_{1n} \quad Q_{2n+1} = \bar{Q}_{2n} \tag{6.2.16}$$

Step 3 Construct the state table from the state equations and output equation.

The possible state transition of D flip-flop occurs at the arriving of the negative edge of the clock signal. Thus, we use $CP = 1$ to represent the negative edge of the clock signal. For D flip-flops, the next state can be solved by eq. (6.2.15) when $CP = 1$. While the next state is the same as the present state when $CP = 0$. Assume that the initial state is $Q_2Q_1Q_0 = 000$; the next state can be derived by the state equations and clock equations. When the first clock signal arrives, $CP_0 = 1$ and thus the next state of FF0 can be calculated by eq. (6.2.16). Since Q_{0n} is a 0, Q_{0n+1} is a 1. Then we turn to determine the next state of FF1. Since Q_0 is changed from LOW to HIGH, $CP_1 = 0$ and thus Q_1 does not change. The next state of FF1 is a 0. When the Q_1 keeps no change, $CP_2 = 0$ and thus Q_2 also keeps no change. So when the first clock signal arrives, the next state of an asynchronous sequential circuit will be $Q_2Q_1Q_0 = 001$. Similarly, you can analyze the next state of an asynchronous sequential circuit when the sequential clock signal arrives. The final state table is shown in Table 6.2.4. In the state table of an asynchronous sequential circuit, the column of clock signal is added to represent if the flip-flop is clocked by negative edge of the CP. If the flip-flop is clocked, $CP = 1$.

Table 6.2.4: A state table.

CP	Q_2	Q_1	Q_0	CP_2	CP_1	CP_0
0	0	0	0	0	0	0
1	0	0	1	0	0	1
2	0	1	0	0	1	1
3	0	1	1	0	0	1
4	1	0	0	1	1	1
5	1	0	1	0	0	1
6	1	1	0	0	1	1
7	1	1	1	0	0	1
8	0	0	0	1	1	1

Step 4 Draw a timing diagram from the state table and deduce the implementing functionality of the given sequential logic circuits.

All *D* flip-flops are triggered at the negative edge of their CP. According to the stable table, a timing diagram is shown in Figure 6.2.6. It can be seen from the timing diagram, the asynchronous sequential circuit in Figure 6.2.5 goes through the states in the following sequence: 000, 001, 010, 011, 100, 101, 110, 111, 000, ... After eight CPs, the state of asynchronous sequential circuit goes back to 000 and then starts a new count cycle again. Therefore, the asynchronous sequential circuit shown in Figure 6.2.6 works as an asynchronous three-bit binary up counter.

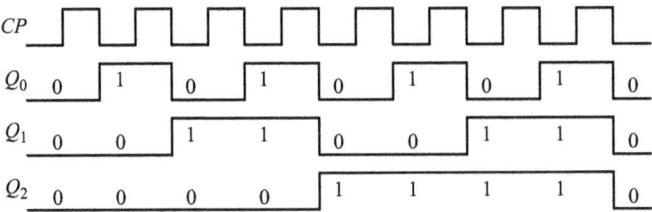

Figure 6.2.6: A timing diagram.

6.3 Design of a sequential logic circuit

The design of sequential logic circuits is also referred to as the synthesis of sequential logic circuits. The requirements of a logic circuit are generally described in words or specification. The design procedure can be regarded a converse procedure of analysis. This section introduces the basic procedure for designing a sequential logic circuit with flip-flops and logic gates according to the word description or specification. *After completing this section, you should be able to design a synchronous sequential logic circuit according to the word description or specification.*

6.3.1 Design of an SSC

The main objective for designing a sequential circuit with flip-flops and logic gates is to reduce the cost, which means that the number of flip-flops and logic gates are used as few as possible for the implementation of a sequential circuit. The basic procedure is as follows.

Step 1 Construct an initial state diagram and initial state table according to the word description or specification.

The state diagram and state table are the basis of designing the sequential logic circuit. The state diagram and state table are a graph and table description that specify the next state and output of the circuit for every possible combination of current state and input, respectively. It is the most important step to derive the correct state diagram and state table according to the word description or specification. By analyzing the given logic requirement of the sequential logic, input variables, output variables, and the required number of states are first determined. States can be denoted by the letters or numbers in the initial state diagram and initial state table. Then determine the outputs and next states for each combination of input and current states and derive the initial state diagram. Finally, convert the initial state diagram to the initial state table.

Step 2 Minimize the number of states in the state table by state reduction (optional).

The initial state table may contain some unnecessary states, which makes the following design more complex. So the initial state diagram or state table should be minimized by state reduction. State reduction is to find equivalent states and eliminate unnecessary states. For each possible combination of input values, if two or more current states have the same next state and the same output, then these states are called the *equivalent states*. A pair of equivalent states can be replaced by a single state. The minimized state table can be obtained by eliminating equivalent states.

Step 3 Encoded states and determine the required number of flip-flops to represent states.

Each state in the minimized state table is assigned to a unique binary code. This process is called *coded state* or *state assignment*. Generally, there are several ways to encode states. Different coded state method results in different complexity of the resulting circuit. Hence, a good coded state method can make synthesis of sequential logic circuits simpler.

A flip-flop can store one-bit binary code, 0 or 1, which can represent two states. So 2^n binary codes can be produced to represent at most 2^n state with n flip-flops. If there are M states in the minimized state table, M binary codes is necessary for encoding M states. The required number of flip-flops, n, at least can be deduced as follows.

$$2^{n-1} < M \leq 2^n$$

The objective to choose a coded state method should be helpful for obtaining the minimized excitation equation of flip-flops. Usually, a natural binary code is selected to encode states and the adjacent binary code is assigned to the adjacent states.

Step 4 Choose the type of flip-flops and derive the excitation equations and output equations.

Different types of flip-flops would lead to different complexity of the resulting synthesis of sequential logic circuits. The aim is to use the number of flip-flops and logic gates as few as possible for implementing a sequential circuit. According to the encoding state table, an excitation table can be deduced according to the function of flip-flop used. The *excitation table* shows the flip-flop excitation input values needed to make the circuit go to the desired next coded state for each combination of coded state and input. Then an excitation equation can be obtained by applying Karnaugh map or Boolean algebra simplification. Alternatively, the output equations and state equations can be directly derived from the coded state table. Then the excitation equations can be derived by comparing the state equations with the characteristic equations of the flip-flops.

Step 5 Check if the synthesis of sequential logic circuits can be self-corrected. In the state assignment, it is a common thing that the code combinations provided by the flip-flops are great than the required number of states. There are some code combinations unused corresponding to unused states. If the circuit is in these unused states, it cannot automatically return to any used states and thus the circuit cannot work normally. That is say, the circuit cannot be self-corrected. If the designing circuit cannot be self-corrected, there are two effective methods to solve this problem. One is to force the circuit entering into one of used states at the beginning of circuit operation. Another is to modify the design and arrange an used state as the next state of the unused state. A new state table can be constructed. Repeat step 4 and new exciting equations can be deduced. As a result, the circuit can be self-corrected.

Step 6 Draw a logic diagram according to the resulting exciting equations and output equations.

Example 6.4 Design a three-bit sequence detector for detecting a sequence input series "011."

Solution

Step 1 Construct the initial state diagram and initial state table.

Sequence detection is a act of recognizing a predefined series of input bits. The bits are input one by one. According to the requirement of detecting the predefined series of inputs "011," we need to see whether they match the given sequences "011" for each three bits that are input. Thus, one sequential input variable, X, and one output variable, Y, are needed. X is the value of the input line in a given clock cycle. The output, Y, goes high for one clock cycle as soon as it receives the third bit that matches a pattern. There are at least four states required to construct the initial state diagram. State A is a waiting state in which the state machine waits for the first input bit "0"; state B represents the case detecting the first bit "0"; state C means the case detecting the continuous two bits "01"; and state D represents the case detecting the continuous three bits "011." The output, Y, goes high for one clock cycle as soon as it receives the third bit that matches a pattern.

When the initial state is state A, the state transitions will depend on whether the input X is a 0 or 1. If the input X is a 1, the next state is still in state A; if the input X is a 0, the state transition will be from state A to state B. This means the first bit "0" occurs. The output Y is a 0.

If in state B and the input X is a 0, the next state does not change with an output of a 0; if the input X is a 1, the continuous two bits "01" are detected and the state will change from state B to state C with the output of a 0.

If in state C and the input X is a 0, the state will turn back state B with an output of a 0; if the input X is 1, the continuous three bits "011" are detected and the state will change from state C to state D with the output of a 1.

If in state D and the input X is a 0, the state will turn back state B and start new sequence detector; if the input X is 1, the state will turn to state A waiting for the first bit "0" with the output of a 0.

According the aforementioned analysis, the initial state diagram for detecting a sequence input series "011" is shown in Figure 6.3.1. The initial state table can be deduced from the initial state diagram, as listed in Table 6.3.1.

Step 2 Minimize the number of states in the state table by state reduction.

The initial state table may contain some unnecessary states, which makes the following design more complex. So initial state diagram or state table should be minimized by state reduction. It can

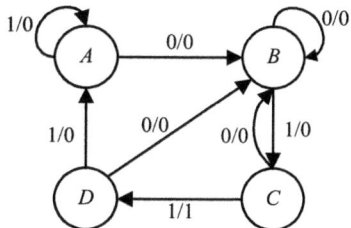

Figure 6.3.1: The initial state diagram.

Table 6.3.1: The state table.

Current State	Next state		Output	
	$X = 0$	$X = 1$	$Y(X = 0)$	$Y(X = 1)$
A	B	A	0	0
B	B	C	0	0
C	B	D	0	1
D	B	A	0	0

be seen from Table 6.3.1 that state A and state D have the same next state and the consistent output for each input value. Thus, these two states can be called as equivalent states. If two or more states are equivalent, they can be represented by one state. That is, state D can be taken place by state A. Note that state A and state B lead to the same output, but they do not lead to the same next state for each input value, and so state A and state B are not equivalent. The minimized state table is shown in Table 6.3.2.

Table 6.3.2: The minimized state table.

Present State	Next state		Output	
	$X = 0$	$X = 1$	$Y(X = 0)$	$Y(X = 1)$
A	B	A	0	0
B	B	C	0	0
C	B	A	0	1

Step 3 Encode states and determine the required number of flip-flops to represent states.

To perform a state assignment, each state should be assigned a unique binary code. Since there are three states, A, B, and C, there are at least two-bit binary codes required. There are several schemes for coded three states with two-bit binary codes. Here, natural binary codes are chosen for coded states A, B, and C. States A, B, and C can be assigned 00, 01, and 10, respectively. So the coded state table can be deduced from Table 6.3.2 to Table 6.3.3. The state of flip-flops is represented by Q_1Q_0. Since one flip-flop can store one-bit binary code, two flip-flops are required to store two-bit binary codes.

Table 6.3.3: The binary coded state table.

	Next state		Output	
Present State	$X = 0$	$X = 1$	$X = 0$	$X = 1$
$Q_1 Q_0$	$Q_1{}^+ Q_0{}^+$	$Q_1{}^+ Q_0{}^+$	Y	Y
00	01	00	0	0
01	01	10	0	0
10	01	00	0	1

Step 4 Choose the type of flip-flops and derive the excitation equations and output equation.

Different types of flip-flops would lead to different complexities of the resulting synthesis of sequential logic circuits. Here, two T flip-flops are chosen for constructing the sequential circuit. In order to derive the excitation equations and output equation, a new style state/output table should be constructed by using the excitable table of T flip-flop and binary coded state stable as listed in Table 6.3.3. The excitation table illustrates the excitation input values of the corresponding flip-flop needed to make the circuit go to the desired next coded state for each combination of coded state and input. Excitation table of T flip-flops can be deduced according to the function of T flip-flop, as shown in Table 6.3.4.

Table 6.3.4: The excitable table of T flip-flop.

Current state	Next state	Input
Q	Q^+	T
0	0	0
0	1	1
1	0	1
1	1	0

A state transition table is shown in Table 6.3.5, which can be deduced by combining Tables 6.3.3 and 6.3.4.

Table 6.3.5: The state transition table.

Input	Present state		Next state		Flip-flop Inputs		Output
X	Q_1	Q_0	$Q_1{}^+$	$Q_0{}^+$	T_1	T_0	Y
0	0	0	0	1	0	1	0
0	0	1	0	1	0	0	0
0	1	0	0	1	1	1	0
1	0	0	0	0	0	0	0
1	0	1	1	0	1	1	0
1	1	0	0	0	1	0	1
0	1	1	0	1	1	0	0
1	1	1	0	0	1	1	0

According to Table 6.3.5, the Karnaugh maps of T_1 and T_0 are shown in Figure 6.3.2. State 11 is the used state and the corresponding two cells are treated as "don't care" terms.

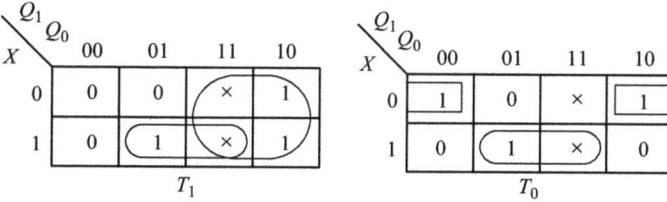

Figure 6.3.2: The Karnaugh map.

The minimized excitation equations can be obtained by simplifying the Karnaugh map:

$$T_1 = Q_1 + XQ_0 \tag{6.3.1}$$

$$T_0 = \bar{X} \cdot \bar{Q}_0 + XQ_0 \tag{6.3.2}$$

Since only one combination of current state and input make the output a 1, the output equation can be directly written as

$$Y = XQ_1\bar{Q}_0 \tag{6.3.3}$$

Step 5 Check if the synthesis of sequential logic circuits can be self-corrected.

In the state assignment, state 11 is unused corresponding to the unused state. If the circuit is in this unused state, it cannot automatically return to one of the used states and thus the circuit cannot be self-corrected. According to the aforementioned simplification process, the value of input T_1 corresponding to two don't care terms, 011 and 111, are both treated as 1; the value of input T_0 are treated as a 0 for the combination of 011 and a 1 for the combination of 111. After state 11, the next state will be state 01 and state 00, which are listed in the last two rows in Table 6.3.5. Both state 00 and state 01 are the used states, so the design circuit can be self-corrected.

Step 6 Draw a logic diagram according to the resulting exciting equations and output equation.

According the excitation eqs. (6.3.1) and (6.3.2), and the output eq. (6.3.3), the logic diagram of a 011 sequence detector is drawn with two T flip-flops and logic gates, as shown in Figure 6.3.3.

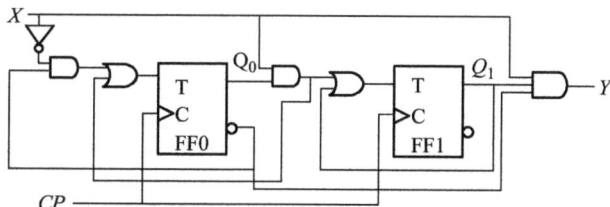

Figure 6.3.3: The logic diagram for detecting a sequence input series "011."

Example 6.5 Design a synchronous counter with J-K flip-flops. The timing diagram for the counter is shown as Figure 6.3.4.

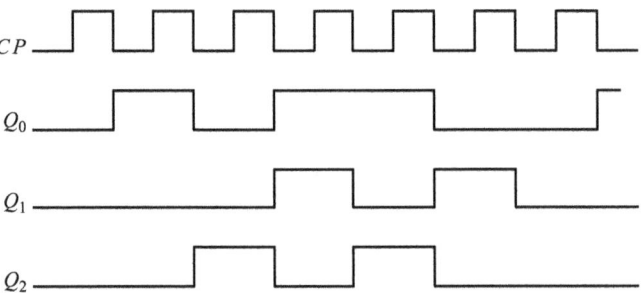

Figure 6.3.4: The timing diagram for Example 6.5.

Solution

Step 1 Construct the state diagram and state table directly from time diagram in Figure 6.3.4. Then determine the number of flip-flops.

The state diagram can be deduced by analyzing the timing diagram in Figure 6.3.4. The initial state is 000. When the falling edge of the first CP arrives, the state transition occurs from state 000 to state 001. Then the state will change from state 001 to state 100 at the falling edge of the second CP. After six CPs, the counter returns to the initial state. Figure 6.3.5 shows the state diagram of the synchronous counter. The state table can be obtained from the state diagram in Table 6.3.6. It can be seen from Table 6.3.6 that the synchronous counter has six used states and thus at least three J-K flip-flops are required.

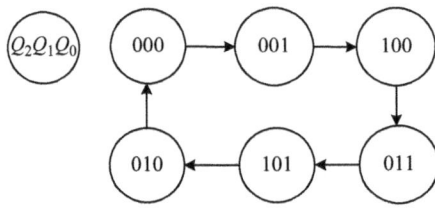

Figure 6.3.5: The state diagram.

Step 2 Derive the excitation equations.

Here, we choose another method to deduce the excitation equations of each flip-flop. First, the state equations can be derived from the state table by Karnaugh map simplification. Karnaugh maps of next state can be obtained from state table, as shown in Figure 6.3.6. Next states, $Q_{2n+1}Q_{1n+1}Q_{0n+1}$, are listed in the cell of Karnaugh map.

In order to obtain the minimized next state equations, the Karnaugh map in Figure 6.3.6 are divided into three Karnaugh maps, as shown in Figure 6.3.7.

By using Karnaugh map simplification in Figure 6.3.7, the minimized state equations can be deduced as follows:

Table 6.3.6: The state table.

Current state			Next state		
Q_{2n}	Q_{1n}	Q_{0n}	Q_{2n+1}	Q_{1n+1}	Q_{0n+1}
0	0	0	0	0	1
0	0	1	1	0	0
0	1	0	0	0	0
0	1	1	1	0	1
1	0	0	0	1	1
1	0	1	0	1	0

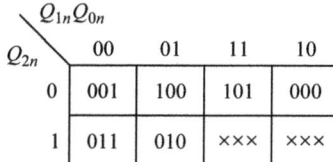

Figure 6.3.6: The Karnaugh map of the next state $Q_{2n+1}Q_{1n+1}Q_{0n+1}$.

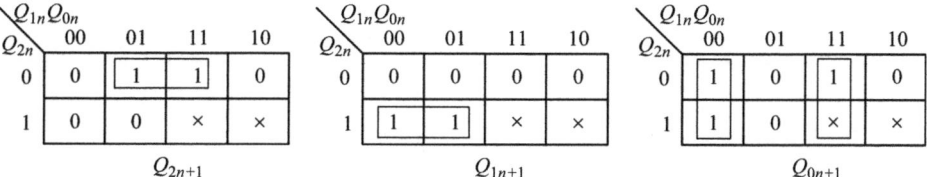

Figure 6.3.7: The Karnaugh map.

$$\begin{cases} Q_{2n+1} = Q_{0n}\bar{Q}_{2n} \\ Q_{1n+1} = Q_{2n}\bar{Q}_{1n} \\ Q_{0n+1} = \bar{Q}_{1n}\bar{Q}_{0n} + Q_{1n}Q_{0n} \end{cases} \tag{6.3.4}$$

Compare each state equation with the characteristic equation of J-K flip-flop, $Q_{n+1} = J\bar{Q}_n + \bar{K}Q_n$; the excitation equations for each J-K flip-flop are deduced as follows:

$$\begin{cases} J_2 = Q_{0n} \\ K_2 = 1 \end{cases}, \quad \begin{cases} J_1 = Q_{2n} \\ K_1 = 1 \end{cases}, \quad \begin{cases} J_0 = \bar{Q}_{1n} \\ K_0 = \bar{Q}_{1n} \end{cases} \tag{6.3.5}$$

Step 3 Check if the synthesis of sequential logic circuits can be self-corrected.

There are two unused states, 110 and 111. If in state 110, the next state is 000, which can be obtained by substituting the current state 110 into state eq. (6.3.4). Similarly, if in state 111, the next state is 001. The state diagram containing all states is shown in Figure 6.3.8. Since both state 000 and state 001 are the used states, the design resulting circuit can be self-corrected.

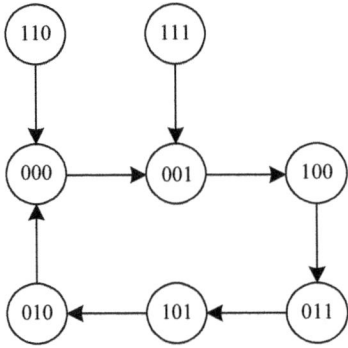

Figure 6.3.8: The state diagram.

Step 4 Draw a logic diagram according to the resulting exciting equations.

According the excitation eq. (6.3.5), the logic diagram of the synchronous counter is implemented with three J-K flip-flops, as shown in Figure 6.3.9.

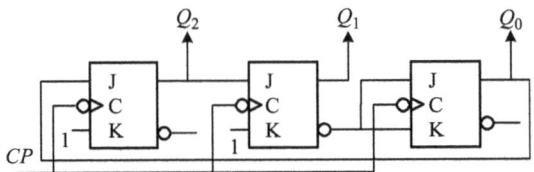

Figure 6.3.9: The logic diagram.

6.3.2 Design of an asynchronous sequential circuit

The design procedure of an asynchronous sequential circuit is similar to that of SSC. Only difference is that there is no global clock signal for all flip-flops in asynchronous circuit, so the CP for each flip-flop should be considered in the design procedure.

Example 6.6 Design an 8421BCD asynchronous up counter with a ripple carry output.

Solution

Step 1 Construct the state table.

Because an 8421BCD counter has ten states represented by ten fixed binary code, the coded state table can be listed directly, as shown in Table 6.3.7. Z is defined as the ripple carry output.

Step 2 Choose the type of flip-flops to be used. Derive the excitation equations and output equation.

The 8421BCD has four-bit binary codes and thus four flip-flops are required. Here, four J-K flip-flops are chosen to construct the 8421BCD asynchronous up counter.

Since there is no global clock signal for all flip-flops in asynchronous circuit, the CP for each flip-flop must be applied as long as the state of flip-flop produce a change .

According to Table 6.3.7, the output Q_1 of the first flip-flop changes state at each CP, so the input CP as its clock input. That is, $CP_1 = CP$ and $J_1 = K_1 = 1$.

The output Q_2 of the first flip-flop changes its state when Q_1 changes from HIGH to LOW, that is, at the negative going edge of Q_1. Thus, Q_1 is set as the input CP of the second flip-flop, $CP_2 = Q_1$.

Table 6.3.7: The state table.

CP	Q_4	Q_3	Q_2	Q_1	Z
0	0	0	0	0	0
1	0	0	0	1	0
2	0	0	1	0	0
3	0	0	1	1	0
4	0	1	0	0	0
5	0	1	0	1	0
6	0	1	1	0	0
7	0	1	1	1	0
8	1	0	0	0	0
9	1	0	0	1	1
10	0	0	0	0	0

Note that at the 10th CP, although Q_1 changes from HIGH to LOW, Q_2 does not change. Thus, J_2 and K_2 input for the second flip-flop should be controlled to make sure that Q_2 does not change at the 10th CP.

The output Q_3 changes its state when Q_2 changes from HIGH to LOW, so Q_2 is set as the input CP of the third flip-flop, $CP_3 = Q_2$ and $J_3 = K_3 = 1$.

The output Q_4 changes state when Q_1 changes from HIGH to LOW, so Q_1 is set as the input CP of the fourth flip-flop, $CP_4 = Q_1$. Note that the J_4 and K_4 input still need to be controlled since Q_4 does not change at its first three CP.

According to aforementioned analysis, Table 6.3.7 can be simplified as Table 6.3.8.

Table 6.3.8: The simplified state table.

CP	Q_4	Q_3	Q_2
0	0	0	0
2	0	0	1
4	0	1	0
6	0	1	1
8	1	0	0

The state transition table is shown as Table 6.3.9.

Table 6.3.9: The state transition table.

CP	Present states			Next states			Excitation inputs			
	Q_{4n}	Q_{3n}	Q_{2n}	Q_{4n+1}	Q_{3n+1}	Q_{2n+1}	J_4	K_4	J_2	K_2
0	0	0	0	0	0	1	0	×	1	×
2	0	0	1	0	1	0	0	×	×	1
4	0	1	0	0	1	1	0	×	1	×
6	0	1	1	1	0	0	1	×	×	1
8	1	0	0	0	0	0	×	1	0	×

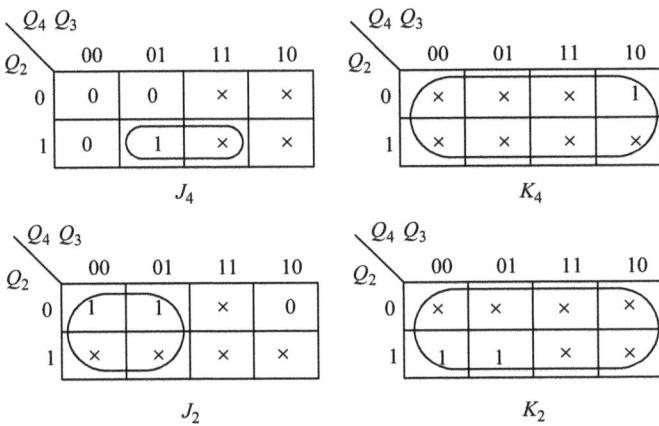

Figure 6.3.10: The Karnaugh map.

According to Table 6.3.9, the Karnaugh maps of the inputs of J-K flip-flops are shown in Figure 6.3.10. Excitation equations can be obtained by Karnaugh map simplification as follows:

$$\begin{cases} J_4 = Q_3 Q_2 \\ K_4 = 1 \end{cases}, \quad \begin{cases} J_2 = \bar{Q}_4 \\ K_2 = 1 \end{cases} \tag{6.3.6}$$

Only when $Q_4 Q_3 Q_2 Q_1 = 1001$, $Z = 1$, the output equation is

$$Z = Q_4 \bar{Q}_3 \bar{Q}_2 Q_1 \tag{6.3.7}$$

Step 3 Check if the 8421BCD asynchronous up counter can be self-corrected.

There are six unused states from 1010 and 1111. Assume that the circuit is in state 1010, when next CP is applied to the clock input of FF1, the next state of the output Q_1 changes from a 0 to a 1 since the values of J and K inputs are both 1. With Q_1 from LOW to HIGH, there is no falling edge of CP applied to the clock input of FF2 and thus the output Q_2 still keeps a 1. Since the output of Q_2 does not change, there is no effective CP applied to the clock input of FF3 and thus the output Q_3 also keeps a 0. With Q_1 from LOW to HIGH, there is no falling edge of CP applied to the clock input of FF4 and thus the output Q_4 does not change. Thus, after state 1010, the next state is state 1011.

Assume that the circuit is in state 1011; when next CP is applied to the clock input of FF1, the next state of the output Q_1 changes from a 1 to a 0 since the values of J and K inputs are both 1. With Q_1 from HIGH to LOW, there is a falling edge of CP applied to the clock input of FF2 and FF4. Therefore, the output Q_2 changes from a 1 to a 0 due to $J_2 = 0$ and $K_2 = 1$ and the output Q_4 changes from a 1 to a 0

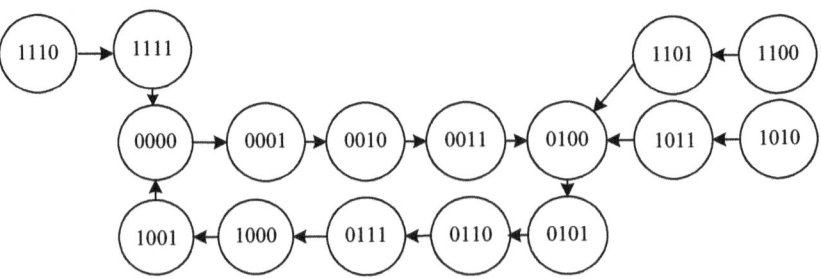

Figure 6.3.11: A state diagram.

since $J_4 = 0$ and $K_4 = 1$. With Q_2 from HIGH to LOW, there is a falling edge of CP applied to the clock input of FF3 and thus the output Q_3 changes from a 0 to a 1 since $J_3 = K_3 = 1$. It is clearly seen that the next state is state 0100 after state 1011. Similarly, you can check the other four unused states. The complete state diagram is shown in Figure 6.3.11. It can be seen from Figure 6.3.11 that the design circuit can be self-corrected.

Step 3 Draw a logic diagram according to excitation equations and output equation.

According the excitation eq. (6.3.6) and output eq. (6.3.7), the logic diagram of the 8421BCD asynchronous up counter is implemented with four J-K flip-flops, as shown in Figure 6.3.12.

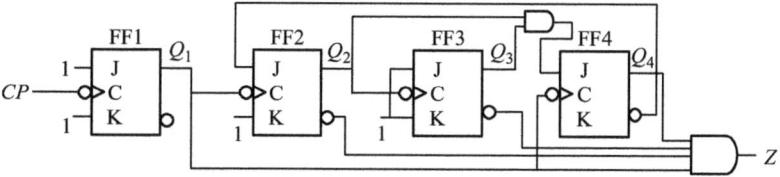

Figure 6.3.12: A logic diagram for Example 6.6.

6.4 State reduction

The complexity of designing a sequential circuit depends on the number of states in the state table. In order to minimize the cost of a design circuit, the states in the state table should be minimized as much as possible and thus the minimized state table can be obtained. In fact, state reduction is to detect the equivalent states. If two or more states are equivalent, they can be replaced by one state. Thus, the equivalent or redundant states can be eliminated from a state table/diagram and the minimized state table can be obtained. This chapter introduces two methods for state reduction: row matching and implication chart.

The objectives of this chapter are to
- Explain the conditions for equivalent states
- Apply row matching method to determine the equivalent states
- Apply implication chart method for state reduction

6.4.1 Row matching

Row matching is the easiest method of state reduction. This method has the advantage of ease use and is generally used to minimize simple state table.

The row matching method uses state equivalence theory. Two states are equivalent if and only if the outputs are consistent and next states are equivalent for each input combination. There are three situations for judging the equivalence of next states, as shown in Figure 6.4.1.

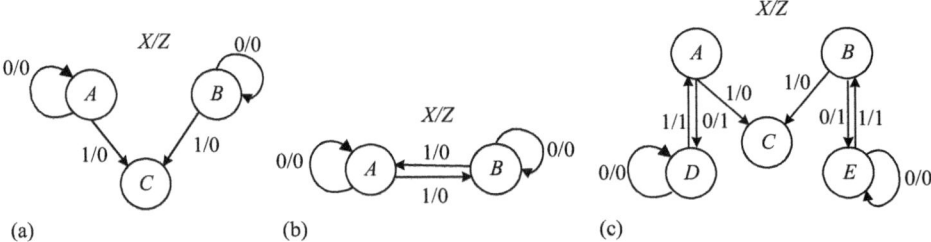

Figure 6.4.1: Illustration of three situations for the equivalence of next states.

The first situation of the equivalence of next states is that next states of two states are the same or the same as their current states for each input combination, as illustrated in Figure 6.4.1(a). For state A and state B, if input X is a 0, their outputs are both 0 and their next states are the same as their current states; if input X is a 1, their outputs are both 0 and their next states are both state C. Therefore, state A and state B are a pair of equivalent states and denoted by $[A, B]$.

The second situation of the equivalence of next states is that next states of two states have crossed each other for each input combination, as shown in Figure 6.4.1 (b). For state A and state B, if input X is a 0, their outputs are both 0, and their next states are the same as their current states; if input X is a 1, their outputs are both 0 and their next states are crossed each other. Therefore, state A and state B are a pair of equivalent states.

The third situation of the equivalence of next states is that next states of two states have implied each other for each input combination, as shown in Figure 6.4.1(c). For state A and state B, if input X is a 1, their outputs are both 0 and their next states are both state C; if input X is a 0, their outputs are both 1. However, next state of state A is state D and that of state B is state E. This means that the equivalence of state A and state B depends on the equivalence of state D and state E. Let us turn to check the equivalence of state D and state E. If input X is a 0, their outputs are both 0 and their next states are the same as their current states; if input X is a 1, their outputs are both 1. However, next state of state D is state A and that of state E is state B. Therefore, whether or not state D and state E is equivalent depends on the equivalence of state A and state B. All in all, the equivalence of state A and state B is determined by the equivalence of state D and state E, while the equivalence of state D and state E is determined by the equivalence of state A and state B. That is, next states have implied each other. For this situation, state A and state B are thought as a pair of equivalent states; state D and state E are also considered as a pair of equivalent states.

Generally, in order to check the equivalence of two states, the first step is to check if the outputs of two states are the same for each input combination. For each input combination, if their outputs are different, then these two states are not a pair of equivalent states; if they are the same, then go to next step. The next step is to check if next states are equivalent. If next states are also equivalent, then two states are equivalent.

According to state equivalence theory, row matching method must check the equivalence for all input combination. Let us take an example of state reduction in Table 6.4.1. It can be seen from Table 6.4.1 that there are totally seven states in the state table. The first step is to check if the outputs are the same for each input combination. For each input combination, the outputs of state A and state C are the same and the outputs of state B, D, E, F, G are the same. This means that state A and state C are possible equivalent and state B, D, E, F, G are possible equivalent. Hence, the next step is to check the equivalence of next states.

Table 6.4.1: The state table.

Present state	Next state		Output	
	$X = 0$	$X = 1$	$X = 0$	$X = 1$
A	B	C	0	0
B	E	C	1	0
C	D	A	0	0
D	E	A	1	0
E	E	C	1	0
F	G	E	1	0
G	F	E	1	0

For state B and state E, if input $X = 0$, their outputs are both 1 and their next state are both state E; if the input $X = 1$, their outputs are both 0 and their next state are both state C. This situation belongs to the first situation of the equivalence of next states. Therefore, state B and state E are a pair of equivalent states expressed as $[B, E]$.

For state F and state G, their outputs are both 0 and their next state are both E if the input $X = 1$; if the input $X = 0$, their outputs are both 1 and their next states are crossed each other. That is, irrespective of whether the circuit starts with state F or state G, it will transform between state F and state G until the input X becomes a 1. When the input $X = 1$, it will go to state E. This situation belongs to the third situation of the equivalence of next states. Therefore, state F and state G are a pair of equivalent states represented by $[F, G]$.

For state A and state C, their outputs are both 0 and their next states cross each other if the input $X = 1$; if $X = 0$, their outputs are both 0 and their next states are state B and state D, respectively. Therefore, whether or not state A and state C is equivalent depends on the equivalence of state B and state D. Let us turn to check state B and state D. For state B and state D, their outputs and next states are both the same when $X = 0$; when $X = 1$, their outputs are the same and their next states are state A and state C, respectively. This means that the equivalence of state B and state D depends on the equivalence of state A and state C, and vice versa. Therefore, state B and state D are a pair of equivalent states and state A and state C are also a pair of equivalent state, which can be expressed as $[A, C]$, $[B, D]$.

Since state B and state D are equivalent, and state B and state E are equivalent, then state D and state E can be regarded as equivalent states. Therefore, state B, state D, and state E are equivalent to each other, which is expressed as $[B, D, E]$.

From the above analysis, there are three pairs of equivalent states listed in Table 6.4.1, which are $[A, C]$, $[F, G]$, and $[B, D, E]$. The equivalent states can be replaced with one state, minimizing the state table. Therefore, state A and state C can be expressed as state A; state F and state G can be denoted by state F; and state B, state D, and state E can be expressed as state B. The minimized state table is shown as Table 6.4.2.

Table 6.4.2: The minimized state table.

Present state	Next state		Output	
	$X = 0$	$X = 1$	$X = 0$	$X = 1$
A	B	A	0	0
B	B	A	1	0
F	F	B	1	0

6.4.2 Implication chart

The row matching method is suitable for minimizing simple state table. It is more difficult to detect equivalent states for a complex state table since there is a large number of states. The *implication chart* uses a graphical cell to systematically detect if the states are equivalent, which is more practical to tackle with the complex states table. Let us take an example to understand the implication charts method.

Example 6.6 Use the implication chart to minimize the state table shown in Table 6.4.3.

Table 6.4.3: The state transition table.

Present state	Next state		Output	
	$X = 0$	$X = 1$	$X = 0$	$X = 1$
1	8	3	0	0
2	8	6	0	0
3	8	1	1	0
4	1	8	1	0
5	4	7	0	0
6	4	2	1	0
7	4	5	1	0
8	5	4	1	0

Solution

Step 1 Construct an implication chart.

The implication chart is a chart in the form of right triangle, as shown in Figure 6.4.2(a). It can be observed from Table 6.4.3 that there are eight states, so we can construct the implication chart as shown in Figure 6.4.2(a). Note that there is no first state in the vertical direction, starting from state 2 to state 8; there is no the final state in the horizontal direction, starting from state 1 to state 7. This organizing style can effectively avoid the repeat comparison. Each cell in the chart represents the pairs of next states corresponding to a pair of states determined by the vertical and horizontal states.

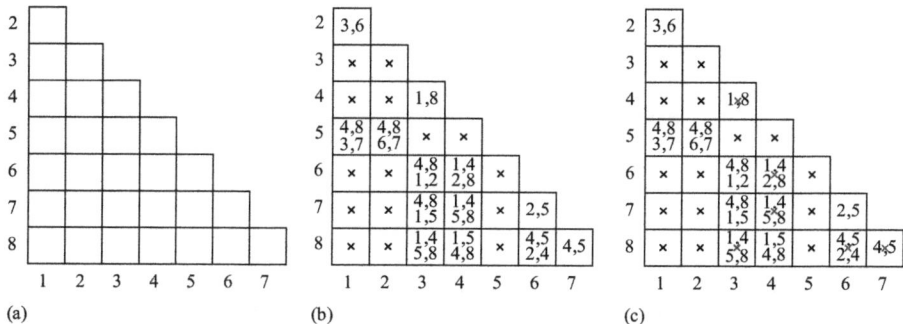

Figure 6.4.2: The implication chart.

Step 2 Fill the implication chart.

The chart is completed by comparing each pair of states in the state table to determine whether they are equivalent. The comparing result can be filled in the corresponding cell. Usually, there are three kinds of comparing results.

The first result is that two states can be directly determined to be a pair of equivalent states, entering a tick "√" in the cells corresponding to these state pairs. For the state table in Table 6.4.3, there is no pair of equivalent states that can be directly determined, so no tick occurs in the cells.

The second result is that two states are directly incompatible. If two states are not equivalent, enter a cross "×" in the cells. For example, state 1 and state 3 do not have the same output for each input value, so they are not equivalent and thus a cross "×" is filled into the corresponding cell.

The third result is that two states are possible equivalent. For each input combination, if two states have the same output, the equivalence of their next states must be determined by the equivalence of other state pairs. For this situation, the implied state pairs represented by (a,b) are listed in the corresponding cells. Let us check state 3 and state 8. For each input value, their outputs are the same. So whether or not these two states are equivalent depends on the equivalence of their next states. If $X = 0$, whether or not state 3 and state 8 are equivalent depends on the equivalence of state 5 and state 8; if $X = 1$, whether or not state 3 and state 8 are equivalent depends on the equivalence of state 1 and state 4. So state pairs (5,8) and (1,4) are the implied equivalence of state 3 and state 8, which is filled in the corresponding cell. Similarly, the implication chart can be completed as shown in Figure 6.4.2(b).

Step 3 Perform the relevant comparison.

In this step, the implied equivalent pairs in the cell must be checked to determine if they are equivalent. If they are equivalent, a tick "√" is filled in the cell; if they are not equivalent, a cross

"×" is filled in the cell. For example, the implied pairs (1,8) corresponding state 3 and state 4 are not equivalent because their outputs are not the same, so a cross "×" is filled in the corresponding cell. The relevant comparison ends until all implied pairs are checked in the cell. Finally, any cell not containing a cross "×" corresponds to an equivalent state pairs, as shown in Figure 6.4.2(c).

Step 4 Merge the equivalent states and minimize the state table.

As shown in Figure 6.4.2(c), each cell without a cross represents a pair of equivalent states. It can be obtained from Figure 6.4.2(c) that all pairs of equivalent states are [1,2], [1,5], [2,5], [3,6], [3,7], [4,8], and [6,7]. With [1,2] and [1,5], state 1, state 2, and state 5 are equivalent, which is represented by [1, 2, 5]; So does [3, 6, 7]. That is, the equivalent pairs are [1, 2, 5], [3, 6, 7], and [4,8]. The resulting minimized state table is obtained as in Table 6.4.4.

Table 6.4.4: The minimized state table.

Present state	Next state		Output	
	$X = 0$	$X = 1$	$X = 0$	$X = 1$
1	4	3	0	0
3	4	1	1	0
4	1	4	1	0

6.5 Summary

1. A sequential logic circuit is a kind of digital circuits, in which the outputs depend on not only the current inputs but also the previous inputs and outputs.
2. A sequential logic circuit is composed of combinational logic circuits and memory elements. Memory elements contain a series of flip-flops.
3. Sequential circuits can be divided into synchronous and asynchronous circuits. In a synchronous circuit, all flip-flops are synchronized by a global clock signal. While in an asynchronous circuit, there is no global clock signal for all flip-flops.
4. A sequential logic circuit is a finite state machine that has a finite state. There are two basic models: Moore model and Mealy model.
5. Any SSC can be described using structural models and behavioral models. The structure models include logic diagram, excitation equations, and output equations, and the behavioral models contain state equations, output equations, state table, and state diagram.
6. Analysis procedure of an SSC includes deriving the exciting equations and output equations, deriving the state equation, constructing the state transition table, drawing the state diagram or timing diagram, and determining the implementing functionality.
7. Design procedure of a synchronous sequential logic circuit includes constructing initial state diagram and state table, minimizing the number of states and encoded states, determining the required number and types of flip-flops, deriving the

excitation equations and output equations, checking whether the synthesis of sequential logic circuits can be self-corrected, and drawing the logic diagram.

8. State reduction refers to the process of eliminating equivalent or redundant states. Two methods for state reduction are row matching and implication chart.

9. The implication chart uses a graphical cell to systematically find equivalences among the states, and is more practical to tackle with the complex states table.

Key terms

Synchronous sequential logic circuit: A sequential logic circuit in which all flip-flops have a global clock.

Asynchronous sequential logic circuit: A sequential logic circuit in which there is no global clock for all flip-flops.

State diagram: A graphic depiction of a sequence of states or values.

State table: A table depiction of the rule of state change.

State machine: Any sequential circuit exhibiting a specified sequence of states.

Characteristic equation: A equation that specifies the flip-flop's next state as a function of its current states and inputs.

State/output table: A table that specifies the next state and output of the circuit for every possible combination of current states and inputs.

Mealy machine: A sequential circuit whose output depends on both current states and inputs.

Moore machine: A sequential circuit whose outputs only depends on current states.

Transition equation: An equation expressing the next states as a function of current states and inputs.

Transition table: A table showing the next coded state for each combination of current coded states and inputs.

Coded state: Process that a particular state is assigned a binary code combination.

Excitation table: A table showing the flip-flop excitation input values needed to make the circuit go to the desired next state for each combination of current states and inputs.

Self-test

6.1 Sequential logic circuit is composed of _____ and _____

6.2 Sequential logic circuit can be divided into two categories:
_____ and _____.

6.3 _____, _____, _____ are three equations to describe the logic function of a sequential logic circuit.

6.4 _____ are Boolean equations that describe the input to the flip-flops.

6.5 The state table of a sequential logic circuit is shown in Table T6.1. The function of this circuit is _____.

Table T6.1

Input	Present State		Next state		Output
X	Q_1	Q_0	Q_1^+	Q_0^+	Y
0	0	0	1	0	1
0	0	1	0	0	0
0	1	0	0	1	0
0	1	1	0	1	0
1	0	1	1	0	0
1	1	0	0	0	1
1	1	1	0	0	0
1	0	0	0	1	0

(a) Modulus-3 up/down counter (b) Modulus-4 up counter
(c) Modulus-4 down counter (d) Modulus-4 up/down counter

6.6 The state table of a sequential logic circuit is shown in Table T6.2. Assume the initial state is S_0, and the input series is X = 01011101, the output series F is _____.

Table T6.2: The state table.

Present State	Next state		Output	
	$X = 0$	$X = 1$	$X = 0$	$X = 1$
S_0	S_0	S_1	1	1
S_1	S_1	S_2	0	0
S_2	S_2	S_0	0	0
S_3	S_2	S_2	0	0

6.7 _____ and _____ are two methods for state reduction.
6.8 If two states have the same output and next states, they are _____.

Problems

6.1 Analyze the implementing functionality of sequential logic circuit, as shown in Figure P6.1. Assume that the initial state for the two flip-flops are both 0.

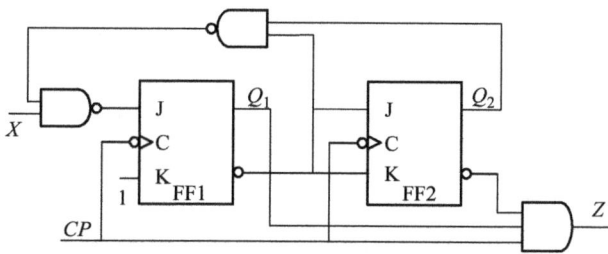

Figure P6.1

6.2 Analyze the sequential logic circuit shown in Figure P6.2. Derive state table and state diagram, and draw output waveform when the input $X = 1011100$. Assume initial state is 0.

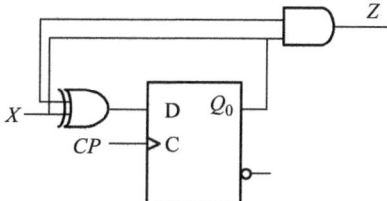

Figure P6.2

6.3 Derive state table and state diagram of sequential logic circuit in Figure P6.3 and determine the modulus of counter. Assume the initial state is $Q_0 Q_1 Q_2 = 000$.

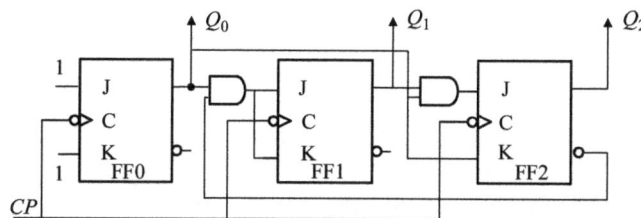

Figure P6.3

6.4 Determine the sequence of counter in Figure P6.4

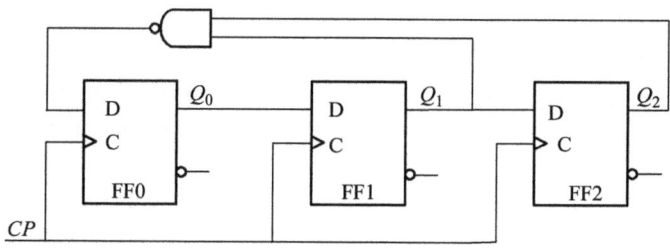

Figure P6.4

6.5 Analyze the logic behaviors of the SSC in Figure P6.5.

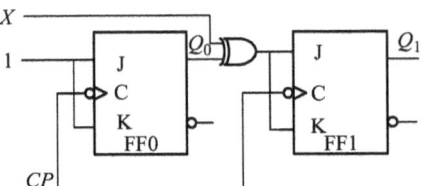

Figure P6.5

6.6 Analyze and complete the logic circuit shown in Figure P6.6(a). The waveform of CP and u_0 are given in Figure P6.6(b).

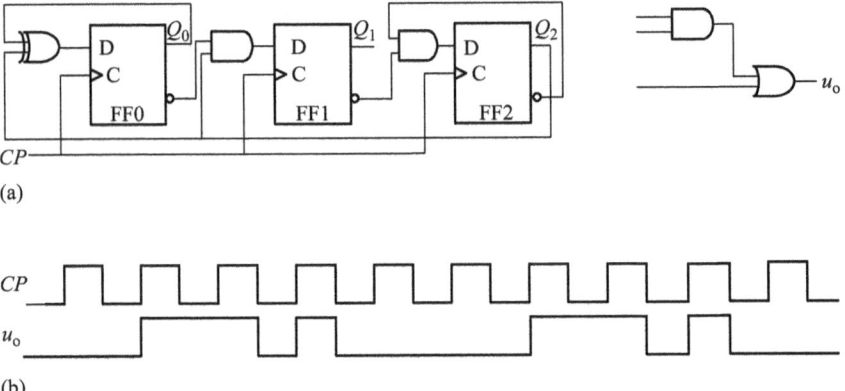

(a)

(b)

Figure P6.6

6.7 Design a counter to produce the following sequence by using J-K flip-flops.
$$00,10,01,11,00,...$$

6.8 Design a sequence detector to detect the sequential input series "111" without overlapping.

6.9 Design a counter with different modulus. When the input X is a LOW, the modulus is 3 and the modulus is 4 when X is a HIGH.

6.10 Design a sequential logic circuit with two input X_1 and X_2 to realize the function that only when two input X_1 and X_2 are the same, if two or more than two continuous CPs are applied, the output is HIGH.

6.11 A sequential signal generator in Figure P6.7 is composed of a counter and a 4-to-1 multiplexer. Analyze the sequence of the counter, list state table, and determine the sequence of the F output by the control of the sequence of the counter.

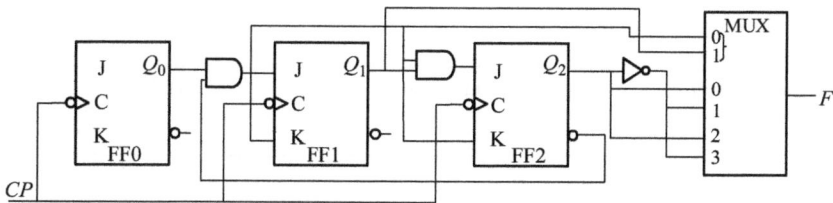

Figure P6.7

6.12 Use row matching method to minimize the state table shown in Table P6.1.

Table P6.1

Present state	Next state		Output	
	$X = 0$	$X = 1$	$X = 0$	$X = 1$
A	A	E	0	1
B	E	C	1	0
C	A	D	1	1
D	F	G	0	1
E	B	C	1	0
F	F	E	0	1
G	A	D	1	1

6.13 Use row matching method to minimize the state table shown in Table P6.2.

Table P6.2

Present State	Next state			Output		
	$X = I$	$X = J$	$X = K$	$X = I$	$X = J$	$X = K$
A	A	B	E	0	1	1
B	B	A	F	0	1	1
C	A	D	E	1	0	0
D	F	C	A	0	1	0
E	A	D	E	0	1	1
F	B	D	F	0	1	1

6.14 Use implication chart to simplify the state table shown in Table P6.3.
6.15 Design a sequence code detector to detect a sequential input series "0010" with an overlapping mode.

Table P6.3

Present State	Next state		Output	
	$X = 0$	$X = 1$	$X = 0$	$X = 1$
A	A	C	0	0
B	D	A	1	0
C	F	F	0	0
D	E	B	1	0
E	G	G	1	0
F	C	C	0	0
G	B	H	1	0
H	H	C	0	0

6.16 Analyze the circuit and draw the output waveforms (Q_0, Q_1, and Q_2) for each flip-flop shown in Figure P6.8. The waveform of CP is given, assume initial state for each flip-flip is a 0.

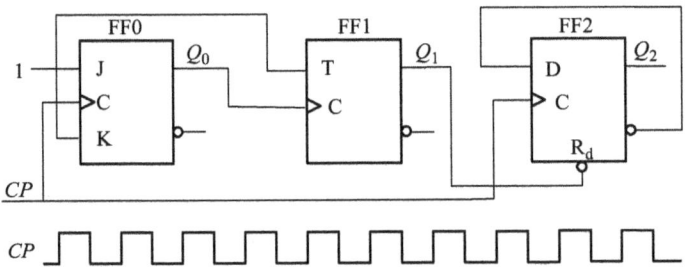

Figure P6.8

6.17 A circuit is shown in Figure P6.9. Draw the output waveforms of Q_0, Q_1, and Z, analyze the relationship between the output Z and the CP. The waveform of CP is given, assume initial state for each flip-flip is a 0.

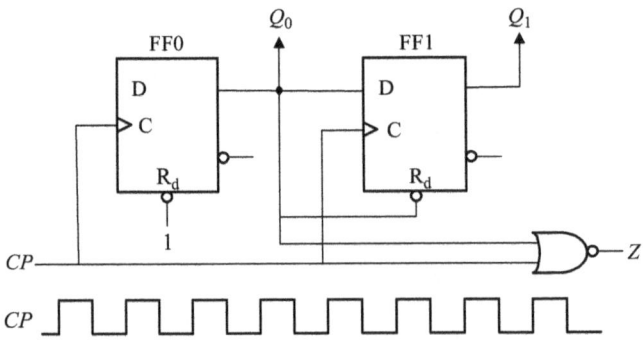

Figure P6.9

6.18 According to Table P6.4, design a counter with J-K flip-flops. X and Y shown in Table P6.4 are the control inputs.

Table P6.4

Present state		XY = 00		XY = 01		XY = 10		XY = 11	
					Next state				
Q_2	Q_1	Q_2^+	Q_1^+	Q_2^+	Q_1^+	Q_2^+	Q_1^+	Q_2^+	Q_1^+
0	0	0	1	1	1	1	1	0	0
0	1	1	0	0	0	1	0	0	1
1	0	1	1	0	1	0	1	1	0
1	1	0	0	1	0	0	0	1	1

7 Counters

A *counter* is usually constructed of a number of flip-flops and cycles through a specified number of states to count the number of clock pulses. There are two types of counters: *asynchronous (ripple) counter* and *synchronous (parallel) counter*. The difference is that asynchronous counters allow some flip-flop outputs to be used as a source of clock for other flip-flops, while synchronous counters apply the same clock to all flip-flops. Counters are widely used components in digital circuits that can be manufactured as specific integrated circuits and can also be incorporated as parts of larger integrated circuits. This chapter introduces asynchronous counters, synchronous counters, cascaded counters, as well as medium-scale integration (MSI) counters and their related applications.

The objectives of this chapter are to
- Understand the difference between an asynchronous counter and a synchronous counter
- Describe the operation of an asynchronous counter
- Describe the operation of a synchronous counter
- Develop counter timing diagrams
- Analyze counter circuits
- Determine the modulus of a counter
- Modify the modulus of a counter
- Analyze the sequence of a counter
- Apply cascaded counters to implement higher modulus
- Explain the logic function of several MSI counters
- Discuss the application of 74LS93, 74LS161, 74LS163, and 74HC190
- Describe how a digital clock operates

7.1 Asynchronous counters

Asynchronous counter is also called *ripple counter* because of the way the clock pulses ripple through the flip-flops. For an asynchronous counter, the first flip-flop is clocked by an external clock, and all subsequent flip-flops are clocked by the output of the preceding flip-flop; hence, the flip-flops do not change states at the same time since they do not have a common clock pulse.

This section introduces the operation of asynchronous binary counters and the 74LS93 four-bit asynchronous binary counter.

The objectives of this section are to
- Describe the operation of a n-bit asynchronous binary counter
- Explain how to design a modulus-n counter
- State the definition of a ripple in relation to counters
- Describe the 74LS93 four-bit asynchronous binary counter

https://doi.org/10.1515/9783110614916-007

7.1.1 Asynchronous binary counters

A two-bit asynchronous binary counter consists of two J–K flip-flops, as shown in Figure 7.1.1. Since the input J and K are both HIGH for each J–K flip-flop, both flip-flops operate at a toggle operation. Therefore, the state transition of each flip-flop depends on whether there is a clock pulse applied on the clock input. Notice that the external clock is connected to the clock input of the first J–K flip-flop (FF0) and thus FF0 changes its state at the falling edge of each clock pulse, but the Q_1 output of FF1 changes only when it is triggered by the falling edge of the Q_0 output of FF0. Because of the inherent propagation delay time through a flip-flop, the transition of the input clock pulse and a transition of the Q_0 output of FF0 can never occur at exactly the same time. Therefore, the two flip-flops are never simultaneously triggered, producing an asynchronous operation.

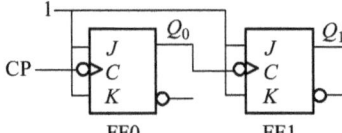

Figure 7.1.1: A two-bit asynchronous binary counter.

Let's analyze the basic operation of the asynchronous counter by applying four clock pulses to FF0 and observing the Q output of each flip-flop. Figure 7.1.2 illustrates the state changes of flip-flops in relation to the clock pulses. Assume that all flip-flops are initially in the RESET state (Q is LOW).

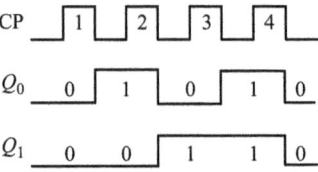

Figure 7.1.2: Timing diagram.

The falling edge of CP1 (clock pulse 1) causes the Q_0 output of FF0 to go HIGH. Due to the Q_0 output from LOW to HIGH, it has no effect on FF1 because a falling transition is required to trigger the flip-flop FF1. After the falling edge of CP1, $Q_0 = 1$ and $Q_1 = 0$. The falling edge of CP2 causes Q_0 to go LOW and thus the flip-flop FF_1 is triggered, causing Q_1 to go HIGH. After the negative-going edge of $CP2$, $Q_0 = 0$ and $Q_1 = 1$. The falling edge of CP3 causes Q_1 to go to HIGH again, which has no effect on FF1. Thus, after the falling edge of CP3, $Q_0 = 1$ and $Q_1 = 1$. The falling edge of CP4 causes Q_0 to go LOW. This makes FF1 triggered, causing Q_1 to go LOW. After the falling edge of CP4, $Q_0 = 0$ and $Q_1 = 0$. The counter is now recycled to its initial state. The term *recycle* refers to the transition of the counter from its final state back to its initial state [3].

It is found that the above two-bit ripple counter has 4 (2^2) different states. If Q_0 represents the least significant bit (LSB) and Q_1 represents the most significant bit (MSB), the counting sequence presents a sequence of binary numbers as shown in Figure 7.1.2 and Table 7.1.1. Since it goes through a binary sequence, the counter in Figure 7.1.1 is a binary counter.

Table 7.1.1: State sequence table.

CP	Q_1	Q_0
0(Initially)	0	0
1	0	1
2	1	0
3	1	1
4(recycle)	0	0

The number of states in a counter is known as its *modulus* (MOD) number. The two-bit binary counter in Figure 7.1.1 has four states; thus it is a MOD-4 counter. Similarly, a binary counter with n flip-flops can have 2^n states; thus, it is a MOD-2^n counter. A MOD-n counter may also be described as a divide-by-n counter. This is because the MSB flip-flop (the furthest flip-flop from the original clock pulse) produces one pulse for every n pulses at the clock input of the least significant flip-flop. Thus, the above counter shown in Figure 7.1.1 is an example of a divide-by-4 counter, which is also called as divide-by-4 frequency divider.

Figure 7.1.3 illustrates a three-bit asynchronous binary counter with three J–K flip-flops triggered by the falling edge of clock pulse. The state sequence for the three-bit binary counter is given in Table 7.1.2. It has 8(2^3) states and counts from 000 (0) to 111(7) for a cycle. It works exactly the same way as a two-bit binary counter mentioned earlier and is an example of divide-by-8 counter. This counter can be easily expanded for higher modulus by connecting additional toggle flip-flops.

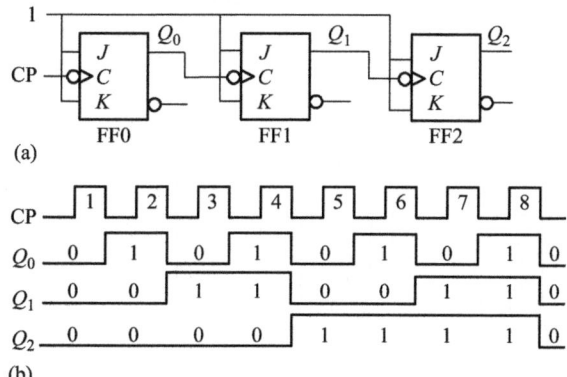

(a)

(b)

Figure 7.1.3: A three-bit asynchronous binary counter: (a) logic diagram; (b) timing diagram.

Table 7.1.2: State sequence table.

CP	Q_2	Q_1	Q_0
0(Initially)	0	0	0
1	0	0	1
2	0	1	0
3	0	1	1
4	1	0	0
5	1	0	1
6	1	1	0
7	1	1	1
8(recycle)	0	0	0

7.1.2 Asynchronous counters with a truncated sequence

The maximum number of states of a binary counter is 2^n, where n is the number of flip-flops used in the counter. Counters can be designed to have a number of states less than the maximum of 2^n. This type of sequence is called a *truncated sequence*.

A common modulus for counters with truncated sequence is ten, also called MOD-10. Counters with ten states in their sequence are called decade counters. A *decade counter* with a count sequence of zero (0000) through nine (1001) is a BCD decade counter because its ten-state sequence produces the BCD code. This type of counter is useful in display applications in which BCD is required for conversion to a decimal readout.

To obtain a truncated sequence, it is necessary to force the counter to recycle before going through all of its possible states. For example, the BCD decade counter must recycle back to the 0000 state after the 1001 state. The MOD-10 counter requires at least four flip-flops because three flip-flops can only afford the maximum of $2^3 = 8$ states. We can use a four-bit asynchronous counter and a few gates to modify its sequence to illustrate the principle of a MOD-10 counter. One way to make the counter recycle after the count of nine (1001) is to decode count ten (1010) with a NAND gate and connect the output of the NAND gate to the direct set inputs of the flip-flops, as shown in Figure 7.1.4.

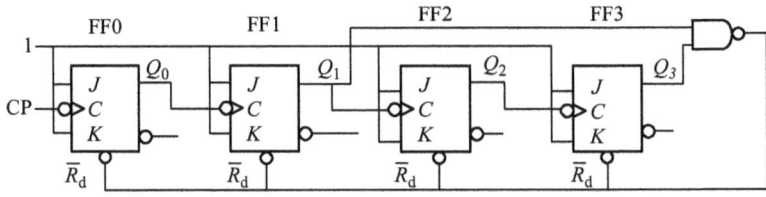

Figure 7.1.4: A decade counter.

Figure 7.1.4 shows that only Q_1 and Q_3 are connected to the NAND gate inputs. This design is an example of *partial decoding*, because from 0000 to 1001, none of the states have both Q_1 and Q_3 HIGH at the same time. When the counter goes into 1010, the NAND output goes LOW and reset all the flip-flops immediately. So after the tenth clock pulse, the stable state is state 0000 rather than state 1010. State 1010 only occurs as a spike. This can be seen in the timing diagram shown in Figure 7.1.5.

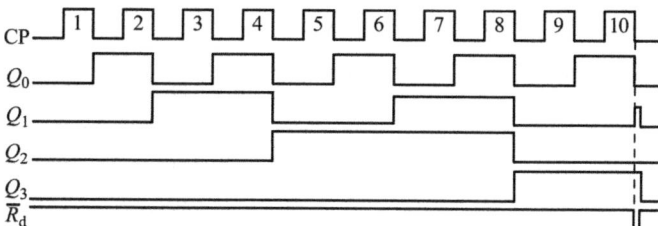

Figure 7.1.5: Timing diagram.

Other truncated sequences can be realized in a similar way, and Example 7.1 shows how to implement a MOD-12 counter.

Example 7.1 Design a MOD-12 asynchronous counter with J–K flip-flops to realize a binary sequence counting from 0000 to 1011.

Solution

To design a MOD-12 asynchronous counter, four J–K flip-flops are required since they can afford any modulus less than or equal to 16 (2^4). The counter counts up from 0000 to 1011, so when it goes to 1011, the next state should be 0000 to make the counter recycle back, as shown in Table 7.1.3.

Normally, the next state after 1011 is 1100, but now it must force the next state changing from 1011 to 0000. Figure 7.1.6 shows the MOD-12 counter, in which a NAND gate partially decodes count

Table 7.1.3: State sequence for the counter in Example 7.1.

CP	Q_3	Q_2	Q_1	Q_0
0(Initially)	0	0	0	0
1	0	0	0	1
2	0	0	1	0
3	0	0	1	1
4	0	1	0	0
5	0	1	0	1
6	0	1	1	0
7	0	1	1	1
8	1	0	0	0
9	1	0	0	1
10	1	0	1	0
11	1	0	1	1
12(recycle)	0	0	0	0

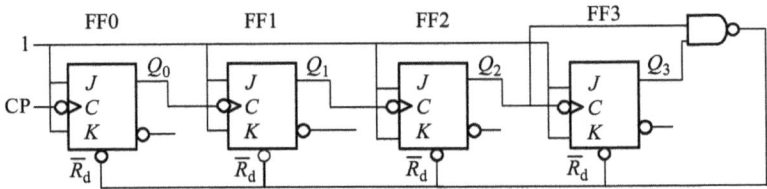

Figure 7.1.6: A MOD-12 counter.

1100 and resets all flip-flops. Thus, when the 12th clock pulse is coming, the counter is forced to recycle from 1011 to 0000.

7.1.3 MSI asynchronous counters

The 74LS93 is an example of a specific MSI asynchronous counter. As shown in Figure 7.1.7, it consists of a J–K flip-flop in toggle operation and a three-bit asynchronous binary counter. This arrangement offers the flexibility to expand. The 74LS93 can be used as a MOD-2 counter if only single J–K flip-flop FF0 is used, or it can be used as a MOD-8 counter if only the three-bit counter portion is used. Except for the two clock inputs, this device also provides asynchronous gated reset inputs $R_{0(1)}$ and $R_{0(2)}$. When both of $R_{0(1)}$ and $R_{0(2)}$ are HIGH, the counter is reset to the 0000 state.

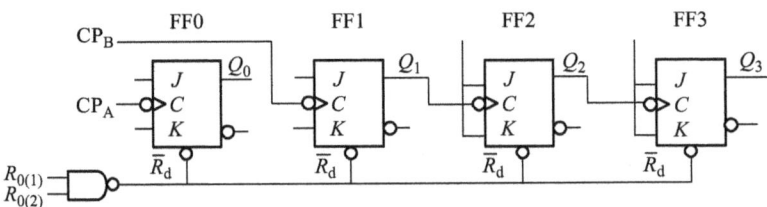

Figure 7.1.7: Internal logic diagram of 74LS93.

Figure 7.1.8 shows the logic symbol of 74LS93. If the Q_0 output is connected to the CP_B input, the 74LS93 can be used as a four-bit binary counter with MOD-16, counting from 0000 through 1111.

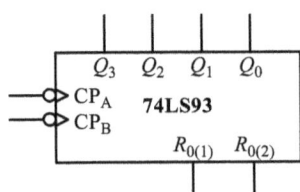

Figure 7.1.8: Logic symbol of 74LS93.

Using the gated reset input, 74LS93 can be used to construct any counter with the MOD less than 16, as mentioned in Example 7.2.

Example 7.2 Construct the MOD-10 counter with 74LS93

Solution

Using the gated reset inputs, $R_{0(1)}$, $R_{0(2)}$, partially decode count 1010. The first step is to construct a MOD-16 counter by connecting Q_0 output to the CP_B input. Then the next step is to determine the connection of reset input. For the MOD-10 counter, the counter counts up from 0000 to 1001, so when it goes to 1001, the next state should be 0000 to make the counter recycle back, as shown in Table 7.1.4. Normally, the next state after 1001 is 1010; it can be observed that Q_0 and Q_3 are both HIGH. Since Q_3 and Q_1 are both HIGH when it counts 1010, the counter can recycle by connecting Q_3, Q_1 with $R_{0(1)}$, $R_{0(2)}$, respectively, as shown in Figure 7.1.9. Notice that an extra state is needed for the asynchronous clear, and the counter must go into the 1010 state for several nanoseconds before recycling.

Table 7.1.4: State sequence table.

CP	Q_3	Q_2	Q_1	Q_0
0(Initially)	0	0	0	0
1	0	0	0	1
2	0	0	1	0
3	0	0	1	1
4	0	1	0	0
5	0	1	0	1
6	0	1	1	0
7	0	1	1	1
8	1	0	0	0
9	1	0	0	1
10(recycle)	0	0	0	0

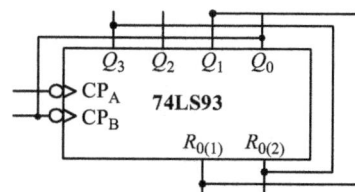

Figure 7.1.9: A 74LS93 connected as a MOD-10 counter.

7.2 Synchronous counters

A *synchronous counter*, also called as a *parallel counter*, is one in which all flip-flops in the counter are clocked at the same time by a common clock pulse. This section introduces synchronous counters and discusses the function and application of several MSI synchronous counters.

The objectives of this section are to
- Describe the operation of an n-bit synchronous binary counter
- Discuss the 74LS161, 74LS163, and 74HC190 MSI counters
- Augment the capacity of MSI counters

7.2.1 Synchronous binary counters

Figure 7.2.1(a) shows a two-bit synchronous counter, which consists of two T flip-flops. As mentioned in Chapter 5, a T flip-flop can be formed by connecting the J and K inputs of a J–K flip-flop together. The T input of FF0 is connected to HIGH and the Q_0 output is connected to the T input of FF1. Assume that the counter is initially in the state 00 ($Q_1 Q_0$); when the positive edge of the first clock pulse is applied, FF0 is in toggle operation and thus Q_0 goes HIGH. While for FF1, at the positive-going edge of CP1, the T input is LOW because there is a propagation delay from the triggering edge of the clock pulse until the Q_0 output actually makes a transition. When the leading edge of the first clock pulse is applied, $T_1 = 0$; thus, the Q_1 output of FF1 keeps the previous state.

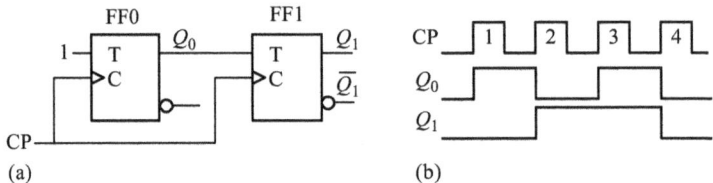

(a) (b)

Figure 7.2.1: A three-bit synchronous counter: (a) logic diagram and (b) timing diagram.

After the first clock, $Q_0 = 1$ and $Q_1 = 0$. When the leading edge of CP2 occurs, Q_0 changes from HIGH to LOW since T_0 is HIGH. Since FF1 has a HIGH on its T_1 input, at the triggering edge of this clock pulse, FF1 is in the toggle operation and Q_1 goes to HIGH. After $CP2$, $Q_0 = 0$ and $Q_1 = 1$.

When the leading edge of CP3 occurs, FF0 toggles again with $Q_0 = 1$ and Q_1 remains HIGH. After the triggering edge of CP3, $Q_0 = 1$ and $Q_1 = 1$.

Finally, at the leading edge of CP4, both Q_0 and Q_1 go to LOW because they have a toggle condition on their inputs, as shown in Figure 7.2.1(b). Thus, the counter has recycled to its initial state. Notice that the propagation delays are normally omitted for simplicity in the timing diagram. However, in high-speed digital circuits, these small delays are important considerations in design and troubleshooting.

Alternatively, you can use the design procedure of the sequential logic (covered in Chapter 6) to design a synchronous binary counter. Let's see Example 7.3.

Example 7.3 Design a three-bit synchronous binary up counter by using T flip-flops.

Solution

The first step to design a counter is to create a state diagram. A state diagram shows the transition of states through which the counter advances when it is clocked. Figure 7.2.2 shows a state diagram for a three-bit synchronous binary up counter.

    The second step is to derive a state table, which lists all the present states, next states, and the inputs of T flip-flop as shown in Table 7.2.1. All possible output transitions are listed. Q and Q^+ represent the present state and the next state.

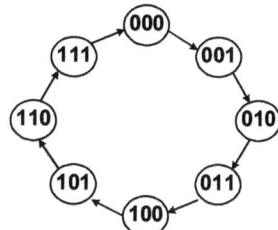

Figure 7.2.2: State diagram for a three-bit synchronous binary up counter.

Table 7.2.1: State table.

Present states			Next states			Inputs		
Q_2	Q_1	Q_0	Q_2^+	Q_1^+	Q_0^+	T_2	T_1	T_0
0	0	0	0	0	1	0	0	1
0	0	1	0	1	0	0	1	1
0	1	0	0	1	1	0	0	1
0	1	1	1	0	0	1	1	1
1	0	0	1	0	1	0	0	1
1	0	1	1	1	0	0	1	1
1	1	0	1	1	1	0	0	1
1	1	1	0	0	0	1	1	1

    The next step is to take use of the Karnaugh map to determine the excitation equation. Figure 7.2.3 shows the Karnaugh maps for the three flip-flops used in the counter.

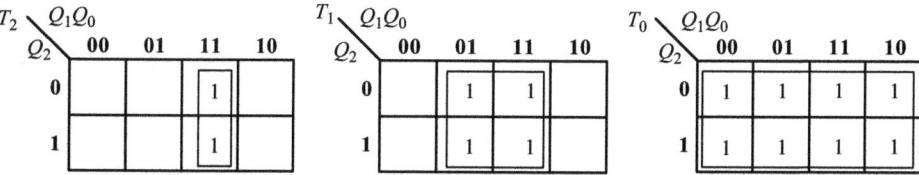

Figure 7.2.3: The Karnaugh maps.

From the Karnaugh maps, excitation equations of three flip-flops are

$$T_2 = Q_1 Q_0, \quad T_1 = Q_0, \quad T_0 = 1 \tag{7.2.1}$$

The last step is to construct the logic diagram, as shown in Figure 7.2.4.

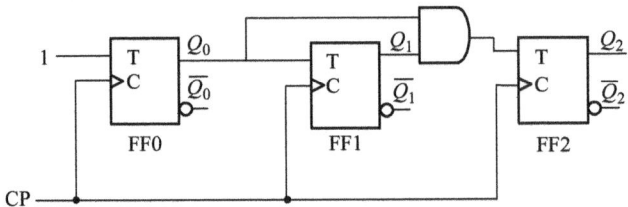

Figure 7.2.4: A three-bit synchronous binary up counter.

7.2.2 Up/down synchronous counter

An up/down counter is a bidirectional counter that is capable of counting either up or down. In order to choose the up or down operation mode, a control line UP/$\overline{\text{DOWN}}$ is used to select the direction of counting. When UP/$\overline{\text{DOWN}}$ is HIGH, the counter counts upward through its sequence, while UP/$\overline{\text{DOWN}}$ is LOW, the counter counts downward across its counting sequence.

Let's take a three-bit binary up/down counter, for example. For up operation mode, the counter advances upward through its sequence, from 000 to 111, and then recycles; for down operation mode, the counter goes through the sequence in the opposite direction, from 111 to 000, and then recycles. So the up/down state sequence table is listed in Table 7.2.2.

Table 7.2.2: Up/down state sequence table.

CP	UP	Q_2	Q_1	Q_0	DOWN
0		0	0	0	
1		0	0	1	
2		0	1	0	
3		0	1	1	
4		1	0	0	
5		1	0	1	
6		1	1	0	
7		1	1	1	

It can be observed from Table 7.2.2 that FF0 toggles on each clock pulse; thus, the T_0 input of FF0 should be

$$T_0 = 1 \tag{7.2.2}$$

For the up sequence, Q_1 changes the state on the next clock when $Q_0 = 1$. For the down sequence, Q_1 changes the state on the next clock when $Q_0 = 0$. Thus, the T_1 inputs of FF1 must be equal to a 1 under the conditions expressed by the following equation:

$$T_1 = (Q_0 \text{UP}) + (\bar{Q}_0 \text{DOWN}) \tag{7.2.3}$$

For the up sequence, Q_2 changes the state on the next clock pulse when $Q_0 = Q_1 = 1$. For the down sequence, Q_2 changes the state on the next clock pulse when $Q_0 = Q_1 = 0$. Thus the T_2 inputs of FF2 must equal to a 1 under the condition expressed by the following equation:

$$T_2 = (Q_0 Q_1 \text{UP}) + (\bar{Q}_0 \bar{Q}_1 \text{DOWN}) \tag{7.2.4}$$

By using the above excitation equations of three flip-flops, the logic diagram of the three-bit up/down binary counter is implemented, as shown in Figure 7.2.5. Notice that the UP/$\overline{\text{DOWN}}$ control input is HIGH for UP and LOW for DOWN.

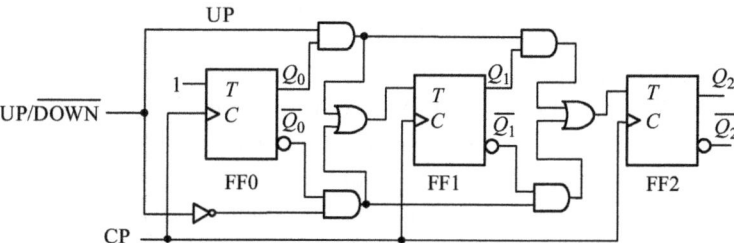

Figure 7.2.5: A three-bit up/down synchronous counter.

7.2.3 MSI synchronous counters

1. Four-bit synchronous binary up counters

The 74LS161 is an example of a four-bit synchronous MSI binary up counter and the corresponding logic symbol is shown in Figure 7.2.6. The functional table of 74LS161 is listed in Table 7.2.3. Except the clock input, it has four control inputs: asynchronous clear (or reset) input $\bar{R}$, synchronous load input $\overline{LD}$, two enable inputs CT_T and CT_P. There are also four flip-flops integrated in the 74LS161 chip. $D_3 D_2 D_1 D_0$ are the four inputs, and $Q_3 Q_2 Q_1 Q_0$ are the four corresponding outputs representing the state of counter, and RCO is the ripple carry output indicating the time that the counter counts to its maximum. The 74LS161 has four operation modes.

An operation mode is *asynchronous clear* or *asynchronous reset*, which is presented in the first row of the functional table. The counter can be asynchronous reset by applying a LOW to the reset input $\bar{R}$. That is, when $\bar{R}$ is an active LOW, the output $Q_3 Q_2 Q_1 Q_0$ is 0000 no matter what the other inputs are.

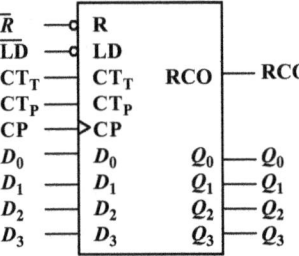

Figure 7.2.6: Logic symbol of 74LS161.

When $\overline{LD}$ is an active LOW, data inputs, $D_3D_2D_1D_0$, are loaded to the outputs, $Q_3Q_2Q_1Q_0$, at the positive edge of the clock pulse; thus, the counter sequence can be started with any four-bit binary number from 0000 to 1111. This operation mode is called as *synchronous load*, which is indicated in the second row of the functional table.

When $\overline{R}$ and $\overline{LD}$ are both HIGH, if at least one of enable inputs (CT_p, CT_T) is an active LOW, the outputs keep the present outputs. This operation mode corresponds to the third and fourth rows in the functional table.

When all control inputs ($\overline{R}$, $\overline{LD}$, CT_p, CT_T) are HIGH, the counter operates as a free-running counter recycling from 0000 to 1111 (0 to 15), which is described in the last row of the functional table. When the counter reaches the state of 1111, which is the last state in its sequence and called the *terminal count* (TC), the RCO goes HIGH. The RCO output can be connected to the enable inputs of the other counters, allowing to cascade several counters to implement the counter with the higher modulus.

In addition to 74LS161, 74LS163 is also a four-bit MSI synchronous binary up counter. By comparing the function table of 74LS161 in Table 7.2.3 and 74LS163 in Table 7.2.4, it can be seen that the only difference between 74LS161 and 74LS163 is the reset operation mode. The 74LS163 has a synchronous reset operation mode. When the reset input $\overline{R}$ is an active LOW, the output $Q_3Q_2Q_1Q_0$ is reset to 0000 only at the positive edge of clock pulse. The logic symbol of 74LS163 is the same as that of 74LS161.

Table 7.2.3: Function table of 74LS161.

				Inputs						Outputs		
CP	$\overline{R}$	$\overline{LD}$	CT_p	CT_T	D_3	D_2	D_1	D_0	Q_3	Q_2	Q_1	Q_0
×	0	×	×	×	×	×	×	×	0	0	0	0
↑	1	0	×	×	D_3	D_2	D_1	D_0	D_3	D_2	D_1	D_0
×	1	1	0	×	×	×	×	×	No change			
×	1	1	×	0	×	×	×	×	No change			
↑	1	1	1	1	×	×	×	×	Count up			

Table 7.2.4: Function table of 74LS163.

	Inputs								Outputs			
CP	$\bar{R}$	$\overline{LD}$	CT_p	CT_T	D_3	D_2	D_1	D_0	Q_3	Q_2	Q_1	Q_0
↑	0	×	×	×	×	×	×	×	0	0	0	0
↑	1	0	×	×	D_3	D_2	D_1	D_0	D_3	D_2	D_1	D_0
×	1	1	0	×	×	×	×	×	No change			
×	1	1	×	0	×	×	×	×	No change			
↑	1	1	1	1	×	×	×	×	Count up			

2. Up/down synchronous counter

74HC190 is an example of an up/down MSI decade counter. Figure 7.2.7 shows the logic symbol of 74HC190. The direction of the count is determined by the level of the up/down input ($\bar{U}/D$). When the up/down input is HIGH, the counter counts down from 1001 to 0000 at the positive edge of each clock pulse; when it is LOW, the counter counts up from 0000 to 1001 at the positive edge of each clock pulse. As the function table in Table 7.2.5, when the load input, $\overline{LD}$, and the enable input, $\overline{CT}$, are both HIGH, the counter keeps the present state with no change. When $\overline{LD}$ is an active LOW, data $D_3D_2D_1D_0$ inputs are loaded to the $Q_3Q_2Q_1Q_0$ output; thus, this device can be preset to any desired BCD digit determined by the state of the data inputs.

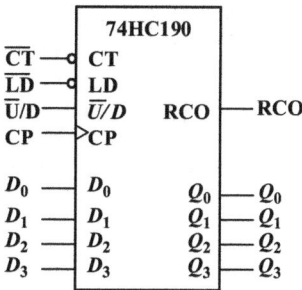

Figure 7.2.7: Logic symbol of 74HC190.

When the terminal counting nine (1001) is reached in the UP mode or when the terminal counting zero (0000) is reached in the DOWN mode, the RCO is HIGH. When the next clock pulse comes, the RCO turns to LOW.

3. Augment the capacity of MSI counters

A four-bit synchronous binary counter can be used to construct any counter with a MOD less than 16. The design method uses clear input or load input to shorten the

Table 7.2.5: Function table of 74HC190.

Inputs								Outputs			
$\overline{CT}$	$\overline{LD}$	$\overline{U}/D$	CP	D_3	D_2	D_1	D_0	Q_3	Q_2	Q_1	Q_0
1	1	×	×	×	×	×	×			No change	
×	0	×	×	D_3	D_2	D_1	D_0	D_3	D_2	D_1	D_0
0	1	0	↑	×	×	×	×			Count up	
0	1	1	↑	×	×	×	×			Count down	

normal sequence and augment the capacity of the counter. Example 7.4 shows the two methods to design a MOD n ($n{\le}16$) counter with one MSI counter.

Example 7.4 Construct a MOD-10 counter with a four-bit binary MSI up counter.

Solution

Both 74LS161 and 74LS163 are four-bit binary MSI up counters. The 74LS161 has an asynchronous reset input: if you use an asynchronous reset input to shorten the normal sequence and design a MOD-10 counter, an extra unstable state (count 10, 1010) is needed to help the counter return to zero. Whereas the 74LS163 has a synchronous reset input: if you use the synchronous reset input to design a MOD-10 counter, a clear signal is produced when the final state (count 9, 1001) is reached, but the counter will be reset at the next clock. Logic diagrams for MOD-10 counter constructed with a 74LS161 and a 74LS163 are shown in Figure 7.2.8. Tables 7.2.6 and 7.2.7 show the corresponding state table.

Another method to construct the MOD-10 counter is to use load input ($\overline{LD}$) to augment the capacity. When the load input $\overline{LD}$ is an active LOW, the data inputs are loaded to the outputs; thus, the counter sequence can be started from any four-bit binary number. Assume that ten states from 0110 to 1111 are used to construct the MOD-10 counter with 74LS161, as shown in Table 7.2.8. The initial state of the counter is set as 0110, and then the counter counts upward; after the tenth clock pulse, the output state is 1111 and the counter reaches the TC. Thus, the RCO is HIGH. The HIGH level emitted by the RCO is fed back to the load input via an inverter, so the load input $\overline{LD}$ is a LOW, making the counter preset and recycle back to the initial state 0110.

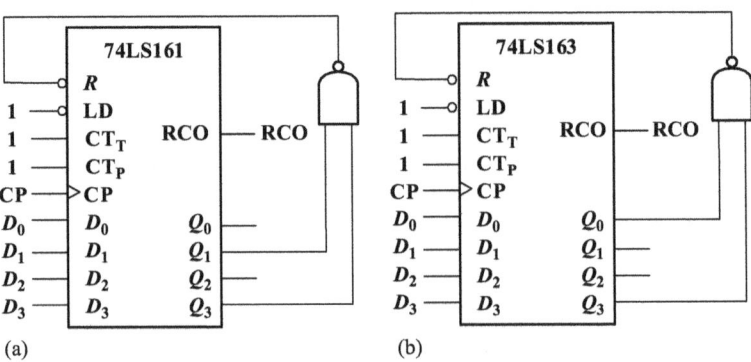

(a)   (b)

Figure 7.2.8: Logic diagram for a MOD-10 counter.

Table 7.2.6: State sequence table.

CP	Q_3	Q_2	Q_1	Q_0	$\overline{R}$
0	0	0	0	0	1
1	0	0	0	1	1
2	0	0	1	0	1
3	0	0	1	1	1
4	0	1	0	0	1
5	0	1	0	1	1
6	0	1	1	0	1
7	0	1	1	1	1
8	1	0	0	0	1
9	1	0	0	1	1
10	1	0	1	0	0

Table 7.2.7: State sequence table.

CP	Q_3	Q_2	Q_1	Q_0	$\overline{R}$
0	0	0	0	0	1
1	0	0	0	1	1
2	0	0	1	0	1
3	0	0	1	1	1
4	0	1	0	0	1
5	0	1	0	1	1
6	0	1	1	0	1
7	0	1	1	1	1
8	1	0	0	0	1
9	1	0	0	1	0

Table 7.2.8: state sequence table.

CP	Q_3	Q_2	Q_1	Q_0	$\overline{R}$
0	0	1	1	0	1
1	0	1	1	1	1
2	1	0	0	0	1
3	1	0	0	1	1
4	1	0	1	0	1
5	1	0	1	1	1
6	1	1	0	0	1
7	1	1	0	1	1
8	1	1	1	0	1
9	1	1	1	1	0

Figure 7.2.9(a) shows the logic diagram for the MOD-10 counter with 74LS161 by selecting the count sequence from 0110 to 1111. Similarly, the first ten states from 0000 to 1001 can be used for implementing a MOD-10 counter with a 74LS161 and the corresponding logic diagram is shown in Figure 7.2.9(b).

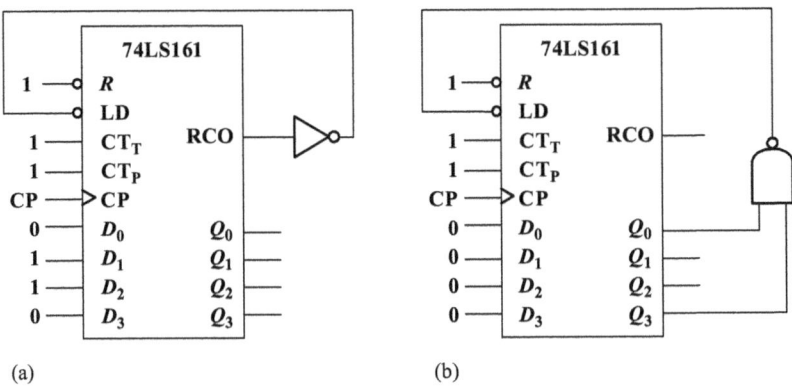

(a)   (b)

Figure 7.2.9: Logic diagrams of MOD-10 counter with 74LS161.

In terms of the above two methods of augmenting the capacity of the counter, notice that whenever using the clear input, you should pay attention to the difference of synchronous clear and asynchronous clear. If using load input, you should pay attention to the initial state and end state of the design sequence, because initial state determines the connection of the data inputs and the end state gives the production method of load signal.

4. Sequence generator

A sequence signal is a serial code with a certain period. A method of constructing a sequence generator is to use the combination of counter and combinational circuit. Let's take a look at the design procedure through an example.

Example 7.5 Design a sequence generator that can generate a sequence code "110001001110."

Solution

Step 1 Design a counter with the modulus which is equal to the length of the sequence code
The length of the given sequence code is 12, so a MOD-12 counter is required to control the code output as the predetermined sequence. Here, a 74LS161 is used to construct a MOD-12 counter with a synchronous load input and the count sequence is from 0100 to 1111.

Step 2 Construct the combinational circuit to implement the sequence code "110001001110"
In order to generate the output of the sequence code "110001001110," one-bit code can be output one by one at the control of the clock pulse. So a combinational circuit is required to implement the output of the sequence code. The outputs of counter act as the inputs of this combinational circuit and an output, F, are needed to output one-bit code at each clock input sequentially. After 12 o'clock

pulse, a group of sequence code "110001001110" is produced and recycles again. The truth table of the combinational circuit is listed in Table 7.2.9.

Table 7.2.9: Truth table.

Q_3	Q_2	Q_1	Q_0	F
0	1	0	0	1
0	1	0	1	1
0	1	1	0	0
0	1	1	1	0
1	0	0	0	0
1	0	0	1	1
1	0	1	0	0
1	0	1	1	0
1	1	0	0	1
1	1	0	1	1
1	1	1	0	1
1	1	1	1	0

This combinational circuit can implement with several methods. Here, an 8-to-1 multiplexer is chosen to implement the logic function given in Table 7.2.9. If $Q_3Q_2Q_1$ are selected for selection input lines, $A_2A_1A_0$, the output expression can be obtained from Table 7.2.9 as follows:

$$F = m_2Q_0 + m_2\bar{Q}_0 + m_4Q_0 + m_6Q_0 + m_6\bar{Q}_0 + m_7\bar{Q}_0 = m_2 + m_4Q_0 + m_6 + m_7\bar{Q}_0 \qquad (7.2.5)$$

The data inputs $D_2 = D_6 = 1$, $D_4 = Q_0$, $D_7 = \bar{Q}_0$, $D_0 = D_1 = D_3 = D_5 = 0$. The logic circuit is shown in Figure 7.2.10.

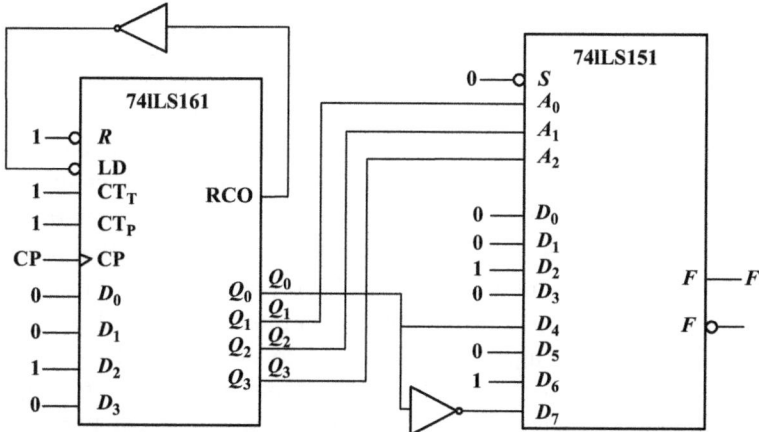

Figure 7.2.10: Circuit diagram for Example 7.5.

7.3 Cascaded counter

Counters can be connected to achieve higher-modulus operation. *Cascading* means that previous output of a counter drives the input of next counter. This section gives a description on cascaded counter.

The objectives of this section are to
- Determine the overall modulus of a cascaded counter
- Use a cascaded counter as a synchronous frequency divider
- Use a cascaded counter to generate a truncated sequence

When operating synchronous counters in a cascaded configuration, it is necessary to use count enable and TC to achieve higher-modulus operation. On some devices, count enable is labeled simply as CT and TC is analogous to RCO on some IC counters.

An example of the cascaded counter in which the two 74HC190s connected in cascade is shown in Figure 7.3.1.

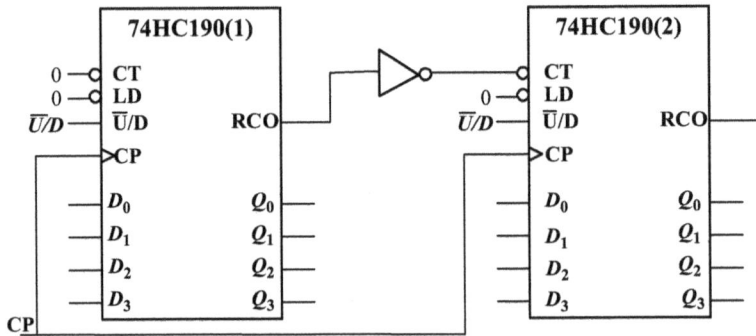

Figure 7.3.1: A MOD-100 counter using two cascade decade counters.

The terminal count (RCO) output of counter 1 is connected to count enable (CT) input of counter 2. Counter 2 is inhibited by a HIGH on its CT input until counter 1 reaches its terminal state (1001) and its terminal count output (RCO) goes HIGH and a LOW is applied on the CT of counter 2. This enables counter 2, so that when the first clock pulse arrives after counter 1 reaches its TC, counter 2 goes from its initial state to the second state. Upon the completion of the entire second cycle of counter 1, counter 2 is enabled again and advances to next state. This count sequence continues. Since they are the decade counters, counter 1 must go through ten cycles before counter 2 completes its first cycle. That is say, for every 10 cycles of counter 1, counter 2 goes through one cycle. Thus counter 2 completes the whole cycle after 100 clock pulses. The overall modulus of these two cascaded counters is $10 \times 10 = 100$.

The cascaded counter can also be regarded as a frequency divider. The cascaded counter shown in Figure 7.3.1 divides the input clock frequency by 100. With this function, cascaded counters are often used to divide a high-frequency clock signal to obtain highly accurate pulse frequencies.

The above discussion has shown how to achieve an overall modulus that is the product of the individual modulus of all the cascaded counters. This can be considered full-modulus cascading. If an application requires a modulus that is less than that achieved by full-modulus cascading, that is, a truncated sequence must be implemented with cascaded counters. To illustrate this method, we use the cascaded counter configuration in Figure 7.3.2. Assume that a modulus 24 counter is required. This circuit uses two 74LS161 four-bit synchronous binary up counters. If the two counters (eight-bit total) are cascaded in a full-modulus arrangement, the modulus is $2^8 = 256$. The technique used in the circuit of Figure 7.3.2 is to preset the cascaded counter to 1 each time it recycles, so that it will count from 1(0000001) to 24(00011000).

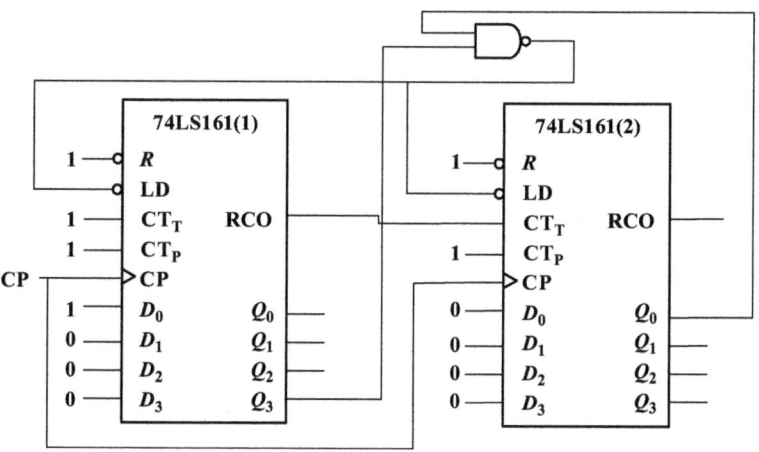

Figure 7.3.2: Modulus 24 counter counts from 1 to 24.

7.4 Counter application

7.4.1 Digital clock

A digital clock is a type of timing equipment to directly display time by digits. It is often incorporated into all kinds of devices such as car, radio, television, microwave oven, standard oven, computer, and cell phone. The digital clock system is made up of crystal oscillator, frequency divider, counter, decoder, LED display circuit, calibrated circuit, and voltage source. Figure 7.4.1 shows the simplified logic diagram of a digital clock system that displays seconds, minutes, and hours.

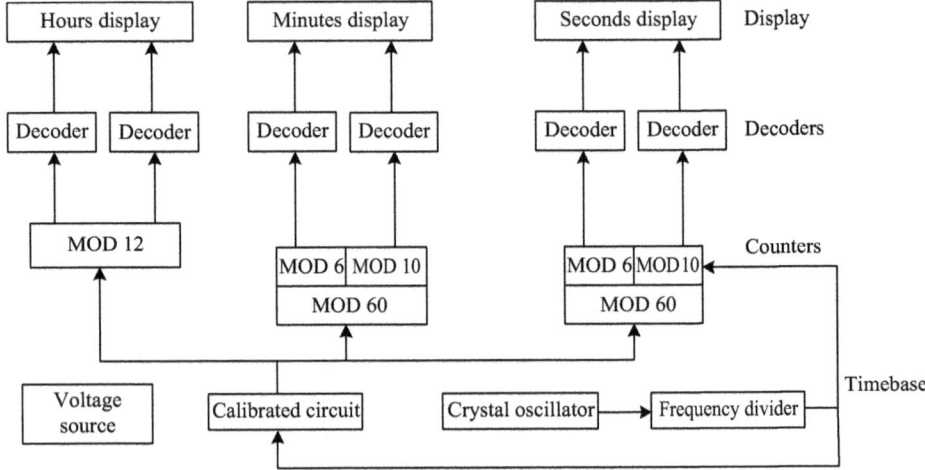

Figure 7.4.1: Simplified logic diagram of a digital clock system.

As the key part of the digital clock, the crystal oscillator can generate an accurate signal with high stability. The high-frequency signal is divided by using a frequency divider to obtain a 1 Hz time base. This 1 Hz time base is used as the clock pulse of the second counter.

To actually see the seconds, two counters produce binary numbers. The MOD-10 counter is used to produce a 0 through 9 sequence on its outputs, while the MOD-6 counter is to produce a 0 through 5 sequence on its outputs. The two counters cascaded as a MOD-60 counter, counts from 0 to 59, and then recycles back to 0. The TC, 59, needs to be decoded to enable the next counter. After each 60 s, the minute section is triggered. And the minute section of the clock looks exactly the same with the second section. Finally, the hour section looks almost the same except the MOD-60 counter is replaced by a MOD-12 counter.

1. Generation of time base signal

To generate a time base signal, CD4060 as an MSI time base generator is generally used. Figure 7.4.2 shows the logic diagram for CD4060. The input of gate 2 (G_2) needs two external resistors, two capacitors, and a quartz crystal to construct a crystal oscillator with a oscillation frequency of 32,768 Hz. F_1–F_{14} are asynchronous binary

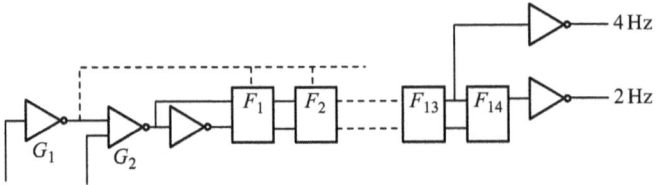

Figure 7.4.2: Logic diagram for CD4060.

counters. The oscillation frequency divided by 2^{14} cascaded binary counters is translated to 2 Hz. The 2 Hz pulse signal can be divided into the 1 Hz pulse as the time base pulse by using the MOD-2 counter. The MOD-2 counter can be constructed by a flip-flop operating at toggle state.

2. Counters

The counters used in the second counter and minute counter are both MOD-60 counters. They can be obtained by cascading a MOD-10 counter with a MOD-6 counter. Figure 7.4.3 shows the MOD-60 counter using two 74LS160 (four-bit counter) in which the first 74LS160 (1) is connected as a MOD-10 counter and the second 74LS160(2) is connected as a MOD-6 counter.

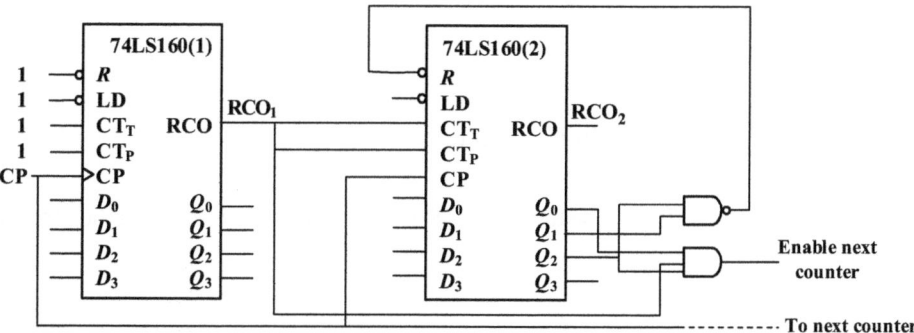

Figure 7.4.3: A MOD-60 counter using two 74LS160s.

The counter used in the hour section is a MOD-12 counter, as shown in Figure 7.4.4. A feedback circuit constructed by G_1 and G_2 is used to help the counter return to zero after one circle. When the output of G_2 is LOW, which means the output of $J–K$ flip-flop and Q_1 of the 74LS160 are both HIGH, the counter will recycle.

3. Decoder and display

We can use a decoder, also named as "binary number to seven-segment display converter", to convert a binary number between 0 and 9 to the appropriate signals to drive a seven-segment LED.

4. Calibrated circuit

The calibrated circuit is used to correct the error of the digital clock and there are many methods to realize a calibrated circuit. The simplest way is adding three switches, S_1, S_2, and S_3 to the second section, minute section, and hour section, respectively. The function of S_2 and S_3 is used to cut off the second signal and use the time base signal to

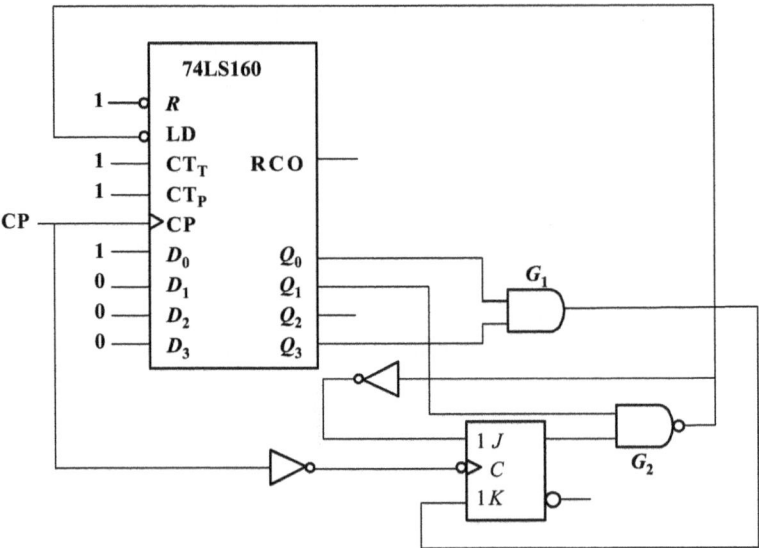

Figure 7.4.4: Logic diagram of a MOD-12 counter using 74LS160.

correct the minute and hour. The function of S_1 is to clear up the counters in the second section.

7.4.2 Implementation of a digital clock with Verilog HDL

1. Generation of a time base signal

```
//Generation of time base signal with a frequency of 1Hz
module clk_1Hz(clk,clk_1);
  input clk;
  output clk_1;
  reg clk_1;
  reg [24:0] count;
  always@(posedge clk)
    begin
      if(count>=24999999)
        begin
          count<=0;
          clk_1<=~clk_1;
        end
      else count<=count+1;
    end
endmodule
```

2. Counters

```verilog
//Counters for minutes and second section
module counter (clk,cnt,out,out_carry);
  input clk;
  input [3:0] cnt;
  output [3:0] out;
  output out_carry;
  reg [3:0] out;
  reg out_carry;
  always@(posedge clk)
  begin
    if(out==cnt)
      begin
        out_carry<=1;
        out<=0;
      end
    else
      begin
        out<=out+1;
        out_carry<=0;
      end
  end
end
endmodule
```

The counting length of the counter can be changed according to the input signal "cnt." For instance, the sentence "cnt_second0=4′b0101" is a MOD-6 counter for the second section.

```verilog
//Counters for hour section
module counter_12(clk,cnt,out);
  input clk;
  input [3:0] cnt;
  output [3:0] out;
  reg [3:0] out;
  always @(posedge clk)
  begin
    if(out==cnt)
        out<=0;
    else
    out<=out+1;
  end
endmodule
```

3. Decoding and display

```verilog
//Decoding for minute and second section
module decode(data,hex);
  input [3:0] data;
  output [7:0] hex;
  reg [7:0] hex;
  always @(data)
    begin
      case(data)
      4'b0000:hex=8'b11000000;
      4'b0001:hex=8'b11111001;
      4'b0010:hex=8'b10100100;
      4'b0011:hex=8'b10110000;
      4'b0100:hex=8'b10011001;
      4'b0101:hex=8'b10010010;
      4'b0110:hex=8'b10000010;
      4'b0111:hex=8'b11111000;
      4'b1000:hex=8'b10000000;
      4'b1001:hex=8'b10011000;
      default:hex=8'b11111111;
      endcase
    end
endmodule
```

```verilog
// Decoding for hour section
module decode_12(data,hex1,hex2);
  input [3:0] data;
  output [7:0] hex1;
  output [7:0] hex2;
  reg [7:0] hex1;
  reg [7:0] hex2;
  always @(data)
    begin
      case(data)
      4'b0000:begin
                      hex1=8'b11000000;
                      hex2=8'b11000000;
                end
      4'b0001:begin
                      hex1=8'b11111001;
                      hex2=8'b11000000;
                end
      4'b0010:begin
                      hex1=8'b10100100;
                      hex2=8'b11000000;
                end
```

```
            4'b0011:begin
                            hex1=8'b10110000;
                            hex2=8'b11000000;
                    end
            4'b0100:begin
                            hex1=8'b10011001;
                            hex2=8'b11000000;
                    end
            4'b0101:begin
                            hex1=8'b10010010;
                            hex2=8'b11000000;
                    end
            4'b0110:begin
                            hex1=8'b10000010;
                            hex2=8'b11000000;
                    end
            4'b0111:begin
                        hex1=8'b11111000;
                            hex2=8'b11000000;
                    end
            4'b1000:begin
                            hex1=8'b10000000;
                            hex2=8'b11000000;
                    end
            4'b1001:begin
                            hex1=8'b10011000;
                            hex2=8'b11000000;
                    end
            4'b1010:begin   hex1=8'b11000000;
                            hex2=8'b11111001;
                    end
            4'b1011:begin   hex1=8'b11111001;
                            hex2=8'b11111001;
                    end
            default:begin
                            hex1=8'b11111111;
                             hex2=8'b11111111;
                    end

        endcase
    end
endmodule
```

```
// Display
module display(clk,hex0,hex1,hex2,hex3,hex4,hex5);//Display with six LEDs defined as
hex1~hex5
    input clk;
    output[7:0] hex0,hex1,hex2,hex3,hex4,hex5;
```

```
wire  [7:0] hex0,hex1,hex2,hex3,hex4,hex5;
wire clk_1;
wire  [7:0] segment0,segment1,segment2,segment3,segment_hour;
wire [3:0] cnt_second0=4'b1001; // counters for second section
wire [3:0] cnt_second1=4'b0101;
wire [3:0] cnt_minute0=4'b1001; // counters for minute section
wire [3:0] cnt_minute1=4'b0101;
wire [3:0] cnt_hour=4'b1011;// counters for hour section
wire second0_carry;
wire second1_carry;
wire minute0_carry;
wire minute1_carry;
clk_1Hz M1(.clk(clk),.clk_1(clk_1));
counter second0(.clk(clk_1),.cnt(cnt_second0),.out(segment0),.out_carry
(second0_carry));
counter second1(.clk(second0_carry),.cnt(cnt_second1),.out(segment1),
       .out_carry(second1_carry));
counter minute0(.clk(second1_carry),.cnt(cnt_minute0),.out(segment2),
       .out_carry(minute0_carry));
counter minute1(.clk(minute0_carry),.cnt(cnt_minute1),.out(segment3),
       .out_carry(minute1_carry));
counter_12 hour(.clk(minute1_carry),.cnt(cnt_hour),.out(segment_hour));

decode  s0(.data(segment0),.hex(hex0));
decode  s1(.data(segment1),.hex(hex1));
decode  m0(.data(segment2),.hex(hex2));
decode  m1(.data(segment3),.hex(hex3));
decode_12 h1(.data(segment_hour),.hex1(hex4),.hex2(hex5));
endmodule
```

The schematic diagram for codes is shown in Figure 7.4.5.

7.5 Summary

1. A counter use of a group of flip-flops to count the number of clock pulses with the number of stable states in its counting cycle.
2. Asynchronous and synchronous counters differ only in the method they are clocked. The maximum clock frequency for an asynchronous counter decreases as the number of bits increases. For a synchronous counter, the maximum clock frequency remains the same, regardless of the number of bits.
3. In asynchronous (ripple) counters, the clock signal is applied to the LSB flip-flop, and all other flip-flops are clocked by the output of the preceding flip-flops.
4. A counter's modulus (MOD) number is the number of stable states in its counting cycle; it is also the maximum frequency-division ratio.

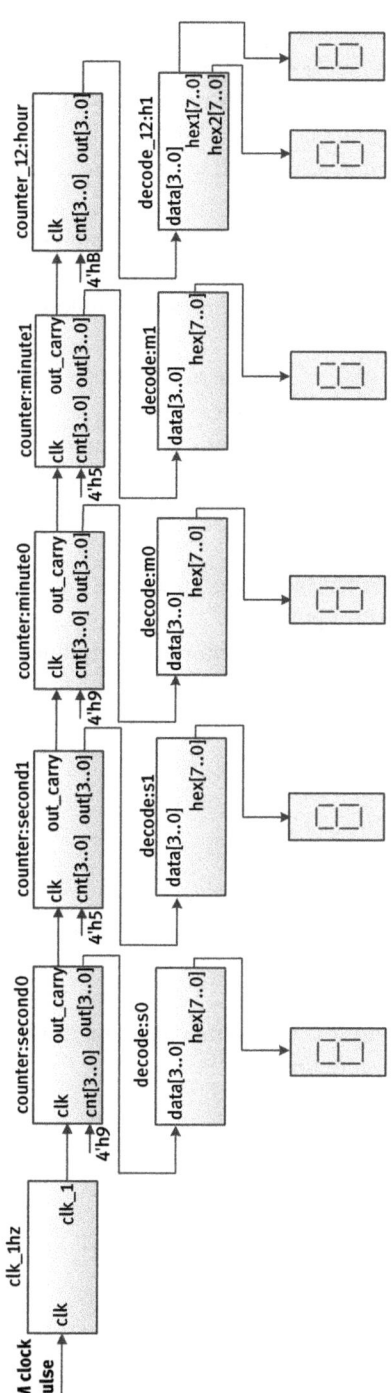

Figure 7.4.5: Schematic diagram.

5. The maximum modulus of a binary counter is the maximum number of possible states and equal to 2^n where n is the number of flip-flops in the counter.
6. One way to modify a counter's MOD number is to add circuitry that forces it to recycle before going through all of its possible states.
7. The 74LS93 is a specific MSI asynchronous counter, which can be configured as a MOD-2, MOD-8, or MOD-16 counter.
8. A decade counter is referred to any MOD-10 counter. A BCD counter is a decade counter that sequences through the ten BCD codes (0–9).
9. In a synchronous (parallel) counter, all of the flip-flops are simultaneously clocked by a global clock input.
10. An up/down counter is a bidirectional counter that is capable of counting either up or down. The 74HC190 is an up/down MSI decade counter.
11. The 74LS161 and the 74LS163 are both four-bit synchronous MSI binary up counters, which have clear (or reset) input, synchronous load input, and two enable inputs. The clear input is asynchronous for 74LS161 and synchronous for 74LS163.
12. By using the counter and combinational circuits, sequence generator can be constructed to generate a serial code in the certain period.
13. Counters can be cascaded to produce a higher-modulus counter and frequency-division. The overall modulus of cascaded counters is equal to the product of the moduli of individual counters.

Key terms

Modulus: The number of states in a counter
Recycle: To undergo transition from final or terminal state back to initial state
Decade: Characterized by ten states or values
Terminal count: The final state in a counter's sequence
Cascade: To connect "end to end" when several counters are connected from the TC output of one counter to the enable input of the next counter

Self-test

7.1 An asynchronous counter is also known as _____ counter

7.2 A four-bit binary counter has a maximum modulus of _____.

7.3 A modulus-12 counter must have _____ flip-flops.

7.4 The modulus of a counter is _____.
    (a) the number of flip-flops

(b) the actual number of state in its sequence

(c) the number of times it recycles in a second

(d) the maximum possible number of state

7.5   An asynchronous counter differs from a synchronous counter in _____.

(a) the number of states in its sequence

(b) the method of clocking

(c) the type of flip-flops

(d) the value of the modulus

7.6   For the circuit shown in Figure T7.1, assume initial state is $Q_2Q_1Q_0 = 000$, and the circuit works as a MOD- _____ counter.

(a) 3   (b) 4   (c) 5   (d) 6

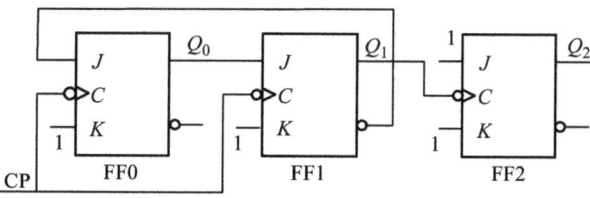

Figure T7.1

7.7   Three cascaded modulus-10 counters have an overall modulus of _____.

7.8   _____ is an example of a specific integrated circuit asynchronous counter.

7.9   The 74LS161 is an example of an integrated circuit _____ bit synchronous binary counter.

7.10 A four-bit binary up/down counter is in the binary state of zero, and the next state in the DOWN mode is _____.

(a) 0001   (b) 1111   (c) 1000   (d) 1100

7.11 The TC of a modulus-13 binary counter with initial state 0000 is _____.

(a) 0000   (b) 1111   (c) 1101   (d) 1100

Problems

7.1   Construct a MOD-5 and a MOD-12 counters with 74LS93.

7.2   For the ripple counter shown in Figure P7.1, sketch the timing diagram including the waveform of the clock, Q_0, Q_1, and Q_2 by applying 16 clock pulses.

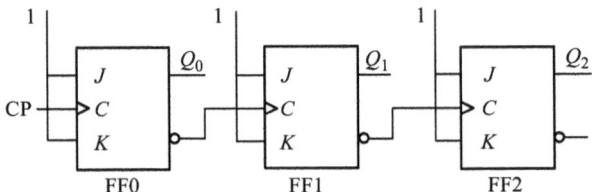

Figure P7.1

7.3 Show how to connect a four-bit asynchronous counter for each of the following moduli using the clear input:
(a) 9   (b) 11   (c) 13   (d) 14   (e) 15

7.4 Design a MOD-4 synchronous counter with D flip-flop.

7.5 Construct a MOD-12 counter with one 74LS161. The initial state is 0100.

7.6 Using asynchronous reset to construct a MOD-9 up counter with one 74LS161.

7.7 The waveforms in Figure P7.2 are applied to a 74LS163 counter. Determine the Q outputs and the RCO waveforms. The inputs are $D_0 = 1$, $D_1 = 1$, $D_2 = 0$, and $D_3 = 1$, $CP_T = 1$.

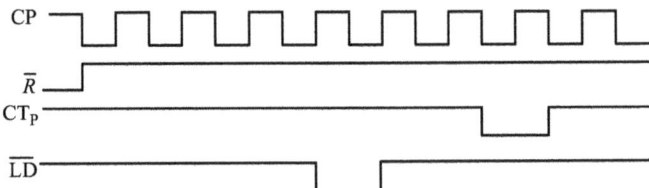

Figure P7.2

7.8 Develop the Q output waveform for a 74HC190 up/down counter with input waveforms in Figure P7.3. Binary 0s are on the data inputs and initial state of counter is 0000.

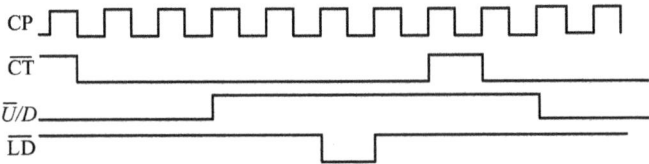

Figure P7.3

7.9 Determine the sequence of the counter shown in Figure P7.4.

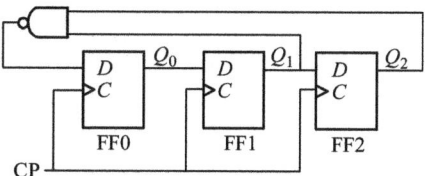

Figure P7.4

7.10 Design a counter to produce the following binary sequence by using J–K flip-flops. 1,4,3,5,7,6,2,1,......

7.11 Determine the modulus of the following counters composed of 74LS161 and 74LS163, as shown in Figure P7.5. List the state sequence table.

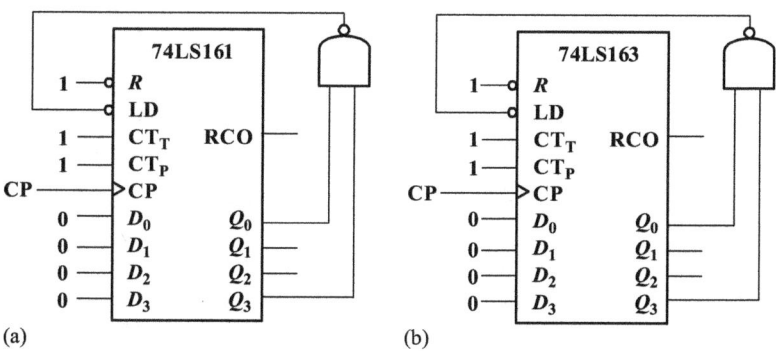

Figure P7.5

7.12 Determine the modulus of the following counters composed of 74LS161 and 74LS163, as shown in Figure P7.6. List the state sequence table.

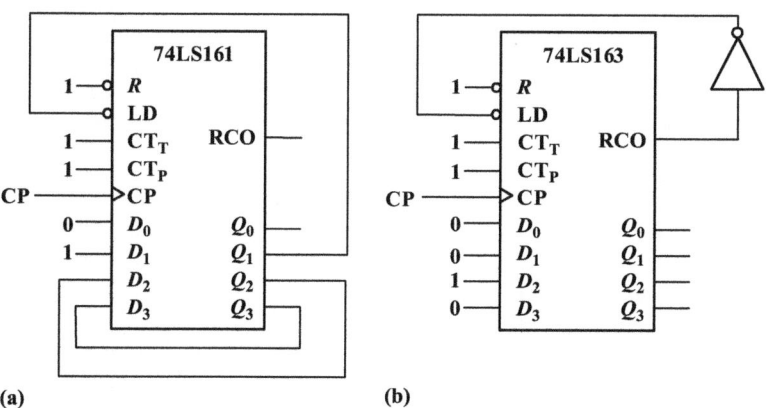

Figure P7.6

7.13 Determine the modulus of the following counters composed of 74LS161 and 74LS163, as shown in Figure P7.7. List the state sequence table.

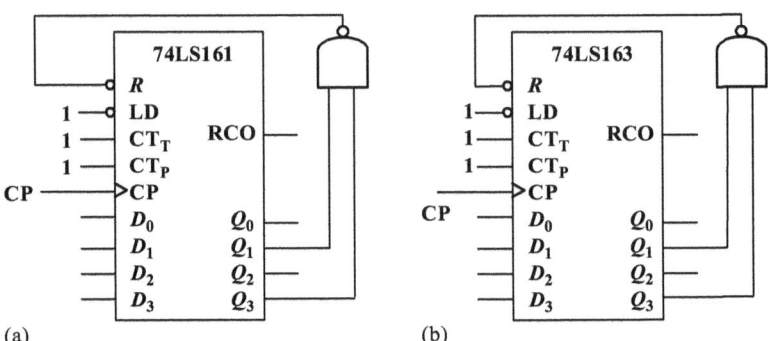

(a)   (b)

Figure P7.7

7.14 Determine the modulus of the following counters composed of 74LS93, as shown in Figure P7.8. List the state sequence table.

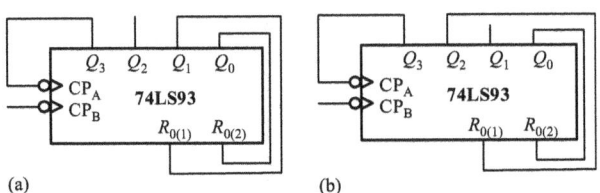

(a)   (b)

Figure P7.8

7.15 Determine the modulus of the following counters composed of 74LS161, as shown in Figure P7.9.

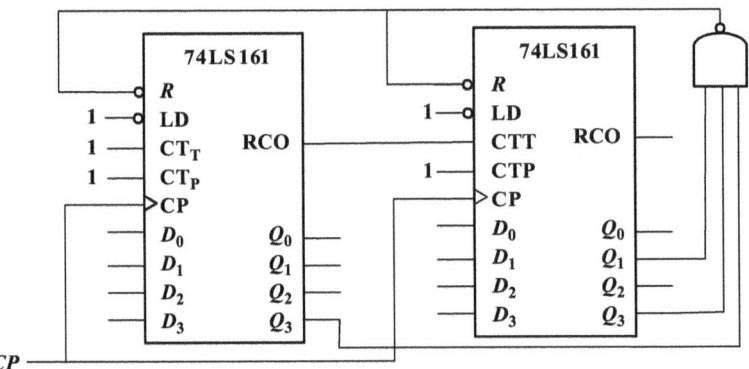

Figure P7.9

7.16 Design a modulus-60 counter with 74LS161s and a few gates.

7.17 Design a modulus-462 counter with 74LS163s and a few gates.

7.18 Design a sequence generator that can generate a sequence code "10110001110101" with a 74LS161, an 8-to-1 multiplexer, and a few gates.

7.19 A circuit is shown in Figure P7.10. The frequency of the clock pulse is 14 kHz.

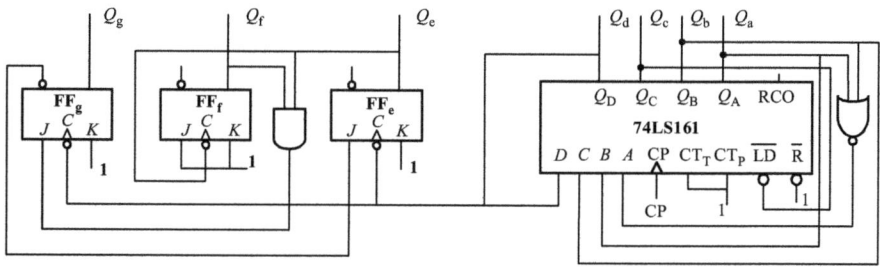

Figure P7.10

(a) List the state table of the counter constructed by the 74LS161 and determine the modulus of this counter.
(b) Analyze the synchronous counter constructed by the flip-flops FF_e, FF_f, and FF_g. Give the state table and draw the state diagram, and determine the modulus of the counter.
(c) Determine the frequency of Q_d and Q_g.

8 Registers and shift registers

Registers and shift registers are commonly used sequential circuits in a digital system. Registers are often used to store binary data or information. Shift register consisting of a group of flip-flops not only stores binary data but also transfers data in a digital system. Especially, shift registers can be applied to construct a specific digital counter. This chapter introduces registers and shift registers. Several applications of shift registers are covered.

The objectives of this chapter are to
- Explain how register stores binary data
- Describe the basic form of data movement in shift register
- Explain the operation of four kinds of shift registers
- Explain the operation of bidirectional shift register
- Use shift registers to construct a ring counter
- Apply shift registers to construct a Johnson counter
- Apply shift registers to construct a sequence generator

8.1 Registers

In a digital system, data are normally stored in a group of bits that represent numbers, codes, or other information. Register is a type of sequential circuits in which binary data and information can be stored. A n-bit register consists of n flip-flops, which is able to store n-bit binary data or information. The storage capability of a register makes it an important type of memory device in digital systems. This section introduces how register store binary data and several MSI registers.

Figure 8.1.1 shows the concept of storing a 0 or a 1 in a D flip-flop. Assume that the D flip-flop is in the SET state and a 0 is applied to the data input; when the triggering edge of the clock pulse is applied, a 0 can be transferred to the corresponding Q output by resetting the flip-flop. Even if the 0 on the input is removed after the clock pulse, the flip-flop remains in the RESET state, thereby storing the 0. A similar procedure applies to the storage of the 1 and can be realized by applying a 1 to the data input when the flip-flop is in the RESET state, as shown in Figure 8.1.1(b).

The storage capacity of a register is total number of bits of digital data it can retain. Each flip-flop in a register represents one bit of storage capacity, so the storage capacity of a register is determined by the number of flip-flops in the register. A common requirement in digital system is to store several bits of data from parallel lines simultaneously in a group of flip-flops. A 4-bit register is composed of four D flip-flops, as illustrated in Figure 8.1.2. Each of the four parallel data lines is connected to the D input of each flip-flop. The clock inputs of the flip-flops are connected

https://doi.org/10.1515/9783110614916-008

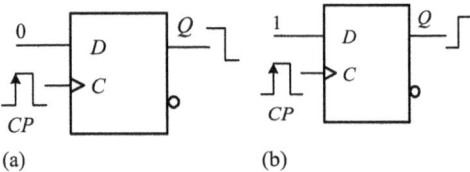

Figure 8.1.1: A flip-flop as a storage element: (a) storing a 0; (b) storing a 1.

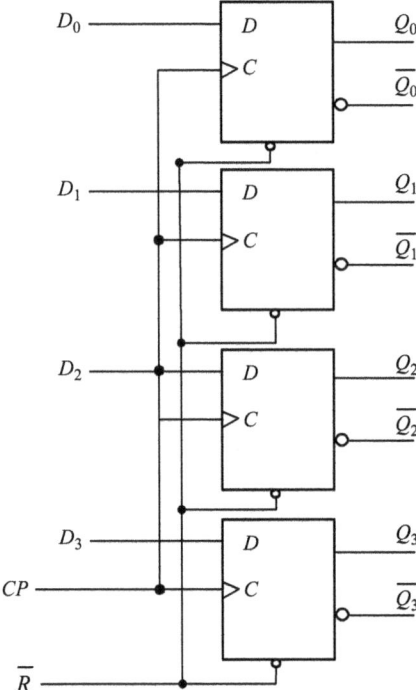

Figure 8.1.2: A 4-bit register for 4-bit parallel data storage.

together so that each flip-flop is triggered by the same clock pulse. Thereby, the parallel data on the D inputs are stored simultaneously into the flip-flops on the positive-going edge of the clock. In addition, the asynchronous reset inputs are connected to a common $\bar{R}$ line. When a LOW is initially applied on the $\bar{R}$ line, all flip-flops can be reset directly and the initial state of the register is 0000.

One of the common used MSI registers is the 74LS175. The 74LS175 is a high-speed Quad D flip-flop with a common clock and an asynchronous clear input, which the internal logic diagram is shown in Figure 8.1.2. The data on the D inputs is stored into the corresponding outputs at the positive edge of clock pulse. Both true and complemented outputs of each flip-flop are provided. Logic symbol of the 74LS175 is shown in Figure 8.1.3.

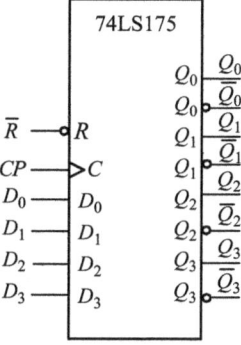

Figure 8.1.3: Logic symbol of 74LS175.

Another commonly used MSI register is the 74LS273. The 74LS273 contains 8 edge-triggered D flip-flops with a common direct reset and a clock input. The data input D immediately appears in the Q output when the positive going edge of clock pulse arrives. The common direct reset is an active LOW input. When a LOW is applied on the common direct reset, all D flip-flop is reset. Normally, the common direct reset must be kept in HIGH, so that the function of D flip-flop can be used. The 74LS273 is widely used for industrial and consumer equipment. The logic symbol of 74LS273 is shown in Figure 8.1.4.

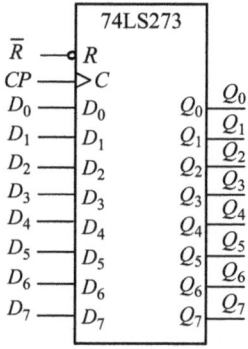

Figure 8.1.4: Logic symbol of 74LS273.

8.2 Shift registers

Data storage and data movement are two basic functions of the register. Shift register has the shift capability, which permits data movement from stage to stage within the register, into, or out of the register upon application of clock pulse. According to the type of data movement, shift registers can be divided into different types: serial in/serial out, serial in/parallel out, parallel in/serial out, parallel in/parallel out shift registers[3].

In this section, the basic types of shift registers are introduced, such as serial in/serial out, serial in/parallel out, parallel in/serial out, and bidirectional shift registers.

The objectives of this section are to
- Explain how data bits enter into a shift register
- Describe how data bits shift through the register
- Explain how data bits are taken out of a shift register

8.2.1 Serial in/serial out shift registers

The serial in/serial out shift register accepts data entry serially and produces the stored information on its output in serial form as well. Figure 8.2.1 shows a serial in/serial out shift register with four D flip-flops, which can store 4-bit binary data. There is a serial data input, D_{in}, and a serial data output, Q_3. The first flip-flop FF0 stores the most significant bit (MSB) and FF4 stores the least significant bit (LSB). When clock pulse is applied, the present state of high bit flip-flops are sent to the input of the next low bit flip-flops as the output of next state.

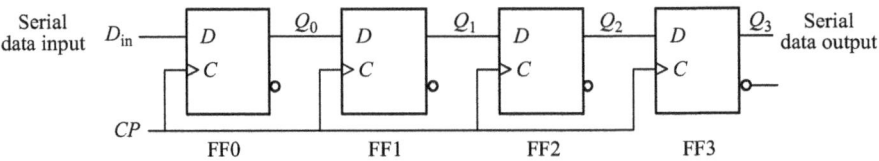

Figure 8.2.1: A serial in/serial out shift register.

If a 4-bit data 1101 is stored into the register, the data is sent to the serial data input serially, beginning with the least significant bit. The first bit 1 is put onto the data input line, making D_{in} = 1 for FF0. When the first clock pulse is applied, FF0 is SET, thus storing the 1.

Then the second bit 0 is applied to the input making D_{in} = 0 for FF0 and $D_1 = Q_{0n} = 1$ for FF1 because the input of FF1 is connected to the Q_0 output. When the second clock pulse is applied, the 0 on the data input is shifted into FF0 causing FF0 to RESET, and the 1 stored in FF0 is shifted into FF1.

The third bit 1 is now put onto the data input and the third clock pulse is applied, this time the 1 is entered into FF0, the 0 stored in FF0 is shifted into FF1 and the 1 stored in FF1 is shifted into FF2.

The last bit 1 is on the data input and a clock pulse is applied, this time the 1 is entered into FF0, the 1 stored in FF0 is shifted to FF1, and the 0 stored in FF1 is shifted to FF2, the 1 stored in FF2 is shifted into FF3. This completes the serial entry of the four bits into the shift register. The stored process of 4-bit data 1101 is summarized in the state table shown in Table 8.2.1. It can be seen that the present output state of

Table 8.2.1: Sate table.

CP	Q_0	Q_1	Q_2	Q_3
1	1	x	x	x
2	0	1	x	x
3	1	0	1	x
4	1	1	0	1

high bit flip-flops are sent to the input of next low bit flip-flops as the output of next state when clock pulse is applied one by one. Because the register shifts begin with LSB, this kind of registers is called as *right shift register*. In contrast, if a shift register begins with MSB, it is called as *left shift register*.

If you want to get the data out of the register, the bits must be shifted out serially and output from the LSB flip-flop. The first bit 1 can be output from Q_3 after the four clock pulse. When the fifth clock is applied, the second bit 0 appears on the Q_3 output. Then the third bit 1 at the sixth clock pulse and the fourth bit 1 at the seventh clock pulse are shifted to the Q_3 output serially. Notice that while all 4-bit data are being shifted out, more bits can be shifted in.

8.2.2 Serial in/parallel out shift registers

A serial in/parallel out shift register is similar to the serial in/serial out shift register. Data bits are input into the register serially by a bit-by-bit mode. The difference is the way in which the data bits output from the register. Therefore, a serial in/parallel out shift register can convert data from serial format to parallel format. If four data bits will shift into the register after four clock pulses, as shown in Figure 8.2.2, the data becomes available simultaneously on the four outputs Q_0 to Q_3.

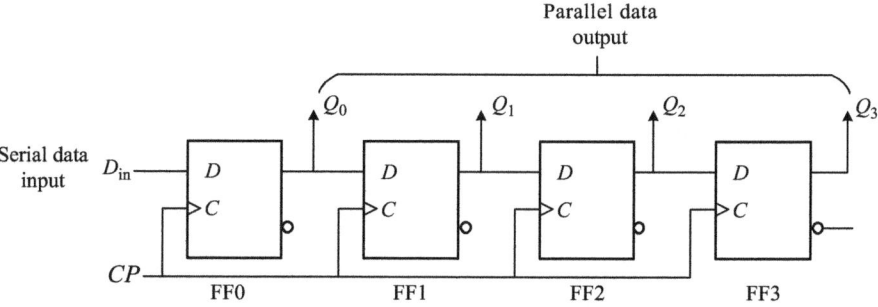

Figure 8.2.2: A serial in/parallel out shift register.

8.2.3 Parallel in/serial out shift registers

For a register with parallel data inputs, the bits enter into the register simultaneously in their respective states occur at parallel lines. Figure 8.2.3 illustrates a parallel in/ serial out shift register. There are four inputs, D_0, D_1, D_2 and D_3, and the SHIFT/$\overline{\text{LOAD}}$ input, which allows four bits of data to *load* in parallel into the register.

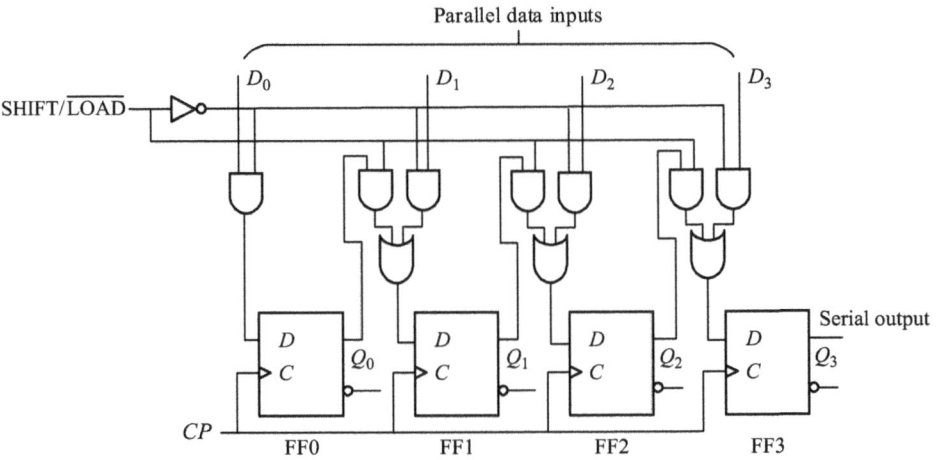

Figure 8.2.3: A parallel in/serial out shift register.

When SHIFT/$\overline{\text{LOAD}}$ input is LOW, each data bit is placed on the D input of its respective flip-flop. When a clock pulse is applied, the flip-flops with $D = 1$ will SET and those with $D = 0$ will RESET, thereby storing all four bits simultaneously.

When SHIFT/$\overline{\text{LOAD}}$ input is HIGH, the data bits shift right from one stage to the next. The OR gates allow either the normal shifting operation or the parallel data-entry operation, depending on which AND gates are enabled by the level of the SHIFT/$\overline{\text{LOAD}}$. The data bits can be taken out from the Q_3 output serially by applying four clock pulses.

8.2.4 Parallel in/parallel out shift registers

The parallel in/parallel out register is shown in Figure 8.2.4. There are four inputs, D_0, D_1, D_2 and D_3, and four outputs, Q_0, Q_1, Q_2 and Q_3, which allows four bits of data to *load* in parallel into the register. When the input bits are applied on the data inputs, they simultaneously appear on the parallel outputs on the positive going edge of the clock pulse.

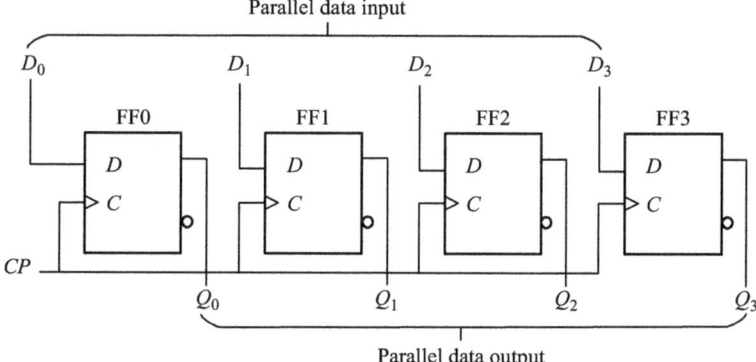

Figure 8.2.4: A parallel in/parallel out shift register.

8.2.5 MSI shift registers

1. A 4-bit right shift register

The 74HC195 is an example of MSI shift registers. It has four parallel inputs, $D_0D_1D_2D_3$, and four parallel outputs, $Q_0Q_1Q_2Q_3$, so the parallel in/parallel out operation can be implemented. Since it also has serial inputs, J and $\bar{K}$, it can be used for serial in/serial out and serial in/parallel out operations. The parallel in/serial out operation can also be implemented by taking the data out from the Q_3 output. A typical logic symbol for 74HC195 is shown in Figure 8.2.5. $\bar{R}$ is an active-LOW asynchronous clear input. When $\bar{R}$ is LOW, the output $Q_0Q_1Q_2Q_3$ is reset to 0000. The $\overline{LD}$ input (also marked as SHIFT/$\overline{LOAD}$) can select the operation mode of shift and load. When the $\overline{LD}$ is an active-LOW, the data on the parallel inputs are synchronously sent to the parallel output on the positive edge of the clock. When $\overline{LD}$ is a HIGH, the stored data will shift right, from Q_0 to Q_3 synchronously with the clock. J and $\bar{K}$ are the serial inputs to the first stage of the register (Q_0), and Q_3 can be used for serial output. The function of 74HC195 is summarized in Table 8.2.2.

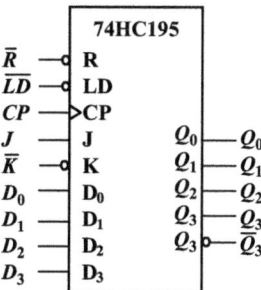

Figure 8.2.5: Logic symbol for 74HC195.

Table 8.2.2: Function table for 74HC195.

Inputs									Outputs			
CP	$\bar{R}$	$\overline{LD}$	J	$\bar{K}$	D_0	D_1	D_2	D_3	$Q_0{}^+$	$Q_1{}^+$	$Q_2{}^+$	$Q_3{}^+$
x	0	x	x	x	x	x	x	x	0	0	0	0
↑	1	0	x	x	D_0	D_1	D_2	D_3	D_0	D_1	D_2	D_3
0	1	1	x	x	x	x	x	x	Q_0	Q_1	Q_2	Q_3
↑	1	1	0	1	x	x	x	x	Q_0	Q_0	Q_1	Q_2
↑	1	1	0	0	x	x	x	x	0	Q_0	Q_1	Q_2
↑	1	1	1	1	x	x	x	x	1	Q_0	Q_1	Q_2
↑	1	1	1	0	x	x	x	x	$\bar{Q}_0$	Q_0	Q_1	Q_2

2. Bidirectional shift register

So far, the registers discussed involved only right shift operations. Each right shift operation has the effect of successively dividing the binary number by two. If the operation is reversed (left shift), this has the effect of multiplying the number by two. *Bidirectional shift register* is one kind of register in which the data can be shift either left or right. This register can be implemented with suitable gate logic that enables the transfer of a data bit from one flip-flop to the next one using right shift mode or left shift mode. The control lines are used to select the shift mode.

The 74HC194 is an example of universal bidirectional MSI shift registers. It is 4-bit multi-function device that can be used in either serial-to-serial, left shift, right shift, serial-to-parallel, parallel-to-serial, or as a parallel-to-parallel multifunction data register, hence named "Universal." Figure 8.2.6 shows logic symbol of the 74HC194. Similar with 74HC195, D_0, D_1, D_2, and D_3 are parallel inputs, Q_0, Q_1, Q_2 and Q_3 are parallel outputs; the active-LOW clear input $\bar{R}$ is an asynchronous clear input. D_{SR} is the serial input of right shift and D_{SL} is the serial input of left shift. M_A and M_B are mode control inputs. When the M_A and M_B inputs are both HIGH, parallel loading, which is synchronous with a positive edge of the clock, is accomplished by applying the four bits of data to the parallel inputs D_0, D_1, D_2, and D_3. When M_A is

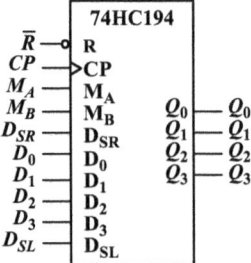

Figure 8.2.6: Logic symbol for 74HC194.

HIGH and M_B is LOW, data shift right synchronously with the positive edge of the clock, and new data are entered at the shift-right serial input (D_{SR}). When M_A is LOW and M_B is HIGH, data shift left synchronously with the positive edge of clock, and new data are entered at the shift-left serial input (D_{SL}). The function of 74HC194 is listed in Table 8.2.3.

Table 8.2.3: Function table for 74HC194.

Inputs										Outputs			
CP	$\bar{R}$	D_{SR}	D_{SL}	M_A	M_B	D_0	D_1	D_2	D_3	Q_0^+	Q_1^+	Q_2^+	Q_3^+
x	0	x	x	x	x	x	x	x	x	0	0	0	0
0	1	x	x	x	x	x	x	x	x		No change		
↑	1	x	x	1	1	D_0	D_1	D_2	D_3	D_0	D_1	D_2	D_3
↑	1	1	x	1	0	x	x	x	x	1	Q_0	Q_1	Q_2
↑	1	0	x	1	0	x	x	x	x	0	Q_0	Q_1	Q_2
↑	1	x	1	0	1	x	x	x	x	Q_1	Q_2	Q_3	1
↑	1	x	0	0	1	x	x	x	x	Q_1	Q_2	Q_3	0
x	1	x	x	0	0	x	x	x	x		No change		

8.3 Application of shift registers

Shift registers have various applications apart from being used as data storage or data movement. They can be used to develop specific counters, convert the data from either serial-to-parallel or parallel-to-serial format, construct sequence code generator. This section introduces a few applications of shift registers.

The objectives of this section are to
- Explain the operation of a ring counter
- Explain the operation of a Johnson counter
- Use the shift register for converting data between a serial and parallel format
- Use the shift register to construct sequence code generator

8.3.1 Shift register counters

The most popular application of shift registers is to construct shift register counter. A shift register counter is a shift register with the serial output connected back to the serial input to produce a special sequence. These devices are often classified as counters because they exhibit a specified sequence of states. Two of the most common types of shift register counters are *ring counter* and *Johnson counter*.

1. The ring counter

A ring counter is a type of counters composed of a type of circular shift register and it has the advantage that decoding gates are not required. The ring counter utilizes one flip-flop for each state in its sequence, a n-bit ring counter cycles through n states.

Figure 8.3.1 shows a 4-bit ring counter that is consisted of four D flip-flops. The output of last flip-flop (FF3) is connected back with the D input of the first flip-flop (FF0). This feedback arrangement produces a characteristic sequence of states, as shown in Figure 8.3.2. If a 1 is preset into the first D flip-flop, and the rest of the flip-flops are cleared. Then, the 1 is always retained in the counter and simply shifted around the ring.

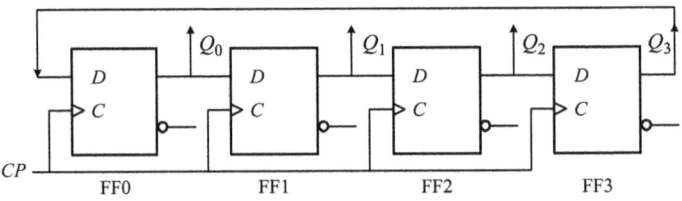

Figure 8.3.1: A 4-bit ring counter.

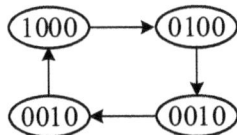

Figure 8.3.2: State diagram.

Figure 8.3.3 illustrates the 74HC195 connected as a 4-bit ring counter. Initially, a LOW level must be applied on the $\overline{LD}$ input and data 1000 is applied to $D_0D_1D_2D_3$ so that the counter begins with the state 1000. After the application of the clock pulse, the 1 continues to circulate through the ring counter, as shown in Table 8.3.1. The timing diagram is shown in Figure 8.3.4.

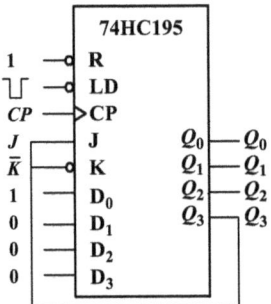

Figure 8.3.3: A 74HC195 connected as a 4-bit ring counter.

Table 8.3.1: 4-bit ring sequence.

CP	Q_3	Q_2	Q_1	Q_0
0	1	0	0	0
1	0	1	0	0
2	0	0	1	0
3	0	0	0	1

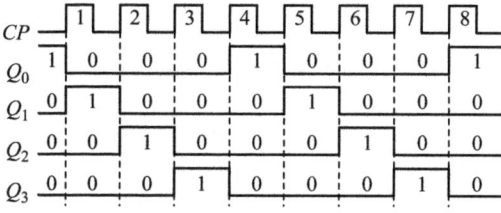

Figure 8.3.4: Timing diagram of a 4-bit ring counter.

2. The Johnson counter

A Johnson counter is a ring counter with an inversion, where the complement of the output of the last flip-flop is connected back to the input of the first flip-flop, so it is also called as *twisted-ring counter*. The implementation of a 4-bit Johnson counter is shown in Figure 8.3.5. The Q output of each flip-flop is connected to the D input of the next flip-flop. The single exception is that the $\bar{Q}_3$ output of the last flip-flop is connected back to the D input of the first flip-flop.

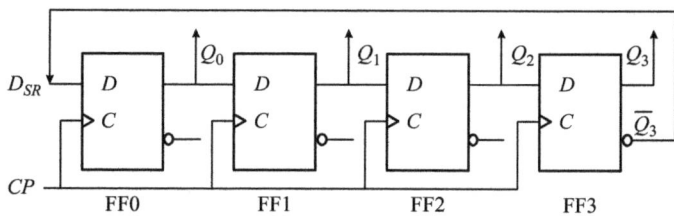

Figure 8.3.5: A 4-bit Johnson counter.

Figure 8.3.6 illustrates the 74HC195 connected as a 4-bit Johnson counter. The $\bar{Q}_3$ is connected to the J and $\bar{K}$ inputs. The initial state of 74HC195 is in RESET state by applying a negative pulse to the asynchronous clear input. This makes the $\bar{Q}_3$ output HIGH. For the first closk pulse, the register performs the shift right operation and Q_0 goes to HIGH. After the fourth clock pulse, Q_0, Q_1, Q_2 and Q_3 are both HIGH, and $\bar{Q}_3$ goes to LOW, so the inputs J and $\bar{K}$ are LOW. With the following clock pluses, Q_0, Q_1, Q_2 and Q_3 turn LOW one by one, as the sequences shown in Table 8.3.2. The counter

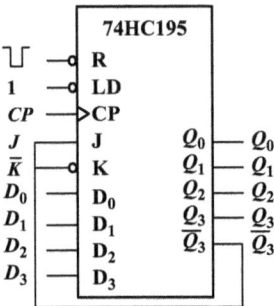

Figure 8.3.6: 74HC195 connected as a 4-bit Johnson counter

will fill up with 1s from left to right and then fill up with 0s again. The modulus of the 4-bit Johnson counter is MOD 8. In general, a Johnson counter will produce a modulus of $2n$, where n is the number of flip-flops used in the counter, so it requires fewer flip-flops than ring counters but more flip-flops than binary counters.

Notice that the 74HC195 consists of four J-K flip-flops, so it can provide at most 16 states, but only 8 states are used for the circuit shown in Figure 8.3.6. You can prove it by yourself that the rest of the 8 states form another counter cycle, which is an unused state cycle for the Johnson counter. The count cycle shown in Table 8.3.2 is regared as the used state cycle. In order to make sure that the counter enter the used state cycle, the counter must be reset at the beginning by applying a negative pulse to the asynchronous clear input.

Table 8.3.2: 4-bit Johnson sequence.

CP	Q_0	Q_1	Q_2	Q_3
0	0	0	0	0
1	1	0	0	0
2	1	1	0	0
3	1	1	1	0
4	1	1	1	1
5	0	1	1	1
6	0	0	1	1
7	0	0	0	1
8(recycle)	0	0	0	0

8.3.2 Frequency divider based on shift register

In viewpoint of clock pulse, a counter is actually a frequency divider. For example, the Johnson counter with MOD 8. If the output pulse is taken out from Q_0 or Q_3, its period is eight times of the period of the clock pulse. That is to say, the frequency of the output pulse is one-eighth as that of clock pulse. Therefore, the Johnson counter with MOD-8 is also called divide-by-8 frequency divider. Figure 8.3.7 shows different frequency dividers by using shift register, in which frequency division signal is taken out from the output representing by arrow.

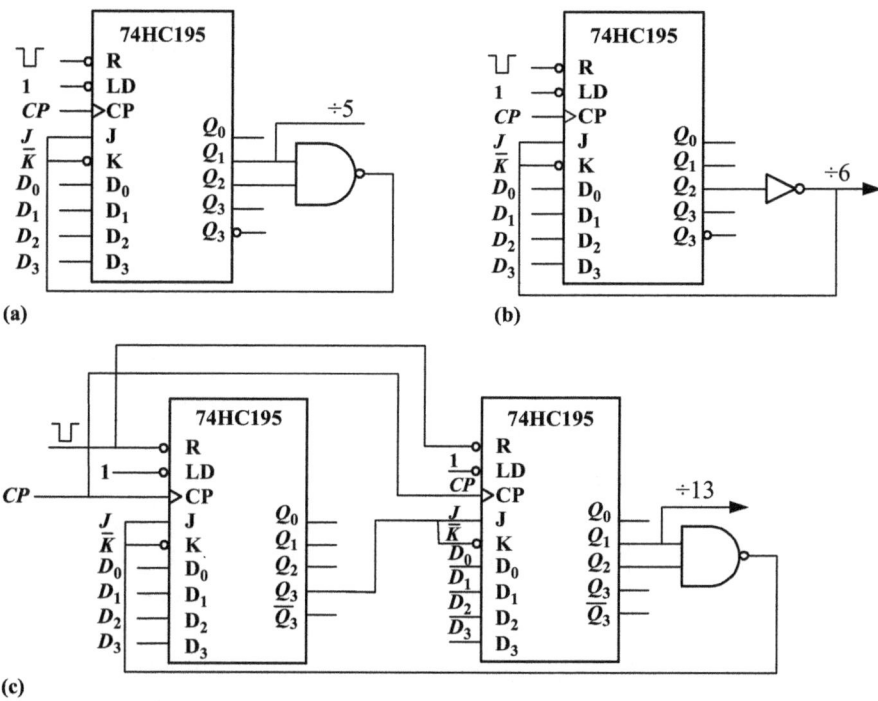

Figure 8.3.7: Frequency divider.

It can be seen from Figure 8.3.6 that the counter with MOD $2n+1$ can be constructed by using a NAND gate with Q_n and Q_{n-1} as its inputs, forming odd number frequency divider. Notice that the frequency divider based on Johnson counter must be reset by applying a negative pulse to asynchronous clear input at the beginning of the operation of the circuit.

8.3.3 Serial/parallel conversion

One of the most common applications of shift registers is to convert parallel data into serial format for transmission or storage, and to convert serial data back to parallel format for processing or displaying. For example, a computer or microprocessor-based system commonly requires incoming data to be in parallel format, thus the requirement for serial-to-parallel conversion.

Figure 8.3.8 shows a 7-bit serial-to-parallel data converter with two 74LS194. The serial input data $(D_6D_5D_4D_3D_2D_1D_0)$ is loaded into the serial input of shift right D_{SR} and the parallel input D_0 of the first circuit 74LS194-1.

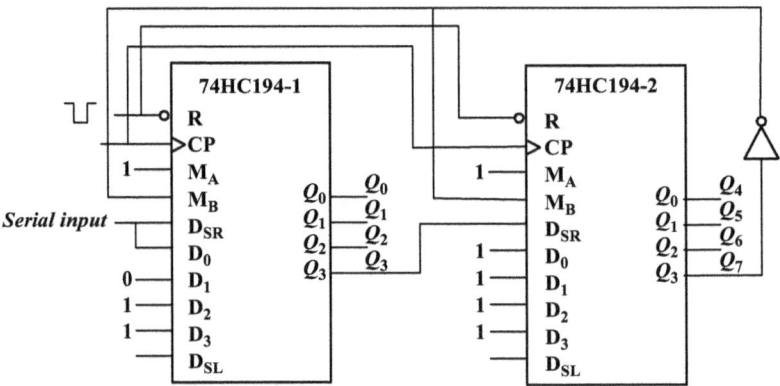

Figure 8.3.8: A 7-bit serial-to-parallel data converter with two 74LS194s.

Table 8.3.3 illustrates the conversion procedure from serial-to-parallel data. It can be seen from Table 8.3.3 that the circuit is RESET with $M_A=M_B=1$ before it starts and then the data is parallel loaded into the shift register, the output data is shown as the second row in the Table 8.3.3. After the positive edge of the first clock pulse is coming, the output data changes to be $D_0 0111111$. With the following clock pulses, the register shifts to the right side. After the positive edge of the seventh pulse clock, the output data is $D_6D_5D_4D_3D_2D_1D_0$, which indicates the circuit accomplishes the serial-to-parallel data conversion.

Table 8.3.3: Transition table for 7-bit serial-to-parallel data converter.

CP	Q_0	Q_1	Q_2	Q_3	Q_4	Q_5	Q_6	Q_7	M_A	$M_B = \bar{Q}_7$	Comments
0	0	0	0	0	0	0	0	0	1	1	Parallel load
1	D_0	0	1	1	1	1	1	1	1	0	Right shift
2	D_1	D_0	0	1	1	1	1	1	1	0	Right shift
3	D_2	D_1	D_0	0	1	1	1	1	1	0	Right shift
4	D_3	D_2	D_1	D_0	0	1	1	1	1	0	Right shift
5	D_4	D_3	D_2	D_1	D_0	0	1	1	1	0	Right shift
6	D_5	D_4	D_3	D_2	D_1	D_0	0	1	1	0	Right shift
7	D_6	D_5	D_4	D_3	D_2	D_1	D_0	0	1	1	Parallel load

Notice that the Q_7 output of 74LS194-2 can be regarded as a flag bit, which indicates the transition state. When Q_7 is HIGH, the circuit is converting the serial input data into parallel format, when Q_7 is LOW, the circuit accomplishes the serial-to-parallel data conversion.

Analogously, Figure 8.3.9 shows a 7-bit parallel-to-serial data converter. The conversion procedure is shown in Table 8.3.4; you can analyze the function of the circuit yourself.

Table 8.3.4: Transition table for 7-bit parallel-to-serial data converter.

CP	74LS194-1				74LS194-2			
	Q_0	Q_1	Q_2	Q_3	Q_0 (Q_4)	Q_1 (Q_5)	Q_2 (Q_6)	Q_3 (Q_7)
0	0	0	0	0	0	0	0	0
1	0	D_0	D_1	D_2	D_3	D_4	D_5	D_6
2	1	0	D_0	D_1	D_2	D_3	D_4	D_5
3	1	1	0	D_0	D_1	D_2	D_3	D_4
4	1	1	1	0	D_0	D_1	D_2	D_3
5	1	1	1	1	0	D_0	D_1	D_2
6	1	1	1	1	1	0	D_0	D_1
7	1	1	1	1	1	1	0	D_0

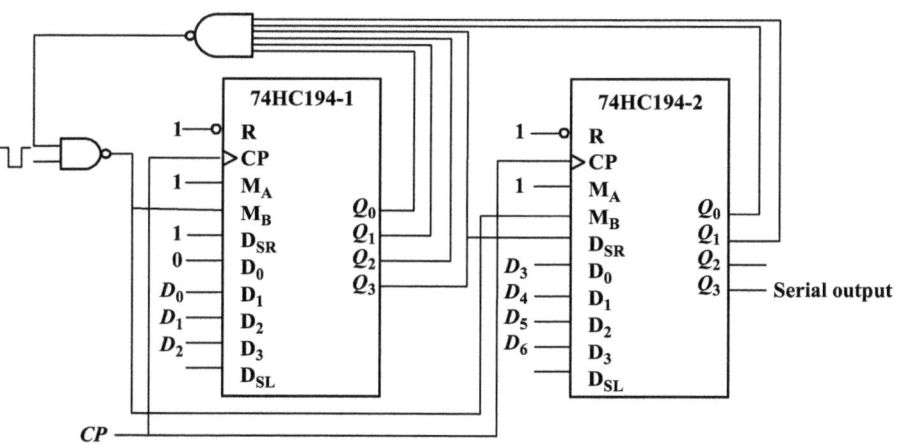

Figure 8.3.9: A 7-bit serial-to-parallel data converter with two 74LS194s.

8.3.4 Sequence generator

A predetermined sequence of binary code can be produced by using a sequence generator. The sequence maybe has indefinite length or a predetermined fixed length. Sequence generators are useful in a wide variety of coding and control applications. A binary counter is a special type of sequence generator, which is already discussed in Chapter 7. In this section, we will discuss the linear feedback shift-register (LFSR) counter to generate sequence outputs.

A *linear feedback shift-register counter* is consisted of an n-bit shift-register and a feedback circuit. A n-bit linear feedback shift-register counter can generate the maximum sequence codes of 2^{n-1}. Such a counter is often called a *maximum-length sequence generator* [40]. For the maximum-length sequence, the occurrence of 0 and 1 in the sequence should be approximately the same.

For the maximum-length sequence generator, the feedback circuit is constructed by using the feedback equation for the n-bit linear feedback shift-register that can be looked up in the table of feedback equation, as shown in Table 8.3.5. With this table, we can design a maximum-length sequence generator by using n-bit shift register.

Table 8.3.5: Feedback equations.

n	$f(Q)$	n	$f(Q)$
1	Q_0	12	$Q_5 \oplus Q_7 \oplus Q_{10} \oplus Q_{11}$
2	$Q_0 \oplus Q_1$	13	$Q_8 \oplus Q_9 \oplus Q_{11} \oplus Q_{12}$
3	$Q_1 \oplus Q_2$	14	$Q_8 \oplus Q_{10} \oplus Q_{12} \oplus Q_{13}$
4	$Q_2 \oplus Q_3$	15	$Q_{13} \oplus Q_{14}$
5	$Q_2 \oplus Q_4$	16	$Q_{10} \oplus Q_{12} \oplus Q_{13} \oplus Q_{15}$
6	$Q_4 \oplus Q_5$	17	$Q_{13} \oplus Q_{16}$
7	$Q_5 \oplus Q_6$	18	$Q_0 \oplus Q_1 \oplus Q_4 \oplus Q_{17}$
8	$Q_1 \oplus Q_2 \oplus Q_3 \oplus Q_7$	19	$Q_{13} \oplus Q_{16} \oplus Q_{17} \oplus Q_{18}$
9	$Q_4 \oplus Q_8$	20	$Q_{16} \oplus Q_{19}$
10	$Q_6 \oplus Q_9$	21	$Q_{18} \oplus Q_{20}$
11	$Q_8 \oplus Q_{10}$	22	$Q_{20} \oplus Q_{21}$

Example 8.1
Construct a maximum-length sequence generator with sequence-code length $S = 7$.

Solution

Step 1 Construct a n-bit linear feedback shift-register (LFSR) counter with 2^{n-1} states which equals to generate sequence code length of 2^{n-1}. So we can get $n = 3$ when the sequence code length $S = 7$ (2^{n-1}).

Step 2 Determine the feedback equation.
From Table 8.3.5, when $n = 3$, the feedback equation is $f(Q) = Q_1 \oplus Q_2$

Step 3 Draw the circuit diagram
A 74HC194 is employed to implement the circuit. Note that once the outputs are all 0s, the next state will be all 0s. In order to avoid appearing all 0s, a term is added to the feedback equation.

$$f(Q) = Q_2 \oplus Q_1 + \bar{Q}_2 \bar{Q}_1 \bar{Q}_0 = Q_2 \oplus Q_1 + \overline{Q_2 + Q_1 + Q_0}$$

The resulting circuit is shown in Figure 8.3.10.

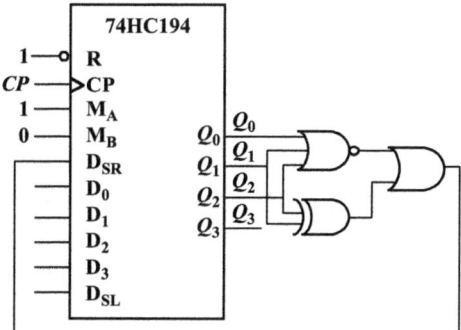

Figure 8.3.10: A maximum-length sequence generator with sequence-code length S = 7.

Another way to avoid all 0s is to use all 0s for producing the load signal. Once all 0s appears, a LOW load signal allows the data inputs 100 to be loaded to the corresponding outputs, as shown in Figure 8.3.11.

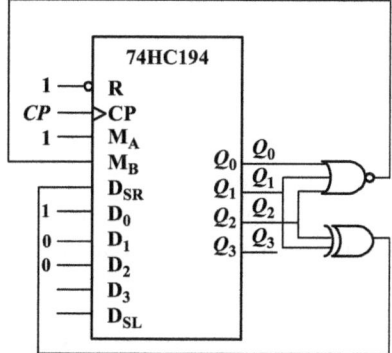

Figure 8.3.11: Sequence generator circuit for Example 8.1.

8.4 Summary

1. Register is a type of sequential circuits in which binary data and information can be stored. A n-bit register consists of n flip-flops, which is able to store n-bits binary data or information.
2. The 74LS175 is a high-speed 4-bit register for 4-bit parallel data storage and the 74LS273 is an 8-bit register with common direct reset and clock inputs.
3. Shift register not only stores binary data but also permits data movement from one stage to another within the register or into or out of the register upon application of clock pulse.

4. The common type of data movement in shift registers are serial in/serial out, serial in/parallel out, parallel in/serial out and parallel in/parallel out shift registers.
5. The 74HC195 is a 4-bit MSI shift right registers with data movement of parallel in/ parallel out, serial in/serial out, serial in/parallel out, parallel in/serial out operations.
6. The 74HC194 is a 4-bit MSI universal bidirectional shift register, which can be used in either serial-to-serial, left shift, right shift, serial-to-parallel, parallel-to-serial, or as a parallel-to-parallel multifunction data register.
7. Shift register counters are shift registers with feedback that implement special sequences. The typical examples are ring counter and Johnson counter.
8. A ring counter is actually a n-bit shift register that recirculates a single 1 continuously, thereby having n states, where n is the number of stages.
9. A Johnson counter is a modified ring counter that has $2n$ states.
10. A maximum-length sequence generator can generate the maximum sequence codes of 2^{n-1} by implementing with one n-bit shift register combined with a feedback circuit.

Key terms

Register: One or more flip-flops that used to store and shift data.
Shift register: A type of register with the shift capability permitting data movement from stage to stage within the register or into or out of the register upon application of clock pulse.
Load: To enter data into a register
Bidirectional: Having two directions. In a bidirectional shift register, the stored data can be shifted right or left.
Serial in/serial out shift register: A type of shift registers that accepts data entry serially and produces the stored information on its output in serial form as well.
Serial in/parallel out shift register: A type of shift registers that accepts data entry serially and produces the stored information on its output in parallel form.
Parallel in/serial out shift register: A type of shift registers that accepts data entry in parallel and produces the stored information on its output in serial form.
Parallel in/parallel out shift register: A type of shift registers that accepts and produces data both in parallel form.
Ring counter: A type of counter composed of a type of circular shift register. The output of the last shift register is fed back to the input of the first register.
Johnson counter: A type of counter in which the complement of the output of the last flip-flop is fed back to the input of the first flip-flop.

Self-test

8.1 Two basic functions of a register are _____ and _____.

8.2 The storage capacity of a register is determined by _____.

8.3 Shift registers can be divided into the following four types, _____, _____, _____, _____.

8.4 To serially shift a byte of data into a shift register, there must be _____.
(a) one clock pulse
(b) one load pulse
(c) eight clock pulses
(d) one clock pulse for each 1 in the data

8.5 To parallel load a byte of data into a shift register with a synchronous load, there must be _____.
(a) one clock pulse
(b) one clock pulse for each 1 in the data
(c) eight clock pulses
(d) one clock pulse for each 0 in the data

8.6 The group of bits 10110101 is serially shifted (right-most bit first) into an 8-bit parallel output shift register with an initial state of 11100100. After two clock pulse, the register contains _____.
(a) 01011110
(b) 10110101
(c) 01111001
(d) 00101101

8.7 The 74HC194 is a _____ shift register.

8.8 A modulus-10 ring counter requires _____ at least
(a) ten flip-flops
(b) four flip-flops
(c) five flip-flops
(d) twelve flip-flops

8.9 A modulus-10 Johnson counter requires _____.
(a) ten flip-flops
(b) four flip-flops
(c) five flip-flops
(d) twelve flip-flops

8.10 A linear feedback shift-register counter is consisted of _____ and _____.

8.11 A n-bit linear feedback shift-register can generate the maximum sequence codes of _____.

8.12 To generate a sequence code of 1110100 with a linear feedback shift register, the minimum number of flip-flops used is _____ .
   (a) 2
   (b) 3
   (c) 4
   (d) 5

Problems

8.1 For the data inputs and clock pulse shown in Figure P8.1, determine the state of each flip-flop in the shift register and draw the $Q_0 Q_1 Q_2 Q_3$ waveforms. Assume that the register contains all 1s initially.

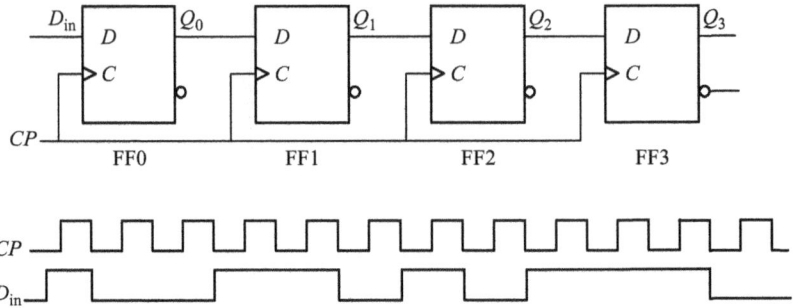

Figure P8.1

8.2 For the data inputs and clock pulse shown in Figure P8.2, sketch a complete timing diagram for all parallel outputs. Assume that the register is initially RESET.

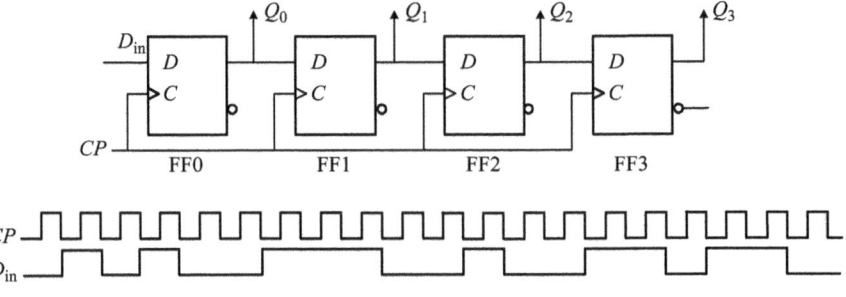

Figure P8.2

8.3   Determine the Q outputs of a 74HC194 with inputs shown in Figure P8.3. Inputs D_0, D_1, D_2 and D_3 are all HIGH.

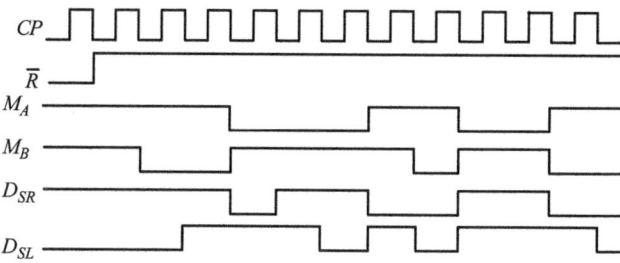

Figure P8.3

8.4   Construct the frequency division of 13 and 7 with 74HC195, respectively.

8.5   Use two 74HC195 shift registers to form an 8-bit shift register. Show the connections.

8.6   Use two 74HC194 4-bit bidirectional shift registers to create an 8-bit bidirectional shift register, show the connections.

8.7   Design an 8-bit ring counter and an 8-bit Johnson counter by using 74HC195, respectively. Determine their modulus.

8.8   Design a 4-bit Johnson counter that can be self-corrected.

8.9   Draw logic diagram for a modulus-18 Johnson counter. Show the timing diagram and write the sequence.

8.10 For the ring counter shown in Figure P8.4, draw the waveforms for each flip-flop output with respect to the clock. Assume that FF0 is initially SET and that the rest flip-flops are RESET. Show at least seven clock pulses.

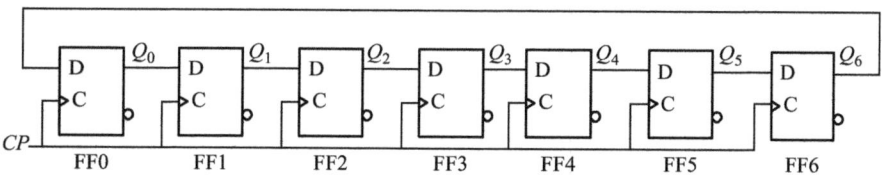

Figure P8.4

8.11 Use 74HC195 4-bit shift register to implement a 16-bit ring counter. Show the connections.

8.12 Determine the modulus of the following frequency divider composed of 74HC195, as shown in Figure P8.5.

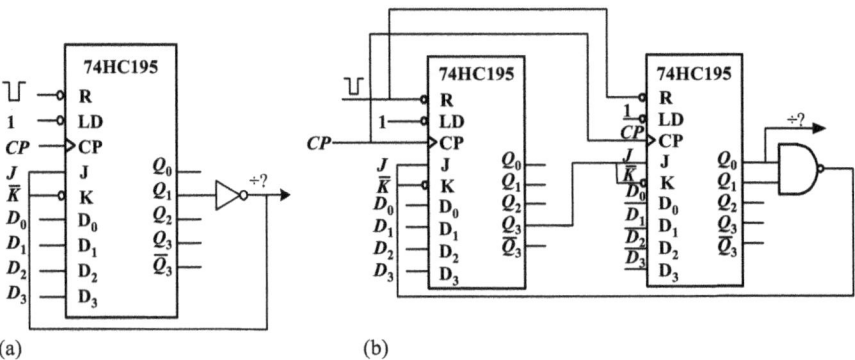

(a)   (b)

Figure P8.5

8.13 Analyze the implementing function of the circuit shown in Figure P8.6.

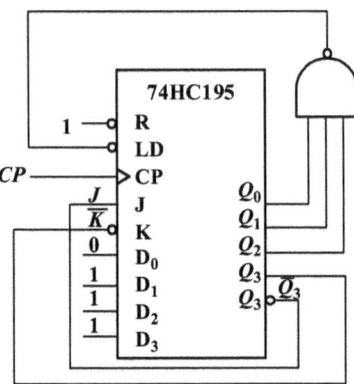

Figure P8.6

8.14 Analyze the implementing function of the circuit shown in Figure P8.7.

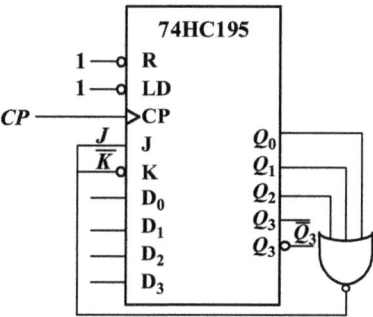

Figure P8.7

8.15 Analyze the implementing function of the circuit shown in Figure P8.8 and list the sequence from the Y output at the trigger of clock pulse.

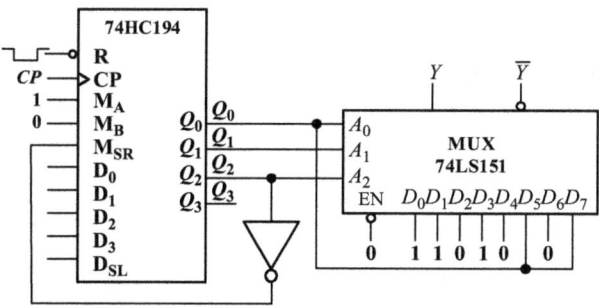

Figure P8.8

8.16 Construct a sequence generator that can generate two groups of sequence code '110101' and '010110' simultaneously by using one 74HC194, a 3-to-8 decoder and a few gates.

8.17 Construct a maximum-length sequence generator with sequence-code length $S=15$ by using 74HC194 and a few gates.

8.18 Construct a maximum-length sequence generator with sequence-code length $S=63$ by using the 74HC194s and a few gates.

9 Semiconductor memory

9.1 Introduction

Memory and storage are essential functions of most digital systems, for example, computer system and microprocessor-based system. They are used to retain binary data or information for a period of time. Generally, memory refers to relatively short-term data retention and storage refers to long-term data retention. A storage device can store a bit or a group of bits as long as necessary. Storage includes magnetic disks (hard drives), optical disks (CDs), and magnetic tapes. Memories contain flip-flops, registers, and semiconductor memories. The flip-flops and registers belong to a small-scale memory. Semiconductor memories are devices typically used for storing large number of data. Semiconductor memory is an electronic data storage device implemented on a semiconductor-based integrated circuit. Most types of semiconductor memories have the property of random access, which means that they take the same amount of time to access any memory location, so data can be efficiently accessed in any random order. There are two types of semiconductor memories. One is called the *read only memory* (ROM) in which the binary data or information are permanently or semipermanently stored and cannot be readily changed. Another is called *random access memory* (RAM) in which binary data or information are temporarily stored and can be easily changed. This chapter mainly introduces semiconductor memories. ROM and RAM are also covered.

The objectives of this chapter are to
– Explain the concept of memory
– Explain what a ROM is and how it works
– Describe the classification of ROMs
– Explain what a RAM is and how it works
– Explain the difference between static RAM (SRAM) and dynamic RAM (DRAM)
– Discuss special types of RAM
– Describe the expansion of ROM and RAM to increase the word length and word capacity

9.2 The concepts of memory

Memory can store binary data or information, which is an important part in a computer and other digital systems. Today, the capacity of semiconductor memory already reaches to the magnitude of order of TB (a terabyte is 1024 GB and a gigabyte is one billion bytes). Such a large memory uses the same operating principles as the smaller one, so we will use the smaller one for illustration to explain the concepts of memory in this chapter.

https://doi.org/10.1515/9783110614916-009

The objectives of this section are to
- Explain how a memory stores binary data
- Explain how a memory is organized
- Define memory cell
- Describe three basic operations: write, read, and addressing operations
- Define what a ROM is
- Define what a RAM is
- Explain the difference between ROM and RAM

9.2.1 Units of data

There are several basic units of binary data in memories. The commonly used units are bit, byte, nibble, and word [5]. A *bit* is the smallest unit of binary data, which represents one-bit binary number. A group of consecutive four bits is called a *nibble*. A nibble is used to represent a BCD or hexadecimal digit. Byte is commonly used for a computer system and other systems. A *byte* contains eight-bit binary numbers, which is the smallest addressable data in memory. Multi-bytes can also be grouped into a word. A *word* is defined as a group of bits or bytes acting as a single entity. In the computer assembly language, a word is directly defined as two bytes involving a high byte and a low byte. High-end (32-bit or 64-bit) microcomputers use double-word and quad-word data structures. These wide data structures are used mainly in highly pipelined and parallel microcomputers.

9.2.2 Memory organization

Memory cell is the smallest storage unit in a memory, which stores either a 1 or a 0. Memory is constructed by the array of memory cells. Let's take a memory with 64 cells for example, as illustrated in Figure 9.2.1. Each block in the memory array represents one memory cell, whose location can be determined by specifying a row and a column or only a row [41].

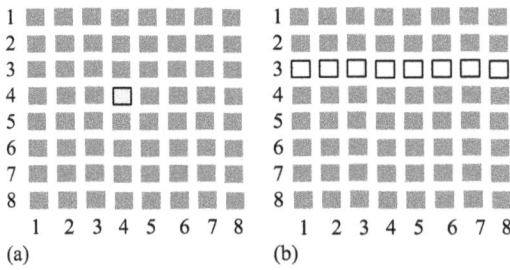

(a) (b)

Figure 9.2.1: A memory array containing 64 cells.

In Figure 9.2.1, the memory with 64 cells is organized as 8×8 array. The location of a unit of data in a memory array is called *address*. If a unit of data is a bit, the address of a bit in the two-dimensional array should be specified by row and column. For example, the address of the white cell is row 4 and column 4 in Figure 9.2.1(a). If a unit of data is a byte (eight bits), the address of a byte in memory array is specified only by a row. In Figure 9.2.1(b), the address of a byte represented by the white block is row 3. So, the address depends on how the memory is organized into the unit of data. For the memory with 64 cells, it can also be organized as 16×4 array.

The *capacity* of a memory is the total number of data units stored. For the bit-organized memory in Figure 9.2.1(a), the capacity is 64 bits. For the byte-organized memory, the capacity is eight bytes. Since each byte contains eight bits, the total capacity is also 64 bits.

9.2.3 Basic memory operations

The basic operations of the memory include *write* operation, *read* operation, and *addressing* operation.

Data in memory is accessed by means of a binary number called an *address code. Addressing* is the process of accessing a specified location in memory. Addressing operation can be realized by placing an address code on a set of lines called *address bus*. The address code through a decoder called *address decoder* is decoded and the specified location in memory is selected. For the single array in Figure 9.2.2, since there are eight rows corresponding to eight words, each word contains eight bits (a byte), the three-bit address codes through the address decoder offers $2^3 = 8$ addresses to access eight locations storing eight words in the memory. If the memory address consists of n-bit address codes, the number of addresses is 2^n, each word containing M bits. Consequently, the capacity of the memory is $2^n \times M$ bits for the memory in Figure 9.2.2.

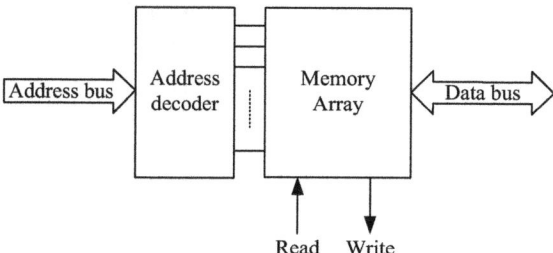

Figure 9.2.2: Block diagram of a single memory.

If a memory contains a large amount of memory cells, memory cells can be organized as a multiple array. Two address decoders, row and column decoders, can be used to access data in memory as shown in Figure 9.2.3. Let's take a simple memory array for example. A memory array containing 64 cells is arranged in a bit as the unit of data, as shown in Figure 9.2.1(a). That is, each word contains one-bit data. In order to access 64 locations, six-bit address codes are required. These six-bit address codes are divided into two groups, three-bit code for row decoder and three-bit code for column decoder. Each location in the memory is determined by row address and column address. Row decoder can select eight rows and column decoder can choose eight columns, so the access locations should be selected simultaneously by rows and columns. The number of lines in the address bus depends on the capacity of the memory. For example, an eight-bit address code can select 2^8 (=256) locations. If each word is eight bits, the capacity of the memory is $2^8 \times 8$. Similarly, a 16-bit address code can select 2^{16} (=65,536) locations in the memory. In personal computers, a 32-bit address bus can select 2^{32} locations, expressed as 4G.

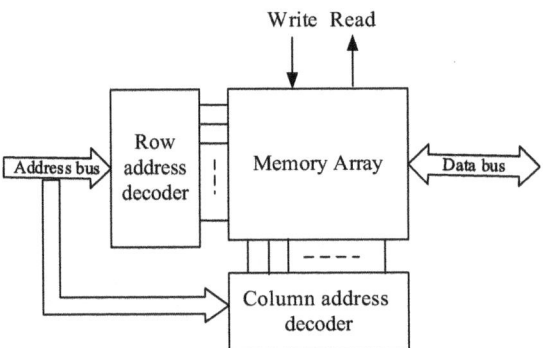

Figure 9.2.3: Block diagram of a multiple array.

The *write operation* is the process of putting the data into a specific address in the memory, while the *read operation* is to copy data from a specific address in the memory. The specific memory address is determined by the address code. Data go into the memory and come out of the memory with a set of lines called *data bus*. Generally, data bus is bidirectional, which means that data can go into the memory and also come out from the memory.

Let's take a practical memory as an example for a better understanding of the concept and operation of the semiconductor memory.

Example 9.1 Design a memory to store the states of eight switches of a circuit system. The state of each switch is represented by a 0 (**LOW**) or a 1 (**HIGH**). The switch changes its state one time each hour. For the continuous eight hours, the states of the switches should be stored each hour in the memory.

Solution

The state of each switch is represented by one bit, a 0 (**LOW**) or a 1 (**HIGH**). So the states of eight switches can be stored by using eight bits, that is, one byte. According to the requirement, the states of the switches should be stored every hour for the continuous eight hours. Totally, there are eight bytes required and thus the capacity of the memory is eight bytes. In order to access eight locations, a three-bit address code is required and thus a 3-to-8 decoder is required to act as the address decoder. The memory array with the address is constructed in Figure 9.2.4.

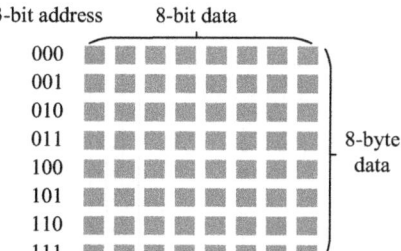

Figure 9.2.4: Memory array.

For the memory array, an 8-bit D latch is used to store the states of eight switches and thus eight 8-bit D latches are required to store the states of the switches each hour for the continuous eight hours. Figure 9.2.5 shows a design memory circuit to store the states of eight switches of a circuit system.

Eight eight-bit latches (74LS373) are used to store eight words; one 3-to-8 decoder (74LS138) is employed as the address decoder to select eight words; $D_0 \sim D_7$ are eight-bit data bus for data inputs; and $A_2 A_1 A_0$ are eight-bit address lines. An active LOW write operation signal, $\overline{WRITE}$, is applied to the "enable input" of the address decoder. When the $\overline{WRITE}$ input is LOW, the decoder is enabled, otherwise, the decoder is disabled.

The write process can be performed as the following steps. When data are to be stored into the memory, the address code stored in the address register is placed on the address *bus,* and the data stored in the data register is placed on the data bus. Then a LOW is applied on the write input and the address decoder is enabled. The address decoder will decode the address and select the corresponding 74LS373 in the specified location. The data on data bus considered as the data inputs of 74LS373 is latched to the output of the enabled 74LS173. That's the end for a write operation.

Figure 9.2.6 shows the timing diagram of a write operation. New data and its storing address must be placed on the data bus and address bus in advance before the active LOW write signal is applied. This time interval between new address (or data) and the active LOW write input must be less than the duration t_s, called the setup time. When the falling edge of the write pulse comes, the address decoder is enabled and decodes the address information. This makes the corresponding 74LS373 enabled and allows the data on data bus loaded into the latch. The new data is latched into the latch after the delay time t_p relative to the falling edge of the write pulse.

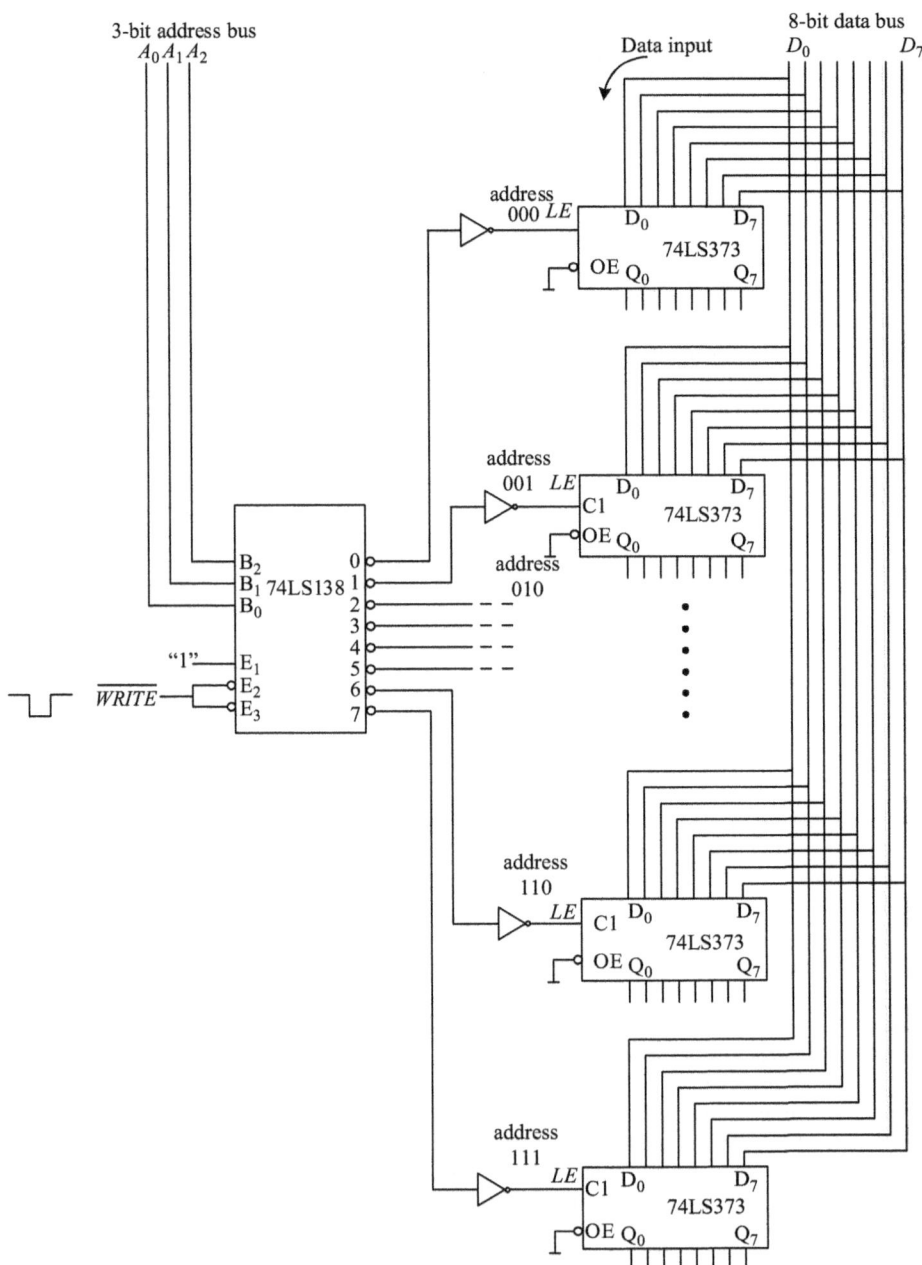

Figure 9.2.5: An 8-byte memory with eight D latches and a 3-to-8 decoder.

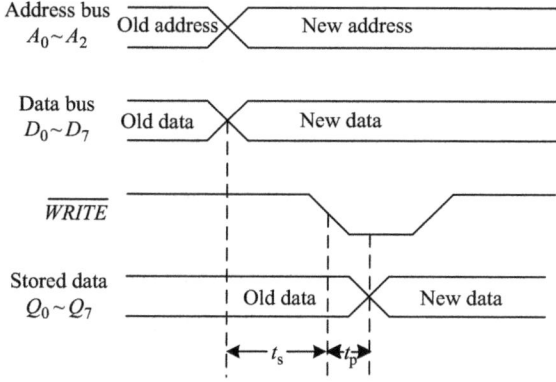

Figure 9.2.6: Timing Diagram of a write operation.

9.2.4 Types of semiconductor memories

There are two types of semiconductor memories. One is read only memory (ROM) and another is random access memory (RAM).

ROMs are used to store the binary data or information permanently or semipermanently. Generally, data can only be read from the ROM. Although some types of data can be written into, the writing process is slow and usually all the data in the chips must be rewritten at the same time. A ROM stores data that are used repeatedly in system, such as tables, conversions, or programed instructions for system initialization and operation. Especially, it can be used to store system software, which must be immediately accessible to the computer, such as the BIOS program using for the start up of computer, and the software (microcode) for portable devices and embedded computers such as microcontrollers. Because ROMs maintain the stored data even if power is turned off, they belong to the nonvolatile memories.

In contrast to ROMs, RAMs have both read and write capacities. Because RAMs lose stored data when the power is turned off, they belong to volatile memories. Although the volatile memory loses its stored data when the power to the memory chip is turned off, it can be faster and less expensive than nonvolatile memory. It is widely used for the main memory in most computers and other digital systems.

9.3 Read only memory

A ROM stores permanent or semipermanent data, which can be read from the memory but either cannot be changed at all or cannot be changed without specialized equipment. In terms of electronic components used as memory cells, semiconductor memory can be divided into two categories: *bipolar memory* and *metal-oxide-semiconductor (MOS) memory* (also called *unipolar memory*). Bipolar memory is generally composed of transistors with the advantage of high speed. But it has some disadvantages of high power consumption, low integration, and high cost, so it is seldom used in large-scale integrated circuits, but often used in the small and medium-sized registers. Compared with bipolar memory, MOS memory is composed of MOS field effect transistor (MOSFET). Although MOSFET has the disadvantage of low speed, it has some distinguished advantages, such as low power consumption, high integration, and low cost. At present, most ROMs use MOSFET as memory cells.

The objectives of this section are to
- Define a ROM
- Describe the classification of ROMs
- Describe the structure of a ROM
- Explain how data are read from a ROM
- Explain the operation modes of ROM chips
- Discuss the application of a ROM

9.3.1 Structure and operational principle of a ROM

Figure 9.3.1 shows the block diagram of a ROM. A ROM contains address input buffer, address decoder, memory array, and data output buffer [42].

The address decoder is a binary decoder with n-bit address inputs and N ($N = 2^n$) outputs, W_0, W_1,..., W_{N-1}. For ROM array, the N input lines, W_0, W_1,..., W_{N-1}, are usually called the *word lines*; the M output lines, D_0, D_1,..., D_{M-1}, are called the *bit lines*; the cross of a word line and a bit line is a memory cell. The number of memory cells represents the capacity of ROM array. The storage capacity of ROM is represented by $N \times M$. The output buffer is used to increase the load capacity of ROM and provides the tristate output in order to connect the ROM with the data bus. Furthermore, it can convert the nonstandard logic levels to the standard logic levels. The input buffer has the similar role as the output buffer to some extent.

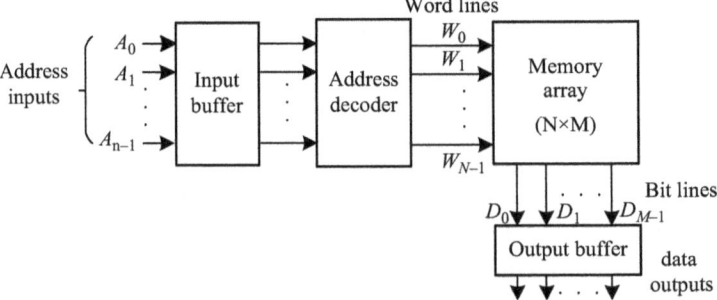

Figure 9.3.1: Block diagram of a ROM.

Let's take a 4×4 ROM as an example. Figure 9.3.2 shows exploded views of a ROM, which can clearly illustrate the concept of the ROM. The ROM array has four word lines and four bit lines. The memory cell in the ROM array is composed of the N-channel enhancement type of MOSFETs, as shown in Figure 9.3.2(a). The address decoder is shown in Figure 9.3.2(b), and its outputs correspond to minterms of the address inputs that can be expressed as

$$W_0 = \bar{A_1}\bar{A_0}, \ \ W_1 = \bar{A_1}A_0, \ \ W_2 = A_1\bar{A_0}, \ \ W_3 = A_1A_0$$

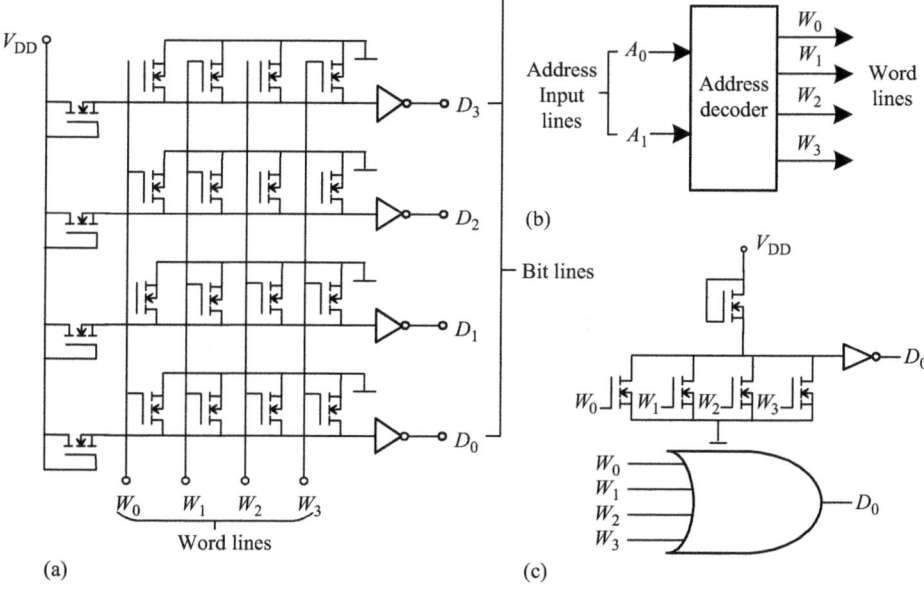

Figure 9.3.2: Relation between word lines and bit lines in ROM array: (a) ROM array; (b) address decoder; (c) logic relation between a bit line and word lines.

There are 16 memory cells (MOS transistors) in the cross of word lines and bit lines. The source of MOS transistor is connected to the ground and the drain is connected to the bit line through an inverter. The presence, or absence, of a connection from a word line to the gate of MOS transistor determines the stored data at that location. If a word line is addressed and appears a HIGH, all transistors with the gate connection to that word line turn on and the output from the drain is a LOW. Then the LOW output passes through an inverter as the output buffer and thus a HIGH is placed on the associated bit lines. That is to say, memory cell stores a 1 when the transistor has its gate connection to the word line, otherwise, a memory cell stores a 0. Assum that the address input is $A_1A_0 = 00$, the word line W_0 is HIGH. So the MOS transistors with the gate connection to the word line W_0 store 1. Others without the gate connection to W_0 store 0. The stored data in the ROM in Figure 9.3.2 are listed in Table 9.3.1.

Table 9.3.1: Stored data in the ROM.

Address lines		Word lines	Stored data			
A_1	A_0		D_3	D_2	D_1	D_0
0	0	W_0	0	1	0	1
0	1	W_1	1	1	1	1
1	0	W_2	0	0	1	1
1	1	W_3	1	0	1	1

The word lines connecting to the bit line D_0 are shown separately in Figure 9.3.2(c). Obviously, the relation between each bit line and word line can be deduced from Table 9.3.1 as follows:

$$D_0 = W_0 + W_1 + W_2 + W_3 = \bar{A}_1\bar{A}_0 + \bar{A}_1A_0 + A_1\bar{A}_0 + A_1A_0$$

$$D_1 = W_1 + W_2 + W_3 = \bar{A}_1A_0 + A_1\bar{A}_0 + A_1A_0$$

$$D_2 = W_0 + W_1 = \bar{A}_1\bar{A}_0 + \bar{A}_1A_0$$

$$D_3 = W_1 + W_3 = \bar{A}_1A_0 + A_1A_0$$

It can be seen from the above four equations that the relation of each bit line D_i (i = 0, 1, 2, 3) and the word line W_j (j = 0, 1, 2, 3) is OR logic; the relation of each word line W_j (j = 0, 1, 2, 3) and the address input (A_1, A_0) is AND logic. Therefore, the relation of bit lines and address inputs is AND-OR logic. The address decoder is equivalent to an AND logic array and the ROM array is equivalent to an OR logic array, therefore the ROM can be equivalent to an AND-OR logic array. In general, AND logic array produces all minterms of the address inputs and thus it is fixed, but OR logic array is programmable.

For simplicity, MOS transistor in the cross of the word line and bit line can be omitted, and a black dot represents the MOS transistor with the gate connection to the corresponding word line. The simplified ROM array diagram of Figure 9.3.2(a) is shown in Figure 9.3.3.

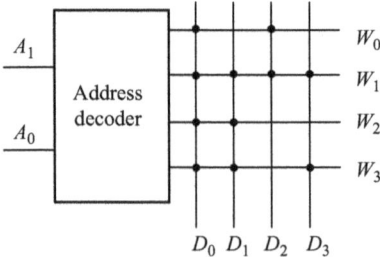

Figure 9.3.3: Simplified ROM array.

9.3.2 Types of ROMs

Classified by write mode, ROM is usually divided into *fixed ROM* (MASK ROM), *one-time programmable ROM* (OTP ROM or PROM), and *multiple times programmable ROM* (MTP ROM). Multiple times programmable ROM can be divided into erasable programmable ROM (EPROM), electrically erasable programmable ROM (EEPROM), and flash memory.

1. Mask ROM

Mask ROM (MROM) is a simple ROM. Its contents are programmed by the integrated circuit manufacturer rather than by the user. Once the memory is programmed, it cannot be changed. The terminology "mask" comes from integrated circuit fabrication, where regions of the chip are masked off during the process of photolithography. Memory cell in a MROM can be either bipolar transistor or unipolar MOS transistor. Figure 9.3.4 shows a MOS ROM cell. A connection between a word line and the gate of a transistor represents a 1 at that location. This can explain that all transistors with the gate connection to that word line turn on and connect HIGH (1) to the associated bit line when the word line is addressed and appears a HIGH. If there is not a connection between a word line and the gate of a transistor, the bit line retains LOW. It utilizes the presence or absence of a transistor connection at a word/bit junction to represent a 1 or a 0. The example of the ROM using MOS transistor as memory illustrated in the previous section is shown in Figure 9.3.2.

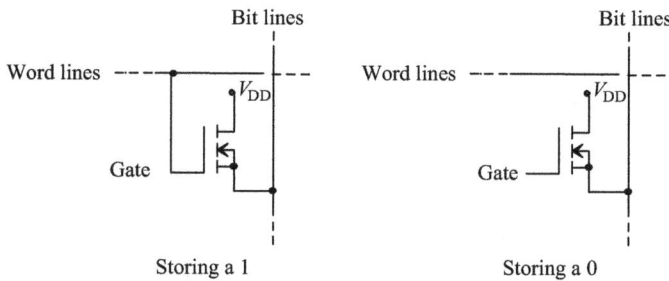

Figure 9.3.4: MOS ROM cells.

Let's take another example of a fixed ROM array using diode as memory cell. A fixed ROM is shown in Figure 9.3.5. Diodes are used to act as memory cells. There is a diode connection between a word line and a bit line, which represents a 1 at that location because when the word line is addressed and appears a HIGH, the diode conducts and connects HIGH on the word line to the associated bit line. If there is not a diode connection between a word line and a bit line, the bit line retains LOW when

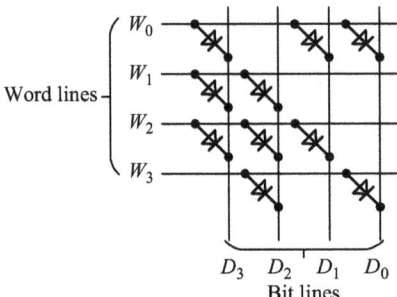

Word lines

W_0
W_1
W_2
W_3

D_3 D_2 D_1 D_0
Bit lines

Figure 9.3.5: A fixed ROM using diode as memory cell.

the word line is addressed. It utilizes the presence or absence of a diode connection at a word/bit junction to represent a 1 or a 0. Therefore, the fixed ROM stores four words and each word has four bits, thus the total capacity is 16 bits. The storing data are listed in Table 9.3.2.

Table 9.3.2: Stored data in the ROM.

Word lines	Stored data			
	D_3	D_2	D_1	D_0
W_0	1	0	1	1
W_1	1	1	0	0
W_2	1	1	1	0
W_3	0	1	0	1

MROM is programmed during manufacture, so it is only used for large production, and new data cannot be rewritten into MROM. The main advantage of MROM is low cost. Since the cost of an integrated circuit strongly depends on its size, MROM is significantly cheaper than any other kind of semiconductor memory. It is widely used in the computer and other digital devices to store system software, which must be immediately accessible to the computer, for example, the BIOS program that starts the computer. However, the one-time masking cost is high, and there is a long turn-around time from design to product phase. Design errors are also expensive. If an error in the data or code is found, the MROM is useless and must be replaced in order to change the code or data.

2. Programmable ROM (PROM)

PROMs are basically the same as MROMs once they have been programmed. The difference is that PROMs are not programmed when they are produced by manufacturer, and they are left to customer and are programmed in field to satisfy the user's demands. A PROM uses some type of fusing process to store bits. There are fusible links in all memory cells after manufacturing process. So the initial storing data in memory cell are all 0s or all 1s. Customers can reprogram PROM to meet their needs by burning the fuse open or leaving the fuse intact to represent a 0 or a 1. Since the fusing process is irreversible, it cannot be changed once programmed.

Figure 9.3.6 shows a 4×4 unprogrammed and programmed MOS PROM array with fusible links, respectively. All memory cells have the fusible links connecting the source of MOS transistor and the associate bit lines after manufacturing process. So the unprogrammed data stored in memory cells are all 1s. If you want to store the data shown in Table 9.3.2, the fusible links in the memory cells storing 0s should be burned open by injecting sufficient current through a fusible links. To write data onto a PROM chip, you need a special device called a PROM *programmer* or PROM *burner*. The process of programming a PROM is sometimes called burning the PROM.

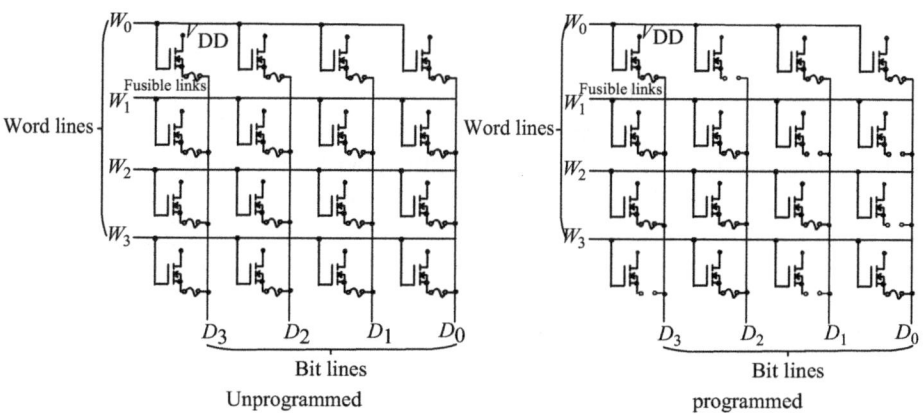

Figure 9.3.6: A 4×4 MOS PROM array.

3. EPROM

EPROM is the abbreviation of erasable programmable read only memory. The stored data in EPROM can be erased and rewritten, but the data in EPROM cannot be rewritten in the circuit board. In order to rewrite the data in EPROM, the EPROM must be removed from the circuit board and exposed to an ultraviolet (UV) light for several minutes to erase the existing data. The IC package has a small transparent "window" in the top to admit the UV light. The UV light clears its contents, making it possible to reprogram the memory. To write and erase an EPROM, you need a special device for programming.

The memory cell of the EPROM uses an MOSFET with the floating gate (FAMOS), as shown in Figure 9.3.7. A floating gate is embedded in highly resistive material (silicon dioxide) and thus the charge contained in it remains unchanged for long

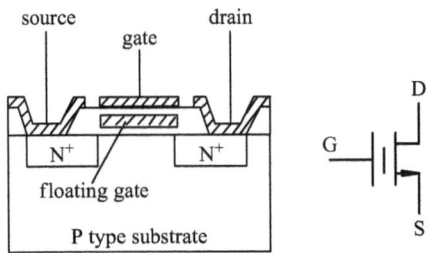

Figure 9.3.7: FAMOS transistor structure and symbols.

periods of time. The data bits can be represented by the presence or absence of a stored gate charge. Erasure of a data bit is a process that removes the gate charge.

4. EEPROM

Although EPROM is a high-density, nonvolatile memory, it can be erased only by removing it from the system and using UV light. It can be reprogrammed only with specialized equipment. This makes the erasing and writing operation complex and inconvenient.

Electrically erasable programmable ROM (EEPROM) is a kind of ROM that can be both erased and programmed with electrical pulse. Because it can be both electrically written into and electrically erased, EEPROM can be rapidly programmed and erased in-circuit for reprogramming.

A floating gate tunnel oxide MOS transistor (Flotox) is used as a memory cell in the EEPROM. The structure of Flotox is shown in Figure 9.3.8. There is a tunnel region formed with a thin oxide layer about 20 nm between the floating gate and the drain of the Flotox transistor. When a strong electric field is applied in this tunnel region, the tunneling effect appears at this region between the floating gate and the drain. As a result, a current path is formed and the floating gate can be charged or discharged.

EEPROM has a more complex cell structure than either ROM or EPROM and so the density is not as high. Due to the lower density, the cost per bit is higher than ROM or EPROM. But it can be reprogrammed in-circuit, so it is much more convenient to use than EPROM.

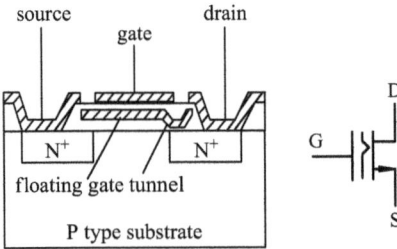

Figure 9.3.8: Flotox transistor structure and symbols.

5. Flash memory

Flash memories are high-density read/write memories that are nonvolatile, which means that data can be stored indefinitely without power [43]. The MOS transistor used in the flash memory is called the flash stacked gate MOS transistor. A storage cell in the flash memory consists of a single flash stacked gate MOS transistor shown in Figure 9.3.9. It has a simpler structure than the Flotox, but its operation is similar to the Flotox. Since the source region is enlarged, this leads the tunnel region to form directly by using the oxide layer between the larger source N^+ and the floating gate. When the data is written into flash memory, the floating gate is charged using hot electron injection; when the data is erased, the floating gate is discharged using high-voltage tunneling effect. A data bit is stored as charge or the absence of charge on the floating gate depending if a 0 or a 1 is stored. Because the charges on the floating gate cannot be leaked, the data can be saved after the power is turned off.

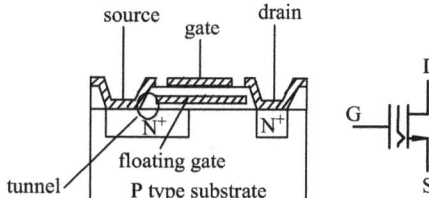

Figure 9.3.9: Structure and symbol of the flash stacked gate MOS transistor.

Flash memory is a variant of EEPROM; data in EEPROM is erased or rewritten in units of byte, but data in flash memory is only erased in units of sector. Usually, the size of flash memory sector is from 256 KB to 20 MB. Except for most features of EEPROM, flash memory also has higher density and higher read/write speed. The simpler structure of flash memory makes it suitable for large-scale integration and highly reliable, and thus the cost can be further reduced.

There are two types of flash memories: NOR-type and NAND-type. NOR-type is very different from NAND-type flash memory. More like RAM, NOR-type flash memory has independent address lines and data lines, but it is more expensive and has smaller capacity. While NAND-type is more like a hard disk, its address lines and data lines are shared as input/output (I/O) lines, and all the information on the hard disk are transmitted through one hard disk lines. Compared with NOR-type flash memory, NAND-type flash memory has lower cost and larger capacity. Therefore, NOR-type flash memory is more suitable for frequently random read and write. The program code can be stored and directly run in the flash memory. NAND-type flash memory is mainly used to store information. The commonly used flash memory products, such as flash drives and digital memory cards, are both NAND-type flash memory.

In addition to different types of flash memory products, flash memory is expanding to other applications. For example, the removable hard disk, equipped in a computer, has replaced the floppy (soft) disk. According to the recent report, flash memory with the capacity of TB will come out in the year of 2018. That means that it can replace the hard disk of computer in the near future. Flash memory has USB interface and supports the hot plug with the advantages of high operation speed and convenient usage.

The ideal memory has high storage capacity, nonvolatility, in-system read/write capability, comparatively fast operation, and cost effectiveness. Flash memory has all these desired characteristics.

ROM is a nonvolatile memory; the data in ROM can be permanently memorized. "Nonvolatile" and "permanent memory" are only relative concepts, that is, data in ROM are not permanently undamaged. The important technical indicator of ROM is the "number of write". Once the rewritten number of ROM reaches or is close to the "number of write", the damage probability of ROM will greatly increase. The average number of write of EEPROM is about 100,000 times and that of the flash is about one million times.

9.3.3 Typical IC ROMs

Among ROMs, EEPROM and flash memory are the most commonly used in digital systems. Here, several typical IC EEPROMs and flash memories are introduced.

1. EEPROM
There are many typical IC EEPROMs, for example, the 2816A/2817A with the capacity of $2k \times 8$ bits, 2864A with $8k \times 8$ bits, 28010 with $128k \times 8$ bits, and many others. Let's take 2864A as an example. Pin diagram of EEPROM Intel 2864A is shown in Figure 9.3.10. It

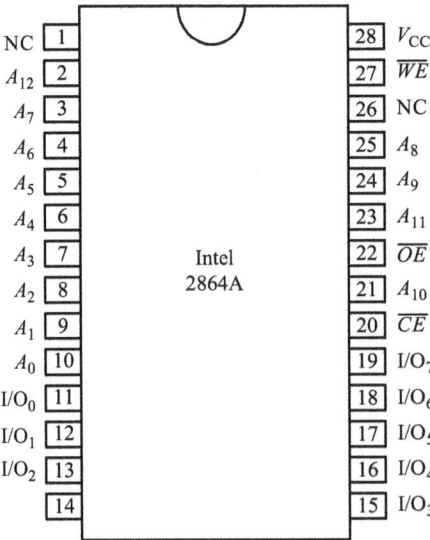

Figure 9.3.10: Pin diagram of EEPROM Intel 2864A.

has 13 address lines (A_0 through A_{12}) that can select 8k (2^{13}) word lines, and each word contains eight-bit data corresponding to eight I/Os (I/O$_0$ through I/O$_7$). There are three control inputs for determining the read/write operation, in which $\overline{CE}$ is an active-LOW input for chip selection, $\overline{WE}$ is also an active-LOW input for enabling write operation, and $\overline{OE}$ is an active-LOW input for enabling read operation. When the chip select input $\overline{CE}$ is HIGH, the access operation is disabled and all I/Os are in high impedance state. When the chip select input $\overline{CE}$ and the write input $\overline{WE}$ are both LOW, the chip performs the write operation. When a 13-bit address code (A_0 through A_{12}) is applied, eight-bit data from the data inputs (I/O$_0$ through I/O$_7$) are written to the memory cells in the specific word line. Similarly, when the chip select input $\overline{CE}$ and the read input $\overline{OE}$ are both LOW, the chip performs the read operation. When a 13-bit address code (A_0 through A_{12}) is applied, eight-bit data appears on the data outputs (I/O$_0$ through I/O$_7$).

The timing diagram of EEPROM 2864A illustrates the write operation and the read operation, as shown in Figure 9.3.11. To write the data to the chip, a LOW is applied on the chip select input and then a valid address code is sent on the address lines. Next, the low level pulse is applied to the $\overline{WE}$ input; the eight-bit data from the inputs (I/O$_0$ through I/O$_7$) can be programmed into a given address after a short delay. To read the data from the chip, a LOW is applied on the chip select input and then a valid address code is sent on the address lines. Next, the LOW level pulse is applied to the $\overline{OE}$ input, the valid output data appears on the outputs (I/O$_0$ through I/O$_7$) from a given address after a short delay.

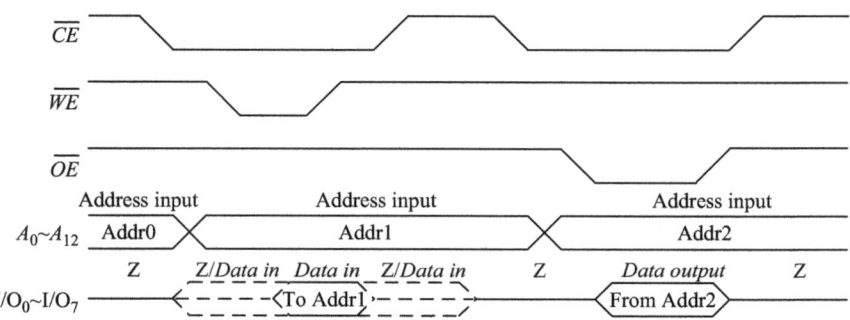

Figure 9.3.11: Timing diagram for illustrating the write operation and read operation.

2. Flash memory

Typical flash memory chips include 29C010 (128 k × 8 = 1 M bits), 29LV020 (256 k × 8 = 2 M bits), 29F040 (512 k × 8 = 4 M bits), etc.

AT29LV020 is a flash memory chip (256 k × 8 = 2 M bits) manufactured by Atmel company. Its internal structure is CMOS type with 3.3V power supply. AT29LV020 has 1024 sectors, in which each sector includes 256 bytes. As a typical NOR-type flash

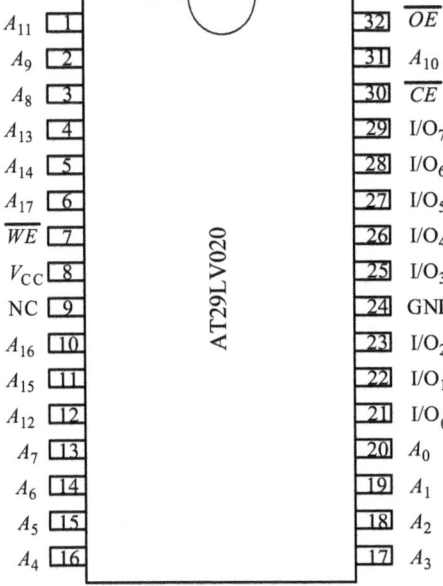

Figure 9.3.12: The Pin diagram for AT29LV020.

memory chip, AT29LV020 has 18 separate address lines (A_0 through A_{17}) that can select 256 k word lines, and each word contains eight data lines corresponding to eight IOs (I/O_0 through I/O_7), as shown in Figure 9.3.12. There are three active-LOW control inputs ($\overline{CE}$, $\overline{WE}$, and $\overline{OE}$) for determining the read/write operation. To write the data into the chip, a LOW is applied on the chip select input and then a valid address code is sent on the address lines. Next the LOW level pulse is applied to the $\overline{WE}$ input, the eight-bit data from the inputs (I/O_0 through I/O_7) can be programmed into a given address after a short delay. To read the data from the chip, a LOW is applied on the chip select input and then a valid address code is sent on the address lines. Next the LOW level pulse is applied to the $\overline{OE}$ input, the valid output data appears on the outputs (I/O_0 through I/O_7) from given address after a short delay. The timing diagram of AT29LV020 illustrates the write operation and the read operation, as shown in Figure 9.3.13.

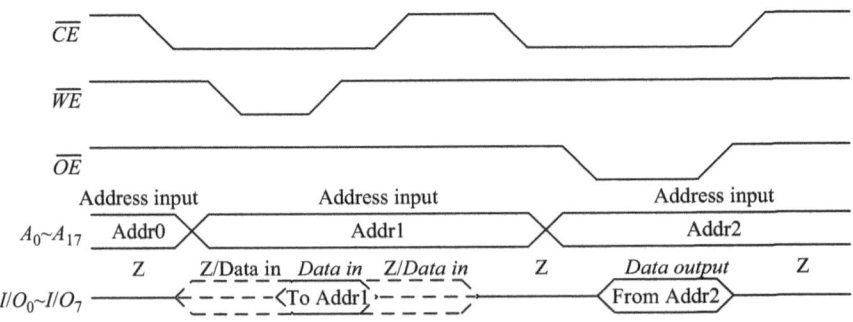

Figure 9.3.13: Timing diagram for illustrating the write operation and read operation.

9.3.4 ROM applications

The main applications of ROM are to store data and program in the computer and other digital systems. But these are just the tip of an iceberg. There are many other applications of ROM. Basically, ROM belongs to the combinational logic circuit and thus any combinational circuit can be implemented by ROM. In addition, the sequential logic circuit can be constructed with ROM by adding some sequential logic components.

1. Design a combinational logic circuit using ROM

In Section 9.3.1, you have learned that the logic relation between each bit line and word line is OR logic and that between each word line and address line is AND logic. Therefore, the logic relations between bit lines and address lines are AND-OR logic. The address decoder is equivalent to an AND logic array producing all minterms; the ROM array is equivalent to an OR logic array producing several sum terms, thus the entire memory is an AND-OR logic array and form the output expression of the sum of minterms. For any combinational logic circuits, its output expression can be transformed into the form of sum of minimum terms. Therefore, the ROM can be used to implement any combinational logic circuits.

Example 9.2 Use ROM to implement a full adder.

Solution

Step 1 Determine the input and output variables and list the truth table.

The full adder has three inputs including the summand A_i, the addend B_i, and the carry input C_{i-1} and two outputs, the sum S_i and the carry output C_i. So the truth table of the full adder is listed in Table 9.3.3.

Table 9.3.3: Truth table of full adder.

A_i	B_i	C_{i-1}	S_i	C_i
0	0	0	0	0
0	0	1	1	0
0	1	0	1	0
0	1	1	0	1
1	0	0	1	0
1	0	1	0	1
1	1	0	0	1
1	1	1	1	1

Step 2 Write the logic expression from the truth table.

Since the output expression of the ROM is the sum of minterms, you can directly deduce the sum of minterms from the truth table in Table 9.3.3.

The output expressions are deduced from Table 9.3.3 as follows:

$$S_i(A_i, B_i, C_{i-1}) = \bar{A}_i \bar{B}_i C_{i-1} + \bar{A}_i B_i \bar{C}_{i-1} + A_i \bar{B}_i \bar{C}_{i-1} + A_i B_i C_{i-1} = m_1 + m_2 + m_4 + m_7 \qquad (9.3.1)$$

$$C_i(A_i, B_i, C_{i-1}) = \bar{A}_i B_i C_{i-1} + A_i \bar{B}_i C_{i-1} + A_i B_i \bar{C}_{i-1} + A_i B_i C_{i-1} = m_3 + m_5 + m_6 + m_7 \qquad (9.3.2)$$

Step 3 Draw logic diagram from the logic expression.

In order to use ROM to implement a full adder, three input variables are arranged as three address lines (A_i, B_i, C_{i-1}) of a 3-to-8 address decoder and output eight word lines representing eight minterms m_0~m_7 of A_i, B_i, C_{i-1}; the ROM array has eight word lines as its address inputs and two bit lines as its outputs representing the sum S_i and the carry output C_i. The ROM array with the capacity of 8 × 2 is needed to implement a full adder, as shown in Figure 9.3.14. From eq. (9.3.1), the sum S_i of the full adder is the sum of the minterms m_1, m_2, m_4, and m_7, so the cross between word lines m_1, m_2, m_4, and m_7 and bit line S_i must have the memory cells connecting the word lines and bit line, and thus the black dots at the corresponding locations represent these programming connections in the simplified ROM array. Similarly, the carry output C_i can be implemented by programming the ROM array.

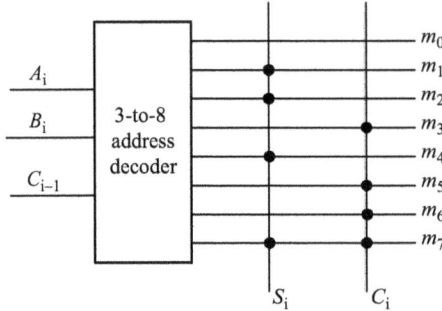

Figure 9.3.14: The ROM array to implement a full adder.

2. Construct the sequence generator using ROM

Sequence generator is used to produce a serial code in the certain period. Sequence codes are widely applied for digital systems, such as radar, communications, remote control, telemetry, and so on. In Chapter 7, you have learned how to design sequence code generator by using the combination of counter and combinational circuit. Since ROM belongs to the combinational circuit, sequence code generator can also be constructed by using the combination of a counter and a ROM.

Example 9.3 Design a sequence pulse generator by using the ROM and the counter to produce the sequence pulse, as shown in Figure 9.3.15.

Solution

Step 1 Design a counter with a modulus that is equal to the length of sequence code.

It can be obviously seen from Figure 9.3.15 that the sequence generator is needed to produce two groups of sequence pulses. One sequence pulse is output from P_1 and another is output from P_2.

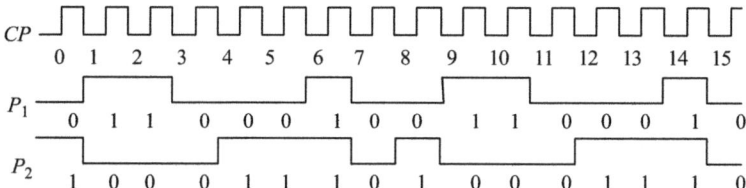

Figure 9.3.15: Timing diagram of a sequence pulse generator.

By checking the P_1 waveform in relation to the clock pulse, you can find that the output sequence of P_1 is 01100010 from initial clock pulse to the seventh clock pulse and then recycle when the eighth clock pulse arrives. Similarly, by checking the P_2 waveform in relation to the clock pulse, you can find that the output sequence of P_2 is 10001110 from initial clock pulse to the seventh clock pulse and then recycle at the arrival of the eighth clock pulse. Therefore, the length of sequence code from P_1 and P_2 are both eight. So a counter of MOD 8 is needed to produce an eight-bit long sequence code. The counter of MOD 8 can be implemented by using a three-bit counter, which has been introduced in Chapter 7. Here, the design process of a three-bit counter is omitted.

Step 2 Construct the combinational circuit to implement the sequence code "01100010" and "10001110".

The outputs of counter Q_0 through Q_2 act as the inputs of this combinational circuit and two outputs, P_1 and P_2, are needed to output the sequence code "01100010" and "10001110", respectively. After eighth clock pulse, a group of sequence codes "01100010' and "10001110" is produced and recycle again. The truth table of the combinational circuit is listed in Table 9.3.4.

Table 9.3.4: Truth table.

Q_2	Q_1	Q_0	P_1	P_2
0	0	0	0	1
0	0	1	1	0
0	1	0	1	0
0	1	1	0	0
1	0	0	0	1
1	0	1	0	1
1	1	0	1	1
1	1	1	0	0

This combinational circuit can be implemented with several methods. Here, the ROM is selected to implement the logic function shown in Table 9.3.4. As you know, the logic expression of the output should be expressed as the sum of minterms when the ROM is used to implement the combinational circuit. So the sum of minterms of the output can be deduced from Table 9.3.4 as follows:

$$P_1 = m_1 + m_2 + m_6, P_2 = m_0 + m_4 + m_5 + m_6 \qquad (9.3.3)$$

where m_i is the minterm of three input variables $Q_2 Q_1 Q_0$.

In order to use ROM to implement the above combinational circuit, three input variables are arranged as three address lines $Q_2Q_1Q_0$ of a 3-to-8 address decoder and output eight word lines representing eight minterms m_0 through m_7 of $Q_2Q_1Q_0$; ROM array has eight word lines as its inputs and two bit lines as its outputs represented by P_1 and P_2. The capacity of the ROM array is 8 × 2. From eq. (9.3.3), put the black dots at the corresponding locations to represent these programming connection in the simplified ROM array. The resulting logic diagram of the sequence pulse generator, which produces the sequence pulse in Figure 9.3.15 by using the ROM and the counter, is shown in Figure 9.3.16.

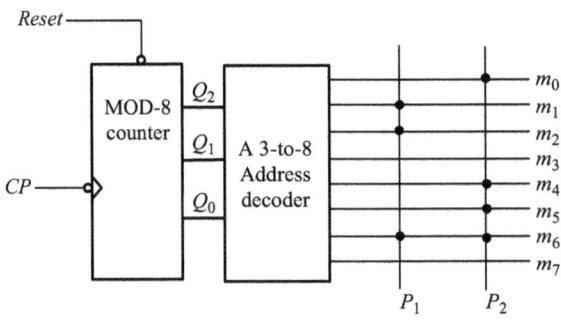

Figure 9.3.16: A sequence pulse generator using the ROM.

3. Construct lookup table using ROM

Suppose we have a ROM circuit written, or programmed, with certain data, such that the address lines of the ROM are served as inputs and the data lines of the ROM are served as outputs, generating the characteristic response of a particular logic function. Theoretically, we could program this ROM chip to emulate whatever logic function we wanted without alternation of any wire connections or gates. For example, if you want to use the ROM to generate the sine wave, the sine wave data should be written into ROM in advance. You can use the angle as the address lines of the ROM served as inputs and the data lines of the ROM served as outputs, generating the value of sine function. A lookup table (LUT) means that definite outputs can be looked up for every given input by constructing the data table with ROM.

Example 9.4 Design the code convertor to convert the four-bit binary code into Gray code by using ROM.

Solution

The code convertor has four inputs (B_3 through B_0) representing four-bit binary code and four outputs (G_3 through G_0) on behalf of four-bit Gray code. So the truth table of the converted code is shown in Figure 9.3.17(a).

When the ROM is used to implement the truth table, we get the address lines of the ROM served as four-bit binary code inputs and the data lines of the ROM served as four-bit Gray code outputs. So the capacity of the ROM array is 2^4 × 4 bits. A 4-to-16 address decoder is needed to provide 2^4 word

Inputs				Outputs			
Binary code				Gray code			
B_3	B_2	B_1	B_0	G_3	G_2	G_1	G_0
0	0	0	0	0	0	0	0
0	0	0	1	0	0	0	1
0	0	1	0	0	0	1	1
0	0	1	1	0	0	1	0
0	1	0	0	0	1	1	0
0	1	0	1	0	1	1	1
0	1	1	0	0	1	0	1
0	1	1	1	0	1	0	0
1	0	0	0	1	1	0	0
1	0	0	1	1	1	0	1
1	0	1	0	1	1	1	1
1	0	1	1	1	1	1	0
1	1	0	0	1	0	1	0
1	1	0	1	1	0	1	1
1	1	1	0	1	0	0	1
1	1	1	1	1	0	0	0

Address inputs of ROM — Data output in specified address

(a)

(b)

Figure 9.3.17: The programmed ROM array: (a) truth table; (b) ROM array.

lines. According to the truth table in Figure 9.3.17(a), the ROM array can be directly programmed to store a 1 by placing a black dot on the cross of the word line and bit line. For example, when the address inputs is 0010, the output is 0011 and thus you can place two black dots on the cross points of the word line W_2 and the bit lines G_1 and G_0; when the address inputs is 0101, the output is 0111 and thus you can place three black dots on the cross points of the word line W_5 and the bit lines G_2, G_1, and G_0. The resulting lookup table for the code convertor is constructed to convert the four-bit binary code into Gray code by using the ROM, as shown in Figure 9.3.17(b).

If you want to construct the lookup table to produce sine wave, square wave, triangular wave, saw tooth wave, and other waveforms, you should store the waveform data into EPROM array by the dedicated programmer. Except that, a counter is needed to implement the scan of the address inputs by driving the clock pulse, and the corresponding outputs from EPROM are the waveform data that are converted into analog signal u_o through the D/A converter. As a result, the required waveform can be produced. Figure 9.3.18 shows the block diagram of the signal generator by using counter and EPROM.

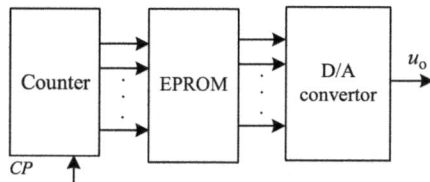

Figure 9.3.18: Block diagram of waveform generator based on EPROM.

ROM stores data that are used repeatedly in system applications, such as tables, conversions, or programmed instructions, for system initialization and operation. ROM keeps the stored data when the power is turned off, therefore ROM is nonvolatile memory.

9.4 Random access memory

Unlike ROM, RAM is also called *read/write memory*, in which data can be written into or read from any selected address in any sequence. When the power is turned off, the stored data in RAM will be lost, so RAM is a volatile memory and is mainly used to store short-term data and temporary programs.

The objectives of this section are to
- Explain the basic structure of RAM
- Explain the difference between static RAM and dynamic RAM
- Describe the SRAM stage cell
- Describe the dynamic RAM stage cell
- Explain the operation modes of the IC RAMs
- Discuss some special RAMs

9.4.1 Basic structure of RAM

The basic structure of RAM contains three parts: address decoder, memory array, and read/write control circuit, as shown in Figure 9.4.1. The address lines A_0 through A_{n-1} are applied on the address decoder and 2^n addresses are produced. The read and write control circuit is used to select the read/write operation. In READ mode, the data stored in a selected address appears on the data output lines. In WRITE mode, the data applied to the data input lines is stored at a selected address. The data input and output lines share the same lines. During READ, they act as output lines. During WRITE, they act as input lines. Therefore, the data lines are bidirectional. If the RAM

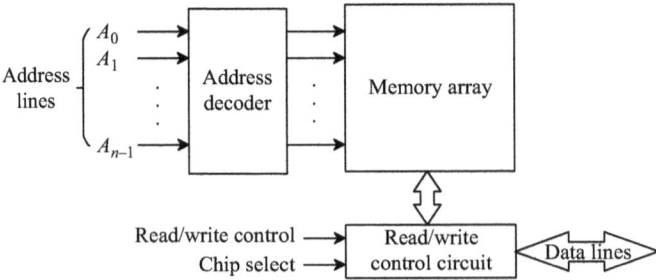

Figure 9.4.1: Basic structure of RAM.

has a larger capacity, dual address decoders containing row decoder and column decoder are adopted in RAM organization to simplify the RAM structure.

9.4.2 Types of RAMs

The two widely used types of RAM are *static RAM (SRAM)* and *dynamic RAM (DRAM)* [44].

SRAM generally uses latches as storage elements and can store data infinitely as long as power is turned on. SRAM is more expensive to manufacture, but is generally faster and requires less dynamic power than DRAM. In modern computers, SRAM is often used as cache memory for the CPU.

DRAM stores a bit of data using a transistor and capacitor pair, which comprises a DRAM cell together. The capacitor holds a high or low charge (1 or 0, respectively), and the transistor acts as a switch that lets the control circuitry on the chip read the capacitor's state of charge or charge it. DRAMs can store much more data than SRAM for a given physical size and cost due to their simple memory cell, thus they are widely used in the system with larger storage capacity. So far, the major memory in the microcomputer uses DRAMs. Due to the use of capacitor as storage elements, data cannot be retained very long and thus the capacitors should be recharged to recover the stored data for a period of time. This recharging process is called *refreshing*. Both SRAMs and DRAMs will lose stored data when power is removed, therefore, they are classified as volatile memories.

The basic types of SRAM are *asynchronous SRAM* and *synchronous SRAM* with a burst feature. The basic types of DRAM are *fast page mode DRAM* (FPM DRAM), *extended data out DRAM* (EDO DRAM), *Burst EDO DRAM* (BEDO DRAM), and *synchronous DRAM* (SDRAM).

9.4.3 Memory cell

1. Static Memory Cell

SRAM generally uses latches as storage elements. A typical SRAM cell with six MOSFETs is shown in Figure 9.4.2. Each bit in a SRAM is stored by a basic S-R latch formed by two cross-coupled inverters that are made up of four MOSFETs, VT_1 through VT_4. This storage cell has two stable states that are used to denote a 0 or a 1. Two additional access transistors VT_5 and VT_6 serve to control the access to a storage cell during read and write operations. When an active level is applied on the row selection line X_i, the transistors VT_5 and VT_6 are turned on, and a data bit (1 or 0) is written into the cell by placing it on the bit line. Similarly, a data bit is read out by taking it off the bit line. In memory array, two gate control transistors VT_7 and VT_8 are

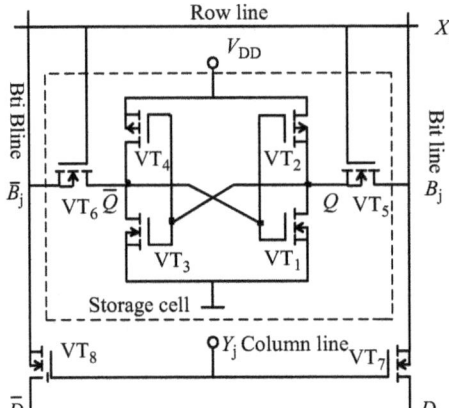

Figure 9.4.2: A SRAM cell with six MOSFETs.

added to control each column of memory cells, and the states of VT_7 and VT_8 are controlled by the column selection line Y_j. If an active level is applied on the column selection line Y_j, VT_7 and VT_8 are turned on, the data D and $\bar{D}$ connect the bit line to be read out or written into.

2. DRAM Memory Cell

Unlike SRAM cell, a DRAM memory cell stores a data bit in a small capacitor rather than in a latch.

A typical DRAM cell consists of a single MOSFET and a capacitor. In this type of cell, the transistor actually acts as a switch. Figure 9.4.3 is a simplified schematic diagram of a single dynamic memory cell. The stored data come from the D_{in} input in the write mode, and the data is output to the D_{out} output in the read mode. The data input D_{in} and the data output D_{out} are independent. In write operation, a write pulse is applied to the gate of MOSFET, causing the MOSFET to turn on. If the D_{in} input is a $1(+ V_{cc})$, the capacitor is charged to the high level and thus a 1 is stored by the capacitor. In read operation, a read pulse is applied to the gate of MOSFET, causing the MOSFET to turn on. If the capacitor stores a 1, the 1 will be sent to the D_{out} output. This is equivalent to reading out the stored data.

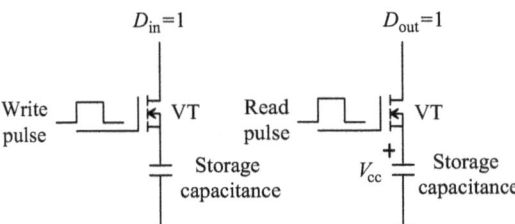

Figure 9.4.3: Simplified DRAM memory cell.

Due to high impedance of gate of MOS transistor, the charge stored in the gate capacitor is not able to discharge within a short time and thus the data can be stored into dynamic memory cell. The charge stored in the capacitor cannot retain very long. It will be discharged slowly, causing stored information disappear. So the dynamic memory must be supplemented by the charge to the gate capacitance for a period of time; this operation is called *refresh*. Therefore, DRAM needs to be equipped with the refresh circuit and the corresponding control circuit, which causes the circuit to be more complex.

DRAM can also be constructed by using three MOS transistors as its storage cell, as shown in Figure 9.4.4. The data is stored in the gate capacitance C_g of the VT_2 transistor, and the voltage on C_g can control the VT_2 to turn on or off. Read control line and write control line are separated. So do read bit line and write bit line. Read control line controls VT_3 transistor and write control line controls VT_1 transistor. The VT_4 transistor is a pre-charged transistor for sharing with a number of memory cells in the same column.

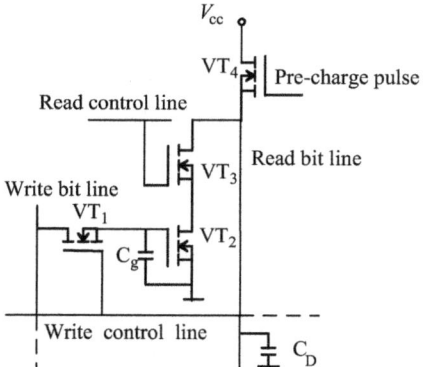

Figure 9.4.4: A dynamic memory cell using three transistors.

In the read operation, the capacitor C_D on the bit line is charged to V_{CC}, and then a HIGH is applied on the read control line causing the VT_3 transistor to turn on. If C_g is charged and the voltage on the C_g exceeds the threshold voltage of the VT_2 transistor, VT_2 and VT_3 are both turned on. As a result, C_D will be discharged to LOW through VT_3 and VT_2. If there is no charge on C_g, VT_2 transistor cuts off and C_D retains HIGH due to lack of discharge path. It can be seen that the level obtained on the read bit line is opposite to the level on the gate capacitance C_g. The data on the read bit line can be sent to the output of the memory via the output amplifier.

In the write operation, the write control line is HIGH and VT_1 transistor is turned on. The data coming from the memory input is transferred to the write bit line. If the data is a 1, the C_g is charged to a high level equivalent to store a 1; if the data is a 0, the C_g is discharged through VT_1 transistor to a low level equivalent to store a 0. So whether or not the stored data is a 1 or a 0 depends on if the C_g has stored the charge.

Because it is inevitable for the charge on C_g to be discharged slowly, the charge on the C_g must be replenished for a period of time to retain the data. The data stored on C_g will be periodically read out to the read control line, performing the inverter operation to control write operation so that the refresh operation is completed by charging the C_g.

Dynamic memory cells store a data bit in a gate capacitor rather than a latch. The advantage of this type of cell is that it is very simple, thus allowing very large memory array to be constructed on a chip at a lower cost per bit. The disadvantage is that the storage capacitor cannot hold its charge for a long time and will lose the stored data bit unless its charge is refreshed periodically. The implementation of refreshing operation requires additional memory circuitry, therefore, DRAM becomes more complex.

DRAM is widely used in digital electronics where low-cost and high-capacity memory are required. One of the largest applications for DRAM is the main memory (colloquially called the "RAM") in modern computers and graphics cards (where the "main memory" is called the graphics memory). It is also used in many portable devices and video game consoles. In contrast, SRAM, which is faster and more expensive than DRAM, is typically used where speed is of greater concern than cost, such as the cache memory in processors.

9.4.4 Typical IC RAMs

There are many typical IC RAMs, such as the traditional SRAM chips 6264 (8k × 8) and 62256 (32k × 8), the new SRAM chips 61LV25608 (256k × 8) and 61LV51216 (512 k × 16), DRAM chips 57V641620 (8MB), etc. Let's take 61LV25616 as an example to introduce the functions of a typical RAM.

IS61LV25616 is high-speed CMOS SRAM of ISSI company with 3.3 V power supply, and the minimum processing time can reach 10 ns. Figure 9.4.5 is the pin diagram of IS61LV25616 (44 feet TSOP package). It has 18 address lines (A_0 through A_{17}) that can select 2^{18} word lines, and each word line contains 16-bit data corresponding to 16 I/Os (I/O$_0$ through I/O$_{15}$), thus the storage capacity is 2^{18} × 16 bits = 256 k × 16 bits. There are three control inputs for determining the read/write operation, in which $\overline{CE}$ is an active-LOW input for chip selection, $\overline{WE}$ is also an active-LOW input for enabling write operation, and $\overline{OE}$ is an active-LOW input for enabling read operation. Additionally, 61LV25616 has two data control inputs, $\overline{UB}$ and $\overline{LB}$, where $\overline{UB}$ is the high eight-bit data control and $\overline{LB}$ is the low eight-bit data control. The use of $\overline{LB}$ and $\overline{UB}$ makes the 61LV25616 chip to be flexibly used for either eight-bit data bus or 16-bit data bus without any external circuit. Table 9.4.1 is a function table for the 61LV25616 chip.

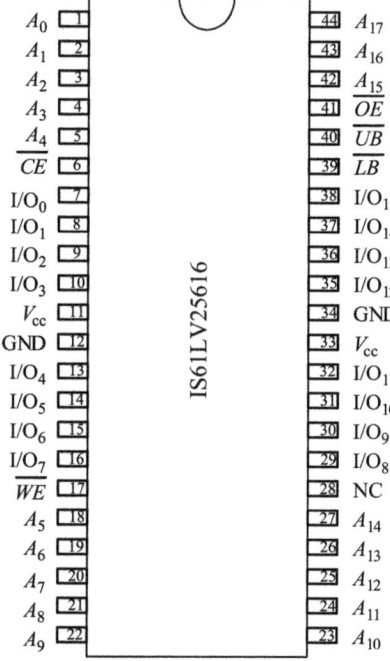

Figure 9.4.5: Pin diagram of IS61LV25616.

Figure 9.4.6 is the timing diagram of IS61LV25616. In write mode, the low eight-bit data and the high eight-bit data can be separately written into the selected addresses Addr1_L and Addr2_U; also, all 16-bit data are simultaneously written into the selected address Addr3. In read mode, the low eight-bit data and the high eight-bit data can be separately read from the selected addresses Addr4 and Addr5; also, all 16-bit data are simultaneously read from the selected address Addr6.

Table 9.4.1: Function Table of IS61LV25616.

Operation mode	$\overline{CE}$	$\overline{WE}$	$\overline{OE}$	$\overline{LB}$	$\overline{UB}$	$I/O_0\text{-}I/O_7$	$I/O_8\text{-}I/O_{15}$
Disable	H	X	X	X	X	Z	Z
Output disable	L	H	H	X	X	Z	Z
	L	X	X	H	H	Z	Z
Read	L	H	L	L	H	D_{OUT}	Z
	L	H	L	H	L	Z	D_{OUT}
	L	H	L	L	L	D_{OUT}	D_{OUT}
	L	L	X	L	H	D_{IN}	Z
Write	L	L	X	H	L	Z	D_{IN}
	L	L	X	L	L	D_{IN}	D_{IN}

Z represents High impedance.

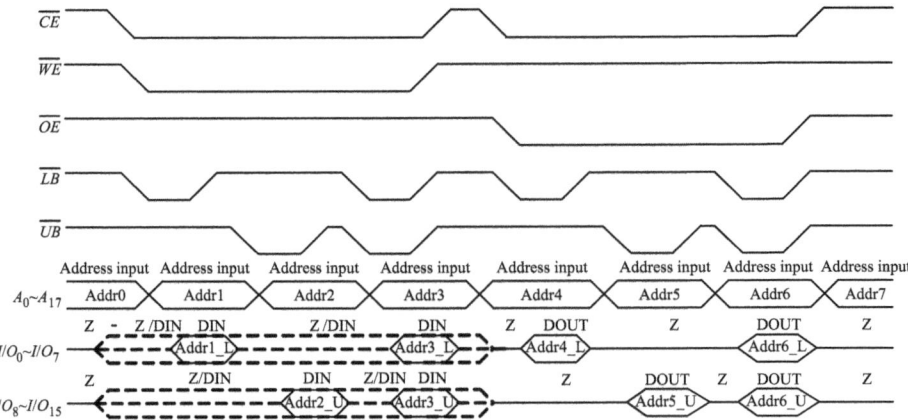

Figure 9.4.6: Timing diagram of the IS61LV25616.

9.4.5 Special RAMs

In addition to general RAMs, there are some RAMs with special structure, such as dual-ported RAM (DPRAM), first in–first out (FIFO), and ferroelectric RAM (FRAM).

1. Dual-ported random access memory

DPRAM is a type of random access memory that allows multiple reads or writes to occur at the same time, or nearly the same time, unlike single-ported RAM that allows only one access at a time.

Generally, single-ported RAM has a set of data, address, and control circuit, while DPRAM has two sets of each of I/O buffers, address decoders, and control lines. It allows two sets of circuits to control a RAM memory array. The typical feature of DPRAM is the storage data for sharing. In Figure 9.4.7, a DPRAM is

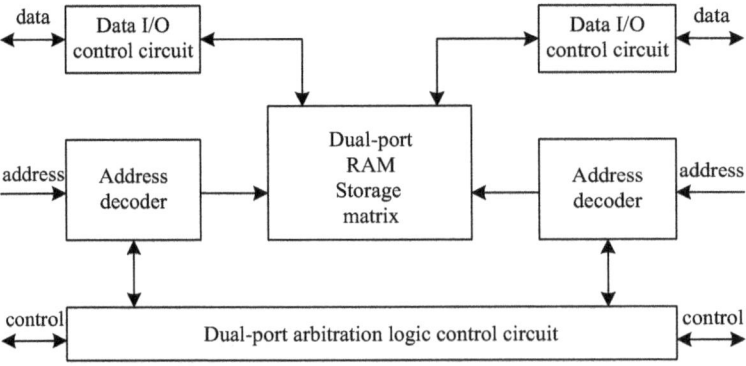

Figure 9.4.7: Block diagram of the DPRAM.

equipped with two separate address decoders and data I/O control circuits, which allow two independent CPUs or controllers to access the memory cells asynchronously at the same time. Due to data sharing, the internal dual-port arbitration control circuit is required for determining access timing sequence and access right of memory cell arrays. The timing of the two sets of circuits is the same as the timing of the ordinary SRAM. The typical DPRAM chips are CY7C130 and CY7C131.

DPRAM is mostly used for video RAM (VRAM) as video memory, allowing the CPU to draw the image while the video hardware is reading it out to the screen.

2. First in–first out (FIFO) memory

This type of memory is formed by an arrangement of shift registers. The term *FIFO* refers to the basic operation of this type of memory, in which the first data written into the memory is the first to be read out.

The difference between ordinary memory and FIFO is that there are no external read and write address lines in FIFO. The address is controlled by the internal read–write pointer, which can automatically add a 1 for each read/write operation. FIFO is easy to use, but its disadvantage is that the data should be written and read in order.

FIFO width refers to the bits by one read and write operation. FIFO width is fixed or optional in a single IC product. If FIFO is implemented by the programmable logic device, the FIFO width (data bits) can be defined by itself. FIFO depth refers to the number of data with specified width that FIFO can store. For an eight-bit FIFO, if the depth is 8, it can store eight-bit data; if the depth is 12, it can store 12 eight-bit data.

According to the operating clock, FIFO can be divided into synchronous FIFO and asynchronous FIFO. The synchronous FIFO means that the read and write clocks are the same clock and thus the read and write operations occur synchronously at the clock edge. Asynchronous FIFO means that the read and write clocks do not use the same clock.

The internal structure of FIFO consists of the dual-port RAM matrix, read and write control circuit, read and write address pointer, and other parts, as shown in Figure 9.4.8. RCLK is the read clock; WCLK is the write clock; $\overline{FF}$ and $\overline{EF}$ are the flag bit of output signal; $\overline{WEN1}$ and WEN2 are the write enable signals; $\overline{REN1}$ and $\overline{REN2}$ are the read enable signals; and $\overline{OE}$ is the output enable signal.

In the READ mode, FIFO reads out data at each clock edge of RCLK. After each read operation, the read pointer of FIFO points to the next read address by automatically adding 1. In the WRITE mode, FIFO writes in data at each clock edge of WCLK. After each write operation, the write pointer of FIFO points to the next write address by automatically adding 1. Read and write pointers are equivalent to read and write address, but this address must be selected continuously.

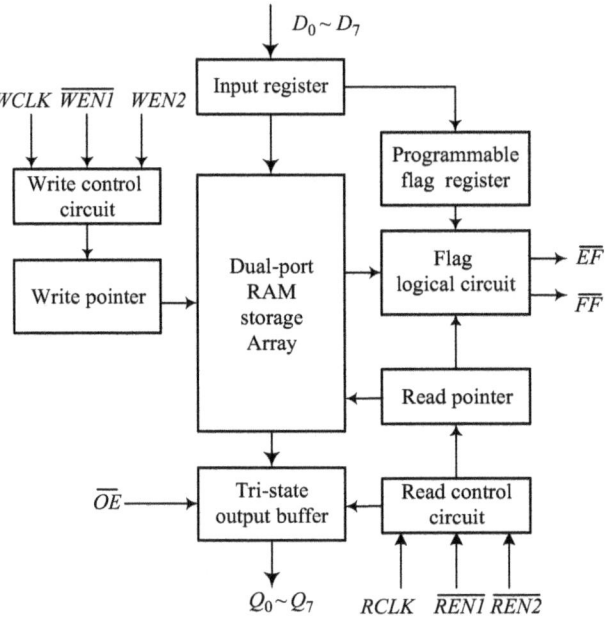

Figure 9.4.8: Schematic diagram of the internal structure of the FIFO.

When the FIFO is full, the write pointer points to the last element of the FIFO and the flag bit logic circuit sends a full flag ($\overline{FF}$) signal; when the FIFO is empty, the read pointer points to the last element of the FIFO and the flag bit logic circuit sends an empty flag ($\overline{EF}$) signal.

Typical FIFO chips include CY7C4201 (256 × 9), CY7C4211 (512 × 9), CY7C4231 (2k × 9), CY7C4251 (8k × 9), etc.

9.5 Memory expansion

If one memory chip does not have enough capacity to store data, the capacity of memory can be expanded by increasing an appropriate number of memory chips. Generally, there are two kinds of expansion methods: word-length expansion and word-capacity expansion.

The objectives of this section are to
- Define the word-length expansion
- Apply the word-length expansion to increase capacity
- Define the word-capacity expansion
- Apply the word-capacity expansion to increase capacity

9.5.1 Word-length expansion

If the word length of a memory is not enough, the number of bits in the data bus must be expanded by increasing the number of memories. This expansion is called as *bit expansion or word-length expansion*. To implement the bit expansion, the corresponding address lines, chip select signal, read and write control signal of all memory chips are connected together, but the data outputs of each chip can be combined as parallel outputs to obtain the required word length. For example, the memory of 1024 × 8 bits can be composed of two RAM chips 2114 with the capacity of 1024 × 4 bits, as shown in Figure 9.5.1. It can be seen from Figure 9.5.1 that the ten-bit address lines are connected together to guarantee the same number of addresses (2^{10}=1024) as each memory; all chip select signals are also connected together so that two memories can be accessed simultaneously. When a ten-bit address is applied on the address lines and a LOW is applied to the chip select signal, an eight-bit data is produced on the data bus and each memory chip provides four-bit data outputs.

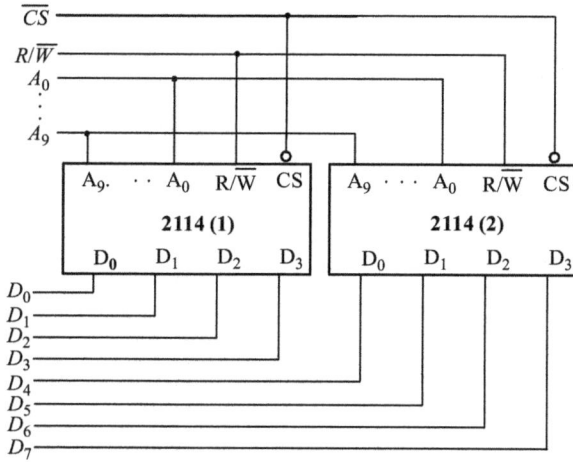

Figure 9.5.1: Expansion of two 1024 × 4 RAMs to a 1024 × 8 RAM.

9.5.2 Word-capacity expansion

If a memory does not have enough word capacity, the number of addresses must be expanded by increasing the number of memories. This expansion is called as *word-capacity expansion*. To implement the word-capacity expansion, the key issue is how to deal with the increased addresses. One alternative method is to use the chip select input, $\overline{CS}$, or to enable inputs act as the increased addresses. For example, the memory of 2048 × 4 bits is composed of two RAM chips 2114 with the capacity of 1024 × 4 bits, as shown in Figure 9.5.2. Each 2114 has ten address bits to select its 1024 addresses. The expanded memory has 2048 addresses and thus need 11 address bits. The eleventh address bit A_{10} is connected to the chip select input of 2114(1) and its complement is connected to $\overline{CS}$ of 2114(2). The ten lower-order address bits (A_0 through A_9) and the read/write control line are connected together. The data lines for the expanded memory remain four-bit width. When the eleventh address bit (A_{10}) is LOW, RAM 2114(1) is selected and RAM 2114(2) is disabled. The ten lower-order address bits (A_0 through A_9) access each address in RAM 2114(1). When the eleventh address bit (A_{10}) is HIGH, RAM 2114(2) is enabled by a LOW on the inverter output and RAM 2114(1) is disabled. The ten lower-order address bits (A_0 through A_9) access each address in RAM 2114(2).

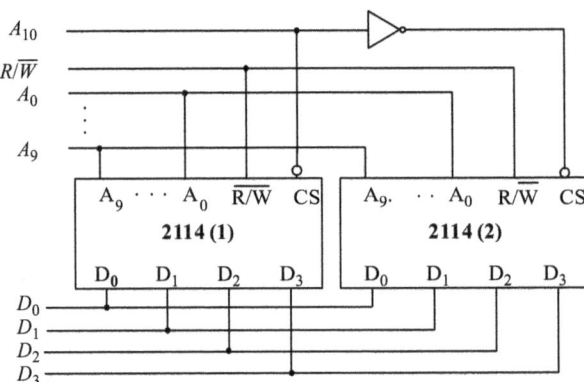

Figure 9.5.2: Illustration of word-capacity expansion.

9.6 Summary

1. Semiconductor memories are electronic data storage devices for storing large number of data on a semiconductor-based integrated circuit. Read only memory (ROM) and random access memory (RAM) are two types of semiconductor memories.

2. All memory devices store data or information by the array of memory cells. Memory cell is the smallest storage unit in a memory, which stores binary logic level of either a 1 or a 0. The size of each binary word that is stored varies depending on the memory device.

3. The location in the memory device where any binary data stored is accessed by another binary number referred to as an address. Each memory location has a unique address.

4. All memory devices operate in the same general way. The basic operations of the memory include write operation, read operation, and addressing operation.

5. ROMs are semiconductor memories to store the binary data or information permanently or semipermanently. They do not lose their data when power is removed from the device, belonging to the nonvolatile memories.

6. MROMs are programmed during the manufacturing process. PROMs are programmed one time by the user. EPROMs are just like PROMs but can be erased using UV light. EEPROMs and flash memory devices are electrically erasable and their contents can be altered after programming.

7. ROMs belong to the combinational logic circuit, so they are not only used to store data and program but also implement any combinational circuit. Moreover, a sequential circuit can be constructed with a ROM by adding some sequential logic components.

8. RAMs are semiconductor memories in which data can be written into or read from any selected address in any sequence. They lose the stored data when the power is turned off, belonging to a volatile memory, so RAMs are mainly used to store short-term data and temporary programs.

9. Static RAM (SRAM) generally uses latches as storage elements. Once the data are stored, they will remain unchanged as long as power is turned on. SRAM is easier to use but more expensive per bit and consumes more power than dynamic RAM.

10. Dynamic RAM (DRAM) stores data by capacitors rather than latches. The simplicity of the storage cell allows DRAMs to store a great deal of data with low cost. Because the charge on the capacitors must be refreshed regularly, DRAMs are more complicated to use than SRAMs.

11. Dual-ported RAM (DPRAM) is a type of RAM with special structure that allows multiple reads or writes to occur at the same time, or nearly the same time, unlike single-ported RAM that allows only one access at a time.

12. First in–first out (FIFO) memory is also a special type of memory formed by an arrangement of shift registers. The term *FIFO* refers to the basic operation of this type of memory, in which the first data bit written into the memory is the first to be read out.

13. When the capacity of one memory chip is not large enough to store data, the capacity of memory can be expanded by word-length expansion or wore-capacity expansion.

Key terms

Memory: The portion of a computer or other digital systems that stores binary data.
ROM: Read only memory; a nonvolatile random access semiconductor memory.
RAM: Random access memory; a volatile read/write semiconductor memory.
Address: The location of a given storage cell or a group of cells in a memory.
Bus: One or more interconnections that interface one or more devices based on a standardized specification.
Byte: A group of eight bits.
Capacity: The total number of data units (bits, bytes, words) that a memory can store.
Memory cell: A single storage element in a memory.

DRAM: Dynamic random access memory; a type of semiconductor memory that uses capacitors and transistors as the storage elements and is a volatile read/write memory.

EPROM: Erasable programmable ROM; a type of semiconductor memory device that typically uses ultraviolet light to erase data.

FIFO: First in–first out memory.

Flash memory: A nonvolatile read/write random access semiconductor memory in which data are stored as charges on the floating gate of a certain type of FET.

PROM: Programmable ROM; a type of semiconductor memory.

Read: The process of retrieving data from a memory.

Write: The process of storing data into a memory.

Self-test

1. The semiconductor memory is divided into _____ and _____.
   (a) EPROM and RAM
   (b) ROM and RAM
   (c) PROM and RAM
   (d) PROM and ROM

2. Two main parts of ROM are _____
   (a) Address decoders and memory array
   (b) Address decoders and flip-flops
   (c) Encoders and counters
   (d) Decoders and counters

3. If an EPROM has eight data lines and 13-bit address lines, then the storage capacity is_____.
   (a) 16 k bytes   (b) 8 bytes   (c) 8 k bits   (d) 64 k bits

4. If a RAM has eight data lines and eight-bit address lines, then its storage capacity is_____.
   (a) 16 k bytes   (b) 8 bytes   (c) 2 k bits   (d) 64 k bytes

5. ROM must store _____ data in the work, the data is_____ after power is off; data can be random read and written into RAM in the work, the data is _____ after power is off.
   (a) In, not lost; lost
   (b) Before, not lost; lost
   (c) Before, no loss; no loss
   (d) Before, lost; lost

6. Which one of the following description of the EPROM is correct?

(a) After programming, you can erase the data with UV light and then rewrite the data.

(b) Its data can be erased with electrical signals.

(c) In the microcontroller system, it is often used as data storage.

(d) After power is off, data is lost.

7. In the following memory, the _____ can ensure that the stored data is not lost after power is off.

   (a) EEPROM   (b) DRAM   (c) SRAM   (d) FIFO

8. FIFO is _____ (first, later) in _____ -(first, later) out _____ (random, read only) memory.

9. DRAM is _____ (nonvolatile, volatile) memory

10. DRAM is a _____ RAM, it _____ (needs, doesn't need) refresh circuit; SRAM is _____ RAM, it _____ (needs, doesn't need) refresh circuit.

Problems

9.1   What are the types of ROM families available?

9.2   What are the types of RAM families available?

9.3   Draw a basic logic diagram for a 512×8 bits static RAM, showing all the inputs and outputs.

9.4   Determine the logic expression of D_3, D_2, D_1, and D_0 for the ROM with capacity of 16×4 bits, as shown in Figure T9.1, in which $A_3A_2A_1A_0$ are four-bit address lines and $D_3D_2D_1D_0$ are four-bit data outputs.

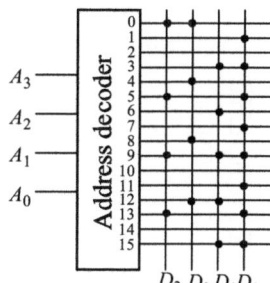

$D_3\,D_2\,D_1\,D_0$ **Figure T9.1**

9.5   Use a 16×4 bits ROM to implement two two-bit binary multiplications ($A_1A_0 \times B_1B_0$). List the truth table and draw the array diagram of memory array.

9.6 A circuit consisting of a three-bit binary up counter and a ROM is shown in Figure T9.2(a).
(a) Write the logic expression of F_1, F_2, and F_3
(b) Draw the waveform of F_1, F_2, and F_3 in relation to the clock pulse. Assume that initial state of the counter is 000.

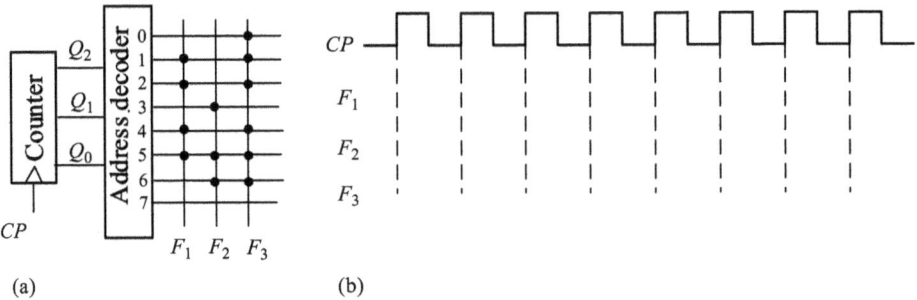

(a)　　　　　　　　　　　(b)

Figure T9.2

9.7 Design a full subtractor using a ROM and draw the programmed ROM array.

9.8 For the ROM array in Figure T9.3, determine the outputs for all possible input combinations and summarize in truth table (white cell is a 1, gray cell is a 0).

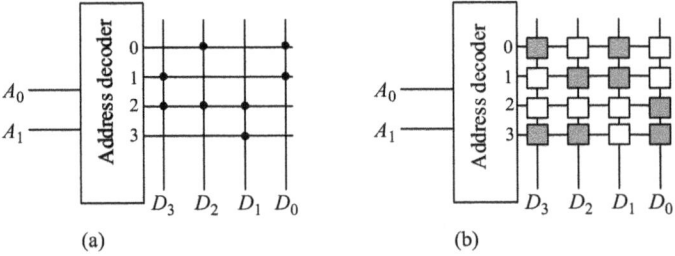

(a)　　　　　　　　　　　(b)

Figure T9.3

9.9 List the truth table for the ROM in Figure T9.4 (white cell is a 1, gray cell is a 0).

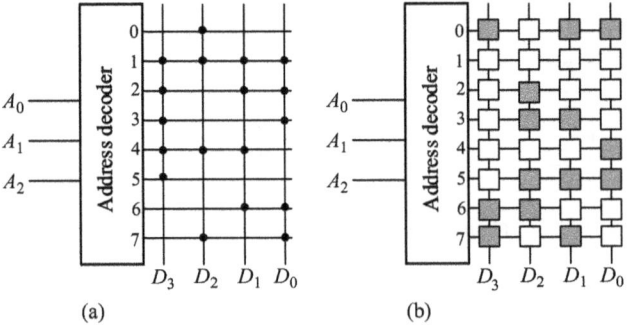

(a)　　　　　　　　　　　(b)

Figure T9.4

9.10 What is the total bit capacity of a ROM that has 14 address lines and eight data output lines?

9.11 Use 16k × 4 DRAMs to construct a 64k × 8 DRAM. Show the logic diagram.

9.12 Use 64k × 1 DRAMs to construct a 256k × 4 DRAM. Show the logic diagram.

9.13 Figure T9.5 shows a ROM array in which the inputs (A_3 through A_0) are the 8421 BCD.
   (a) List the truth table of the ROM array.
   (b) Explain what code the outputs Y_3 Y_2 Y_1 Y_0 are.

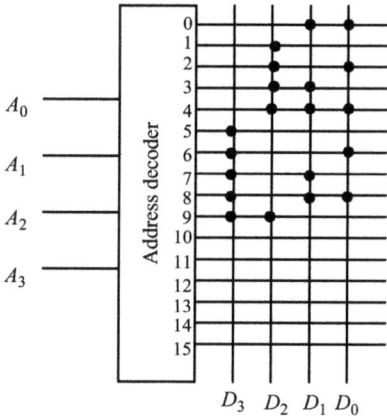

Figure T9.5

9.14 Use an EPROM to implement the following logic function and draw a programmed ROM array.

$$\begin{cases} F_1 &=& \bar{A}\bar{B}\bar{C} + \bar{B}C + AB \\ F_2 &=& \bar{A} + \bar{B} + \bar{C} \\ F_3 &=& \bar{A}\bar{B} + AB \\ F_4 &=& A + B + C + \overline{ABC} \end{cases}$$

9.15 Use EPROM to design a code convertor to convert 8421 BCD to Excess-3 code. Draw the programmed ROM array.

9.16 Use EPROM and a 74161 four-bit binary up counter to design a sequence generator that can produce the sequence code "1001110111" and "0001001011" simultaneously. Draw the logic diagram.

9.17 Use EPROM and a 74161 four-bit binary up counter to design a sequence generator that can produce the sequence code "A356789FB456." Draw the logic diagram.

9.18 Use EPROM to construct a character "H" generator, draw the programmed ROM array.

10 Programmable logic device

10.1 Introduction

Programmable logic devices (PLDs) are one type of electronic components used to build reconfigurable digital circuits. Unlike a logic gate, which has a fixed function, a PLD has an undefined function at the time of manufacture. Before applying PLD to a circuit, it must be programmed, that is, reconfigured. The earlier PLDs were *programmable read-only memory* (PROM) and *programmable logic array* (PLA), which were manufactured in the earlier 1970s. PROM has a fixed AND gate array linked to a programmable OR gate array. While the AND gate array and OR gate array in PLA are both programmable.

Besides PROM and PLA, *programmable array logic* (PAL) and *generic array logic* (GAL) are two major types of *simple programmable logic devices* (SPLDs). PAL has a programmable AND gate array connected to a fixed OR gate array. Compared with PLA, PAL omit the programmable OR array and thus its architecture is simpler, which makes the PAL faster, smaller and cheaper. However, PAL belongs to *one-time programmable* (OTP) device, which could not be updated and reused once programmed, confining the application of PAL. GAL, invented by Lattice Semiconductor in 1985, is a revised version of the PAL. GAL has the same logical properties as the PAL does, but GAL can be erased and reprogrammed. Although SPLDs have the simple structure and the flexible design, PALs and GALs are available only in small size, equivalent to a few hundred logic gates. This makes them difficult to implement large scale and complex logic circuits.

Complex *programmable logic devices* (CPLDs) can be used to implement larger scale logic circuits. For most practical purposes, a CPLD can be thought as multiple PLDs linked by programmable interconnections in a single chip, which can replace hundreds of thousands, even several millions of logic gates. The higher density of CPLD allows you to implement either more logic functions or more complicated designs. When PALs were being developed into GALs and CPLDs, a separate stream of development was happening. This type of device is based on gate array technology, called as the *field-programmable gate array* (FPGA). The term "field-programmable" means the device is programmed by the customer, not the manufacturer. FPGA was first launched by Xilinx Company in 1985. Different from the architecture of CPLD, FPGA consists of many independent programmable logic modules that can be flexibly connected to each other. This architecture of FPGA is much more flexible than that of a CPLD, so FPGA is meant for more complex designs and often used to implement all hardware design in place of a processor-plus-software solution.

With rapid development of PLDs, several hardware description languages (HDLs) are developed to help the PLD programmer. PALASM, Advanced Boolean

https://doi.org/10.1515/9783110614916-010

Expression Language (ABEL) and CUPL are frequently used for low-complexity devices, while Verilog and VHDL are popular higher-level description languages for more complex devices. Now, VHDL and Verilog are more popular, even for low-complexity designs. Today the boundary between hardware and software becomes more and more blurred. Hardware engineers create the bulk of their new digital circuitry by using programming languages such as VHDL and Verilog. A quiet revolution is taking place. Over the past few years, the density of the average PLD has begun to skyrocket. The maximum number of gates in an FPGA is currently over the order of millions and doubling every 18 months. Meanwhile, the price of these chips is dropping quickly. Many types of programmable logic are available. Some system designers try to eliminate the processor and software altogether instead of choosing an alternative hardware-only design. In addition to this incredible difference in size, there are also much variations in architecture. This chapter introduces the most common type of PLD and highlights the most important features of each type.

The objectives of this chapter are to
- Describe the basic structures and features of PALs and GALs
- Explain the basic structures and features of PLAs
- Explain how a macrocell works
- Describe the basic structures and features of CPLDs
- Describe the differences between CPLDs and FPGAs
- Explain the basic operation of the look-up table (LUT)
- Draw a basic software design flow for PLDs

10.2 Simple Programmable Devices (SPLDs)

Besides PROM and PLA, *programmable array logic* (PAL) and *generic array logic* (GAL) are two major types of *simple programmable logic devices* (SPLDs). PAL is *one-time programmable* (OTP) device, which could not be updated and reused once programmed. GAL is an improvement version based on PAL. GAL has the same logical properties as the PAL but can be erased and reprogrammed. The basic structure of both PAL and GAL is a programmable AND array linked to a fixed OR array, which is a basic sum of the product of the architecture.

The objectives of this section are to
- Explain the architecture of PAL and GAL
- Apply PAL or GAL to implement the sum of product expression
- Explain the simplified PAL/GAL logic diagram
- Describe a basic PAL/GAL macrocell
- Explain the difference of PAL and GAL

10.2.1 The PAL

Programmable array logic (PAL) is a family of programmable logic device used to implement logic functions in digital circuits introduced by Monolithic Memories, Inc. (MMI) in 1978. **A PAL consists of a programmable AND array connected to a fixed OR array.** The programmable OR array is omitted in the architecture of PALs, and a bipolar fuse process technology are used as the memory cell in a programmable AND array of PALs. All these make the PALs faster, smaller and cheaper. But PALs belong to one-time programmable (OTP) devices, which could not be updated and reused after initial programmed.

In Chapter 4, you have learned that any combinational logic function can be expressed in the sum-of-product (SOP) form. The PAL structure allows the implementation of logical expressions of any SOP form with a certain number of variables. Figure 10.2.1 shows a simple PAL structure that includes three inputs and one output. There are four AND gates forming a program AND array. Each row is connected to the input of an AND gate and each column is linked to an input variable or its complement. There is a fuse called as a memory cell connecting the row and the column. If there is a fuse link, the input variables or complemented variables can be an input of the corresponding AND gate. In the initial PAL chip, all fuse links exist at each cross point between each row and each column. Only if the PAL is programmed, some fuse links are burned on. Therefore, according to the requirement of combinational logic circuit, you can determine which fuse links should be retained to connect the desired variables or complemented variables as the inputs of the AND gate, and you can select which fuse links should be burned on by programming the AND array of PAL. The resulting desired product term is formed from the AND gates. The outputs of AND gates are applied on the OR gate, and thus the resulting SOP output can be obtained. Most of PALs has multiple inputs and multiple outputs.

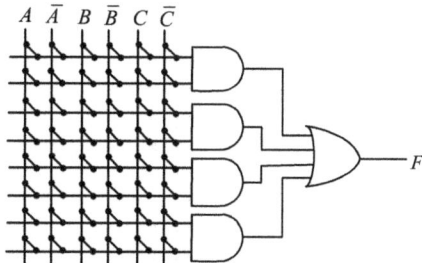

$A\ \bar{A}\ B\ \bar{B}\ C\ \bar{C}$

F

Figure 10.2.1: Basic AND/OR structure of a PAL.

Figure 10.2.2 is an example of the programmed PAL to implement the SOP expression. The top AND gate generates a product term $A\bar{B}$, the next top AND gate generates a product term $\bar{A}\bar{C}$, the next bottom AND generates a product term $\bar{A}B$, and

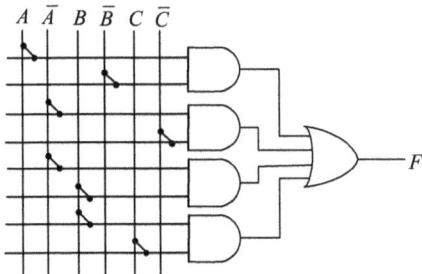

Figure 10.2.2: A PAL programmed to implement a SOP expression.

the bottom AND gate produces a product term BC. It can be seen that the fuses are left intact to connect the desired variables or complemented variables to the input of AND gate, and the fuses are burned off where the variables and complemented variables are not used in a given product term. The resulting output from the OR gate is the SOP expression as

$$F = A\bar{B} + \bar{A}\bar{C} + \bar{A}B + BC$$

10.2.2 Simplified notation for PLD diagrams

The practical PAL has many AND gates and OR gates and is capable of handling multiple variables and their complements. Most of the PAL diagrams on the data sheet use simplified notations to make the schematic diagram much clearer.

1. Buffer notation
In PLD, the input buffer is often used to provide an enough driving capacity, which allows to connect multiple inputs of AND gates. Usually, a triangle symbol represents a buffer that gives both the variable and its complement as its outputs, as shown in Figure 10.2.3.

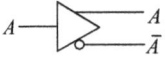

Figure 10.2.3: Buffer notation.

2. AND gate and OR gate notation
Generally, there are a large number of programmable interconnection lines in the PLD chip. Each AND gate in the PLD has multiple inputs. The typical logic diagrams for PLD use a special AND gate symbol to represent a multi-input AND gate, as shown in Figure 10.2.4 (a). The multi-input AND gate has a single input line with a slash and a digit representing the actual number of input lines. Sometimes a slash and a digit can be omitted. A black solid dot "•" at the cross point represents a fixed connection

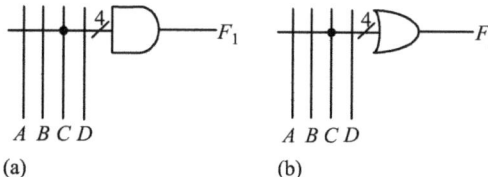

Figure 10.2.4: Simplified notation in PLD: (a) AND gate; (b) OR gate.

(a)   (b)

and a "×" at the cross point indicates a programmable connection. No mark indicates no connection. Similarly, the multi-input OR gate can be represented by the same way as the multi-input AND gate, as shown in Figure 10.2.4(b). The corresponding traditional logic diagram is shown in Figure 10.2.5.

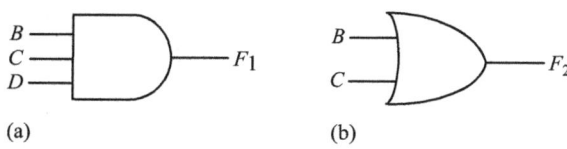

(a)   (b)

Figure 10.2.5: Traditional logic diagram: (a) AND gate; (b) OR gate.

The output logic expressions of AND gate and OR gate in Figure 10.2.4 are expressed as

$$F_1 = BCD, F_2 = B + C$$

Figure 10.2.2 can be redrawn with the simplified notation in Figure 10.2.6 (a). Sometimes the buffer can also be omitted; variables and its complements are directly labeled in the simplified logic diagram, as shown in Figure 10.2.6 (b).

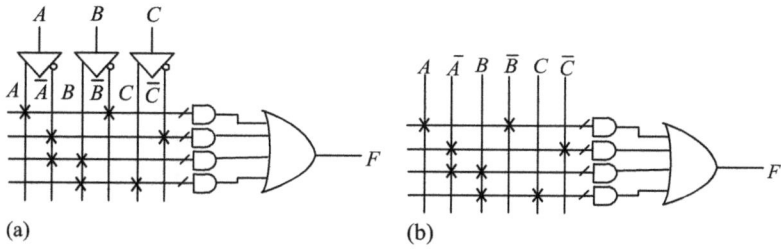

(a)   (b)

Figure 10.2.6: Simplified logic diagram of the programmable PAL in Figure 10.2.2.

Example 10.1 Show how a PAL is programmed for the following 4-variable logic function.

$$F = A\bar{B}C\bar{D} + A\bar{B}\bar{C}D + \bar{A}B\bar{C} + BC.$$

Solution
The programming array is shown in Figure 10.2.7. The intact fuse links are represented by the "×". The absence of a "×" means that the fuse is open.

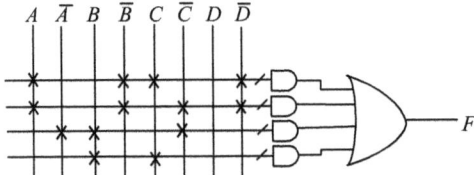

Figure 10.2.7: Logic diagram of programmable PAL for Example 10.1.

10.2.3 The GAL

Generic array logic (GAL) is an improvement version of the PAL. It has the same logical properties as the PAL but can be erased and reprogrammed repeatedly, which belongs to multiple times programmable device. The basis difference is that a GAL uses a reprogrammable memory cell, for instance, a floating gate MOSFET replacing fuse link in Figure 10.2.1. **A GAL usually consists of input buffer, programmable AND logic array, fixed OR gate array, and output logic macrocell.**

Figure 10.2.8 shows a partial logic diagram in a GAL. The actual GAL structure is more complex than it. IN_1 and IN_2 are two input variables, which pass through the buffers to form the variables and their complements as the inputs of the AND array. The AND array is programmable, the OR array is fixed. The programmable AND array produces the product terms that is the AND of input variabes and their complements labeled with a "×". Then the product terms pass through the OR gate and a SOP expression can be produced.

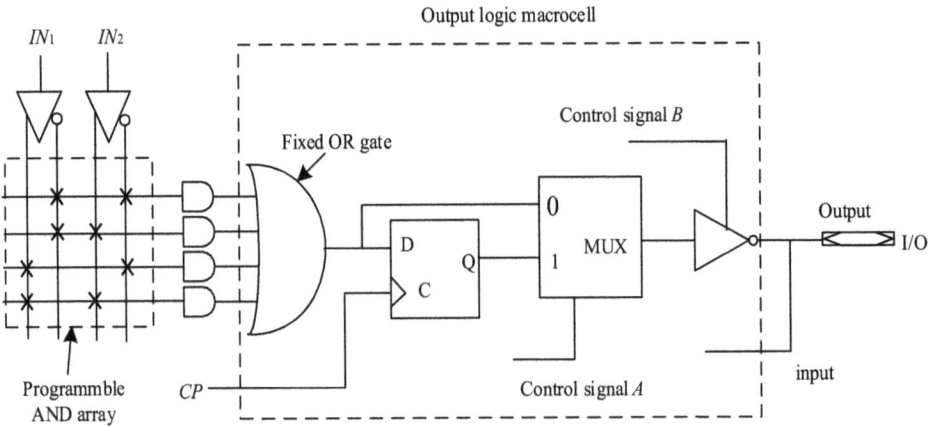

Figure 10.2.8: Partial logic diagram in a GAL.

The internal control signal A is used to control the output of a 2-to-1 MUX (multiplexer). If A is 0, the output of MUX is determined by the output of a fixed OR gate, which is called as the combinational output mode. If A is a 1, the output of the MUX comes from the output of flip-flop, which is called as the sequential output mode.

The control signal B controls a tristate buffer. If B is a 1, the tristate gate is enabled, the I/O pin is selected as the output pin; if B is a 0, the tristate buffer is disabled, and the I/O pin is selected as the input pin. The state of control signal B is generated automatically by computer programming.

In essence, a GAL is a PAL that can be erased and reprogrammed. It has the same type of AND/OR organization as a PAL.

10.2.4 PAL/GAL general block diagram

Figure 10.2.9 shows a general block diagram of PAL and GAL. Both of them have the programmable AND array and the fixed OR array [21]. The main difference between a PAL and GAL is the reprogrammable process technology E^2CMOS and flexible output logic macrocell (OLMC) is used in the output part of a GAL. Since there are flip-flops in an OLMC, GAL can implement both combinational logic and sequential logic by programming the OLMC.

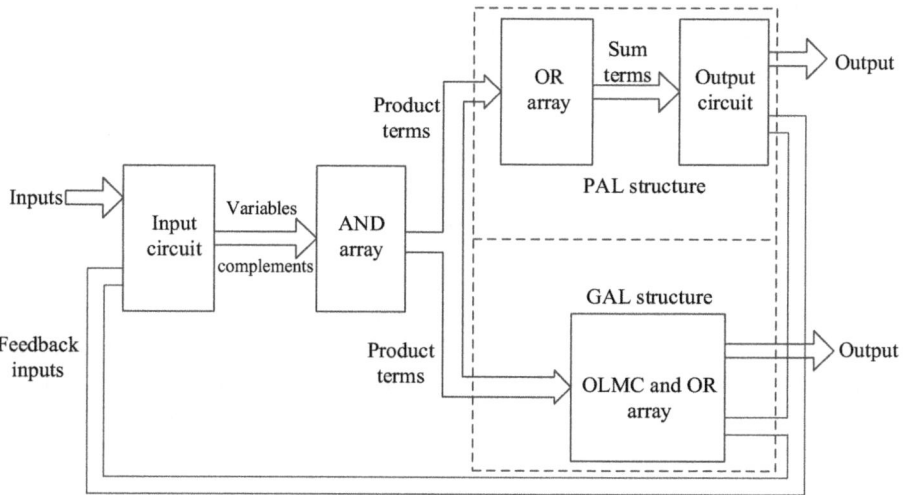

Figure 10.2.9: A general block diagram of PAL or GAL.

10.2.5 Macrocell

A macrocell is usually composed of an OR gate and some associated output logic. **The various macrocells are not the same in complexity depending on the particular**

type of GAL. For example, a GAL chip, type No. GAL16V8, where 16 refers to the number of the inputs, 8 represents the number of the outputs, and **V** refers to the flexible form of the output. There are eight same OLMCs in a GAL16V8. Figure 10.2.10 shows the functional block diagram of GAL16V8, and the internal logic of one OLMC. The XOR(n) signal of each macrocell controls the output polarity. When XOR(n) =1, the XOR gate is equivalent to an inverter. The output of OR gate passes through XOR gate and an inverter output buffer and then the final output is the active high level. When XOR(n) = 0, the output XOR gate is the same as its input. It continues to pass through the inverter output buffer and the final output is active-LOW output. There are two global bits, SYN and AC_0, control the mode configuration for all macrocells to select three global OLMC configuration modes possible: simple, complex and registered. The control bit, SYN, selects the output mode of GAL and the use of clock input (CLK) and enable input ($\overline{OE}$). If SYN=1, all outputs are the combinational output without using any flip-flop. The clock input (CLK) and enable input ($\overline{OE}$) can be configured as the common inputs; if SYN=0, at least one OLMC is configured as the sequential output. At this situation, the CLK input and $\overline{OE}$ input must be used for the clock pulse input and the enable control of the tristate gate rather than for common inputs. $AC_{1(n)}$ of each macrocell controls the input/output configuration. All control signals of the OLMC include SYN, AC_0, $AC_{1(n)}$, XOR(n), which are corresponding to programmable bits in structure control word. The structure control word can be configured by software complier. According to the programmable bits in the structure control word, the GAL can be configured as five basic operation modes as follows.

1. Dedicated input configuration
Dedicated input configuration is shown in Figure 10.2.11. At this configuration, the control bits are set as SYN=1, AC_1=1, AC_0=0, making the tristate inverter disconnection and the OLMC configured as dedicated input.

2. Dedicated combinational output configuration
Combinational output configuration is shown in Figure 10.2.12. At this configuration, the control bits are set as SYN=1, AC_1=0, AC_0=0, making the tristate inverter connection and the OLMC configured as combinational output. At this situation, the clock input (CLK) and enable input ($\overline{OE}$) can be configured as the common inputs.

3. Combinational Input/Output (I/O) configuration
Combinatorial I/O configuration is shown in Figure 10.2.13. At this configuration, the control bits are set as SYN=1, AC_1=1, AC_0=1, the macrocell is configured as output only or I/O functions. When it is used as an input, the tristate inverter is disabled, and the input goes to the buffer that is connected to the AND array. The clock input (CLK) and enable input ($\overline{OE}$) can be configured as the common inputs.

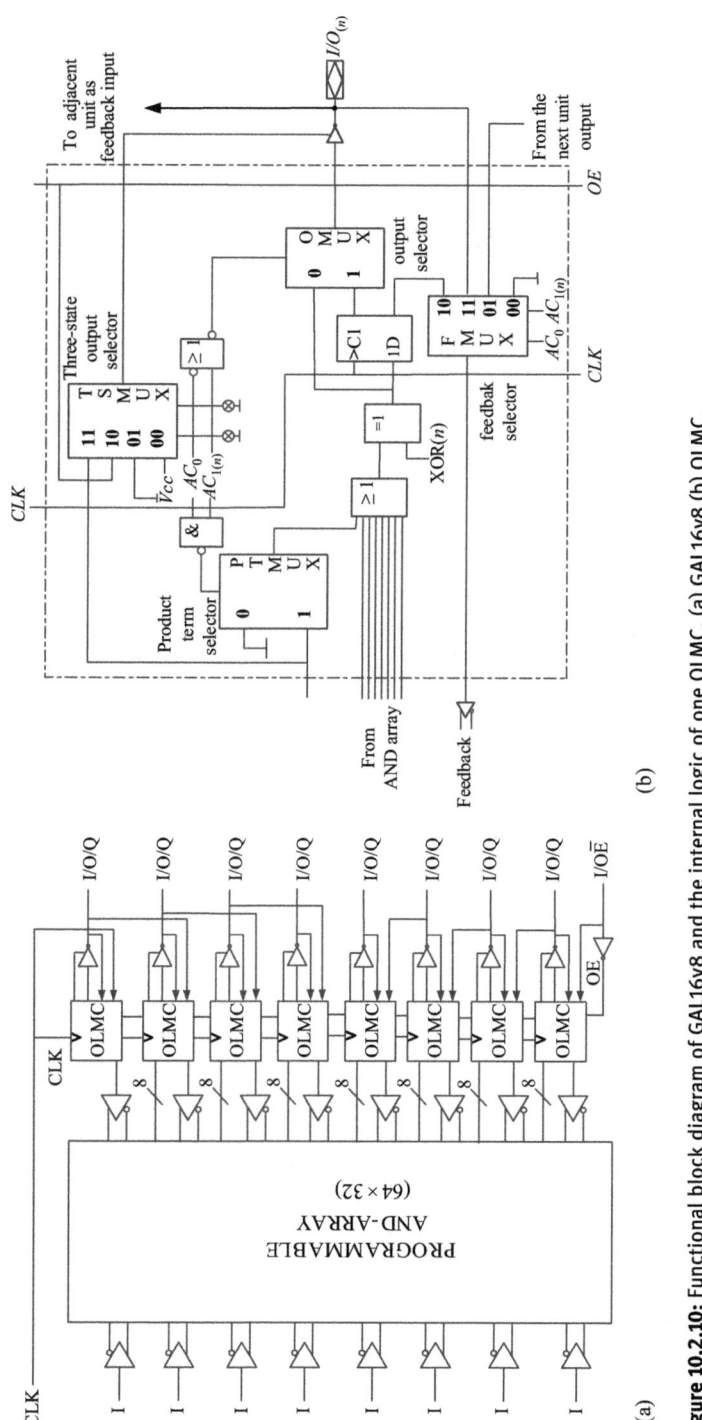

(a)

(b)

Figure 10.2.10: Functional block diagram of GAL16v8 and the internal logic of one OLMC. (a) GAL16v8 (b) OLMC.

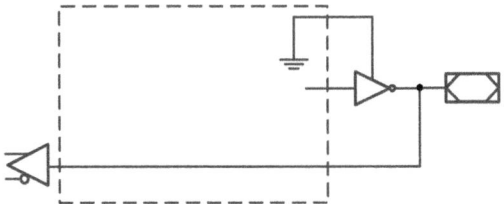

Figure 10.2.11: The OLMC configured as dedicated input.

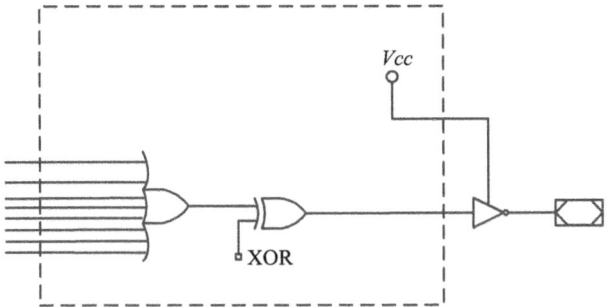

Figure 10.2.12: The OLMC configured as combinatorial output.

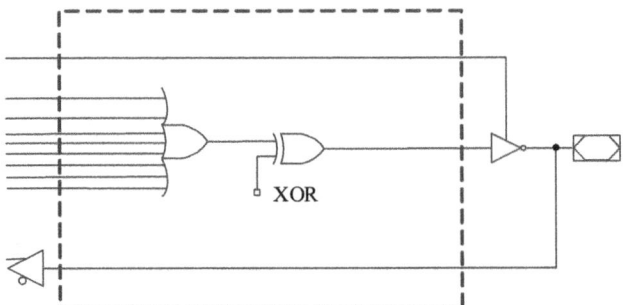

Figure 10.2.13: The OLMC configured as combinational I/O.

4. Registered configuration

Registered configuration is shown in Figure 10.2.14. At this configuration, the control bits are set as $SYN = 0$, $AC_1 = 0$, $AC_0 = 1$, then macrocell is configured as sequential output. The clock input (CLK) and enable input ($\overline{OE}$) must be configured as the clock pulse input and enable input for the registered output rather than common inputs.

5. Registered combinational I/O configuration

Registered combinational I/O configuration is shown in Figure 10.2.15. At this configuration, the control bits are set as $SYN=0$, $AC_1=1$, $AC_0=1$, the macrocell is configured as combinational I/O output. When it is used as an input, the tristate inverter is

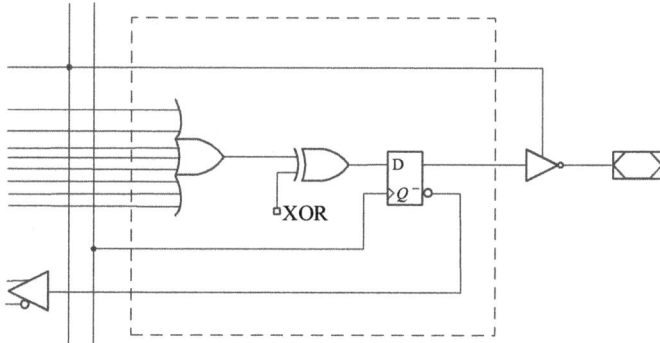

Figure 10.2.14: The OLMC configured as register mode.

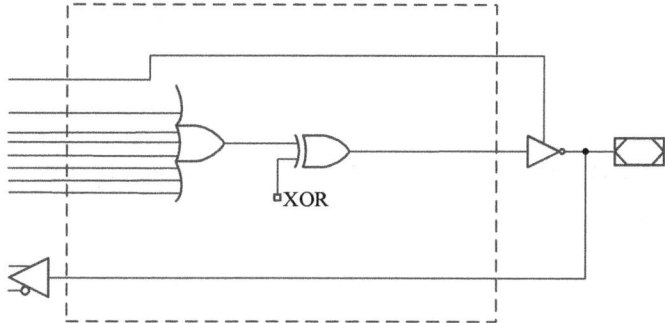

Figure 10.2.15: The OLMC configured as registered Combinational I/O

disabled, and the input goes to the buffer that is connected to the AND array. Although the logic diagram is similar with that of combinational I/O configuration, the clock input (CLK) and enable input ($\overline{OE}$) must be configured as the clock pulse input and enable input rather than common inputs.

10.2.6 Features of the PAL and GAL

PAL and GAL both consist of the programmable AND logic array and fixed OR gate array. The memory cells in PAL use bipolar fuse, thus PAL has simple structure and high speed. But PAL belongs to one-time programmable device, which limits the application of PAL. GAL is an innovation of the PAL. With the floating gate MOS as memory cell, GAL is erasable and reprogrammable, making prototyping and design change easier for engineers. This brings many advantages of the GAL over the PAL. Due to the adoption of the advanced CMOS technology, GAL consumes lower power about 50% to 75% reduction in comparison to PAL. Although the speed of GAL is about half or a fourth of PAL, GAL can be erased and reprogrammable. It has high speed electrical erasure (<100 ms) and can be reprogrammable in the ratio up to 100%.

Usually, a GAL can be reprogrammable near one hundred times. By configuring OLMC, GAL can be configured as a sequential form to assure the programming process tested in the ratio up to 100%. In addition, the stored data can be retained over two decades due to the use of E^2COM process technology. These high performances make GAL widely applied for DMA control, state machine control, high speed graphics processing, and standard logic speed upgrade.

10.3 CPLD

As IC technology advances, there is naturally great interest in creating larger PLD architecture to take advantage of higher chip density. CPLD packages multiple SPLDs into a single chip to get more capacity for larger-scale logic designs. The fully programmable AND/OR array and a bank of macrocells are combined in a CPLD. Macrocells are functional blocks that perform combinatorial or sequential logic, and also have the added flexibility for variables or their complements, along with varied feedback paths. This section introduces the traditional CPLD architecture, remembering that CPLDs may be slightly different in architecture and parameters, such as density, process technology, power consumption, voltage and speed.

The objectives of this section are to
– Explain the basic structure of CPLD
– Describe how to generate the product terms in the CPLD

10.3.1 CPLD architecture

CPLD architecture is mainly composed of multiple SPLD arrays, a programmable interconnection and multiple I/O control blocks [45]. Each SPLD is called a logical array block (LAB), sometimes using other names such as function blocks, logical blocks, or generic blocks. Programmable interconnections are often referred to as programmable interconnect array (PIA), which is also called advanced interconnect matrix (AIM) or similar names by other manufacturers, for example, Xilinx Inc. Although CPLD products manufactured by different companies have their own characteristics, they have the similar architecture.

Let's take an EPM7128S chip in MAX 7000s as an example to illustrate CPLD architecture. The EPM7128S is a typical CPLD with high density and high performance, which is manufactured by ALTERA Company. It has totally 84 pins with PLCC package and the corresponding pin arrangement as shown in Figure 10.3.1. The pin description of EPM7128S is listed in Table 10.3.1. Among them, four

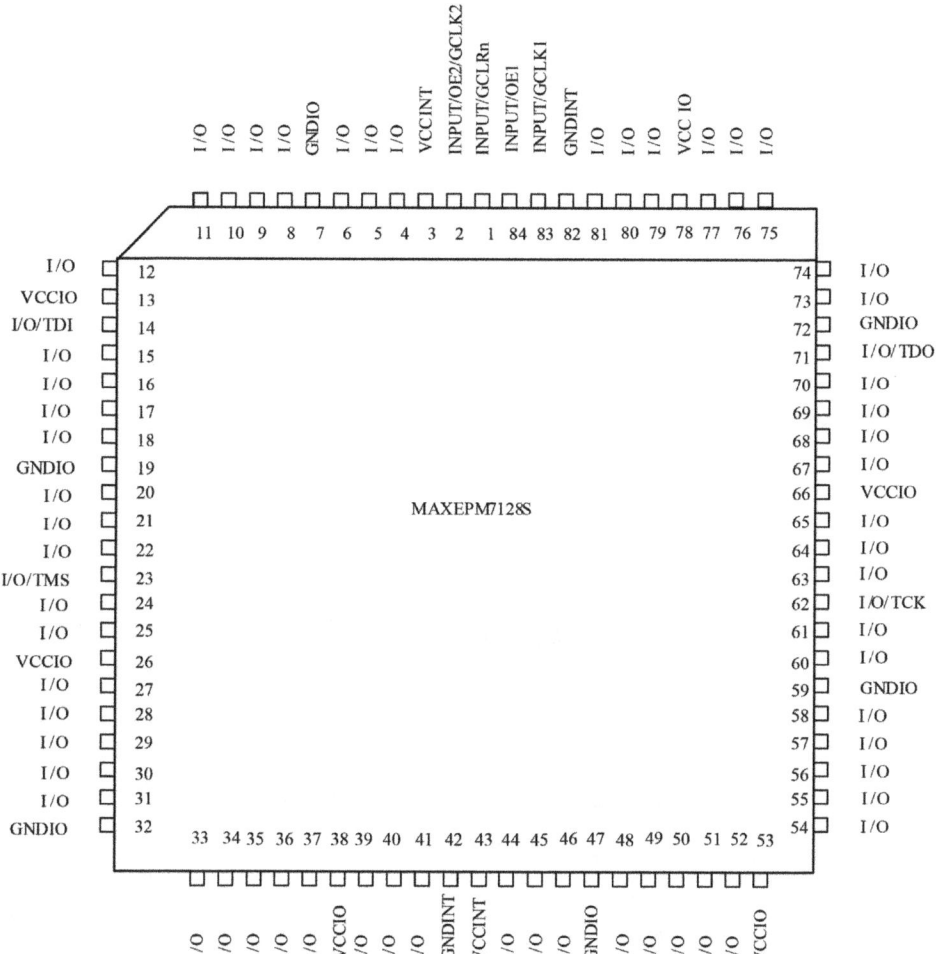

Figure 10.3.1: 84-pin PLCC package pin-out diagram.

specified pins including TDI, TMS, TCK, and TDO are used not only for in-system programming but also for the chip test. There are also four dedicated inputs that can be used as general-purpose inputs or as high-speed, global control signals (clock, clear and two output enable signals) for each macrocell and I/O pin. Figure 10.3.2 shows the architecture of MAXEPM7128S.

In EPM7128S, there are eight LABs, one PIA and multiple I/O control blocks. Figure 10.3.2 only shows 4 LABs. Each LAB consists of 16 macrocells, and LABs can be connected together via PIA. PIA is a programmable global bus structure, which can connect all LABs, to connect the general inputs, I/O control blocks and macrocells. The internal interconnection between LABs can be programmed with

Table 10.3.1: Pin description of MAXEPM7128S.

Pin names	Pin No.	Pin description
INPUT/GCLK1	83	Input/global clock 1
INPUT/GCLRn	1	Input/global clear
INPUT/OE1	84	Input/output enable control1
INPUT/OE2/GCLK2	2	Input/output enable control2/global clock 2
TDI	14	Programming data input
TMS	23	Programming mode selection
TCK	62	Programming clock
TDO	71	Programming data output
GNDINT	42,82	Ground of 5V dc supply voltage
GNDIO	7,19,32,47,59,72	Ground of input/output
VCCINT(5.0V Only)	3,43	5V dc supply voltage
I/O	4,5,6,···	68 input/output pins

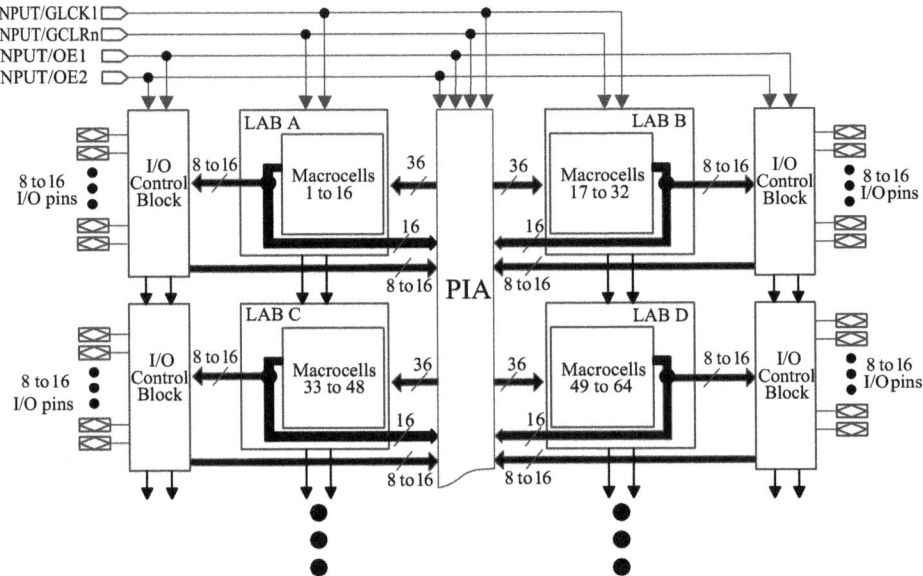

Figure 10.3.2: Block diagram of EPM7128S.

software. CPLD can be programmed to implement more complex logic functions. These complex logic functions are SOP structures based on a single LAB (actually SPLD). The inputs can be connected to any LAB, and the outputs can be inter-connected via the PIA to any other LABs.

10.3.2 Macrocell

Each LAB in EPM7128S has 16 macrocells. Macrocell can be individually configured for either sequential or combinational logic operations. Figure 10.3.3 shows a macrocell of EPM7128S device. The macrocell mainly consists of three functional blocks: the logic array, the product-term select matrix, and the programmable register. The logic array contains a small AND programmable gate array with five AND gates, an OR gate, a XOR gate for choosing the output polarity, and a product term selection matrix used to connect the outputs of AND gate to the OR gate. The other associate logic and the programmable registers can be programmed as input, combined logic output or register output [3].

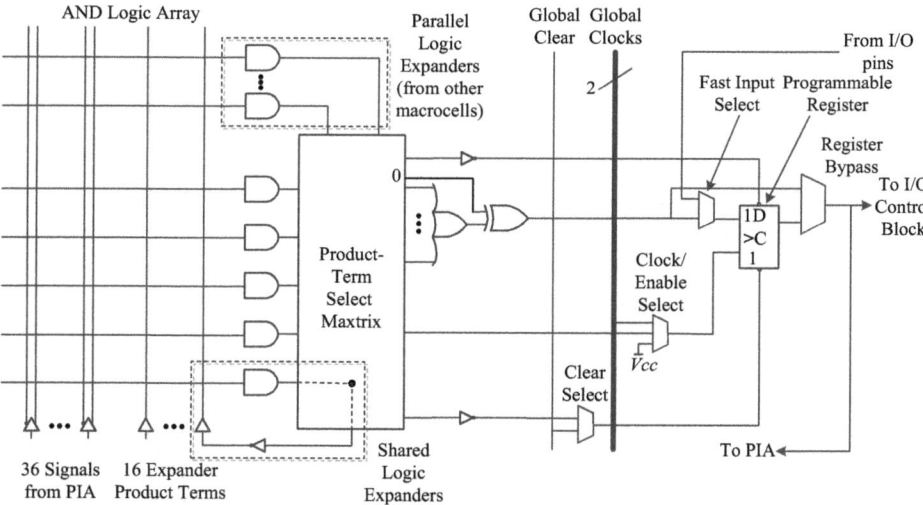

Figure 10.3.3: EPM7128S device macrocell.

Combinational logic is implemented in the logic array, which provides five product terms per macrocell. The product-term select matrix allocates these product terms for using as either primary logic inputs (to the OR and XOR gates) to implement combinatorial function, or as secondary inputs to the macrocell's register clear, preset, clock and clock enable control functions. Two kinds of expander product terms ("expanders") are available to supplement the lack of macrocell logic resources: one is shareable expanders, which are inverted product terms fed back into the logic array; another is parallel expanders, which are product terms borrowed from adjacent macrocells. These two expanders will be introduced in detail later. The Altera development system automatically optimizes product-term allocation according to the logic requirements of the design.

For registered functions, each macrocell flip-flop can be individually programmed to implement D, T, J-K, or S-R operation with programmable clock control. The flip-flop can be bypassed for combinatorial operation. During design entry, the designer specifies the desired flip-flop type; Altera development software then selects the most efficient flip-flop operation for each registered function to optimize resource utilization.

Each register also supports asynchronous preset and clear function. As shown in Figure 10.3.3, the product-term selection matrix allocates product terms to control these operations. Although the product-term-driven preset and clear of the register are active high, active-low control can be obtained by inverting the signal within the logic array. In addition, the clear function of registers can be individually driven by the active-low dedicated global clear pin

This macrocell in CPLD is slightly different from that of the SPLD because it contains a portion of the programmable AND gate array and a product selection matrix. As shown in Figure 10.3.3, the inputs of five AND gates come from PIA, and the output of AND gate is sent to the product selection matrix. The product of the bottom AND array is inverted and sent back to the programmable array for the extension of sharing with other macrocells. This parallel expander entry allows the use of other product terms in the adjacent macrocells to extend an SOP expression. The product term selection matrix is an array of programmable connections, which connects the selection outputs from the AND gate array and the expander entry to the OR gate.

10.3.3 Expander product terms

Although most logic functions can be implemented with the five product terms available in each macrocell, the more complex logic functions require additional product terms. The architecture of EPM7128S allows both shareable and parallel expander product terms ("expanders") that provide additional product terms directly to any macrocell in the same LAB. These expanders help you ensure that logic is synthesized with the fewest logic resources to obtain the fastest speed.

1. Shareable expander

Each LAB in an EPM7128S chip has 16 shareable expanders and each macrocell has one sharable expander. The sharable expander term is an inverted product term that is fed back into the logic array to increase the number of product terms for building complex logic function. Figure 10.3.4 shows how shareable expanders can feed multiple macrocells.

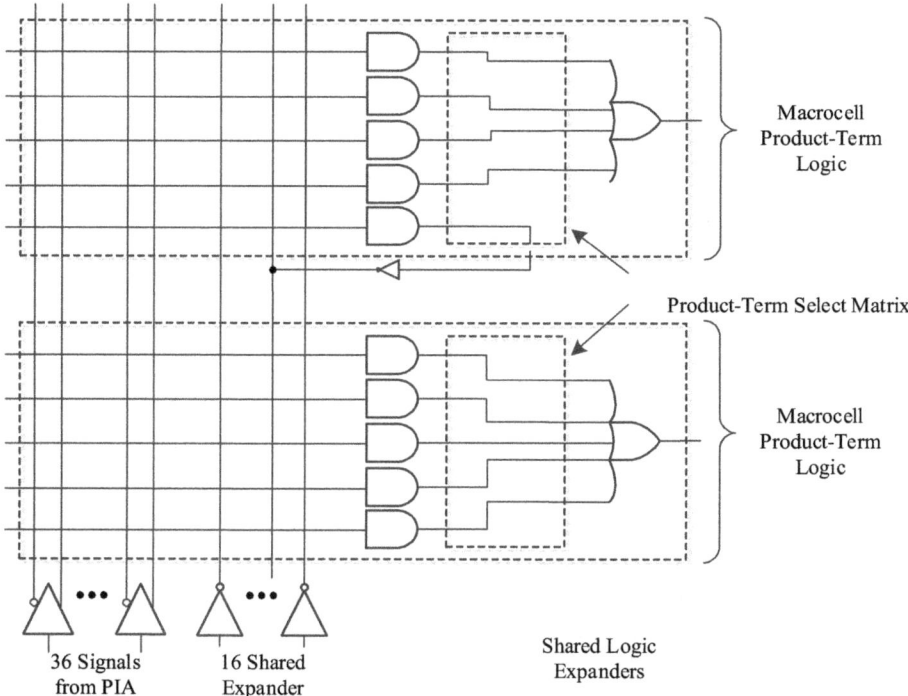

Figure 10.3.4: Shared expander.

Each shareable expander can be shared by any or all macrocells in the LAB. Each macrocell can produce five product terms, and these product terms are generated from the AND gate array. If a macrocell requires more than five product terms to get the SOP output, the macrocell can use an extension of adjacent macrocells in the same LAB.

Assume that a design requires a SOP expression containing six product terms, Figure 10.3.5 illustrates that the product term of adjacent macrocell is used to increase an SOP output. The unused macrocell 2 generates a shared expander $(E + F)$ which connects to the fifth AND gate in macrocell 1, resulting in a SOP expression with 6 product terms.

2. Parallel expander

Parallel expander is another way to increase the number of macrocell product terms. Parallel expanders are unused product terms that can be allocated to a neighboring macrocell to implement fast and complex logic function. Parallel expanders allow up to 20 product terms to directly feed the macrocell OR logic, with five product terms provided by the macrocell and 15 parallel expanders provided by neighboring macrocells in the LAB. Figure 10.3.6 shows how parallel expanders can be borrowed from a neighboring macrocell.

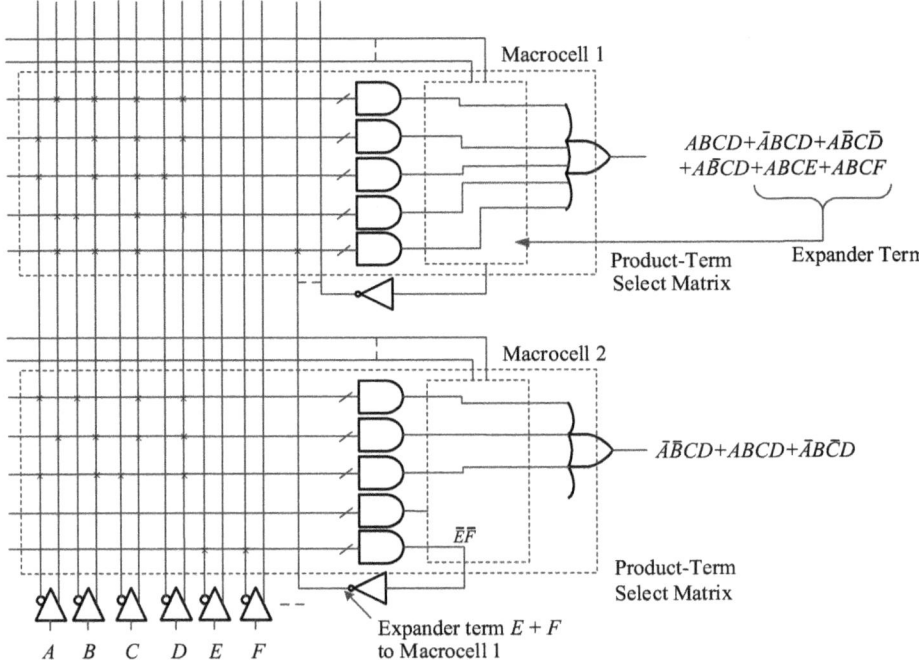

Figure 10.3.5: Sketch of using a shared expander term from adjacent macrocell to increase an SOP expression.

Figure 10.3.7 shows how a macrocell borrows a parallel expander from neighboring macrocell to increase the SOP output. The macrocell 2 uses three product terms from the macrocell 1 to generate an 8-item SOP expression.

10.3.4 Programmable interconnect array

Logic is routed between LABs via the programmable interconnect array (PIA). This global bus is a programmable path that connects any signal source to any destination on the device. All EPM7128S dedicated inputs, I/O pins, and macrocell outputs feed the PIA, which make the signals available throughout the entire device. Only the signals required by each LAB are actually routed from the PIA into the LAB. Figure 10.3.8 shows how the PIA signals are routed into the LAB. An EEPROM cell controls one input to a 2-input AND gate, which selects a PIA signal to drive into the LAB.

10.3.5 I/O control blocks

The I/O control block allows each I/O pin to be individually configured for input, output, or bidirectional operation. All I/O pins have a tristate buffer that is individually

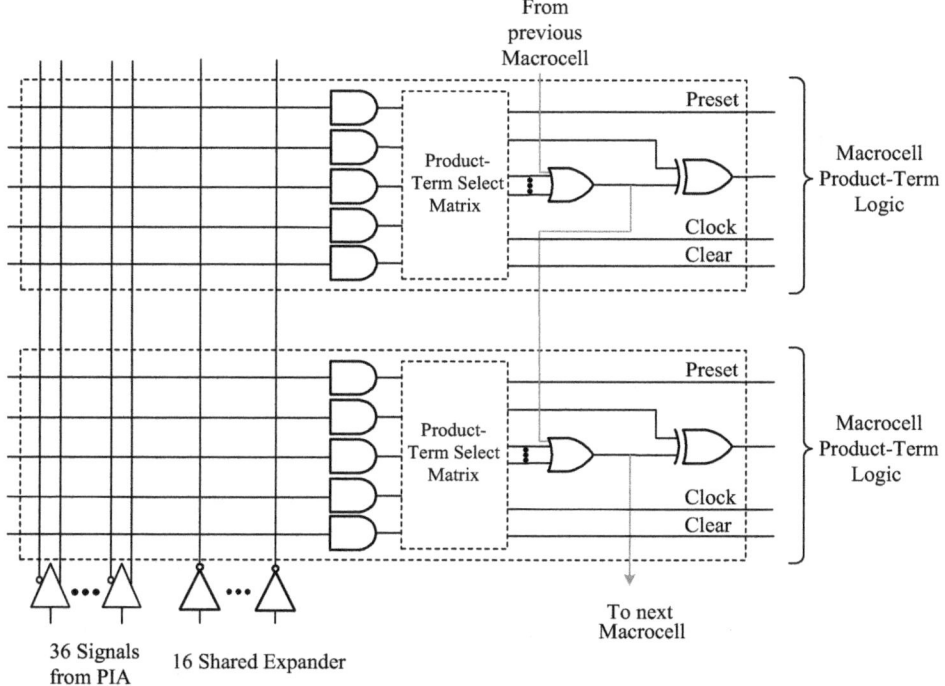

Figure 10.3.6: Parallel Expanders.

controlled by one of the global output enable signals or directly connected to ground or V_{CC}. Figure 10.3.9 shows the I/O control block of the EPM7128S. The I/O control block of EPM7128S device has two global output enable signals that are driven by two dedicated active-low output enable pins (OE1 and OE2). The I/O control block has six global output enable signals that are driven by the trues or complements of two output enable signals, or a subset of the I/O pins, or a subset of the I/O macrocells.

When the tristate buffer control is connected to the ground, the output is tristated (high impedance) and the I/O pin can be used as a dedicated input. When the tristate buffer control is connected to V_{CC}, the output is enabled. The output buffer for each I/O pin has an adjustable output slew rate that can be configured for low-noise or high-speed performance. A faster slew rate provides high-speed transition for high-performance system. However, these fast transition may introduce noise transients into the system. A slow slew rate reduces system noise, but adds a nominal delay of 4 to 5 ns. For MAX 7000S devices, each I/O pin has an individual EEPROM bit that controls the slew rate, allowing designers to specify the slew rate on a pin-by-pin basis.

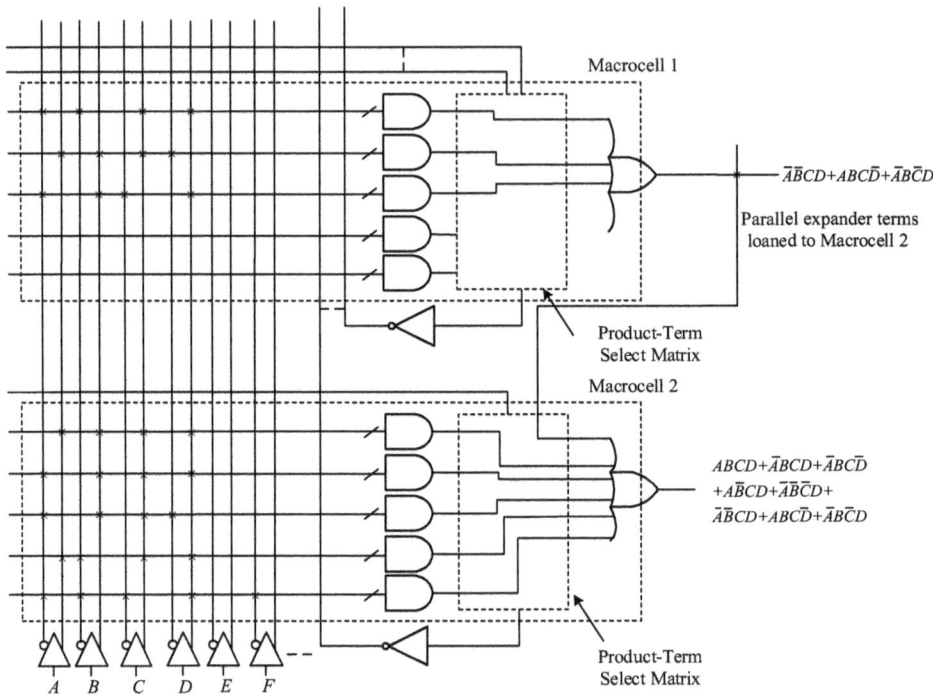

Figure 10.3.7: Sketch of using parallel expander terms from another macrocell to increase an SOP expression.

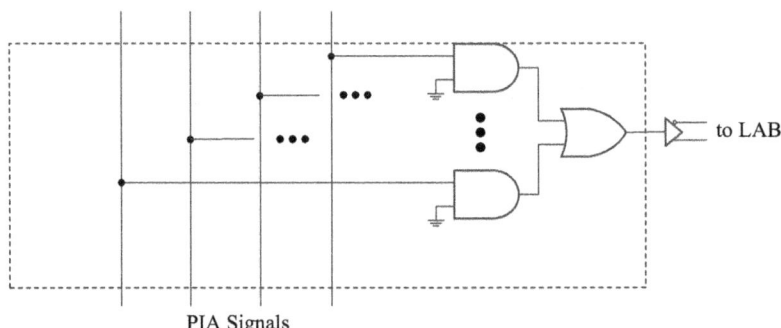

Figure 10.3.8: PIA routing.

10.3.6 PLA

As you have learned earlier, the structure of CPLD refers to the organization and arrangement of internal units. Some PLDs' structures use PLA-based structures instead of the PAL-based structures that are discussed above.

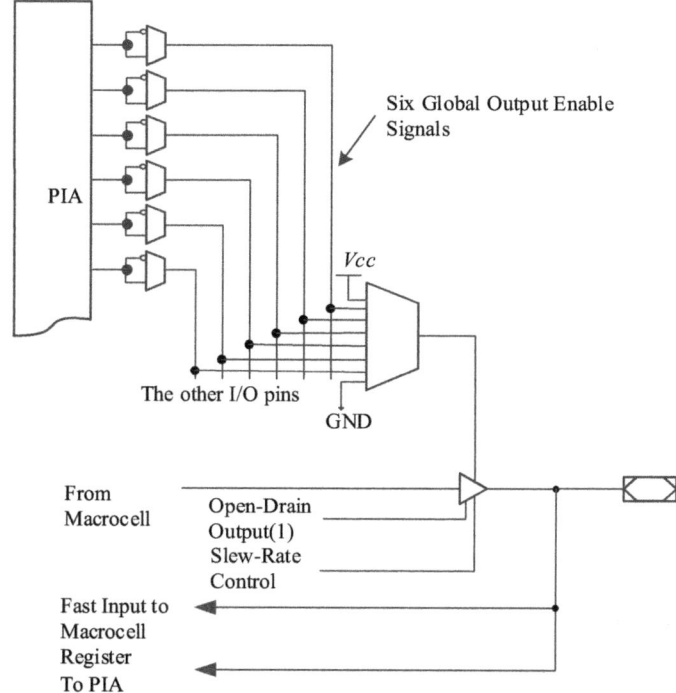

Six Global Output Enable
Signals

PIA

Vcc

The other I/O pins

GND

From
Macrocell

Open-Drain
Output(1)
Slew-Rate
Control

Fast Input to
Macrocell
Register
To PIA

Figure 10.3.9: I/O control block in EPM7128S.

Figure 10.3.10 compares a simple PAL structure with a simple PLA structure. PAL has a programmable AND array followed by a fixed OR array to produce an SOP expression, as shown in Figure 10.3.10 (a). PLA has a programmable AND array followed by a programmable OR array, as shown in Figure 10.3.10 (b).

10.3.7 Combination mode and register mode

When a macrocell is programmed to produce a SOP combinatorial logic function, the logical unit in the data path is marked with dash squares and double lines as shown in Figure 10.3.11, as we can see, the only one multiplexer is used and the flip-flop is bypassed.

When a macrocell is programmed to implement the register mode, the SOP output of combinational logic is sent to the input of the register and the global clock is used as clock-drive signal. The logical unit in the data path is marked with the dashed square and double lines in Figure 10.3.12. It can be seen that the four multiplexers are used in the figure, and the flip-flop is valid.

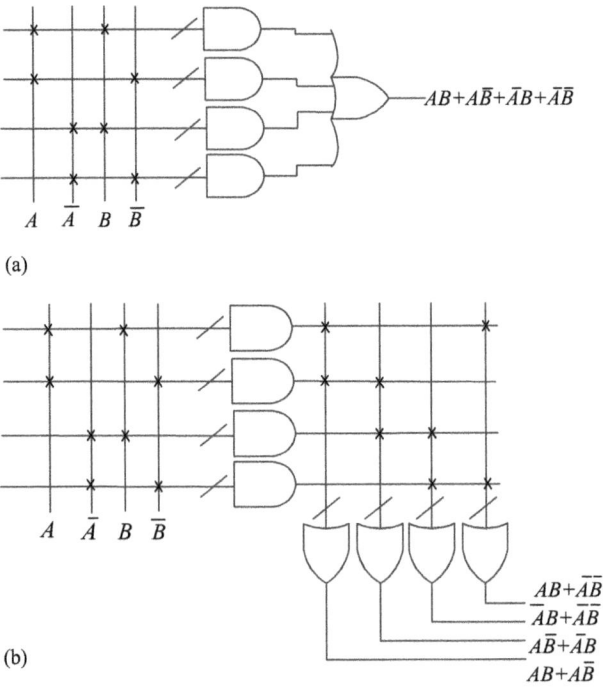

(a)

(b)

Figure 10.3.10: Comparison of a basic (a) PAL-type array and (b) PLA -type array.

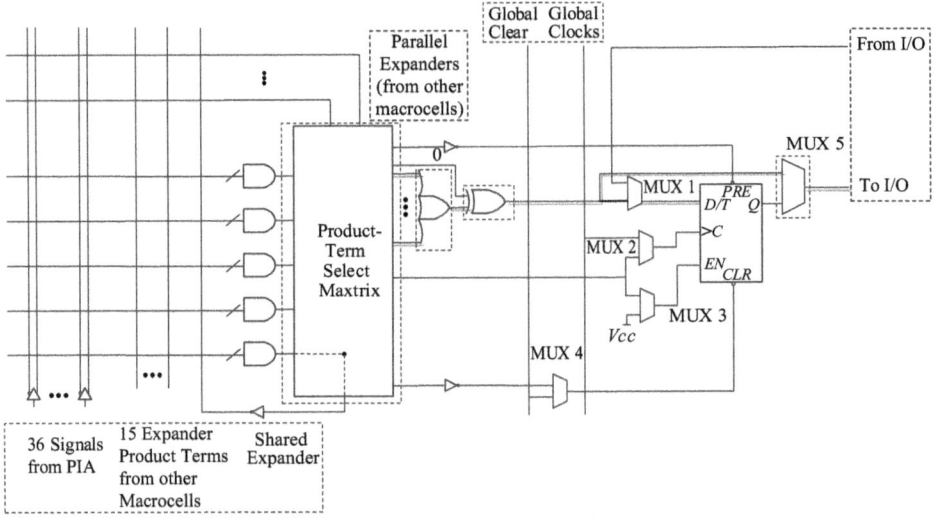

Figure 10.3.11: A macrocell configured for generation of an SOP logic function, dash squares and double lines indicate the data path.

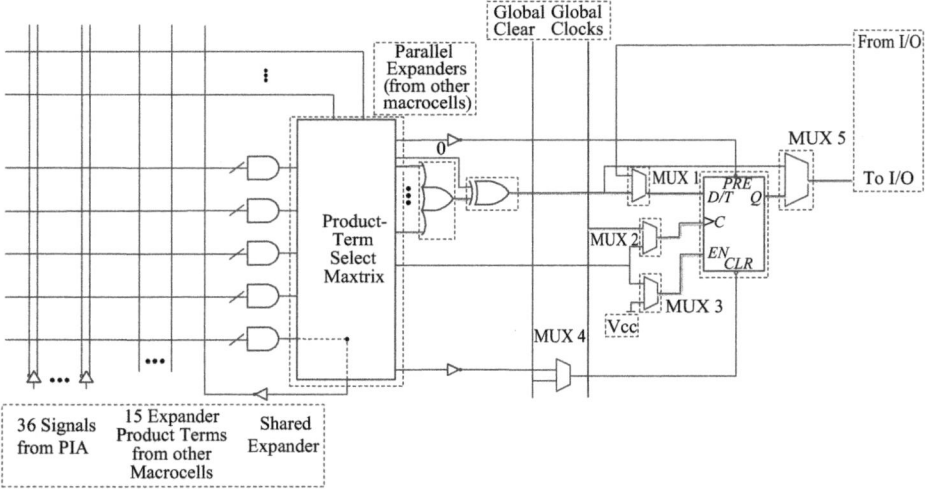

Figure 10.3.12: A macrocell configured for generation of a registered logic function, dashed square and double lines indicate data path.

Figure 10.3.13 shows a complete macrocell containing the flip-flop. The SOP output of the OR gate is reversed by the XOR gate, and then we can get the function in the form of product of sum (POS). For the input of XOR gate, if the input is 1, the XOR gate output is reversed; if the input is 0, the XOR output is directly output (SOP form). MUX1 selects the XOR output or the input from the I/O terminal as the input of flip-flop. MUX2 can be programmed to implement a global clock signal or a clock

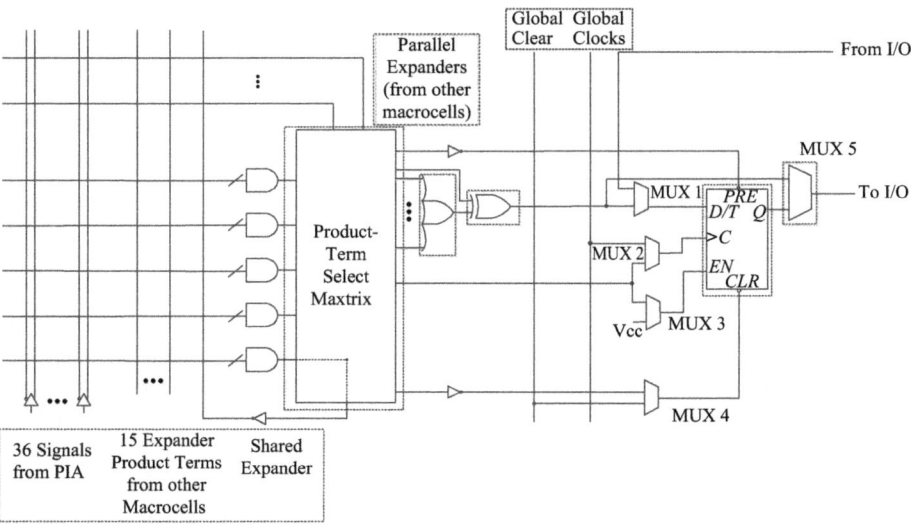

Figure 10.3.13: A CPLD macrocell.

signal based on a product term as a clock signal of the flip-flop. MUX3 can be programmed to implement a high level or a product term for enabling the flip-flop. MUX4 can be chosen as global clear or the product term clear. MUX5 is used to select the bypass flip-flop to connect the output of combinational logic to I/O, or connect the output of the register to the I/O. The flip-flop can be programmed to implement D, T flip-flop or J-K flip-flop.

10.3.8 In-system programmability (ISP)

MAX 7000S devices are in-system programmable via an industry-standard 4-pin Joint Test Action Group (JTAG) interface (IEEE Std. 1149.1-1990). The MAX 7000S architecture internally generates high programming voltage required to program EEPROM cells, allowing in-system programming with only a single 5.0 V power supply. During in-system programming, the I/O pins are tristated and pulled-up to eliminate board conflicts. The pull-up value is nominally 50kΩ. ISP simplifies the manufacturing flow by allowing devices to be mounted on a printed circuit board with standard in-circuit test equipment before they are programmed. MAX 7000S devices can be programmed by downloading the information via in-circuit testers (ICT), embedded processors, or the Altera MasterBlaster, ByteBlasterMV, ByteBlaster, BitBlaster download cables. Now, the ByteBlaster cable is obsolete and is replaced by the ByteBlasterMV cable, which can program and configure 2.5V, 3.3V, and 5.0V devices. Programming the devices after placed on the board, which eliminates lead damage on high-pin-count packages due to device handling and allows devices to be reprogrammed after a system has already shipped to the field. For example, product upgrades can be performed in the field via software or modem.

During in-system programming, instructions, addresses, and data are shifted into the MAX 7000S device through the TDI input pin. Data is shifted out through the TDO output pin and compared against the expected data.

10.3.9 Features of the CPLD

Compared with PAL and GAL with low density, the CPLD has many advantages as follows.
1. The CPLD has high density. A CPLD consists of multiple SPLD arrays with programmable internal connections. Some CPLD chip contains the logic gates reaching up to the order of millions. In addition, CPLD has a large amount of I/O pins to avoid the fact that the I/O pins have been completely occupied before the logic elements exhausted in the low-density devices.
2. The CPLD has high speed and low power consumption. The designer can program each individual macrocell in a CPLD device for either high-speed or low-power

operation. As a result, speed-critical paths in the design can run at high speed, while the remaining paths can operate at reduced power. Some CPLDs, for instance MAX 7000 devices, offer a power-saving mode that supports low-power operation. This feature allows total power dissipation to be reduced by 50% or more, because most logic applications require only a small fraction of all gates to operate at maximum frequency.

3. CPLD has the capacity of in-system programmability. CPLD can be programmed in-system by using an industry-standard 4-pin Joint Test Action Group (JTAG) interface (IEEE Std. 1149.1-1990). This can avoid physical loss caused by plugging in and pulling out the programmed chip for specified programmer. The CPLD architecture internally generates high programming voltage required to program EEPROM cells, allowing in-system programming with only a single 5.0 V power supply. During in-system programming, the I/O pins are tristated and pulled-up to eliminate board conflicts.

4. The internal logic of the CPLD can be tested and programmed. Most CPLDs contain the built-in JTAG boundary-scan test (BST) circuitry. This makes the internal logic of CPLD testable, allowing for not only the troubleshooting but also determining the error location.

5. The CPLD has the sharable product terms. The most product terms of PAL and GAL is fifteen; while CPLD allows providing up to 32 product terms for each macrocell by parallel expander and sharable expander.

6. CPLD has asynchronous clock and clear function. PAL and GAL must use the same clock and can construct synchronous sequential logic circuit. However CPLD can construct both asynchronous and synchronous sequential logic circuit.

7. CPLD has programmable security bit for protection of proprietary designs and also provide open-drain output option allowing the wired-OR operation.

10.4 Field programmable gate arrays (FPGAs)

A *field-programmable gate array* (FPGA) is an integrated circuit designed to be configured by a customer or a designer after manufacturing – hence "field-programmable". Unlike CPLD using PAL/GAL arrays in architecture, FPGA contains a large number of programmable logic blocks that are individually smaller than a PLD. They are distributed to form a square array across the entire chip in a sea of programmable interconnections, and a hierarchy of reconfigurable interconnection that allow the blocks to be "wired together", like many logic gates that can be internally wired in different configurations. Therefore, the equivalent gates in a typical FPGA are many times as that in a typical CPLD. The logic generation unit of the FPGA is usually smaller in size and greater in amount than that of the CPLD.

Xilinx Company invented FPGAs. This section we will use one of their popular families, the XC4000 series, to illustrate FPGA architecture and highlight the most important features of FPGA.

The objectives of this section are to
– Describe the basic structure of FPGA
– Explain the difference between FPGA and CPLD
– Discuss look up tables (LUTs)
– Describe FPGA core

10.4.1 FPGA architecture

The programmable logic blocks in FPGA architecture are called *configurable logic blocks* (CLBs). Generally, FPGA consists of configurable logic block (CLB), input/output block (I/OB), programmable interconnection (PI) and programmable switch matrix (PWM), as shown in Figure 10.4.1.

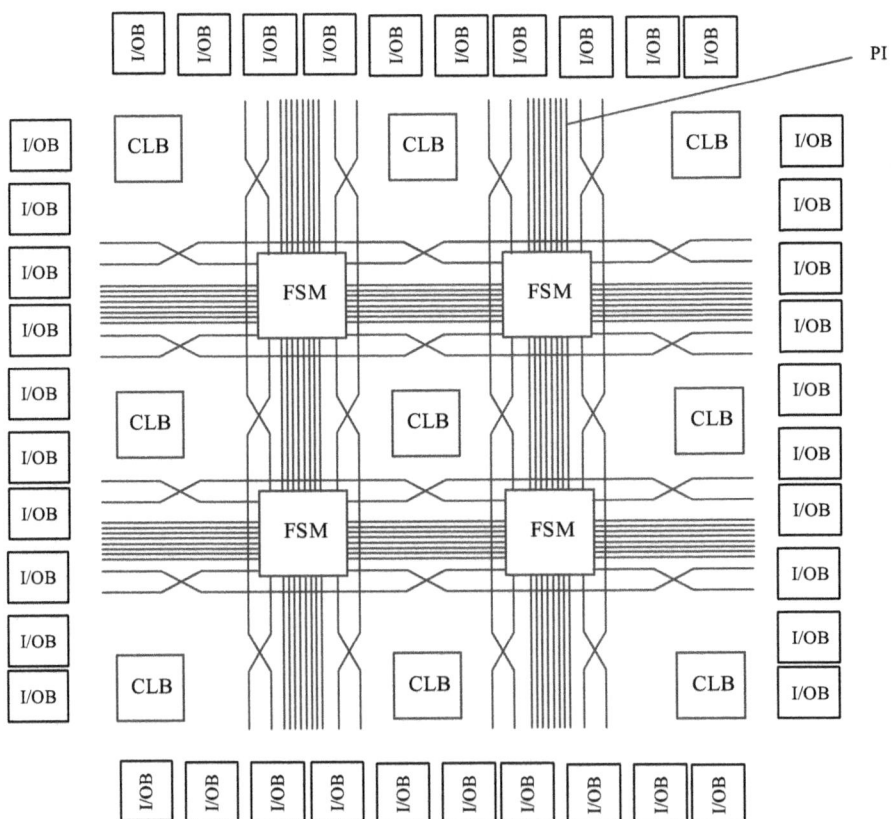

Figure 10.4.1: The structure of FPGA.

CLB in the FPGA is simpler than the LAB in the CPLD, but the number of CLBs in the FPGA are usually more than that of LAB. When the structure of CLB is relatively simple, the FPGA structure is called the *fine-grained structure*; when the CLB volume is relatively large and more complex, the FPGA structure is called the *coarse-grained structure*. I/O blocks are arranged the outside on the perimeter of the structure and provide a single selectable input, output or bidirectional access to the outside world. Programmable switch matrix of programmable interconnection provides the inter-connections between different CLBs and connections to inputs and outputs.

10.4.2 Configurable logic block (CLB)

Configurable logic block implement most of the logic in an FPGA. Since an FPGA can have lots of CLBs, it is important for you to understand it. Figure 10.4.2 shows the internal structure of an XC4000 series CLB [46]. The CLB contains three logic function generators, two D flip-flops, multiple programmable multiplexers represented by the trapezoidal boxes, and RAMs that are not shown in Figure 10.4.2.

1. Combinational and sequential logic

The most important programmable elements are the logic function generator F, G and H. Two 4-input function generators (F and G) offer unrestricted versatility. Four independent inputs are provided to each of two function generators (F1–F4 and G1–G4). These function generators, with outputs labeled F' and G', are individually capable of implementing any arbitrarily defined Boolean function of four inputs or fewer inputs. The third function generator, labeled H', can implement any Boolean function of its three inputs. Two of these inputs can select the output of F' and G' functional generator. Alternatively, one or both of these inputs can come from outside the CLB (H_2, H_0). The third input must come from outside the block (H_1). The CLB can, therefore, implement certain functions of up to nine variables.

Let's think about how to build a universal function generator for 4-input logic function. It's a hard problem if you think about it at the gate level, but pretty easy if you think about it from another point of view. Any 4-input logic function can be described by its truth table, which has 16 rows corresponding to 16 input combinations. If we apply the function's four input bits to the memory's address lines, its data output is the value of the function for that input combination. That is to say, the function generators are implemented as memory *look-up table* (LUT). A LUT (look-up table) is a type of memory that is programmable and can be used to generate SOP expression of combinational logic. In fact, the LUT can implement the same function as the PAL does.

Generally, a LUT structure contains 2^n memory cells; n is the number of input variables. For example, a LUT with four input variables as the address lines can

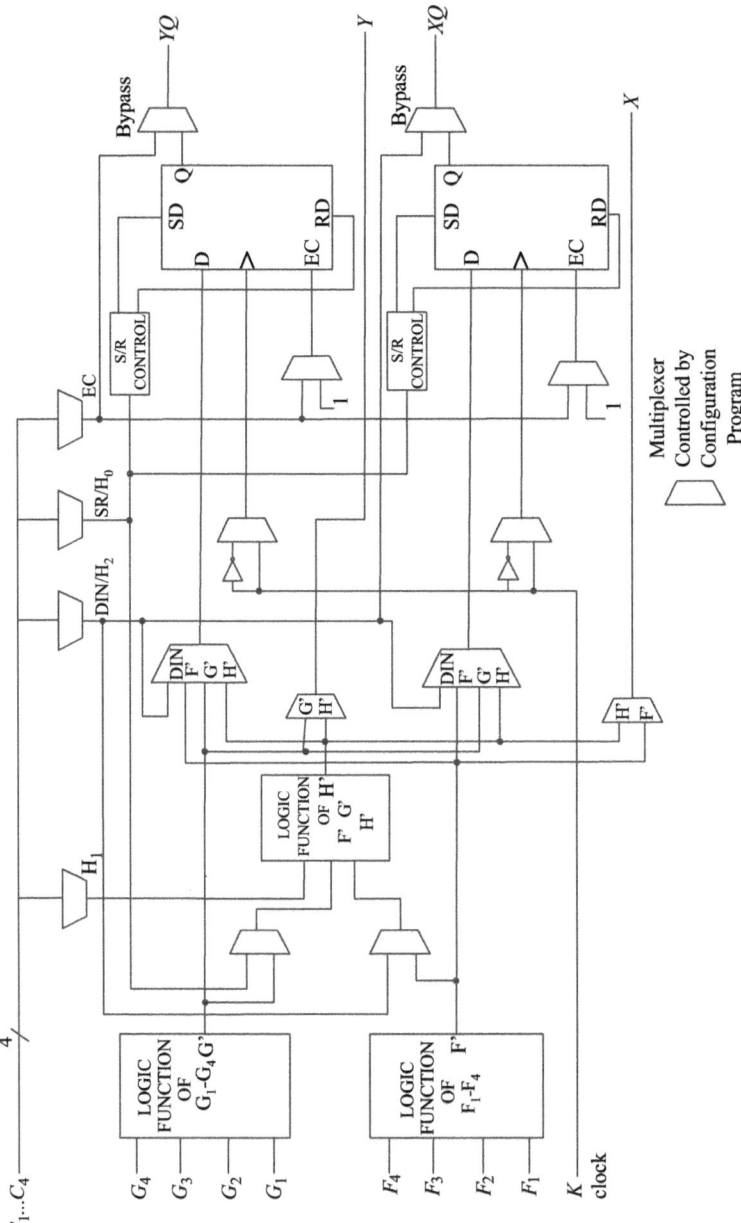

Figure 10.4.2: Simplified Block Diagram of XC4000 Series CLB.

offer 16 addresses to select 16 memory cells at most, thus a LUT with four input variables can form an SOP expression with 16 product terms at most. When you program 1s and 0s into different memory cells in a LUT, a SOP expression is formed, in which each 1 represents the corresponding product term in the SOP expression and each 0 indicates the corresponding product term that does not appear in the SOP expression. For example, Figure 10.4.3 shows a programmed LUT with four input variables. There are four memory cells programmed as 1s, so the resulting SOP expression is.

$$\bar{A}_3\bar{A}_2\bar{A}_1\bar{A}_0 + \bar{A}_3\bar{A}_2A_1A_0 + A_3\bar{A}_2\bar{A}_1\bar{A}_0 + A_3\bar{A}_2A_1A_0$$

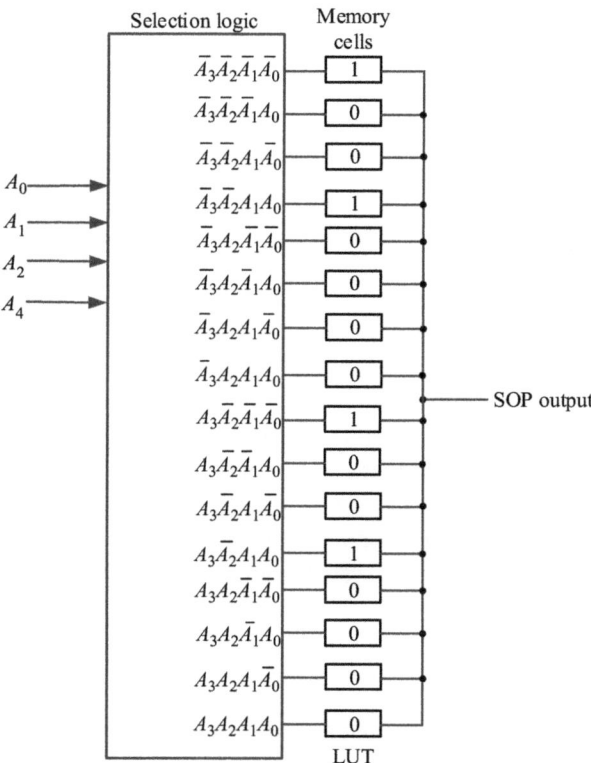

Figure 10.4.3: Example of the basic concept of an LUT programmed for a particular SOP output.

The above logic function of four inputs can be implemented by function generator F or G. Signals from the function generators can exit the CLB on two outputs. F' can be connected to the X output. G' can be connected to the Y output.

There are two D flip-flops in one CLB. Sequential logic can be implemented by combing D flip-flops and combinational logic. The outputs of two flip-flops can be output from two independent outputs YQ and XQ.

2. Fast Carry Logic

Each CLB contains dedicated arithmetic logic for the fast generation of carry logic and borrow signals. This extra output is passed on to the function generator in the adjacent CLB. The carry chain in XC4000E devices can run either up or down. At the top and bottom of the columns where there are no CLBs above or below, the carry is propagated to the right, as shown in Figure 10.4.4.

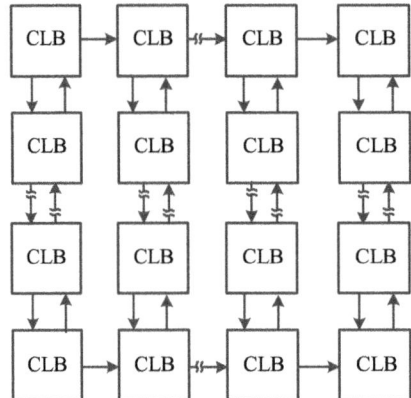

Figure 10.4.4: Available XC4000E carry propagation paths.

The carry chain is independent of normal routing resources. Dedicated fast carry logic greatly increases the efficiency and performance of adders, subtractors, accumulators, comparators and counters. It also opens the door to many new applications involving arithmetic operation, where the previous generations of FPGAs were not fast enough or not efficient. There are two typical applications: high-speed address offset calculations in microprocessors or graphics systems, and high-speed addition in digital signal processing.

3. RAM

The function generators in any CLB can be configured as RAM arrays in the following sizes:
- Two 16 × 1 RAMs. F and G are used as SRAMs with independent address and write-data inputs. However, they share a common write-enable input.
- One 32 × 1 RAM. The same four address bits are used for F and G, and the fifth address bits is applied to H function generator for the write-enable circuitry to select F and G as the upper and lower halves of the memory.
- Synchronous or asynchronous. For write operation, the SRAMs above can be configured to have normal asynchronous latching behavior, or they can be configured to occur on a designated edge of the clock signal.

- One 16 × 1 dual-port SRAM. The two sets of address inputs are used to independently read and write different locations in the same SRAM. Only synchronous write operations are supported in this mode.

In these modes, function inputs F1–F4 and G1–G4 supply address, other CLB inputs H0-H2 provide data inputs and the write-enable signal, and data outputs are produced by the F and G function generator and can be loaded into two D flip-flops. The outputs of two flip-flops can be output from two independent outputs YQ and XQ. Furthermore, the F' output of the F function generator can be connected to the X' output; the G' output of the G function can be connected to the Y' output.

The versatility of the CLB function generators significantly improves system speed. In addition, the design-software tools can deal with each function generator independently. This flexibility improves the usage of cells.

10.4.3 Input/output block (I/OB)

Programmable I/OBs are arranged the outside on the perimeter of the structure, as shown in Figure 10.4.1. It can be flexibly programmed to connect the device pins and different logic interfaces of the internal circuit. An I/O pin can be used for input or output or both. The structure of the XC4000 I/OB is shown in Figure 10.4.5. The main

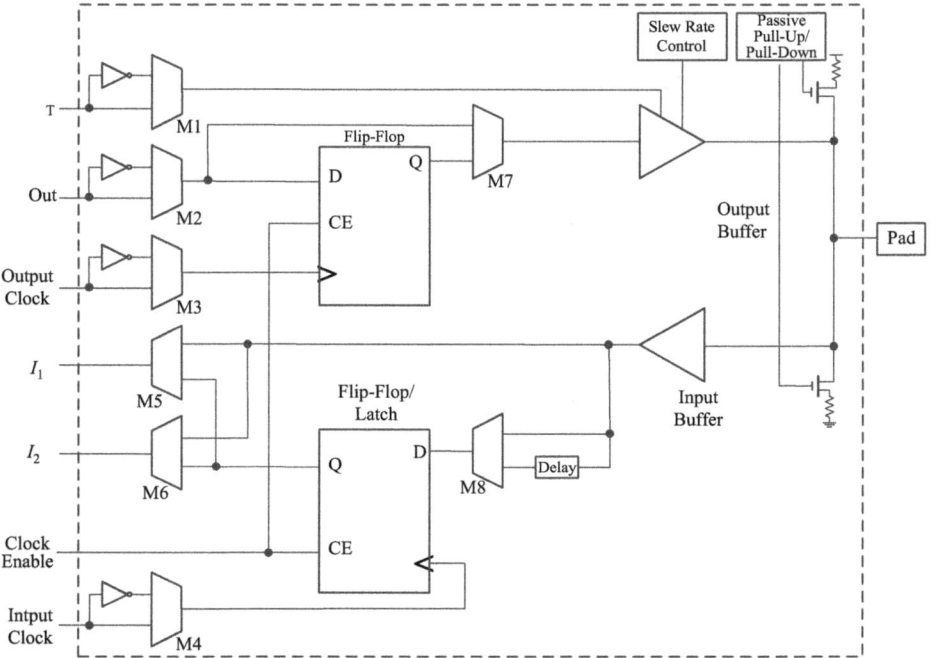

Figure 10.4.5: Simplified block diagram of XC4000E I/OB.

parts include control circuit, combinational I/O circuit, sequential I/O circuit and other associated circuits.

The XC4000 I/OB has more "logic" controls. In particular, its input and output paths contain D flip-flops selected by multiplexers M5-M7. The output enable (T) controls the tristate output buffer, allowing the data output can be transferred to the output pin. Multiplexer M1 can select the activate level (LOW or HIGH) to enable the tristate buffer. The clock enable (CE) controls the clock of D flip-flop. When it is a LOW, the flip-flops are disabled and thus only combinational input or output circuit can be built. The input clock and the output clock are individually control the input and the output D flip-flops. The D flip-flops triggered by the rising edge or the falling edge are determined by the multiplexer M3 and M6. An input register can be programmed as either an edge-triggered flip-flop or a level-sensitive latch. This capability makes the FPGA highly synthesis compatible.

Each I/OB can be configured as combination logic or sequential logic. The main difference of combination logic and the sequential logic is determined whether or not the flip-flops are used. Figure 10.4.6 shows the typical configurations of input and output.

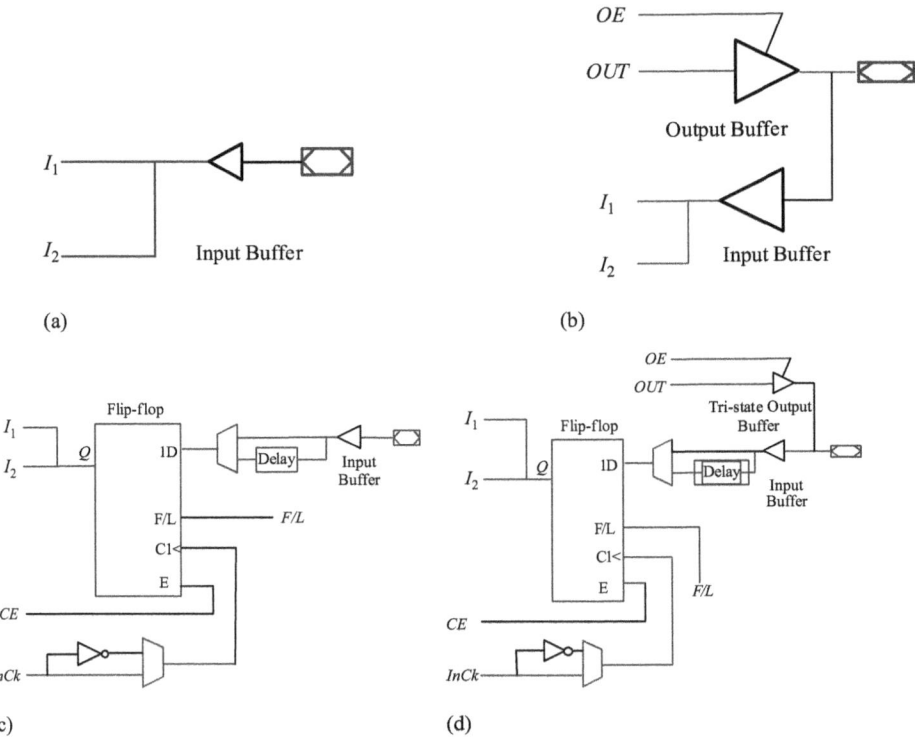

Figure 10.4.6: Four typical configurations of input and output: (a) combinational input; (b) combinational input/output; (c) register input; (d) register input/output.

Other associated circuits include slew-rate control and the pull-up/pull-down circuit. The I/OB output buffers have a slew-rate control, which is by default reduced to minimize power bus transients when switching noncritical signals. For critical signals, we attach a FAST attribute or property to the output buffer or flip-flop. Programmable pull-up and pull-down resistors are useful for tying unused pins to V_{cc} or Ground to minimize power consumption and reduce noise sensitivity.

10.4.4 Programmable interconnect

The XC4000 programmable interconnect architecture is a fascinating example of a structure that provides rich, symmetric connectivity in a small silicon area.

In an FPGA, each CLB is embedded in the interconnect structure, which is really just wires with programmable connections to them. Wires are not really "owned" by any one CLB and they are distributed among the CLBs and between the CLBs and I/OBs. All internal connections are composed of metal segments with programmable switching points and switching matrices to implement the desired routing. A structured hierarchical matrix of routing resources is provided to achieve efficient automated routing.

Different types of the chips have different interconnection line types. According to the length of interconnection lines, the internal lines in XC4000E can be divided into several types. The first type of lines is single-length lines, as shown in Figure 10.4.7. *Single-length lines* provide the greatest interconnect flexibility and fast routing between adjacent blocks. There are eight vertical and eight horizontal

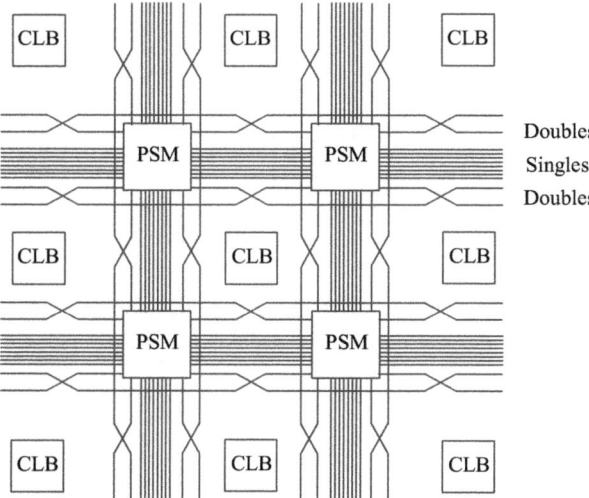

Figure 10.4.7: Single- and double-length lines, with PSMs.

single-length lines associated with each CLB. These lines connect the switching matrices that are located in every row and column of CLBs. The second is double-length lines also shown in Figure 10.4.7. The *double-length lines*, which are twice as the length of the single-length lines, consist of a grid of metal segments. They run past two CLBs before entering a switch matrix. Double-length lines are grouped in pairs with the switch matrices staggered, so that each line goes through a switch matrix at every other row or column of CLBs. There are four vertical and four horizontal double-length lines associated with each CLB. These lines provide faster signal routing over intermediate distances, while retaining routing flexibility. Double-length lines are connected by way of the programmable switch matrices. The third is longlines as shown in Figure 10.4.8. *Longlines* form a grid of metal interconnect segments that run the entire length or width of the array. Each longline is split into two segments by a PSM and form two individual routing regions. The input of the CLB can be driven directly by the longlines and the output of the CLB can be connected to the longlines through tristate output buffer or single-length lines. These longlines are intended for high fan-out, time-critical signal nets, or nets that are distributed over long distances. The interconnection between single-length lines and longlines can be controlled by programmable points. There is no connection between double-length lines and longlines.

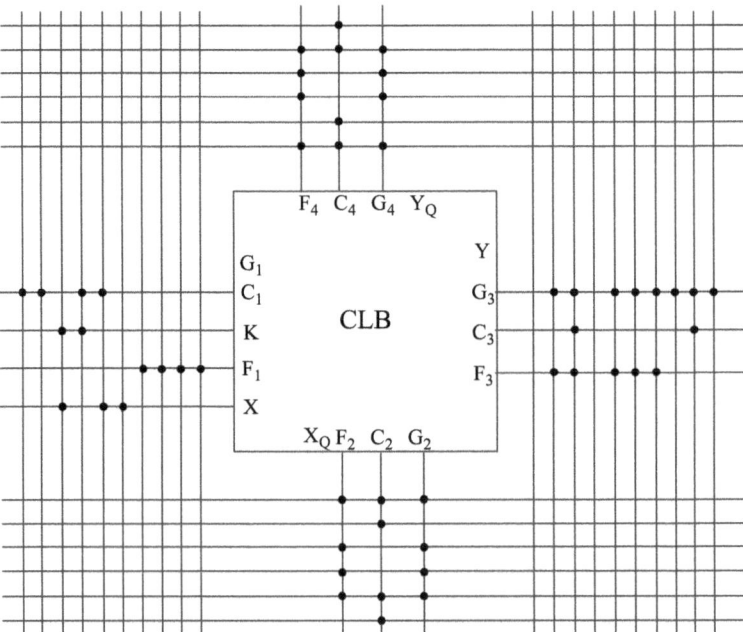

Figure 10.4.8: Longlines in XC4000E.

The horizontal and vertical single-and double-length lines intersect at a box called a programmable switch matrix (PSM). Each switch matrix consists of programmable pass transistors used to establish connections between the lines, as shown in Figure 10.4.9. For example, a single-length signal entering on the right side of the switch matrix can be routed to a single-length line on the top, left, or bottom sides, or any combination. If multiple branches are required, similarly, a double-length signal can be routed to a double-length line on any or all of the other three edges of the programmable switch matrix.

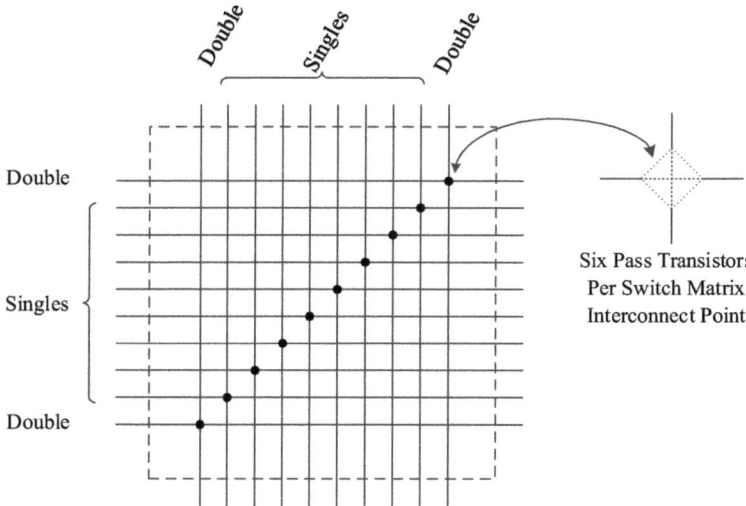

Figure 10.4.9: Programmable Switch Matrix (PSM).

10.4.5 Storage cell in FPGA

The storage cell in FPGAs adopts antifuse technology and SRAM technology. FPGAs based on antifuse technology are nonvolatile, while FPGAs based on SRAM technology are volatile. Therefore, SRAM-based FPGAs should be reconfigured each time when power is turned back on or they use an external memory with data transfer controlled by a host processor.

10.4.6 FPGA core

As discussed in previous section, FPGA is actually like "blank board," the end users can program FPGA to implement any logic design. The hard-core logic is part of the logic in the FPGA and is embedded by the manufacturer according to specific functions and cannot be rewritten. For example, if the customer needs a small microprocessor to be a part of the system design, the microprocessor can be programmed by the customer and then implanted in the FPGA, or can be achieved with the hard-core provided by the manufacturer. Soft-core is the embedded functions with programmable features.

One of the advantages of hard-core approach is that hard-core takes less space of the FPGA than that of the user's on-site programming for implementing the same design. Therefore, hard-core reduces the space in the chip and saves the user's development time. Furthermore, the hard-core has been fully tested. The disadvantage of the hard-core is that specific function has been fixed and the user can only use the existing hardware which is not allowed to be changed later.

Hard-core is often adopted to implement functions that frequently used in digital systems, such as microprocessors, standard I/O interfaces and digital signal processors. Multiple hardcore functions can be programmed into an FPGA. Figure 10.4.10 illustrates the concept of a hard-core that is surrounded by configurable logic implemented by user programming. Since this hard-core is embedded into the user's programming logic, this is a basic embedded system.

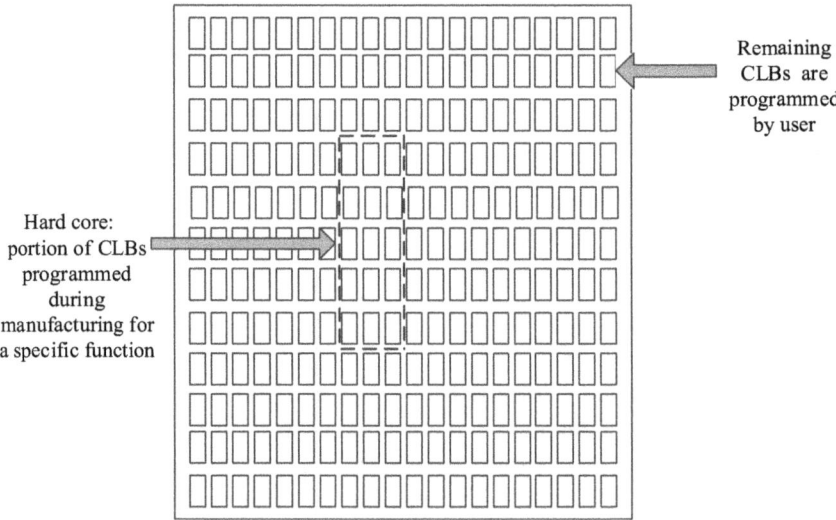

Figure 10.4.10: Basic idea of a hard-core function embedded in an FPGA.

10.4.7 Features of FPGA

Compared with CPLD, FPGA has the following features.

1. The programming unit of most FPGA adopts SRAM as memory, so FPGA can be programmed infinitely. Because SRAM belongs to the volatile component, the

information within the chip will be lost once power-down; therefore, each time the power is turned on, FPGA need to be configured again.

2. FPGA is different from typical CPLD architectures. In FPGA, the size of CLB is smaller than the LAB in CPLD. This makes the number of CLB much greater than that of LAB. Therefore, FPGA has much larger density than CPLD. In addition, the number of flip-flops in the FPGA is greater than that of the CPLD, which makes the FPGA perform better than CPLD in the implementation of sequential circuit.

3. FPGA has various programmable interconnection lines, which surround the CLBs. There is various signal paths from one CLB to another, thus the logic occupancy rate in FPGA can reach to a higher level than that in CPLD. But the speed of the system based on FPGA cannot be predicted accurately.

4. The power consumption of CPLD is generally between 0.5 and 2.5 W. However, FPGA has very lower power consumption from 0.25 mW to 5 mW, so FPGA is usually called zero-power device.

5. Hard-core is a part of the logic embedded in the FPGA, which is embedded into the FPGA by the manufacturer to provide specific functionality and not allowed reprogramming. Soft-core is embedded in part of the logic of the FPGA, it has some programmable characteristics.

10.5 The programming process

An SPLD, CPLD, or FPGA can be thought of as a "blank slate" on which you implement a specified design by using a certain process. This process requires a software development package installed on a computer to implement a circuit design in the programmable chip. These software packages are in category of software known as computer-aided design (CAD) software. In this section, the programming process in terms of design are briefly introduced. Tutotials for Altera Quartus II and Xinlinx ISE are provided on the website. You can download them for reference.

Before you begin to design functionality into a CPLD or a FPGA, you should have four necessary things: a computer or work station, the development software run on windows-based PCs as well as workstation, a PLD chip, and a way to connect the PLD chip to computer. For different PLD chip manufactured by different company, there are the corresponding development software. Electronic design automation (EDA), also referred to as electronic computer-aided design (ECAD), is a category of software tools for designing electronic systems such as integrated circuits. Now Xilinx and Altera Inc. are two main manufacturers to offer high desity CPLD and FPGA. They provide the EDA tool including Altera Quertus II and Xilinx ISE. There are demo software versions that can be downloaded from their website. The PLD chip can be programmed through connecting a computer via cable by using either the programming fixture in which the chip is inserted or the development board on which the chip is mounted. After the software has been installed on your computer, you can start to design.

The programming process is usually called as design flow. The design flow diagram for implementing a logic design in a PLD chip is shown in Figure 10.5.1.[3].

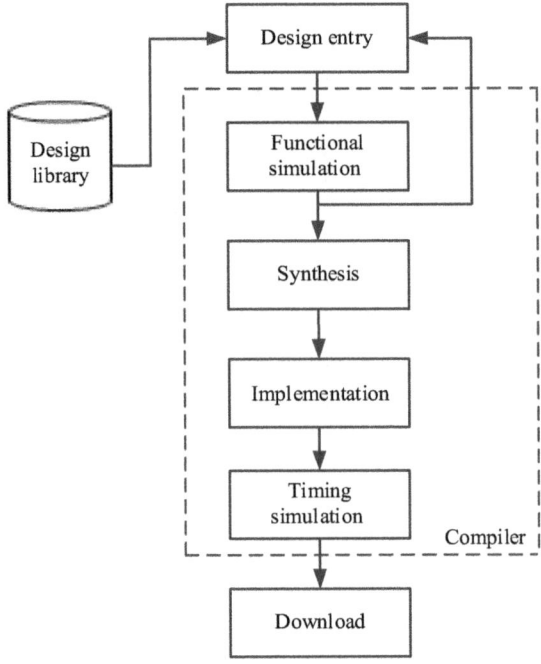

Figure 10.5.1: Basic Programming flow block diagram.

1. Design entry

When you perform design entry, you do not need to know the information about the target device, that is, design entry is device independent. You can enter your designs by using text entry, graphic entry, or state machine entry in the aided design application software enviorment. Text-based entry is finished by using hardware description languages (HDLs) such as VHDL, Verilog or AHDL. Verilog has been introduced in Chapter 4. Figure 10.5.2 shows the design of the logic function $X = \bar{A}B + \overline{BC}$ by using verilog HDL description. Graphic entry usually refers to schematic entry. Schematic entry allows you to place symbols of logic gates and other logic functions from a library on the screen and connect them as required. Figure 10.5.3 shows the schematic design of $X = \bar{A}B + \overline{BC}$ by logic diagram. State machine entry requires specification of both the state through which a sequential logic circuit progressed and the conditions that produce each state change.

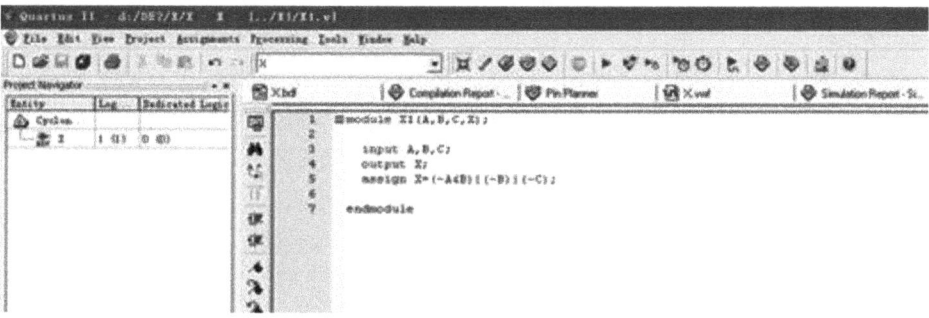

Figure 10.5.2: Design entry with Verilog.

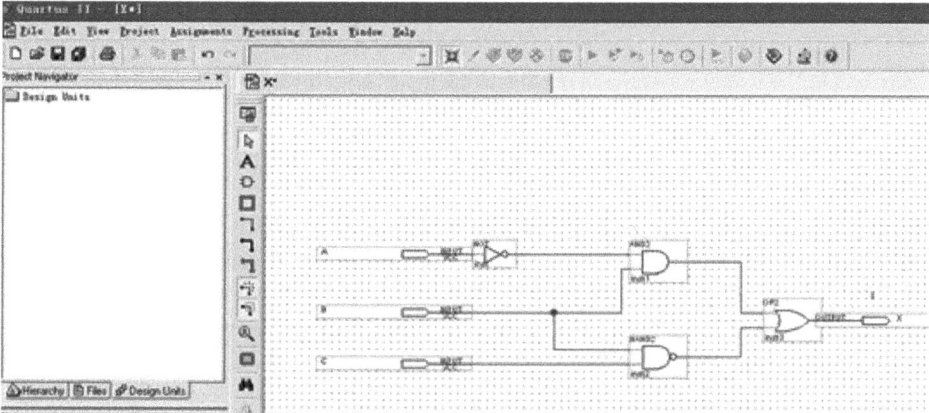

Figure 10.5.3: Schematic entry.

Once a design has been entered, it is compiled. A *compiler* is a program that controls the design flow process and translates source code into object code in a format that can be logically tested or downloaded to a target device. The source code is created during design entry and the object code is final code that actually makes the design to be implemented in the programmable device.

2. Funtional simulation

Funtional simulation is to verify if the design can implement the required logic function by checking the relation of the outputs and the inputs. A waveform editor is used to funtional simulation. The errors found by the simulation would be corrected by returnning back to design entry and making the appropriate changes. Figure 10.5.4 shows the output waveform in relation to the inputs for logic function $X = \bar{A}B + \overline{BC}$. It can be seen from the waveform diagram that the output waveform is correct for the given inputs, verifying that the logic function implemented is correct.

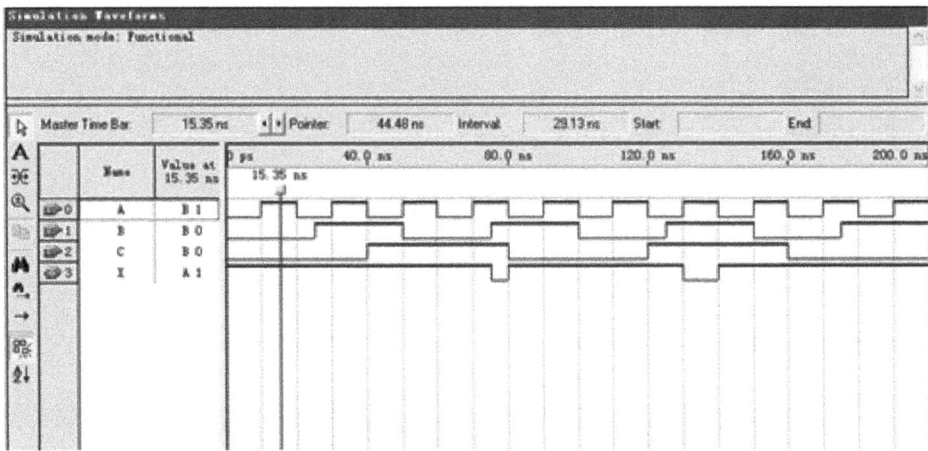

Figure 10.5.4: The output waveform in relation to the inputs after the functional simulation.

3. Synthesis

Synthesis is used to translate the design into a netlist, which is also device independent. During symthesis, the design is optimized in terms of minimizing the number of gates, replacing logic elements with other logic elements that can perform the same function more effectively, and eliminating any redundant logic. The final output from the synthesis phase is a netlist that describes the optimized version of the logic circuit.

4. Implementation

Implementation is used to map the logic structures described by the netlist into the real structure of the given target device being programmed. Note that the implementation process is device dependent. You must select the target device and assign pin configurations. The implementation process is called *fitting* or *place and route* and thus produce an output called a *bitstream*.

5. Timing simulation

This step comes after the design is mapped into the given target device. The timing simulation is basically used to confirm if there are no design error or timing problems due to propagation delays that will affect the overall operation. The above functional simulation is to verify if the design circuit can work properly from a logic point of view and the specification of the target device has no been considered. But for the timing simulation, the target device must be selected. Similar to functional simulation, the waveform editor can be used to observe the result of the timing simulation.

6. Download

After the functional simulation and timing simulation have been verified that the design circuit operates correctly. Once a bitstream has been produced for a given programmable device, it has to be downloaded to the device to implement the software design in hardware. Some programmable devices have to be installed in a special piece of equipment called a device programmer or on a development board. Other types of devices can be programmed while in a system programming (ISP)—using a standard JTAG interface. Some devices are volatile, the bitstream data must be stored in a memory and reloaded into the device after each reset or power-on. Also, the contents of an ISP device can be manipulated or upgraded while it is operating in a system. This is called "on-the-fly" reconfiguration. Figure 10.5.5 shows the basic concept of downloading.

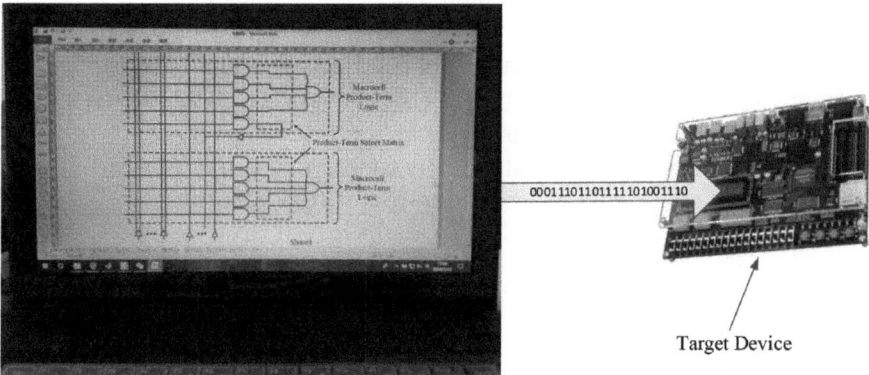

Target Device

Figure 10.5.5: Downloading a design to the target device.

10.6 Summary

1. PLD is a semi-custom IC, which can be programned by users to realize the required logic relationship between inputs and outputs. Accoding to the internal stucture, PLD can be categoried as SPLD, CPLD and FPGA.
2. SPLDs includes PROMs, PLAs, PALs, and GALs.
3. The PAL is one-time programmable (OTP) device, which could not be updated and reused once programmed. It consists of a programmable AND array that connects to a fixed OR array.
4. The GAL has the same logical properties as the PAL does, but GAL can be erased and reprogrammed.

5. The macrocell is usually composed of an OR gate and some associated output logic in PAL or GAL. Macrocell can be set to two modes: combinatorial mode and register mode.
6. CPLD is a complex programmable logic device that is mainly composed of multiple SPLD arrays, a programmable interconnection and multiple I/O control blocks.
7. Each SPLD array in a CPLD is called a logical array block (LAB).
8. CPLDs are based on the AND/OR logic to realize combnational logic function.
9. Unlike CPLD using PAL/GAL arrays in architecture, FPGAs contain a large number of programmable logic blocks and have much greater density than CPLDs.
10. FPGA mainly consists of configurable logic block (CLB), input/output block (I/OB), programmable interconnection (PI) and programmable switch matrix (PWM). The CLB in the FPGA is simpler than the LAB in the CPLD, but the number of CLB in the FPGA is usually more than that of LAB in the CPLD.
11. FPGAs are based on LUT architecture. LUT stands for look-up table, which is a type of memory that is programmable and used to generate SOP expression.
12. The hard-core is part of the logic in the FPGA and is embedded by the manufacturer according to specific functions and cannot be rewritten.
13. The programming process is usually called as design flow.

Key Terms

PLD: A programmable logic device produced as a semi-custom integrated circuit.
PROM: A form of digital memory where the setting of each bit is locked by a fuse or antifuse; one type of ROM (read-only memory).
PAL: A type of SPLD that belongs to one-time programmable device and has a programmable AND array connected to a fixed OR array.
GAL: A type of SPLD that is similar to a PAL except that it uses the reprogrammable storage cells, such as a floating gate MOSs, instead of fuses.
ISP: A kind of PLD programming method that a PLD chip can be programmed in the installed circuie system, rather than requiring the chip to be programmed prior to installing it into the system.
SPLD: Simple programmable logic devices.
CPLD: Complex programmable logic device(CPLD) that contains multiple SPLD and programmable interconnection.
FPGA: Field programmable gate array(FPGA) that use LUT as logic element.

OTP: One-time programmable(OTP) that means the circuit can only be programmed once.

PIA: Programmable interconnect array (PIA) that offer the route between LABs.

AIM: Advanced Interconnect Matrix(AIM).

LUT: Look up table (LUT) that is a type of memory that is programmable and used to generate SOP expression.

CLB: Configurable logic blocks(CLB) that is a important parts of a FPGA.

Self-test

10.1 Two types of SPLD are_____.
   (a) CPLD and PAL
   (b) PAL and FPGA
   (c) PAL and GAL
   (d) GAL and SRAM

10.2 The structure of PAL is_____.
   (a) a programmable AND gate array and a programmable OR gate array
   (b) a programmable AND gate array and a fixed OR gate array
   (c) a fixed AND gate array and a programmable OR gate array
   (d) a fixed AND gate and OR gate array

10.3 The organization of the macrocell is_____.
   (a) a fixed OR gate and other associated logic
   (b) a programmable OR gate array and other associated logic
   (c) a fixed AND gate and other associated logic
   (d) a fixed AND gate and OR gate array with flip-flop

10.4 Which of the following devices is not a PLD device_____.
   (a) CPLD   (b) GAL   (c) ROM   (d) FPGA

10.5 The term LAB stands for_____.
   (a) logic AND blocks
   (b) logical array blocks
   (c) final judgment bit
   (d) logical assembly block

10.6 In which of the following modes the macrocell is configured to generate SOP functions_____.
   (a) combnational mode
   (b) parallel mode
   (c) hosting mode
   (d) sharing mode

10.7 Typical macrocells include_____.
(a) gate, multiplexer and flip-flop
(b) gate and shift register
(c) gray code counter
(d) fixed logic array

10.8 Based on the complexity of configurable logic blocks (CLBs) in FPGAs, they can be divided into_____.
(a) volatile or nonvolatile
(b) programmable or reproducible
(c) fine-grained or coarse-grained
(d) platform type or embedded type

10.9 Volatile FPGAs are usually based on_____.
(a) fuse technology
(b) non-fuse technology
(c) EEPROM technology
(d) SRAM technology

10.10 Which of the following FPGAs with embedded logic function cannot be programmed_____.
(a) Volatile   (b) Platform type   (c) Hard-core   (d) Soft-core

10.11 Hard core design is usually designed and developed by FPGA vendors, and owned by FPGA manufacturers. These designs are called_____.
(a) intellectual property
(b) special logic
(c) user design
(d) IEEE standard

10.12 Text input as a logical design_____.
(a) must use logical symbols
(b) must use HDL
(c) can only use boolean algebra
(d) must use specific code

10.13 In the functional simulation, the user must specify_____.
(a) the specific target device
(b) output waveform
(c) input waveform
(d) HDL

Problems

10.1 How does a PLA differ from a PAL?

10.2 Explain how a programmed polarity output in a PAL works.

10.3 Basically, what does a macrocell contain?

10.4 Describe how a CPLD differs from an SPLD.

10.5 When a macrocell is configured to produce an SOP function, which mode is it in?

10.6 How does an FPGA differ from a CPLD?

10.7 Describe a LUT and discuss its purpose?

10.8 What is a FPGA core?

10.9 Determine the Boolean output expression for the simple PAL array in Figure P10.1. The "×" represent connected links.

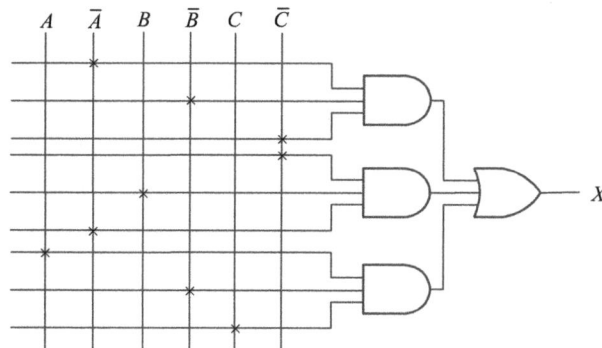

Figure P10.1

10.10 Show how the PAL-type array in Figure 10.2 should be programmed to implement each of the following SOP expressions. Use "×" to indicate a connected link.
(a) $Y = A\bar{B}C + \bar{A}B\bar{C} + ABC$
(b) $Y = A\bar{B}C + \bar{A}\bar{B}C + \bar{A}BC$

10.11 Determine the product term for the AND gate in a CPLD array shown in Figure P10.3 (a). If the AND gate is expanded, as shown in Figure P10.3 (b), determine the SOP output.

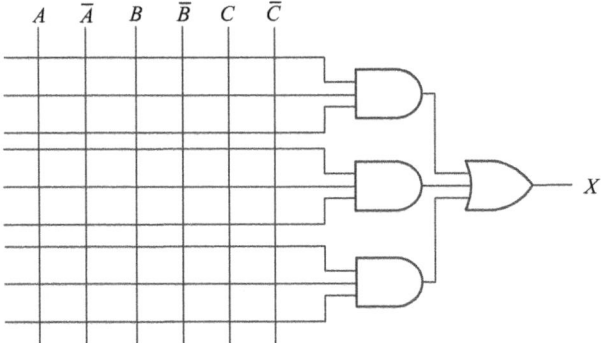

Figure P10.2

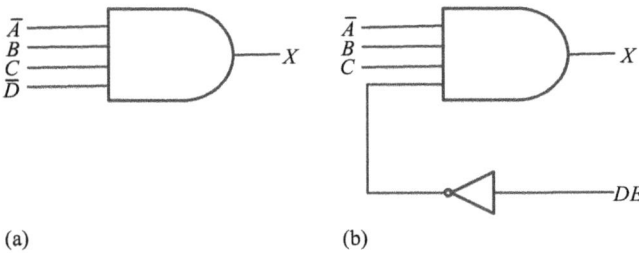

(a)   (b)

Figure P10.3

10.12   Determine the output of the macrocell logic in figure P10.4 if $AB\bar{C}D + \bar{A}BCD$ is applied to the parallel expander input.

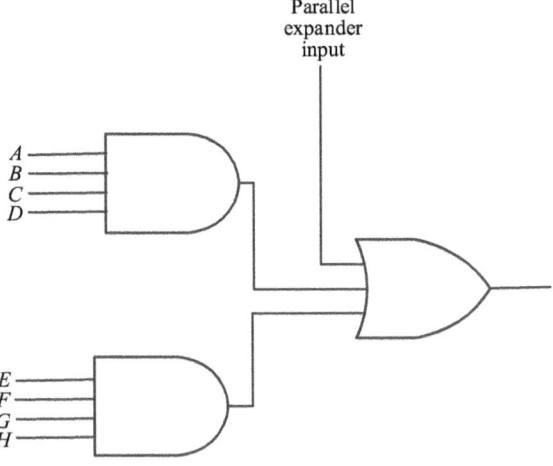

Figure P10.4

10.13 Determine the output of the array in Figure P10.5. The "×" represents connected links.

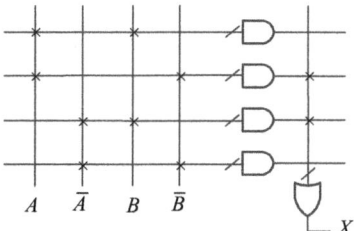

Figure P10.5

10.14 Modify the array in Figure P10.5 to produce an output
$X = \bar{A}\bar{B}\bar{C} + \bar{A}B\bar{C} + ABC + A\bar{B}C$.

10.15 Show a logic module configured in the normal mode to produce one 4-variable SOP function and one 2-variable SOP function.

10.16 Determine the final SOP output function for the logic module shown in Figure P10.6.

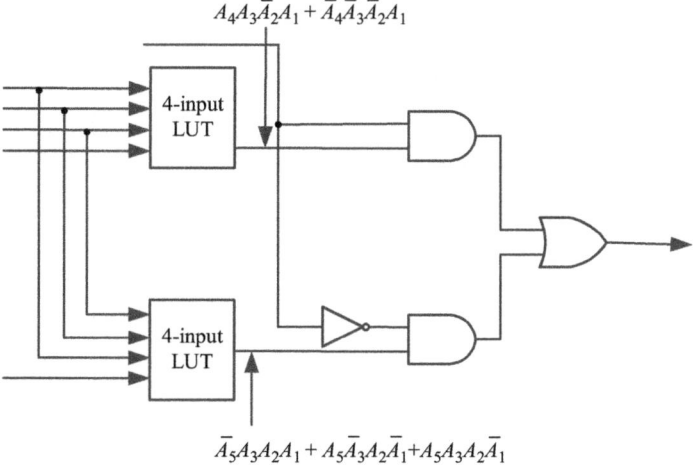

Figure P10.6

10.17 Determine the output expression of the LUT for the internal conditions shown in Figure P10.7.

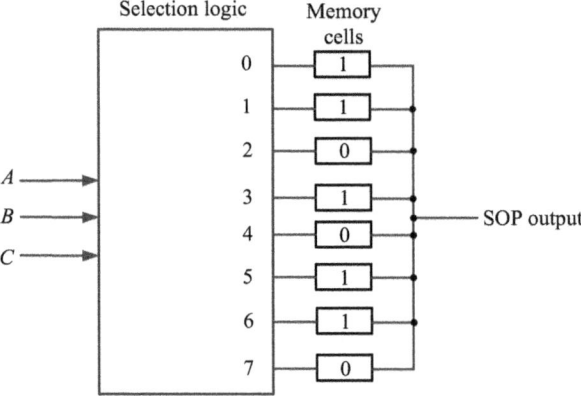

Figure P10.7

10.18 Show how to reprogram the LUT in Figure P10.8 to produce the following SOP output:

$$\bar{A}B\bar{C} + A\bar{B}\bar{C} + ABC$$

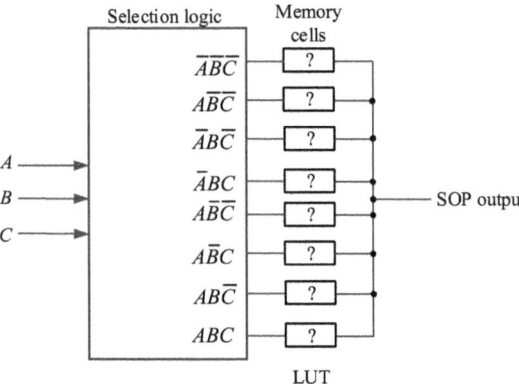

Figure P10.8

10.19 Show how to reprogram the LUT in Figure P10.9 to produce the following SOP output: $Y = A\bar{B}CD + \bar{A}B\bar{C}D + A\bar{B}CD + \bar{A}\bar{B}C\bar{D}$.

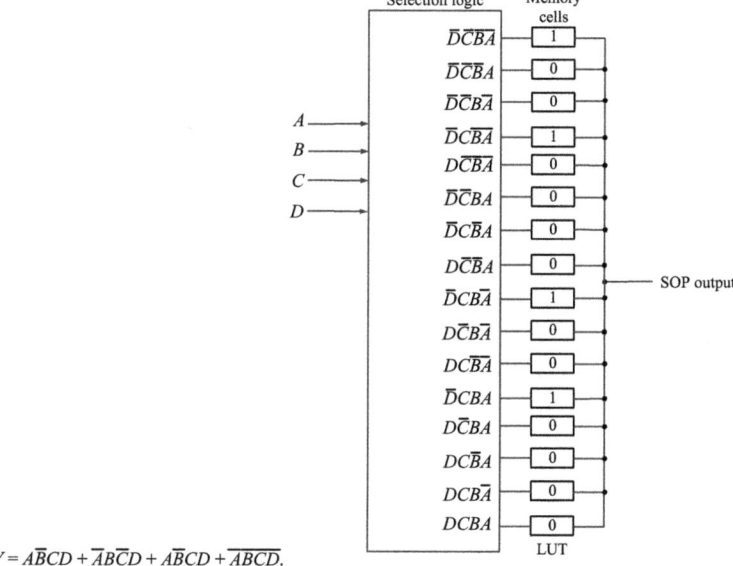

$Y = A\overline{B}CD + \overline{A}B\overline{C}D + A\overline{B}CD + \overline{ABCD}.$

Figure P10.9

10.20 Use the PLA and D flip-flop to design a 4-bit Johnson ring counter.

11 Analog-to-digital and digital-to-analog converter

11.1 Introduction

With the rapid development of digital technology, digital technology has already been applied to a wide range of areas besides computer system. Such applications include communication systems, radar, navigation and guidance systems, military systems, medical instrumentation, industrial process control, and many others. However, digital circuits and digital systems deal with digital quantity. From physics point of view, a digital quantity is the one having a discrete set of values. Most things that can be measured quantitatively in nature with analog form. In order to process these analog quantities with digital technique, it is necessary to convert the analog quantity to a digital one. The device that converts an analog signal to a digital signal is called the *analog-to-digital converter* (ADC). Moreover, most electronic instruments are driven by analog signals and thus the processed digital quantity must be converted back to analog signal to drive the electronic equipment. The circuit that converts the digital signal to an analog signal is called the *digital-to-analog converter* (DAC). This chapter first introduces the basic concepts and then the operating principles of DAC and ADC. Several typical integration DAC and ADC chips and their applications are also covered.

The objectives of this chapter are to
- Explain how analog signals are converted to digital forms
- Describe the sample process
- State the purpose of digital-to-analog conversion
- Explain the operating process of several types of DACs
- State the purpose of analog-to-digital conversion
- Explain the operating process of several types of ADCs

11.2 Digital-to-Analog Converter (DAC)

In electronics, a digital-to-analog converter (DAC, D/A, or D-to-A) is a circuit that converts a digital signal into an analog signal. DAC is an important part of a digital processing system. After the digital data are processed, they need to be converted back to analog form. In fact, there are several types of DACs. This section mainly introduces two types of DACs: binary-weighted-input DAC and the $R/2R$ ladder DAC.

The objectives of this section are to
- Explain how binary-weighted input DAC works
- Explain how the $R/2R$ ladder DAC works
- Understand the accuracy representation of the DAC

https://doi.org/10.1515/9783110614916-011

11.2.1 Basic concepts of ADC and DAC

In real world, most of the physical parameters involved are analog signals, such as temperature, angle, speed, pressure, and so on, which have infinite continuous values. To use the digital processing technique, these physical parameters must be converted into their corresponding continuous voltage or current signals using the sensor, then the voltage or current signals are converted into discrete digital signals so that it can be processed by digital signal process (DSP) system. The circuit that converts an analog signal to a digital signal is called the ADC. After digital signals are processed, digital signals should be converted back to analog signals to drive the actuator completing the required operations. The circuit that converts a digital signal to an analog signal is called the DAC. The measuring and controlling systems with analog-to-digital and digital-to-analog conversion can be roughly sketched by the block diagram as shown in Figure 11.2.1.

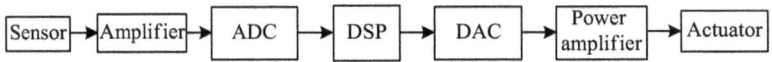

Figure 11.2.1: A general block diagram of the measuring and controlling systems.

The sensor first converts physical parameters into electrical signals, which are amplified by the amplifier and sent to ADC. Then, ADC converts the amplified signals into digital signals and sends them to the DSP system through which the digital signals are processed and sent to DAC to generate analog signals for driving the actuator.

For example, noisy sound signals are picked up by the microphone and are changed into analog (voltage or current) signals. Then, the analog signals are amplified by the voltage or current amplifier to a certain magnitude and then sent to the ADC, converting the analog signal into the digital signal. Usually, the acquired sound signals contain noises, which would affect the quality of sound signals. The noisy sound signals are processed by the DSP system to remove or reduce the noise in the sound signals. Next, the sound signals, without noise, are sent to the DAC to convert digital signals back to analog signals. In succession, analog signals should be amplified by the power amplifier to drive the loud-speaker (or actuator) playing the noiseless (clean without noise) analog sound signals.

11.2.2 Digital-to-analog conversion method

The DAC is used to convert an n-bit binary number, $D_{n-1}D_{n-2}\cdots D_1D_0$, to the corresponding analog signal represented by A. In Chapter 2, you have already seen that any n-bit binary number can be expanded by the sum of the digits after each digit has been multiplied by its weight.

$$D = D_{n-1} \times 2^{n-1} + D_{n-2} \times 2^{n-2} + \cdots + D_1 \times 2^1 + D_0 \times 2^0 = \sum_{i=0}^{n-1} D_i \times 2^i \qquad (11.2.1)$$

The equivalent decimal number, D, can be directly obtained by using eq. (11.2.1) and thus the corresponding analog value should be proportional to the input digital signal, which can be expressed as

$$A = KD = K \sum_{i=0}^{n-1} D_i \times 2^i \qquad (11.2.2)$$

where the conversion parameter K is constant.

The conversion features of a DAC are illustrated in Figure 11.2.2. Note that the output analog quantities are not continuous and they only have the discrete specified values corresponding to the input digital quantities. There is a basic quantitative unit corresponding to the least significant bit (LSB) of the input binary number. The analog signals are K times basic quantitative unit.

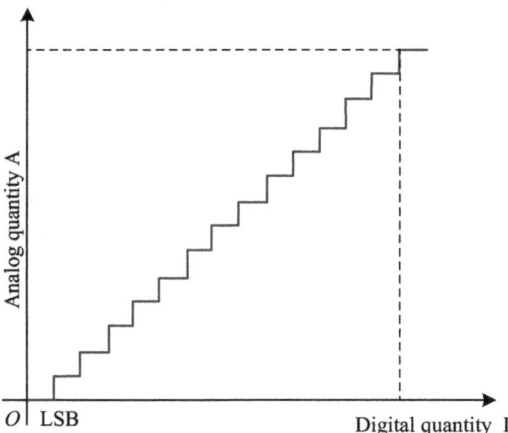

Figure 11.2.2: Illustration of the conversion characteristics of a DAC.

The key issue of the DAC is how to implement the sum of the digits after each digit has been multiplied by its weight so that the digital signal represented by $D_{n-1}D_{n-2}\cdots D_1D_0$ can be converted into an analog signal A with a certain proportional relation. Therefore, resistor networks, analog switches, and amplifier can be used to construct the DAC.

11.2.3 Binary-weighted-input DAC

A method of DAC uses a resistor network with resistor values representing the weights of input bits of binary number. Figure 11.2.3 shows a four-bit binary-weight-input DAC .

Each input resistor may or may not have current, depending entirely on whether the input bit is a 1. The low level is 0 V and the high level is represented by the voltage V. If the input level is LOW (binary 0), the current is 0. If the input level is HIGH (binary 1), the current can be determined by the input resistor value.

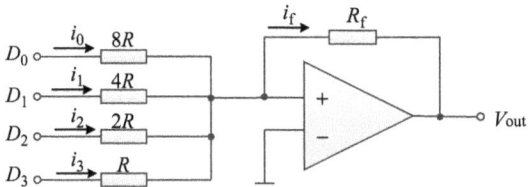

Figure 11.2.3: A four-bit DAC with binary-weighted inputs.

Since the inverting input level is 0 V (virtual ground), the noninverting input level is approximately equal to the inverting input level for the op-amp with negative feedback. Thus, the noninverting input level is also 0 V so the current of each path can be deduced as

$$i_0 = \frac{V}{8R}D_0 = \frac{V}{2^3R}D_0, i_1 = \frac{V}{4R}D_1 = \frac{V}{2^2R}D_1,$$

$$i_2 = \frac{V}{2R}D_2, i_3 = \frac{V}{R}D_3 = \frac{V}{2^0R}D_3$$

Since the current entering the inverting input of the op-amp is approximately equal to zero, the current i_f going through the feedback resistor R_f can be expressed as

$$i_f = i_0 + i_1 + i_2 + i_3 = \frac{V}{2^3R}(D_3 2^3 + D_2 2^2 + D_1 2^1 + D_0 2^0) = \frac{V}{2^3R}\sum_{i=0}^{3}D_i 2^i$$

The output voltage V_{out} can be deduced as

$$V_{out} = -i_f R_f = -\frac{VR_f}{2^3R}\sum_{i=0}^{3}D_i 2^i \tag{11.2.3}$$

Thus, the output voltage is proportional to the sum of the digits after each digit has been multiplied by its weights.

The input resistor value is inversely proportional to the binary weight of the corresponding input bit. The lowest value resistor (R) corresponds to the highest binary weight input (2^3). The other resistor values are multiples of $2R$, $4R$, and $8R$, representing the binary weights of 2^2, 2^1, and 2^0, respectively. The input current is also proportional to the binary weight, since the sum of the input current flows through R_f, so the output voltage is proportional to the sum of the digits after each digit has been multiplied by its weights.

The disadvantage of this type of DAC is the use of multiple different resistor values. An eight-bit converter requires eight resistors ranging from R to $128R$ in the binary-weighted steps. In order to convert accurately, the resistor requires accuracy of $1/255$ (less than 0.5%). This makes the mass production of this DAC very difficult.

Example 11.1 Determine the output waveform of DAC in Figure 11.2.4(a). The input four-digit sequence waveforms are shown in Figure 11.2.4(b), where the input D_0 is the LSB.

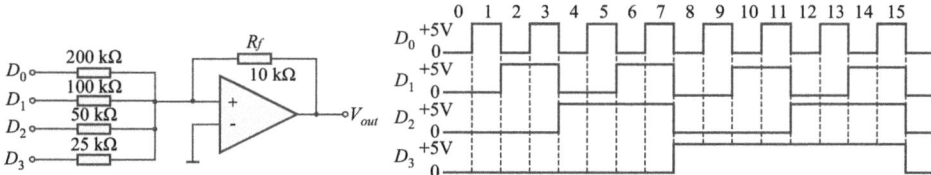

Figure 11.2.4: (a) A four-bit DAC with binary-weighted inputs, (b) the input four-bit sequare waveforms.

Solution

First, the input current for each weight is obtained. Since the inverting input of the op-amp is 0 V (virtual ground), the binary 1 represents the voltage of 5 V, so the current flowing through each resistor is equal to 5 V divided by the corresponding resistance values. Since the input current of the op-amp's inverting input is about zero, all current flows across the feedback resistor R_f. Therefore, the current flows across the resistor R_f can be obtained as follows:

$$i_f = \frac{V}{2^3 R} \sum_{i=0}^{3} D_i 2^i$$

Because the voltage drop across the resistor R is equal to the output voltage, the output voltage is negative relative to the virtual ground, which can be expressed as

$$V_{out} = -i_f R_f = -\frac{V R_f}{2^3 R} \sum_{i=0}^{3} D_i 2^i = -0.25 \sum_{i=0}^{3} D_i 2^i$$

From Figure 11.2.4(b), the first input code is 0000; the corresponding output voltage is 0 V. The second input code is 0001; the corresponding output voltage is −0.25 V. The third input code is 0010; the corresponding output voltage is −0.5 V. Then the input code becomes 0011; the corresponding output voltage is −0.75 V. The output of each adjacent binary code is increased by −0.25 V for this continuous binary sequence from 0000 to 1111 at the inputs; the output is a stair waveform from 0 to −3.75 V with a stair step of −0.25 V, as shown in Figure 11.2.5. The basic quantitative unit is −0.25 V. The output voltage is the product of digital quantity and basic quantitative unit.

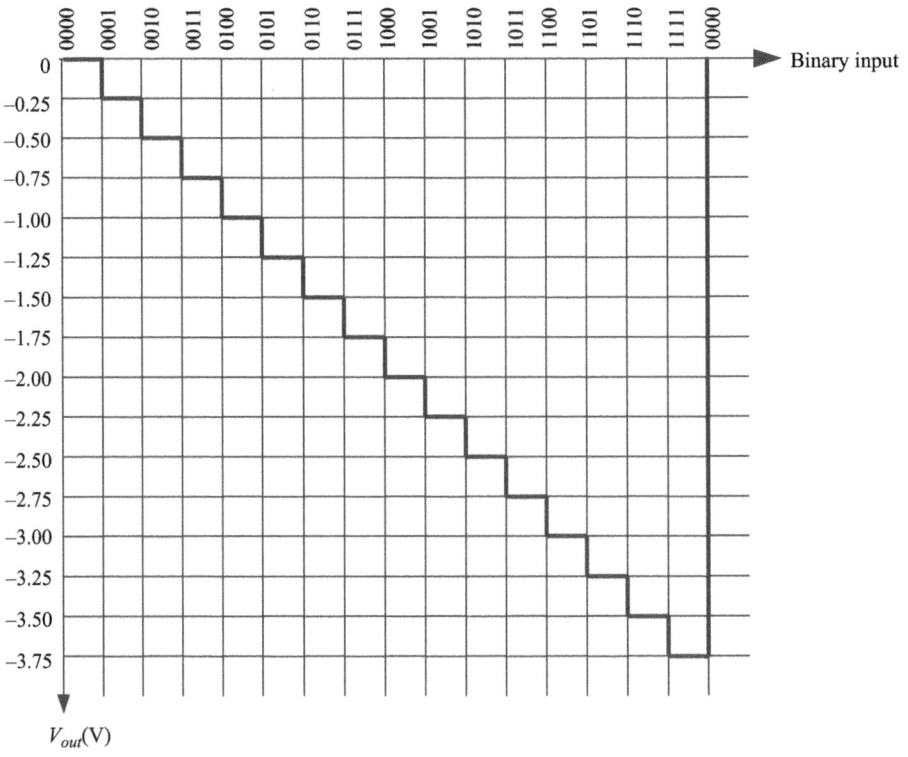

Figure 11.2.5: Output of the DAC in Figure 11.2.4.

11.2.4 The R/2R ladder DAC

The $R/2R$ ladder DAC is another commonly used DAC. The $R/2R$ ladder resistor network is used to implement the weights of input bits of binary number.

Figure 11.2.6 shows a four-bit $R/2R$ ladder DAC. The $R/2R$ ladder DAC consists of an $R/2R$ resistor network, a reference voltage V_{REF}, a group analog switches that can be controlled by the input bits of digital signal, and an integrated op-amp that can convert the current to the voltage. The $R/2R$ resistor network contains only two resistor values, R and $2R$. Since only two resistor values are required, this method can overcome the disadvantage of multiple resistor values for the binary-weighted resistor network DAC, which is greatly convenient for design and manufacture of the integrated circuit. The positions of analog switches S_3–S_0 are controlled by input digitals D_3–D_0, respectively. When the input digital bit is 1, the corresponding analog switch connects the branch to the inverting input of the op-amp; otherwise,

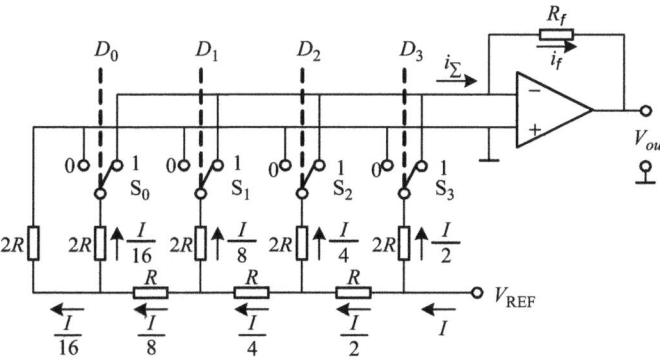

Figure 11.2.6: The $R/2R$ ladder DAC.

the branch is connected to the noninverting input (to ground) of the op-amp. Op-amp is connected to a negative feedback form with the noninverting input directly connected to ground and the inverting input is a "virtual ground." In fact, irrespective of the position the analog switch is connected to, it is equivalent to a "ground."

From the leftmost side of the resistor network to the rightmost of Figure 11.2.6, $2R$ and $2R$ are first connected in parallel to get the resistor value of R, and then R and R are connected in series to get $2R$, repeating the former equivalent process until the rightmost of the resistor network. The final equivalent resistor value between the reference voltage and ground equals to R and thus the total current I supplied by the reference voltage can be deduced as follows:

$$I = \frac{V_{\text{REF}}}{R}$$

As long as V_{REF} is fixed, the total current I is constant. The currents flowing through each branch from right to left are divided by two, so the currents flowing through each branch are $I/2^1$, $I/2^2$, $I/2^3$, and $I/2^4$, respectively. When input bit of the digital signal is a 1, the current flows to the inverting input of the op-amp; otherwise, it flows to the noninverting input of the op-amp. Hence, the current i_0 can be expressed by the sum of all currents flowing to the inverting input as follows:

$$i_\Sigma = \frac{I}{2^1}D_3 + \frac{I}{2^2}D_2 + \frac{I}{2^3}D_1 + \frac{I}{2^4}D_0 = \frac{I}{2^4}(D_3 \times 2^3 + D_2 \times 2^2 + D_1 \times 2^1 + D_0 \times 2^0)$$

$$= \frac{I}{2^4}\sum_{i=0}^{3} D_i \times 2^i$$

Since the input current of the op-amp's inverting input is about zero, the current i_Σ flows through the feedback resistor R_f, that is, $i_f = i_\Sigma$. Because the voltage drop across the resistor R_f is equal to the output voltage, the output voltage is negative relative to the virtual ground, which can be expressed as

$$V_{out} = -i_f R_f = -i_\Sigma R_f = -\frac{IR_f}{2^4}\sum_{i=0}^{3}D_i 2^i = -\frac{V_{REF}R_f}{2^4 R}\sum_{i=0}^{3}D_i 2^i \quad D_i \in (0,1)$$

The input bits of the digital signal can be expanded to the n-bit binary number, $D_{n-1}D_{n-2}\cdots D_1 D_0$, the corresponding analog output can be expressed as

$$V_{out} = -\frac{V_{REF}R_f}{2^n R}\sum_{i=0}^{n-1}D_i 2^i \quad D_i \in (0,1) \tag{11.2.4}$$

Assuming that the $V_{REF}=10$ V, $n = 4$, $R_f = R$, and the input digital signal varied from 0000 to 1111, the corresponding analog voltage can be solved by eq. (11.2.4) as listed in Table 11.2.1.

Table 11.2.1: DAC conversion table.

Inputs				Output
D_3	D_2	D_1	D_0	V_{out} / V
0	0	0	0	0.000
0	0	0	1	−0.625
0	0	1	0	−1.250
0	0	1	1	−1.875
0	1	0	0	−2.500
0	1	0	1	−3.125
0	1	1	0	−3.750
0	1	1	1	−4.375
1	0	0	0	−5.000
1	0	0	1	−5.625
1	0	1	0	−6.250
1	0	1	1	−6.875
1	1	0	0	−7.500
1	1	0	1	−8.125
1	1	1	0	−8.750
1	1	1	1	−9.375

Figure 11.2.7 shows the analog output V_{out} versus the digital inputs, which increases from 0000 to 1111. Each successive binary input increases the output voltage by −0.625 V, so for this particular straight binary sequence from 0000 to 1111 on the inputs, the output is a stair step waveform going from 0 to −9.375 V.

The AD7524 is a typical CMOS low-power eight-bit integrated DAC based on an $R/2R$ ladder network manufactured by Analog Devices in the United States. Its internal circuit diagram and pin diagram are shown in Figure 11.2.8.

The DC supply voltage V_{DD} is from + 5 to +15 V; data inputs (DB_7 through DB_0) can be the TTL/CMOS level; $\overline{CS}$ is the chip select signal; $\overline{WR}$ is the write control signal; V_{REF} is a reference voltage that can be either positive or negative value; and

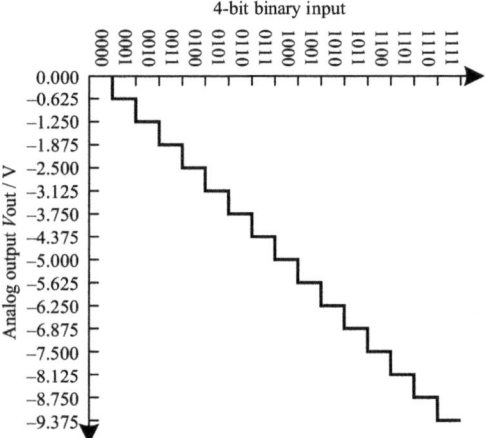

Figure 11.2.7: Output of DAC for four-bit binary input increases from 0000 to 1111.

OUT_1 and OUT_2 are two current output. Generally, OUT_1 is connected to the inverting input of external op-amp, and OUT_2 is connected to the noninverting input of external op-amp; R_{FB} is the access terminal for a feedback resistor. The output of the external op-amp can be connected to pin 16 using an internal feedback resistor R_{FB} or can also be connected to pin 1 using an external feedback resistor.

The typical application of AD7524 is shown in Figure 11.2.9. The output of the external op-amp is connected directly to pin 16 using the internal feedback resistor R_{FB} (10 kΩ). When the chip select signal $\overline{CS}$ and the write control signal $\overline{WR}$ are both active LOW, the AD7524 operates at the write state. In this case, the data from data inputs (DB_7 through DB_0) are loaded into the data latch and then converted to the analog output voltage. The relation between the output voltage and the data input is expressed as

$$V_{\mathrm{out}} = - \frac{V_{\mathrm{REF}}}{2^8} \sum_{i=0}^{7} DB_i \times 2^i \tag{11.2.5}$$

11.2.5 Performance characteristics of DACs

1. Conversion precision

There are two ways to represent the conversion precision of DAC: *resolution* and *conversion error*.

(1) Resolution

There are two representation methods for the resolution of DAC. First, the number of bits in the digital input is used to represent the resolution. The DAC with n-bit resolution outputs the maximum 2^n different analog voltages. The more the number of bits in the digital input, the smaller the basic quantitative unit of the analog quantity is and thus higher the resolution of the DAC is.

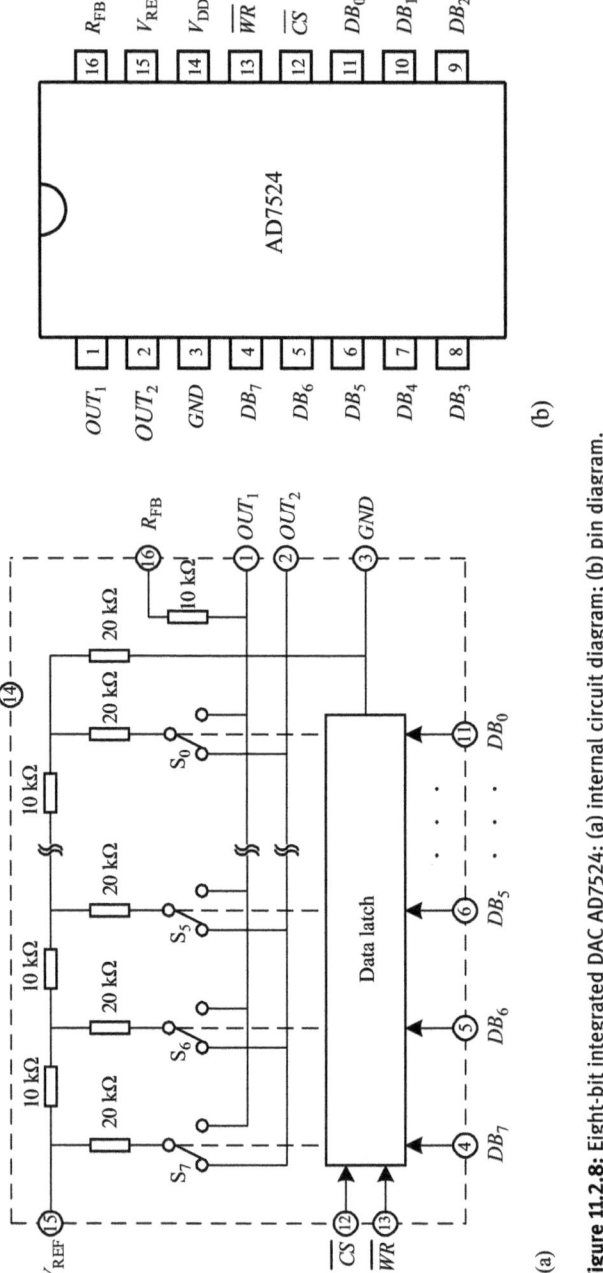

(a)

(b)

Figure 11.2.8: Eight-bit integrated DAC AD7524: (a) internal circuit diagram; (b) pin diagram.

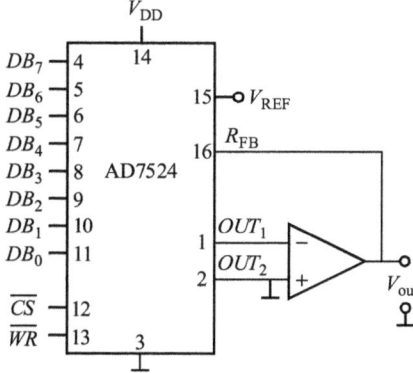

Figure 11.2.9: Typical practical circuit of AD7524.

In addition, the resolution of DAC can also be defined as the ratio of the minimum output voltage U_{LSB} to the maximum output voltage U_m, which can be expressed as

$$\text{Resolution} = \frac{U_{\text{LSB}}}{U_m} = \frac{-\frac{V_{\text{REF}}}{2^n} \cdot 1}{-\frac{V_{\text{REF}}}{2^n} \cdot (2^n - 1)} = \frac{1}{2^n - 1} \tag{11.2.6}$$

where U_{LSB} refers to the corresponding analog output voltage when only LSB input is a 1 and the remaining bits are 0.

For example, the resolution of ten-bit DAC is

$$\frac{1}{2^{10} - 1} = \frac{1}{1023} \approx 0.000978$$

Equation (11.2.6) characterizes the sensitivity of DAC to the minimum digital input. For the DAC, the more the number of bits in the data input is, the smaller the resolution is. As a result, DAC has a higher conversion precision and more sensitive to the digital input.

(2) Conversion Error

The conversion error is used to describe the conversion accuracy that the DAC can actually achieve. The conversion error can be expressed as a percentage of the full scale of the output voltage, or as multiples of the minimum output voltage U_{LSB}. For example, the conversion error is $U_{\text{LSB}}/2$, indicating that the absolute error of the output analog voltage is equal to one-half of the minimum output voltage U_{LSB}. The conversion error is divided into static error and dynamic error. The reasons for the static error are the instability of the reference power V_{REF}, the zero drift of the op-amp, the internal resistor and voltage drop during the analog switch on, and the deviation of the resistor values in the resistor network. Dynamic error is the additional error generated in the dynamic process of conversion, which is due to the influence of the distribution parameters in the circuit.

2. Conversion speed

There are two ways to access the conversion speed of DAC. The first is the settling time t_{set} to represent the conversion speed. The settling time is defined as the time it takes a DAC to settle within $\pm U_{LSB}/2$ of its final value when a change occurs in the input code, as shown in Figure 11.2.10. If the DAC only contains a resistor network and analog switch, its typical settling time is less than or equal to 0.1 µs; if the DAC also contains a reference power and a sum op-amp except for the resistor network and analog switched, its shortest settling time t_{set} is about 1.5 µs.

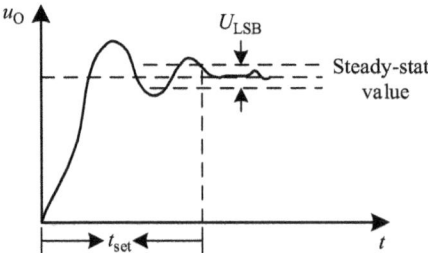

Figure 11.2.10: DAC settling time.

The second is to use the conversion rate S_R to represent the conversion speed. When a change occurs in the input code, the maximum rate of the output voltage changing into the new steady output voltage is called as *conversion rate*. This parameter is similar to the slew rate of the op-amp.

Example 11.2 Determine the resolution of the following DACs, expressed by a percentage. (a) an eight-bit DAC (b) a 12-bit DAC

Solution
For the eight-bit DAC, the resolution can be solved by using eq. (11.2.6) as follows:

$$\frac{1}{2^8 - 1} \times 100\% = \frac{1}{255} \times 100\% = 0.392\%$$

For the 12-bit DAC, the resolution is

$$\frac{1}{2^{12} - 1} \times 100\% = \frac{1}{4095} \times 100\% = 0.0244\%$$

11.2.6 Types of conversion errors

Take four-bit DAC for an example to illustrate several DAC conversion errors, as shown in Figure 11.2.11. The four-bit DAC has 15 discrete steps and each graph with an ideal step slope is given for comparison with the fault output [47].

*Nonmonotonicity:*The reversal step in Figure 11.2.11(a) represents nonmonotonicity, which is also a form of nonlinearity. In the specific case, this error occurs because the 2^1 bit in the binary code is always a 0, that is, the short circuit causes the input stuck in the LOW.

Nonlinearity: Figure 11.2.11(b) illustrates the differential nonlinearity, where the step amplitudes are lower than the corresponding step of a given input code. This kind of output may be due to a faulty input resistor and results in the 2^2 bit having an insufficient weight. If the binary weights are higher than that it should be, the steps with amplitudes will be higher than normal.

High or low gain: Figure 11.2.11(c) shows the output errors caused by high or low gain. In low gain, all step amplitudes are lower than the ideal one. In high gain, all step amplitudes are higher than the ideal one. This error is generally caused by a feedback resistor fault in the op-amp.

Offset error: Figure 11.2.11(d) shows the offset error. Note that when the input is 0000, the output voltage is not 0 V and the offset in the conversion process is the same for each step. The situation is usually caused by op-amp errors.

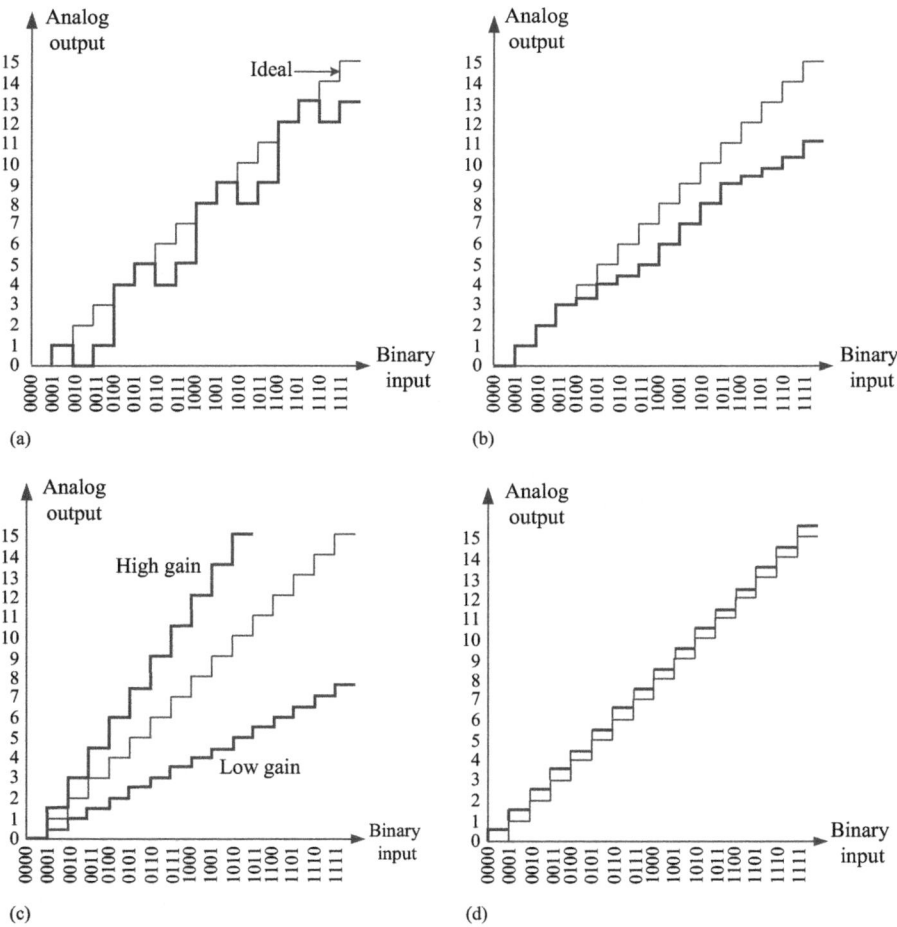

Figure 11.2.11: Illustrations of several digital-to-analog conversion errors.

Example 11.3 When a continuous four-bit binary sequence is added to the input of the DAC, the observed output is shown in Figure 11.2.12. Identify the type of error and give a method of isolating the fault.

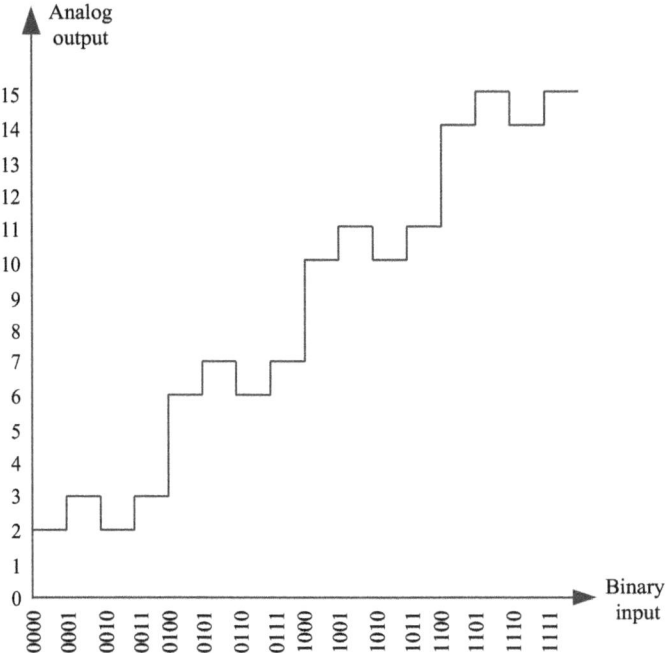

Figure 11.2.12: The observed output wavefoms of DAC.

Solution

In this case, the DAC is nonmonotonic. Analysis of the output reveals that the device is converted in the following sequence, rather than actually the input binary sequence added.

0010, 0011, 0010, 0011, 0110, 0111, 0110, 0111, 1010, 1011, 1010, 1011, 1110, 1111, 1110, 1111

Obviously, the position of 2^1 is kept at a high level. In order to find the problem, we first detect the bit input pin to the device. If it changes states, then the error is inside the DAC and it should be replaced. If the external pin does not change its states and is always high, check for an external short to +V that may be caused by a solder bridge somewhere on the circuit board.

11.3 Converting analog signals to digital

The input analog signals must be converted to digital forms in order to process signals using digital techniques. The ADC is the device that converts analog signal to the digital in a certain proportion. The ADC digitizes the analog signal in the time and amplitude, including four steps: sampling, holding, quantization, and encoding. This section introduces the method of converting an analog signal to a digital one.

The objectives of this section are to
- Explain how to convert an analog signal to digital
- Describe what sampling, holding, quantization, and encoding are
- Explain the Nyquist frequency
- Explain how to improve the conversion accuracy

11.3.1 Sampling

Usually, an analog signal has the infinite continuous values. It is impossible to convert all analog values to digital signals. Therefore, the analog signal must be discrete in time and amplitude so that they can be represented by digital signal. Sampling is a process to use enough number of discrete values at point on a waveform and the discrete values will essentially represent the shape of waveform, as shown in Figure 11.3.1. The more the number of sample points, the higher the accuracy is. Sampling is performed by acquiring the value of the analog signal at a certain time interval, *T*, which is called the *sampling interval* or the *sampling period*.

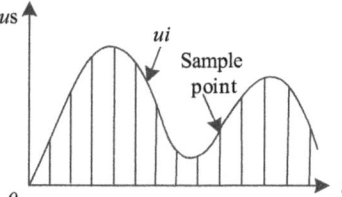

Figure 11.3.1: Analog signal sample.

In order to accurately represent the original analog signal, it is required that the sampling period must satisfy the Nyquist sampling theorem. The Nyquist sampling theorem defines a sample frequency f_{sample} that must be at least twice the highest frequency component $f_{a(max)}$ of the analog signal. In other words, the maximum frequency $f_{a(max)}$ of the analog signal is not greater than half of the *sample frequency* f_{sample}, where $f_{a(max)}$ is called the *Nyquist frequency* and is expressed by the following equation. *In practice, the sample frequency is greater than the twice as the highest analog frequency.*

$$f_{sample} > 2f_{a(max)} \qquad (11.3.1)$$

In digital audio system, the sample rates used are 32, 44.1, or 48 kHz. The rate of 44.1 kHz is often used for audio CDs and tapes. According to the Nyquist frequency, the sample frequency must be twice higher than the audio signal. Therefore, 44.1 kHz can be used to sample the signal with the maximum frequency about 22 kHz.

11.3.2 Hold the sample value

The second step is the hold operation after the sample pulse. The hold operation is to keep the sampling value constant for a period of time so that ADC has enough time to convert the sampling value into the corresponding digital signal. The process of sample-and-hold operation is shown in Figure 11.3.2. The initial analog signal is sampled at the control of sampling pulse and become a series of impulses, each representing the amplitude of the signal at a given instant time. Each sampling impulse must be kept constant for a certain time interval that usually equals to a sampling period. After the hold operation, the resulting signal waveform appears a "stair step waveform." The outline of the stair steps is close to the initial analog input waveform.

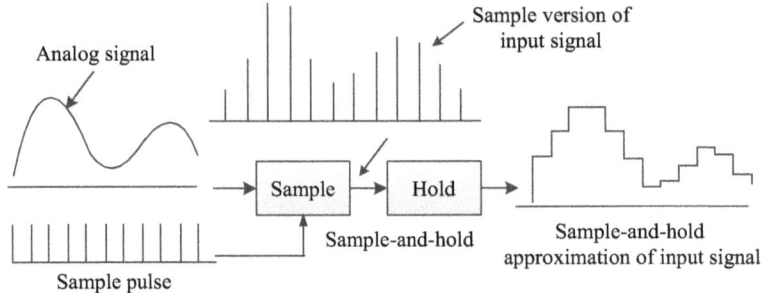

Figure 11.3.2: Illustration of a sample-and-hold operation.

Usually, the sample-and-hold operation can be implemented within a circuit called as sample-and-hold circuit. Figure 11.3.3 shows a kind of sample-and-hold circuit including two noninverting input op-amps, in which u_I is the analog input signal, u_O is the output signal after sample-and-hold operation, and the sampling switch is controlled by the control signal S. The capacitor C is used to retain the input analog quantity. The op-amp A_1 is used to provide high input impendence for the analog input signal, and A_2 is to isolate the capacitor C and offer the enough output current. When the switch is closed, the capacitor C is charged, which corresponds to the sampling process. While the switch is opened, the charge in the capacitor can be retained because the input impendence of the op-amp A_2 is very high, which corresponds to the holding sample value. In fact, the input impendence of the op-amp A_2 is not infinitely high, so the charge in the capacitor has a drop with a small magnitude.

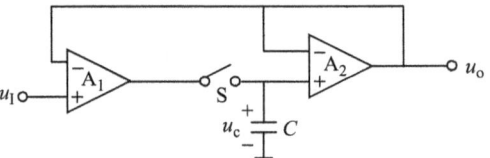

Figure 11.3.3: An example of the sample-and-hold circuit.

11.3.3 Quantization and encoding

Quantization refers to the process of representing a continuously analog signal with a discrete set of points. Usually the value of analog signal can have any value between two limits, but a discrete set of points only represent a limited number of possible values and information. Hence, the values between the points are lost. This unavoidably creates error called as quantization error. *Quantization error* is the difference between the actual sampled value and the quantized value. In order to improve the performance of the ADC process, much of efforts are related to minimizing these errors. *Encoding* refers to the process of using a group of binary code to represent the quantized value. The number of the quantized levels depends on how many bits in the binary code. The more the bits that are used to represent a sampled value, the more accurate the representation is.

Let us take a three-bit ADC as an example. The three-bit ADC converts the analog signal to a three-bit binary code. Assume that the analog signal is between 0 and 1 V. Because the three-bit binary code can only represent eight (2^3) possible values, the input analog signal should be uniformly quantized with eight quantization levels: 0 V, 1/8 V, 2/8 V, 3/8 V, 4/8 V, 5/8 V, 6/8 V, and 7/8 V, as shown in Figure 11.3.4(a). Each quantized level corresponding to a segment of analog voltage region can be encoded by using a unique three-bit binary code. If the analog sampled value is between 0 and 1/8 V, the quantized level is 0 V encoded as 000; if the analog sampled value is between 1/8 and 2/8 V, the quantized level is 1/8 V encoded as 001; and the rest can be done in the same manner. Due to the difference between the actual sampled value and the quantized value, the maximum quantization error is 1/8 V. This quantization process is called *uniform quantization*. In order to minimize the quantization error, an optional method is adopted for nonuniform quantization. The input analog signal between 0 and 1 V should be quantized with eight quantization levels: 0 V, 2/15 V, 4/15 V, 6/15 V, 8/15 V, 10/15 V, 12/15 V, and 14/15 V, as shown in Figure 11.3.4(b). The first quantized level of 0 V represents the analog sampled value between 0 and 1/15 V, which is different from the other quantized levels. The maximal quantization error in this region is 1/15 V. The second quantized level of 2/15 V represents the analog sampled value between 1/15 and 3/15 V. The maximal quantization error in this region is also 1/15 V; the rest can be done in the same manner. Although the voltage range of the first quantized region is different from that of others, the maximal quantization error is the same for all quantized segments. This quantization process is called as *nonuniform quantization* . Because 1/15 V is less than 1/8 V, the nonuniform quantization can further reduce the quantization error. Obviously, the more the bits that are used to represent a sampled value, the less the maximal quantization error is and thus the more accurate the representation is.

To illustrate the quantization process of the analog signal, an analog waveform is quantized into four levels (0–3), where two bits are required for encoding four levels. As shown in Figure 11.3.5, the sample intervals in the horizontal axis are labeled by

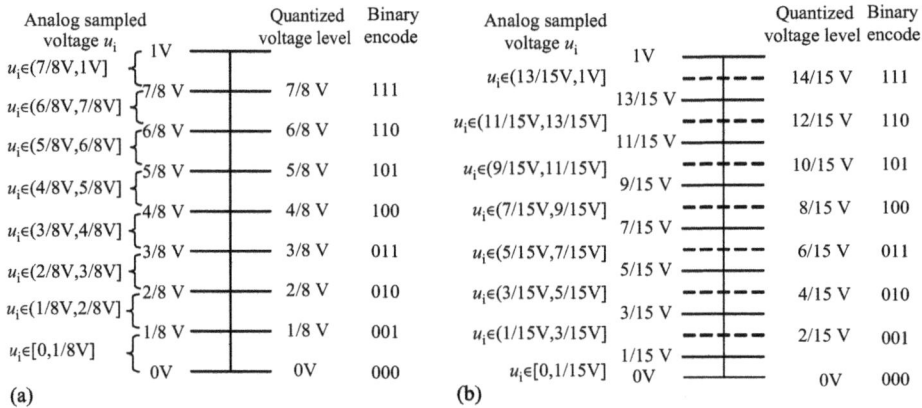

Figure 11.3.4: The quantization process of a three-bit ADC: (a) uniform quantization and (b) nonuniform quantization.

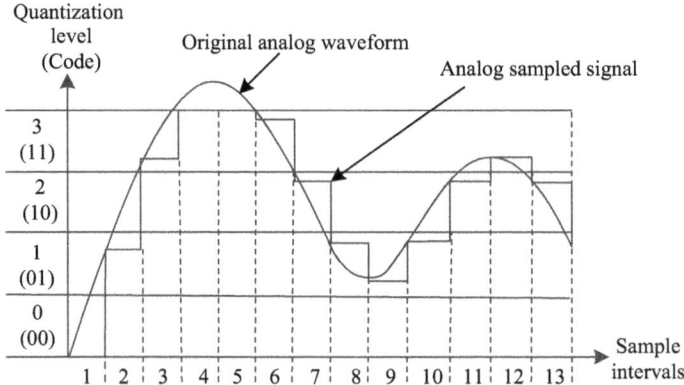

Figure 11.3.5: Sample-and-hold output waveform with four quantization levels. The original analog waveform is also shown for reference.

numbers; the quantization level in the vertical axis is represented by a two-bit binary code. The sampled data remain unchanged throughout the sample period. The data are quantized as adjacent lower levels as shown in Table 11.3.1.

The waveform shown in Figure 11.3.6 is reconstructed by using two-bit digital code; this process is called encoding. It can be seen from Figure 11.3.6 that the accuracy of the reconstructed signal waveform is quite low if only two-bit binary code is used to represent the sample value.

In order to improve the conversion accuracy, more bits are selected to encode the analog signal. Figure 11.3.7 shows the quantization process and Figure 11.3.8 shows the corresponding reconstruction process with 16 levels, that is, four-bit binary code.

Table 11.3.1: Two-bit quantization for the waveform in Figure 11.3.5.

Sample interval	Quantization level	Code
1	0	00
2	1	01
3	3	11
4	3	11
5	3	11
6	3	11
7	2	10
8	1	01
9	1	01
10	1	01
11	2	10
12	3	11
13	2	10

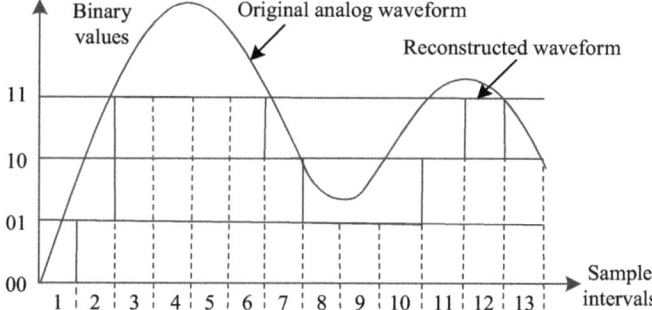

Figure 11.3.6: The reconstructed waveform of Figure 11.3.5 with four quantization levels. The original analog waveform is also shown for reference.

The waveform in Figure 11.3.8 is the original waveform reconstructed by using four-bit binary codes. It can be seen that the result is more approximate to the original waveform than that by only using two-bit digital codes. However, the reconstructed waveform is still deviating from the original waveform. This illustrates that the higher accuracy is achieved with more bits. For most of the integrated ADCs, the bits in the output digital code are from 8 to 24 bits, and the sample-and-hold circuit is often integrated inside the ADC chip.

11.4 Analog-to-digital converter (ADC)

Analog-to-digital conversion is the process by which an analog signal is converted into a digital signal. The conversion involves quantization of the input, so it

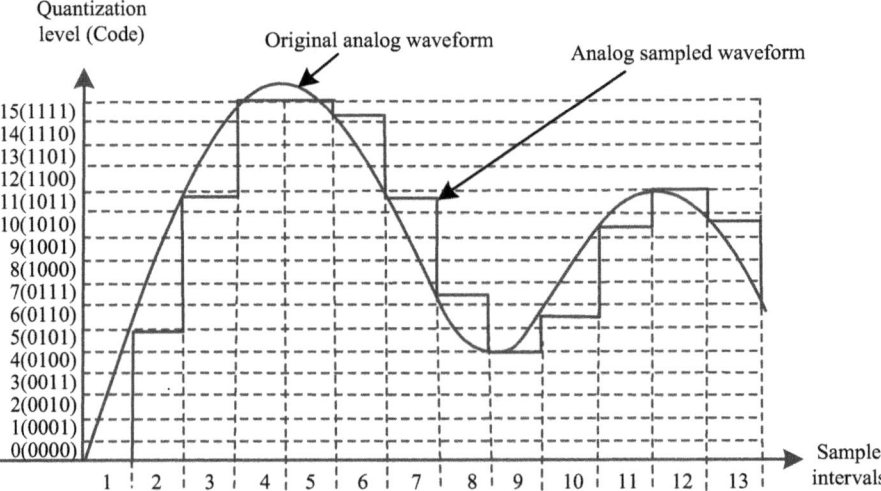

Figure 11.3.7: Sample-and-hold output waveform with 16 quantization levels. The original analog waveform is also shown for reference.

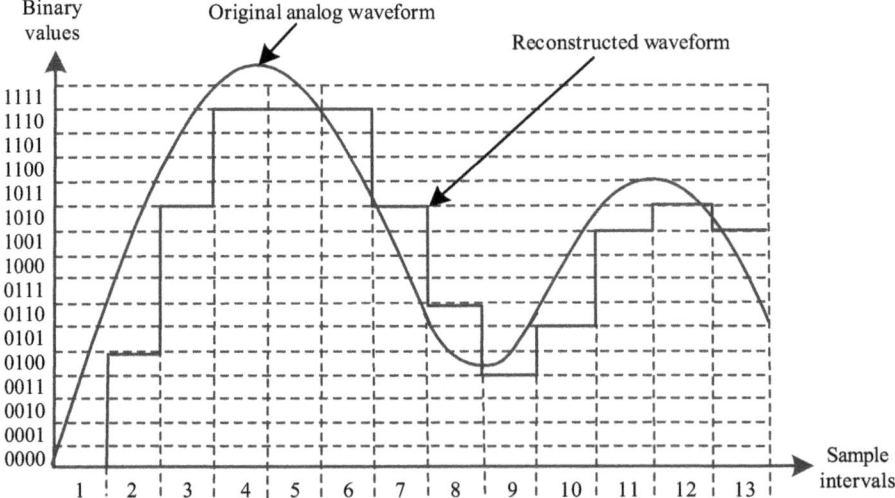

Figure 11.3.8: The reconstructed waveform of Figure 11.3.7 with 16 quantization levels. The original analog waveform is also shown for reference.

necessarily introduces a small amount of error or noise. Furthermore, instead of continuously performing the conversion, an ADC performs the conversion periodically, limiting the allowable bandwidth of the input signal. This chapter introduces several commonly used types of ADCs including the flash ADC, successive-approximation ADC, dual-slope ADC, and sigma-delta ADC.

The objectives of this section are to
- Describe the basic function of the ADCs
- Explain how the flash ADCs works
- Describe the operation of successive-approximation ADC
- Discuss how dual-slope ADC works
- Explain how sigma-delta ADC works
- Discuss performance characteristics of ADCs

ADC is to convert the analog sampled value of the sample-and-hold circuit to a series of binary codes that represent the amplitude of analog signal at each sample time. The sample-and-hold process keeps the amplitude of the analog signal constant between sample pulses, so ADC can complete analog-to-digital conversion by using a constant value instead of time-varying analog signal values during a conversion interval, which corresponds to the time between sample pulses. Figure 11.4.1 shows the basic function of an ADC circuit that performs analog-to-digital conversion. The sample intervals are indicated by dashed lines.

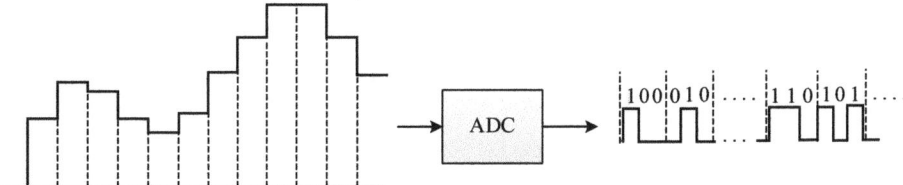

Figure 11.4.1: Basic function of an ADC.

11.4.1 Flash (simultaneous) ADC

Figure 11.4.2 show a three-bit flash ADC converting analog signal, u_I, to a series of three-bit binary codes, D_2 through D_0. The three-bit flash ADC consists of a resistive voltage-divider, seven comparators, seven flip-flops, and a priority encoder. The analog signal, u_I, is between 0 and + V_{REF}. Eight resistors are used to construct a resistive voltage-divider, which divides V_{REF} into eight voltage region and provides seven reference voltages to seven comparators. The corresponding reference voltage value is labeled in Figure 11.4.2. In fact, the resistive voltage-divider plays a role of quantizing the analog signal between 0 V and + V_{REF} into eight quantization levels: 0, 2/15 V_{REF}, 4/15 V_{REF}, 6/15 V_{REF}, 8/15 V_{REF}, 10/15 V_{REF}, 12/15 V_{REF}, and 14/15 V_{REF}. This quantization process is a nonuniform quantization, with the maximal quantization error of 1/15 V_{REF}. Seven comparators are adopted to compare input analog voltage with the reference voltage to determine which voltage region the input analog voltage belongs to, that is, judge which quantization level the input analog voltage should be converted to. When the conversion clock pulse arrives, the outputs of

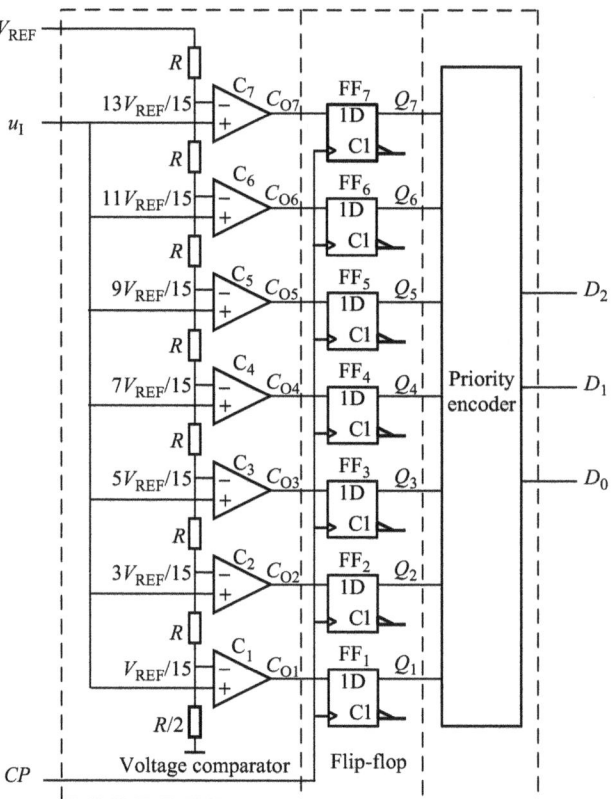

Figure 11.4.2: The three-bit flash ADC.

seven comparators (C_{O7}–C_{O1}) are sent to the output of seven D flip-flops (Q_7–Q_1), and then passed to the priority encoder. The resulting output from the priority encoder is a three-bit binary code, which corresponds to the quantization level of the input analog signal.

A comparator is composed of an op-amp with open loop. The inverting input of op-amp is a reference voltage provided by the resistor voltage-divider and the non-inverting input is the input analog voltage. If the input analog voltage is greater than its reference voltage, the output of the op-amp is a HIGH. Otherwise, the output is a LOW. The priority order of the priority encoder is from input Q_7 (the highest order) to input Q_1 (the lowest order) and the activate level is HIGH. Therefore, the conversion table of a flash ADC is listed in Table 11.4.1.

The flash ADC is to compare the input analog voltage with the reference voltage by using the specified high-speed comparators. A three-bit flash ADC converter uses 7 (2^3–1) comparators. A four-bit flash ADC needs 15 (2^4–1) comparators. In general, an n-bit flash ADC needs 2^n–1 comparators. The number of bits used in the ADC determines its resolution. Therefore, the flash ADC with high resolution requires a large number of

Table 11.4.1: Conversion table of three-bit flash ADC.

Input analog voltage	Outputs of comparators							Binary code		
u_I	C_{O7}	C_{O6}	C_{O5}	C_{O4}	C_{O3}	C_{O2}	C_{O1}	D_2	D_1	D_0
$0 \leq u_I < V_{REF}/15$	0	0	0	0	0	0	0	0	0	0
$V_{REF}/15 \leq u_I < 3V_{REF}/15$	0	0	0	0	0	0	1	0	0	1
$3V_{REF}/15 \leq u_I < 5V_{REF}/15$	0	0	0	0	0	1	1	0	1	0
$5V_{REF}/15 \leq u_I < 7V_{REF}/15$	0	0	0	0	1	1	1	0	1	1
$7V_{REF}/15 \leq u_I < 10V_{REF}/15$	0	0	0	1	1	1	1	1	0	0
$10V_{REF}/15 \leq u_I < 11V_{REF}/15$	0	0	1	1	1	1	1	1	0	1
$11V_{REF}/15 \leq u_I < 13V_{REF}/15$	0	1	1	1	1	1	1	1	1	0
$13V_{REF}/15 \leq u_I < V_{REF}$	1	1	1	1	1	1	1	1	1	1

comparators, which make the integrated circuit scale difficult to be enlarged. The main advantage of the flash ADC is the fast conversion because of high throughput measured in samples per second.

There are many monolithic integrated flash ADC products such as AD10012 (8-bit), AD10002 (8-bit), and AD10020 (10-bit) manufactured by the AD Company.

Example 11.4 Determine the binary code output of the three-bit flash ADC with uniform quantization level for the input signal and conversion clock pulse as shown in Figure 11.4.3. Assume the reference voltage V_{REF} = +8V.

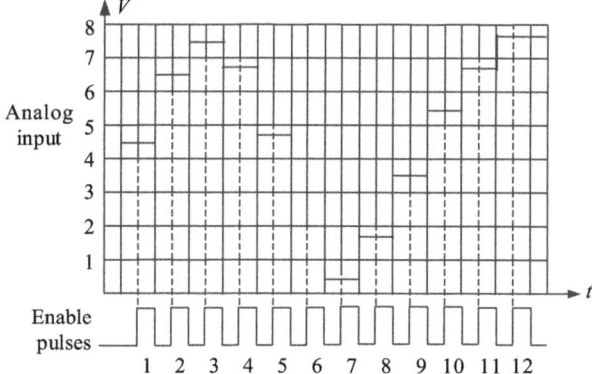

Figure 11.4.3: Sampling of values on a waveform for converting to binary code.

Solution
Since the input analog signal is between 0 and 8 V and the three-bit code have eight states, eight uniform quantization level is 0 V, 1 V, 2 V, 3 V, 4 V, 5 V, 6 V, and 7 V, which correspond to eight voltage region, that is, [0 V, 1 V], (1 V, 2 V], (2 V, 3 V], (3 V, 4 V], (4 V, 5 V], (5 V, 6 V], (6 V, 7 V], and (7 V, 8 V]. Each voltage region has the unique binary code. Therefore, the resulting binary code sequence is listed as follows and the corresponding waveforms is shown in Figure 11.4.4 associated with the clock pulse.

100,110,111,110,100,010,000,001,011,101,110,111

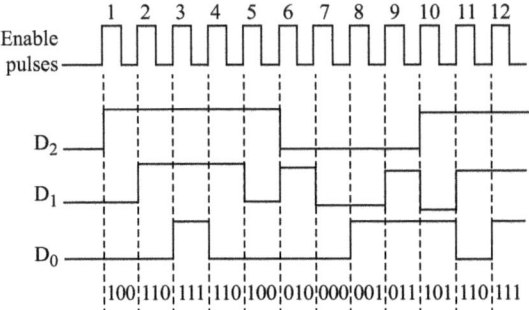

Figure 11.4.4: The final digital output for sample-and-hold values.

11.4.2 Successive-approximation ADC

1. Operating principle

The successive-approximation ADC is one of the most widely used ADCs. Figure 11.4.5 shows the typical successive-approximation ADC circuit that consists of four main subcircuits: a sample-and-hold circuit, a voltage comparator, a successive-approximation register (SAR) , and an n-bit internal DAC. A sample-and-hold circuit is to acquire the input sampled voltage, u_I. The analog voltage comparator compares the input voltage u_I to the output of the internal DAC and outputs the comparison result to the SAR. A SAR is designed to supply an approximate digital code of the input voltage u_I to the internal DAC. The internal reference DAC, for comparison with V_{REF}, supplies the comparator with an analog voltage equal to the digital code output of the SAR.

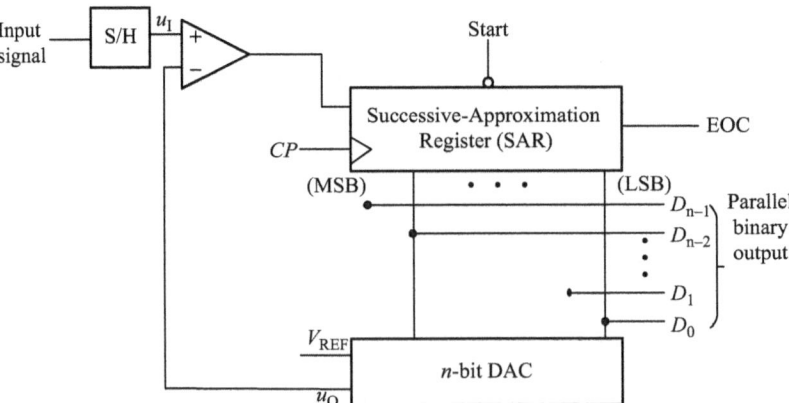

Figure 11.4.5: Schematic diagram of n-bit successive-approximation ADC.

The conversion process of the successive-approximation ADC is similar to the process of weighting object with a balance. Note that the digit at different position is equivalent to apply different weights on a balance. The SAR is initialized so that the most significant bit (MSB) is equal to a digit "1". This code is fed into the DAC, which then supplies the analog equivalent of this digital code ($V_{REF}/2$) into the comparator circuit for comparison with the sampled input voltage. If this analog voltage exceeds u_I, the comparator causes the SAR to reset this bit; otherwise, the 1 of the bit is left. Then the next bit is set to a 1 and the same test is done, continuing this binary search until every bit in the SAR has been tested. The resulting code is the digital approximation of the sampled input voltage and is finally output by the SAR at the end of conversion (EOC). The system starts from the MSB, then the next MSB, then the next, and so on. When all the bits in the DAC have been compared, the conversion cycle is completed.

In order to better understand the operation of successive-approximation ADC, we take an example of an eight-bit conversion, which has the reference voltage (V_{REF}) of 8 V and input analog voltage of 5.27 V. Table 11.4.2 lists the output of the DAC at different bit input.

Table 11.4.2: The output of the DAC corresponding to different bit input.

Inputs								Output
D_0	D_7	D_6	D_5	D_4	D_3	D_2	D_1	u_o
1	0	0	0	0	0	0	0	4V
0	1	0	0	0	0	0	0	2V
0	0	1	0	0	0	0	0	1V
0	0	0	1	0	0	0	0	0.5V
0	0	0	0	1	0	0	0	0.25V
0	0	0	0	0	1	0	0	0.125V
0	0	0	0	0	0	1	0	0.0625V
0	0	0	0	0	0	0	1	0.03125V

When a negative pulse is applied on the start input, the ADC begins to work and all bits in the SAR is cleared at the beginning of each cycle. The conversion starts with the MSB in SAR; let the 2^7 bit equals to a 1 in the first conversion clock pulse and the output of DAC is half of the reference voltage, that is, 4 V. The output of DAC u_O is fed back to the inverting input of a comparator and compared with the input voltage u_I. Because u_O is less than u_I, the output of the comparator is HIGH; the 1 in MSB should be retained in the SAR. In the second clock pulse, let the 2^6 bit equal to a 1. The output of the DAC is 6 V because there is a 1 on the 2^7 bit input and on the 2^6 bit input: 4 V + 2 V = 6 V. Since this output of 6 V is greater than the input of 5.27 V, the output of the comparator become LOW, making the 2^6 bit reset in the SAR. In the third clock pulse, let the 2^5 bit equal to a 1. The output of the DAC is 5 V because there is a 1 on the 2^7 bit input and on the 2^5 bit input: 4 V + 1 V = 5 V. Since this output of 5 V is less than the input of 5.27 V, the output of the comparator becomes HIGH, making the 1 in 2^5 bit retained in the SAR. The test can be done in the same manner until all eight bits have been tried, thus completing the conversion cycle. The resulting binary code in the register is 10101000 that corresponds to the analog voltage of 5.25 V, and there is 0.02 V error compared with 5.27 V. The timing waveform generated during conversion is shown in Figure 11.4.6. When one conversion cycle is completed, the next conversion cycle begins and repeats the aforementioned process. Note that the SAR should be reset at the beginning of each cycle.

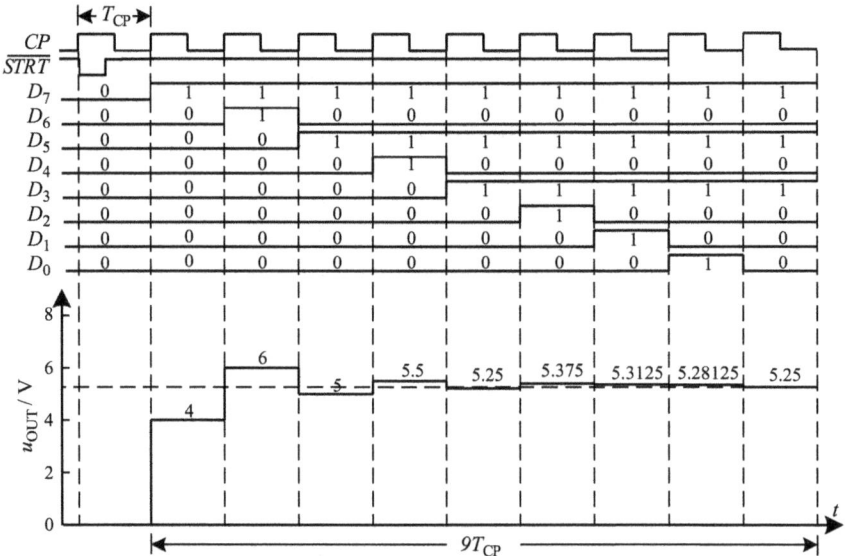

Figure 11.4.6: Timing waveform of one conversion cycle for a successive-approximation ADC.

2. Typical integrated successive-approximation ADC

The ADC0809 is a typical integrated ADC for data acquisition component, which is a monolithic CMOS device with an eight-bit ADC, eight-channel multiplexing analog switches, and microprocessor compatible control logic. The eight-bit ADC uses successive approximation as the conversion technique. The converter features a high impedance chopper stabilized comparator, a 256R voltage divider with analog switch tree and a SAR. The eight-channel multiplexer can directly access any of the eight-single-ended analog signals. The device eliminates the need for external zero and full-scale adjustments. Easy interfacing to microprocessors is provided by the latched and decoded multiplexer address inputs and latched TTL tristate outputs. The ADC0809 offers high speed, high accuracy, minimal temperature dependence, excellent long-term accuracy, and repeatability, as well as consumes minimal power. These features make this device ideally suited for applications from process and machine control to consumer and automotive applications. The internal block diagram and dual-in-line package with 28 pins are shown in Figure 11.4.7.

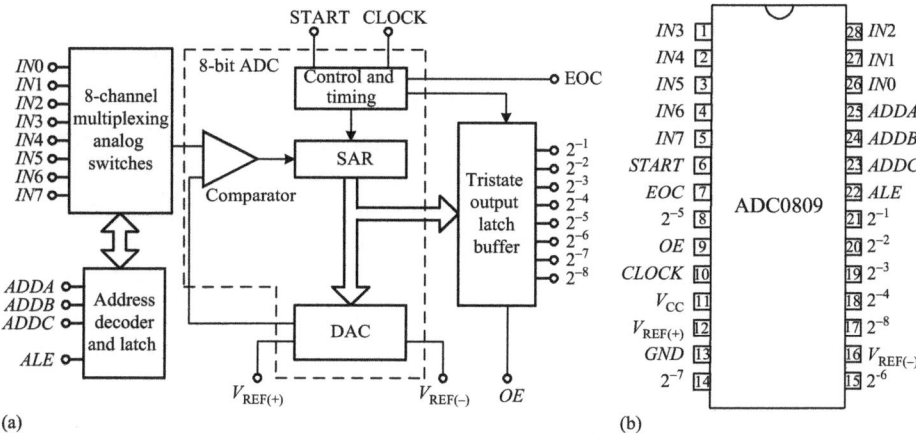

Figure 11.4.7: An eight-bit integrated ADC ADC0809: (a) internal block diagram and (b) pin diagram.

(1) Multiplexer

The device contains an eight-channel single-ended analog signal multiplexer. A particular input channel is selected by using the address decoder. Table 11.4.3 lists the input states for the address lines to select any channel. The address is latched into the decoder on the low-to-high transition of the address latch enable signal.

Table 11.4.3: Channel selection.

Selected analog channel	Address lines			Selected analog channel	Address lines		
	ADDC	ADDB	ADDA		ADDC	ADDB	ADDA
IN0	0	0	0	IN4	1	0	0
IN1	0	0	1	IN5	1	0	1
IN2	0	1	0	IN6	1	1	0
IN3	0	1	1	IN7	1	1	1

(2) The eight-bit ADC

The eight-bit ADC is the heart of ADC0809. The converter is designed to give fast, accurate, and repeatable conversions over a wide range of temperatures. The converter is partitioned into three major sections: the 256*R* ladder network, the SAR, and the comparator. The 256*R* ladder network approach is chosen over the conventional *R/2R* ladder because of its inherent monotonicity, which guarantees no missing digital codes. Monotonicity is particularly important in closed-loop feedback control systems. A nonmonotonic relationship can cause oscillations that will be catastrophic for the system. Additionally, the 256*R* network does not cause load variations on the reference voltage.

(3) Control and timing

The SAR is reset on the positive edge of the start conversion pulse (START) into the ADC. The conversion begins on the falling edge of the start conversion pulse. A conversion in process will be interrupted by receiving a new start conversion pulse. Continuous conversion may be accomplished by tying the EOC output to the START input. If used in this mode, an external start conversion pulse should be applied after power up. EOC will go low between 0 and 8 clock pulses after the rising edge of start conversion. When the conversion is finished, EOC goes high and the resulting digital code is latched into the "tristate output latch."

(4) Power line and others

V_{CC} is +5 V power supply and *GND* is connected to the ground. $+V_{REF}$ and $-V_{REF}$ are reference voltage inputs, which are used to supply reference voltages to DAC. $+V_{REF}$ is often connected with V_{CC} and $-V_{REF}$ is often connected to the ground. *CLOCK* is the clock input for ADC0809 to provide a 640 kHz clock pulse for successive comparisons.

11.4.3 Dual-slope ADC

1. Operating Principle

Dual-slope ADC is commonly used in digital voltmeters and other types of measurement instruments. It is a kind of indirect converter. Generally, it converts the input analog signal into an intermediate variable "time", then quantizes and encodes the time to obtain digital signal. This kind of ADC is all called the voltage–time conversion. Because the integrator generates two different ramps, one with the unknown analog input voltage V_I and another with a known reference voltage $-V_{REF}$. Hence, this ADC is called as *dual-slope ADC*. Since two integral processes are required in one conversion cycle, this ADC is also called as *dual-integral ADC*. The block diagram of the dual-slope ADC is shown in Figure 11.4.8.

When the dual-slope conversion starts, the binary counter is initially reset; the output of integrator is reset to 0 V. The switch S_1, which is controlled by control logic, is switched to the positive input voltage V_I, so the positive input voltage V_I is applied to the inverting input of the integrator through the resistor *R*. Assume that the positive input voltage is constant for a period of time T_1. Since the inverting input of the integrator is a

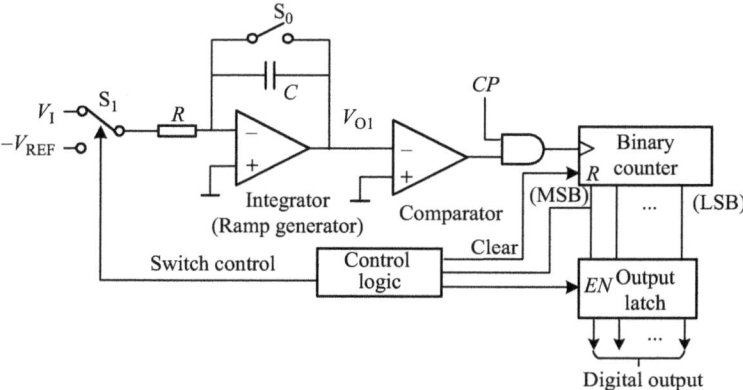

Figure 11.4.8: Block diagram of the dual-slope ADC.

virtual ground, the capacitor can be charged by the V_I through the input resistor R. Because the charging current is fixed, the charge on the capacitor C will increase linearly and thus a negative-going linear ramp is on the output of the integrator. At the same time, the output of comparator is HIGH and the clock is passed through the AND gate driving the binary counter to count up. The negative ramp continues for a fixed time period T_1, which is determined by a count detector for the time period T_1. At the end of the fixed time period T_1, the ramp output of integrator can be deduced as

$$V_{O1} = \frac{1}{C}\int_0^{T_1}(-\frac{V_I}{R})dt = -\frac{T_1}{RC}V_I \tag{11.4.1}$$

When the counter reaches the fixed count at time period T_1, it will be reset and the control logic switches the integrator input to a negative reference voltage $(-V_{REF})$. Now the ramp generator starts with the initial value V_{O1} and increases in positive direction until the output voltage of the integrator reaches 0 V; at the same time, the counter is advanced. When V_{O1} reaches 0 V, the output of comparator becomes LOW (i.e., logic 0) and thus AND gate is deactivated. This results in no further clock are applied through AND gate. Now, the conversion cycle end and the positive ramp voltage can be deduced as

$$|V_{O1}| = \frac{T_2}{RC}V_{REF} = \frac{T_1}{RC}V_I \tag{11.4.2}$$

Therefore,

$$T_2 = \frac{T_1}{V_{REF}}V_I \tag{11.4.3}$$

Figure 11.4.9 shows two linear integral processes, where T_1 is the charging time interval in which the voltage on capacitor increases until the MSB of the counter is a 1, and T_2 is the discharging time interval during which the capacitor is discharged until

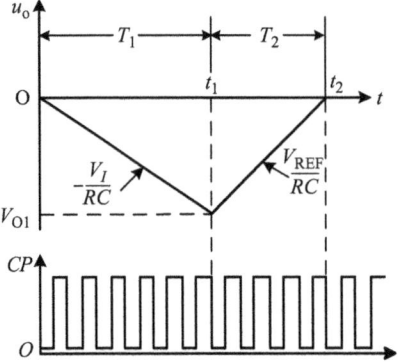

Figure 11.4.9: Waveform of dual-slope ADC.

the voltage on the capacitor is 0 V. Assume that the period of the clock pulse is represented by T_{cp}, then the time interval T_1 can be expressed as

$$T_1 = 2^n T_{CP} \qquad (11.4.4)$$

where 2^n is the count value of the binary counter when the MSB of the counter is a 1.

The actual conversion of analog voltage V_I into a digital count occurs during time interval T_2. The binary counter gives the corresponding digital value for time interval T_2. The clock is connected to the counter at the beginning of T_2 and is disabled at the end of T_2. So the time interval T_2 can be expressed as

$$T_2 = D T_{CP} \qquad (11.4.5)$$

where D is the count digital output of the counter at the end of T_2.

By substituting eqs. (11.4.4) and (11.4.5) into eq. (11.4.3), the digital output of the counter can be deduced as follows.

$$D = \frac{T_1}{T_{CP} V_{REF}} V_I = \frac{2^n}{V_{REF}} V_I \qquad (11.4.6)$$

The digital output D of the counter is proportional to the input voltage V_I. Now the voltage-to-time conversion is complete.

Compared with the successive-approximation ADC, the dual-slope ADC can only respond to the average of the input signal because of the existence of integrator. Therefore, the outstanding advantages of dual-slope ADC are stable operating performance and strong anti-interference ability. From the aforementioned analysis, it can be seen that as long as the integrator's time constants of the two integrations are equal, the counting result of the counter is independent of RC. Thus, the requirement of the circuit for RC accuracy is not strict, and the structure of the circuit is relatively simple. The dual-slope ADC belongs to the low-speed ADC, with the single conversion time of 1–2 ms; however, the successive-approximation ADC can reach 1 μs. In many industrial control systems, the millisecond conversion time is enough.

2. Typical Integrated Dual-slope ADC

There are many IC products of dual-slope ADCs, such as ICL7107, 7109, 5G14433, and so on. Figure 11.4.10 shows the block diagram of a typical dual-slope ADC chip (ICL7107). ICL7107 is high performance, low power, the $3^1/_2$ digits ADC, including seven segment decoders, display drivers, a reference, and a clock. It can directly drive an instrument size common anode light-emitting diode (LED) display. Four BCD digit readings can be directly displayed by using four seven-segment LED display. The maximal reading is 1999, in which the 3 refers to the rear three 7-segment display can display ten BCD digits from 0 to 9 and 1/2 refers to the headmost one 7-segment display can only display 1 or no display. ICL7107 can directly drive the common anode LED displayer and can display the analog voltage directly by decimal numbers from 0000 to 1999.

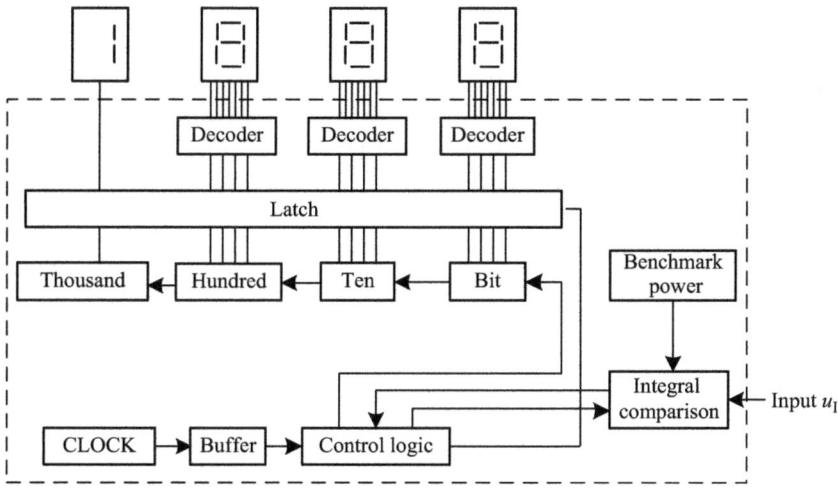

Figure 11.4.10: A block diagram of dual-slope ADC ICL7107.

The advantages of this dual-slope ADC are that *it can achieve high accuracy with fewer components, and can achieve strong anti-interference performance for the general input, which is a direct current (DC) or a slow changing DC.* Dual-slope ADC is widely used in digital measuring instruments, panel of industrial control cabinets, automobile instruments, and so on.

11.4.4 Sigma-delta ADC

Sigma-delta is a widely used conversion method in analog-to-digital conversion[39], especially in the area of communication using audio signals. In the conventional ADC, an analog signal is sampled with a sampling frequency and subsequently quantized by using the multilevel quantizer. This process introduces quantization noise. Sigma-delta ADC is based on *delta modulation* , which quantizes the difference (increase or decrease) between two consecutive samples and the change in the signal (delta) is encoded, rather than the absolute value like in the other ADCs. The result is a stream of pulses, as opposed to a stream of numbers. Delta modulation is a one-bit quantization method. In sigma-delta modulation, the accuracy of the modulation is improved by passing the digital output through a one-bit DAC and adding (sigma) the resulting analog signal to the input signal (the signal before delta modulation), thereby reducing the error introduced by the delta modulation.

The output of the delta modulator is a single-bit data stream, that is, a stream of pulses. The number of 1s in a stream of pulses represents the level or amplitude of the input signal. When the number of 1s exceeds the given number of clock cycles, the number exceeded determines the amplitude during that interval. The maximum number of 1s corresponds to the maximum value of the positive input voltage. Half of the maximum number of 1s corresponds to input voltage of zero. No 1s, that is, all 0s correspond to the maximum value of the negative input voltage. Figure 11.4.11 illustrates the quantization process in a simplified way. For example, suppose that if the input signal is a positive maximum, there will be 4096 1s occurring during the interval. Then when the input signal is zero, there are 2048 1s during the interval; when the input signal is negative maximum, there is no 1s during the interval. For the signal level between the negative maximum and the positive maximum, the number of 1s is proportional to the level of the input signal.

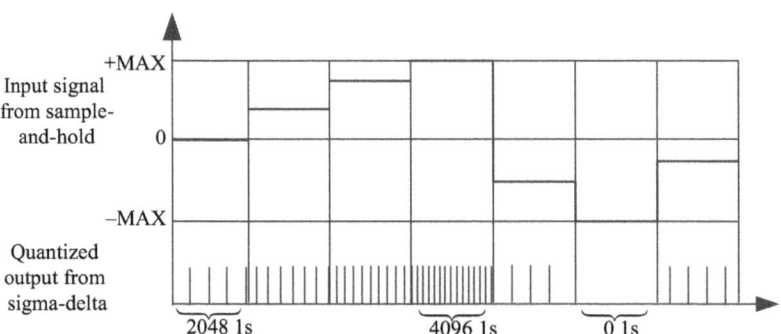

Figure 11.4.11: A simplified illustration of sigma-delta ADC.

The basic block diagram of the dashed rectangle shown in Figure 11.4.12 completes the conversion shown in Figure 11.4.11. The analog input signal and the analog signal from the converted quantized bit stream from the ADC in the feedback loop are applied to the summation (Σ) point. The difference (Δ) signal out of the Σ is integrated. The one-bit ADC increases or decreases the number of 1s according to the change of the difference signal. The purpose of this is to keep the quantized signal that is fed

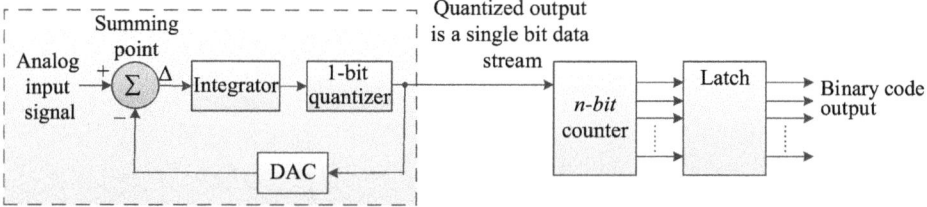

Figure 11.4.12: A partial function block diagram of a sigma-delta ADC.

back equals to the incoming analog signal. The one-bit quantizer is essentially a comparator followed by a latch.

To accomplish the sigma-delta conversion, a special method is used to convert a single-bit stream into a series of binary codes, as shown in Figure 11.4.12. The counter counts the 1s in the quantized data stream for successive intervals. The code in the counter then represents the amplitude of the analog input signal for each interval. These codes are shifted out into the latch for temporary storage. The output of the latch is a series of n-bit codes that completely represent this analog signal.

11.4.5 Analog-to-digital conversion errors

When analyzing the ADC errors, a four-bit conversion is used to illustrate the basic principle, assuming that the test input is an ideal linear ramp signal[47].

Missing codes: Figure 11.4.13(a) shows that the binary code 1001 that does not appear on the output of ADC. Note that the signal 1000 stays for two cycles, and then the output jumps directly to 1010. In a flash ADC, a failure of one of the op-amp comparators may cause errors of missing coding.

Incorrect codes: Figure 11.4.13(b) shows that several binary codes from the ADC output are incorrect. The results of the analysis show that 2^1 bit has been stuck in LOW and resulted in errors.

Offset: The offset is shown in Figure 11.4.13(c). In this case, the analog output voltage of the ADC is always higher than the actual value.

11.5 Summary

1. DAC with n bits divides a range of analog values (voltage or current) into 2^n steps. The size or magnitude of each stairs step is the analog equivalent weight of the LSB. This is called the resolution or step size.
2. DAC usually consists of reference voltage or current, the resistor networks, analog switches, and amplifier. Two kinds of DACs are the binary-weighted-input DAC and $R/2R$ ladder DAC.

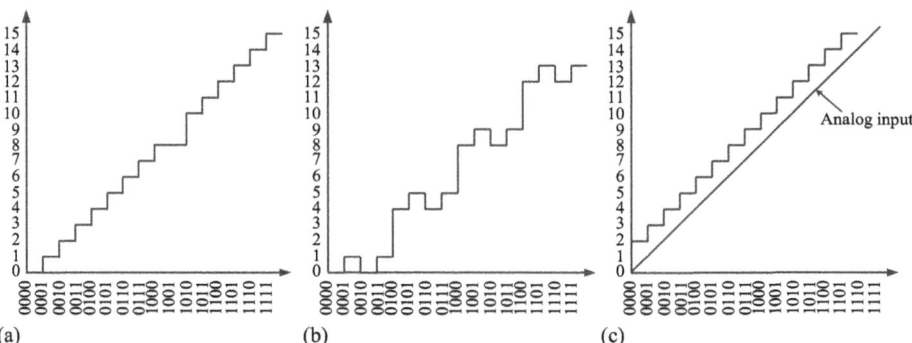

Figure 11.4.13: Illustrations of analog-to-digital conversion errors: (a) missing code; (b) incorrect codes; and (c) offset.

3. The binary-weighted-input DAC uses a resistor network where resistor values represent the weights of the input bits of the binary number.

4. The $R/2R$ ladder DAC uses a resistor network with only two resistor values with R and $2R$, which is greatly convenient for design and manufacture of integrated circuit and overcomes the disadvantage of multiple resistor values for binary-weighted-input DAC.

5. The ADC is the device that converts an analog signal to a digital one in a certain proportion. Analog-to-digital conversion is generally divided into four steps: sampling, holding, quantization, and encoding.

6. Sampling is the process to use an enough number of discrete values at point on a waveform and the discrete values that will essentially represent the shape of waveform. In practice, the sample frequency is at least greater than the twice as the highest analog frequency.

7. The hold operation is to keep the sampling value constant for a period of time so that ADC has enough time to convert the sampling value into the corresponding digital signal.

8. Quantization refers to the process of representing a continuous analog signal with a discrete set of points called as quantized levels that only represent a limited number of possible values. The difference between the actual sampled value and the quantized value is called as quantization error.

9. Encoding refers to the process of using a group of binary code to represent the quantized value. The number of the quantized levels depends on how many bits in the binary code. The more bits that are used to represent a sampled value, the more accurate is the representation.

10. The development of ADC and DAC is facilitated by the widespread use of digital systems such as microprocessors and computers in a variety of instrumentation and control systems. Accuracy and conversion speed are the two important parameters of ADC and DAC that are the main factors determining the accuracy and operating speed of the entire system.

11. The DAC can convert the digital signal to analog signal. The resistor network of $R/2R$ ladder DAC contains only two resistor values with R and $2R$. Since only two resistors are required, this method overcomes the disadvantage of multiple resistor values for binary-weighted-input DAC, which is greatly convenient for the design and manufacture of the integrated circuit.

12. Four types of ADCs are flash ADC, successive-approximation ADC, dual-slope ADC, and sigma-delta ADC.

13. Flash converters use analog comparators and an encoder to assign a digital value to the analog input. These are the fastest converters because the only delays involved are propagation delays.

14. A successive-approximation converter has a constant conversion time and is probably the most common general-purpose converter.

15. Dual-slope ADC is a kind of indirect converter. Generally, it converts the input analog signal into an intermediate variable–time and thus the conversion speed is slow.

16. Sigma-delta ADC is based on delta modulation, which quantizes the difference (increase or decrease) between two consecutive samples and the change in the signal (its delta) is encoded, rather than the absolute value like in the other ADCs.

Key terms

Analog-to-digital converter (ADC): A circuit that converts analog signal to a digital signal.

Digital to analog converter (DAC): A circuit that converts the digital signal back to a analog signal.

Sample: The process of taking a sufficient number of discrete values at points on a waveform that will define the shape of waveform.

Quantization: The process of converting an analog signal in a quantized level represented by a binary code.

Nyquist frequency: The highest signal frequency that can be sampled at a specified sampling frequency; a frequency equal to or less than half the sampling frequency.

Missing codes: The binary codes disappearing on the output of ADC.

Offset: The phenomenon that the analog output voltage of the ADC is always higher or lower than the actual value.

Resolution: The number of bits in the digital input; the ratio of the minimum output voltage to the maximum output voltage.

Conversion error: A percentage of the full scale of the output voltage, or as multiples of the minimum output voltage.

Self-test

11.1 An ADC refers to an _____
(a) alphanumeric data code
(b) analog-to-digital converter
(c) analog device carrier
(d) analog-to-digital comparator

11.2 A DAC refers to a _____
(a) digital-to-analog computer
(b) digital analysis calculator
(c) data accumulation converter
(d) digital-to-analog converter

11.3 In a digital representation of voltages using an eight-bit binary code, how many values can be defined?
(a) 16   (b) 64   (c) 128   (d) 256

11.4 In a digital reproduction of an analog curve, accuracy can be increased by_____
(a) sampling the curve more often
(b) sampling the curve less often
(c) decreasing the number of bits used to represent each sampled value
(d) (a), (b), and (c)

11.5 In a binary-weighted DAC, the resistors on the inputs_____
(a) determine the amplitude of the analog signal
(b) determine the weights of the digital inputs
(c) limit the power consumption
(d) prevent loading on the source

11.6 A four-bit $R/2R$ ladder DAC uses_____
(a) one resistor value
(b) two resistor values
(c) three resistor values
(d) four resistor values

11.7 Sampling of an analog signal produces_____
(a) a series of impulses that are proportional to the amplitude of the signal
(b) a series of impulses that are proportional to the frequency of the signal
(c) digital codes that represent the analog signal amplitude
(d) digital codes that represent the time of each sample

11.8 According to the sampling theorem, the sampling frequency should be_____
(a) less than half the highest signal frequency
(b) greater than twice the highest signal frequency
(c) less than half the lowest signal frequency
(d) greater than the lowest signal frequency

11.9 Generally, an analog signal can be reconstructed more accurately
with_____
(a) more quantization levels   (b) fewer quantization levels
(c) a higher sampling frequency   (d) a lower sampling frequency
(e) either answer (a) or (c)

11.10 The resolution of a 6-bit DAC is_____
(a) 63%   (b) 64%   (c) 15.8%   (d) 1.58%

11.11 For a four-bit DAC, the LSB is_____
(a) 6.25% of full scale   (b) 0.625% of full scale
(c) 12% of full scale   (d) 1.2% of full scale

11.12 A flash ADC use_____
(a) counters   (b) op-amps
(c) an integrator   (d) flip-flops
(e) answers (a) and (c)

11.13 A dual-slope ADC uses_____
(a) a counter   (b) op-amps
(c) an integrator   (d) a differentiator
(e) answers (a) and (c)

Problems

11.1 Determine the resolution of the following DACs, expressed by a percentage.
(a) a four-bit DAC   (b) an eight-bit DAC
(c) a ten-bit DAC   (d) an 24-bit DAC

11.2 Determine the output waveform of the binary-weighted input DAC in Figure
P11.1 (a) for a given input waveforms in Figure P11.1(b).

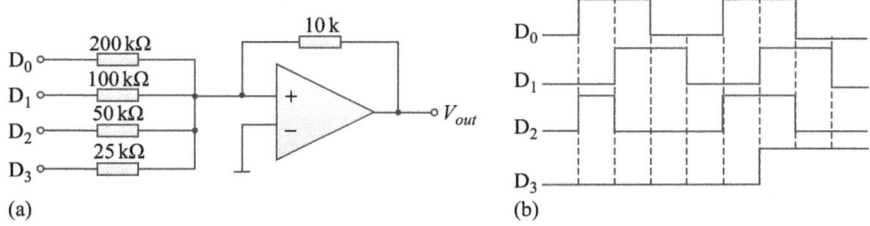

Figure P11.1

(a) when the sequence of four-bit numbers in part
(b) is applied to the inputs. Assume that a low level is 0 V and a high level is +5 V.

11.3 Determine the output of the $R/2R$ ladder DAC in Figure P11.2 when the digital input is $(0FDA)_{16}$. Assume that $V_{REF} = 10$ V.

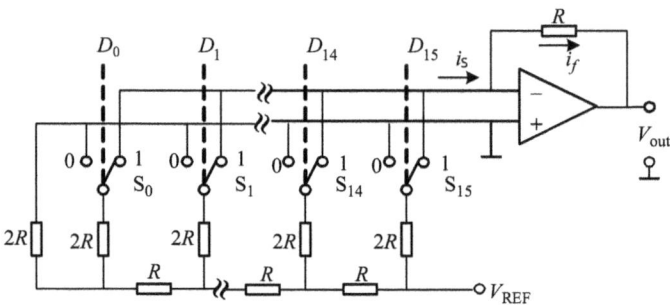

Figure P11.2

11.4 Analyze the implementing function of a circuit that consists of an AD5724 and an op-amp, as shown in Figure P11.3, where V_I is analog input voltage and V_{out} is the output voltage.

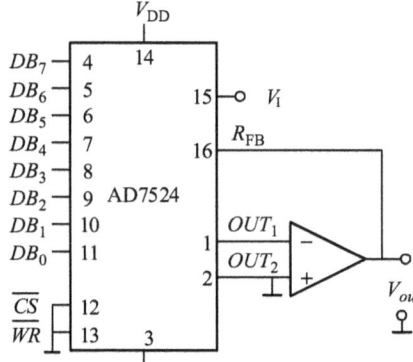

Figure P11.3

11.5 Figure P11.4 shows an $R/2R$ ladder DAC, where the outputs of the counter are connected to the digital inputs. The state sequence of the counter is listed in Table P11.1. Complete Table P11.1 by filling the corresponding output of DAC and draw the waveform of output V_{out} in relation to the clock pulse. Assume that the high level is 8 V, the low level is 0 V, and the initial state (Q_1Q_0) of the counter start from the state 00.

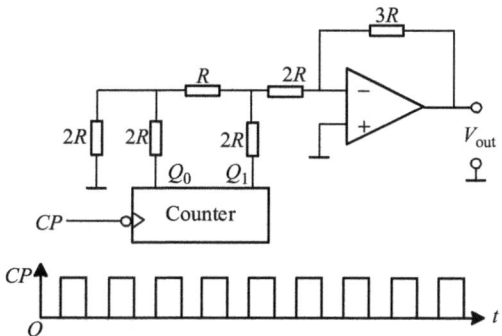

Table P11.1 State sequence table

Q_1	Q_0	V_{out}/ V
0	0	
0	1	
1	0	
1	1	
1	0	
0	1	
0	0	

Figure P11.4

11.6 Explain the operating principle and draw the waveform of output voltage u_O for the logic circuit shown in Figure P11.5. The stored data in EPROM 2716 are given in Table P11.2.

Table P11.2: Stored data in EPROM 2716.

Address inputs				Data contents			
A_3	A_2	A_1	A_0	D_3	D_2	D_1	D_0
0	0	0	0	0	0	0	0
0	0	0	1	0	0	1	0
0	0	1	0	0	0	0	0
0	0	1	1	0	0	1	0
0	1	0	0	0	1	0	0
0	1	0	1	0	0	0	0
0	1	1	0	0	0	0	0
0	1	1	1	0	0	1	0
1	0	0	0	0	1	0	0
1	0	0	1	0	1	1	0
1	0	1	0	0	0	0	0
1	0	1	1	0	0	1	0
1	1	0	0	1	1	0	0
1	1	0	1	0	1	1	0
1	1	1	0	0	0	0	0
1	1	1	1	0	0	0	0

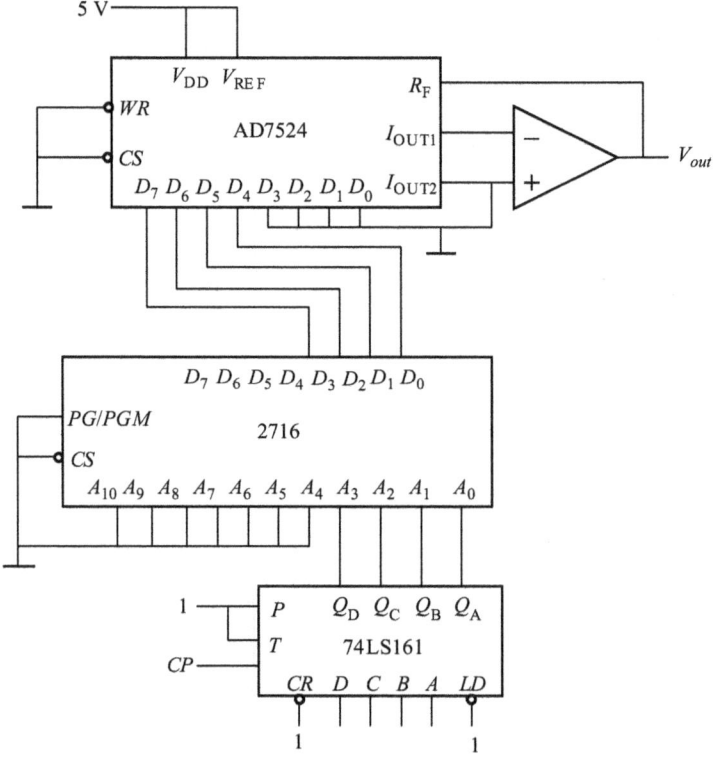

Figure P11.5

11.7 Determine the binary code output of the three-bit flash ADC circuit in Figure 11.4.2 when the input signal is applied to its input at the clock pulses shown in Figure P11.6. Assume that the reference voltage V_{REF} = +8 V.

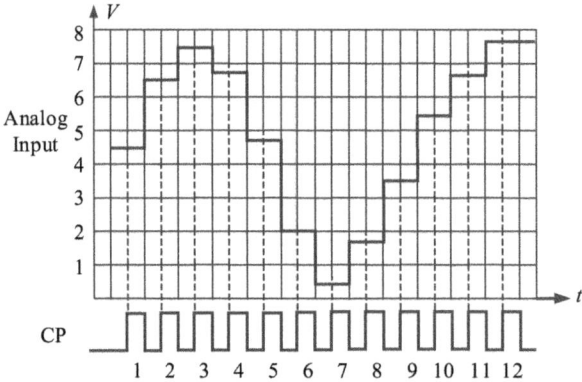

Figure P11.6: Sample values on a waveform for conversion to binary code.

11.8 Show the output of the sample circuit when the waveform shown in Figure P11.7 is applied to a sample circuit and the sampled period is a 1 ms interval. Assume a one-to-one correspondence between the input and output.

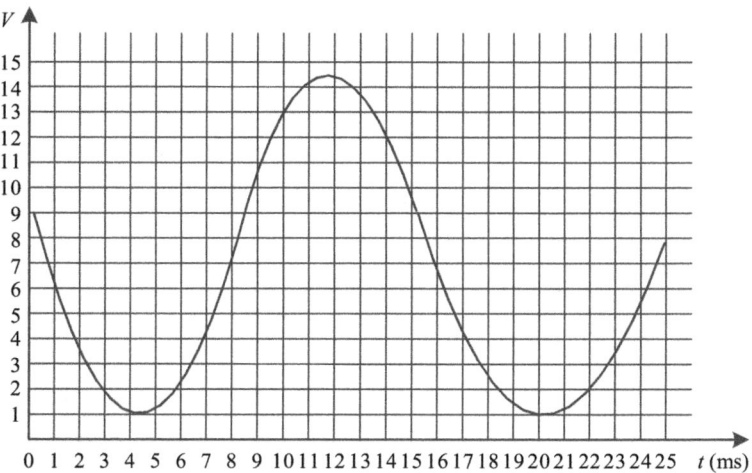

Figure P11.7

11.9 The output of the sample circuit in Problem 11.8 is applied to a hold circuit. Show the output of the hold circuit.

11.10 If the output of the hold circuit in Problem 11.9 is quantized using two bits, what is the resulting sequence of binary codes.

11.11 When using four-bit quantization, repeat Problem 11.9.

11.12 How many comparators are required to form an eight-bit flash ADC?

11.13 Determine the binary output code of a three-bit flash ADC for the analog input signal in Figure P11.8.

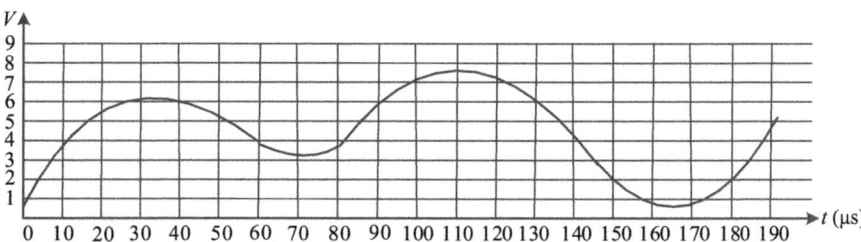

Figure P11.8

11.14 For a certain two-bit successive-approximation ADC, the maximum ladder output is +8 V. If a constant voltage of + 6 V is applied to the analog input, determine the sequence of binary states for the SAR.

11.15 Repeat Problem 11-14 for a four-bit successive-approximation ADC.

11.16 Figure P11.9 shows the timing waveform of one conversion cycle for a certain eight-bit successive-approximation ADC with the clock frequency of 250 kHz. Determine the time required for one conversion cycle and the resulting digital output.

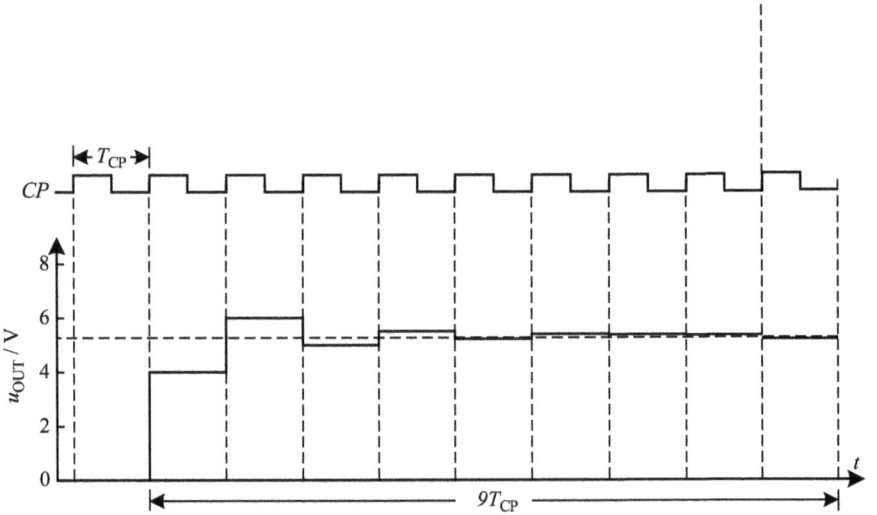

Figure P11.9

11.17 The maximum output voltage u_{omax} equal to 12.276 V for a certain ten-bit successive-approximation ADC with the clock frequency of 500 kHz. Determine the resulting digital output and the requiring time for one conversion cycle when the input voltage u_I is 4.32 V.

11.18 A dual slope ADC is shown in Figure P11.10.
(a) Analyze its operating principle.
(b) If the maximum of the unknown input voltage V_I is 2 V and the minimum voltage distinguished is 0.1 mV, determine what the maximum count of binary counter is and how many bits is required in binary counter.
(c) Determine the sampling time T_1, taking the clock frequency as 200 kHz.
(d) Determine the integral time constant RC when the maximal output voltage of integrator is 5 V at the input voltage of 2 V and clock frequency of 200 kHz.

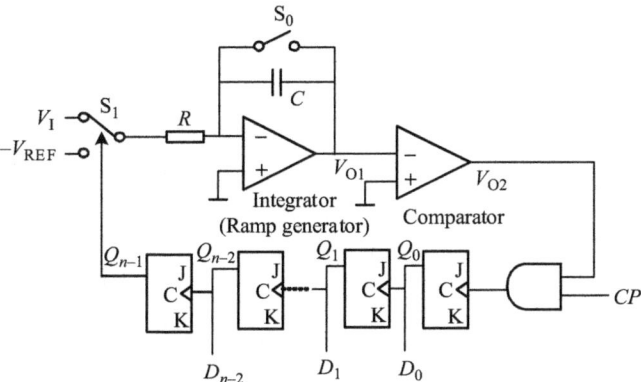

Figure P11.10

11.19 A circuit is shown in Figure P11.11, where 74LS190 is an 8421BCD up/down counter and AD5724 is eight-bit DAC chip.
(a) Explain the function of the circuit.
(b) Determine the maximum of the output.
(c) Sketch the output waveform of u_o in relation to the clock pulse.

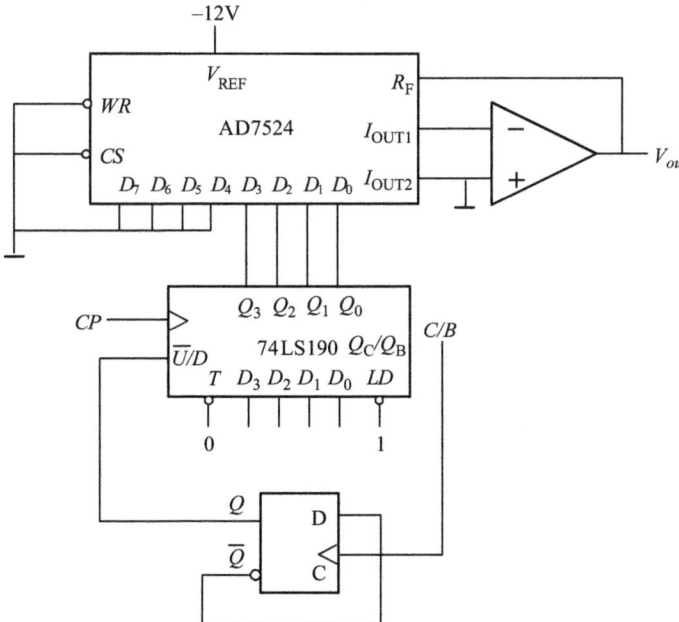

Figure P11.11

11.20 Analyze the function of a circuit that consists of eight-bit ADC0809, four-bit binary counter, and RAM with capacity 16 × 8, as shown in Figure P11.12.

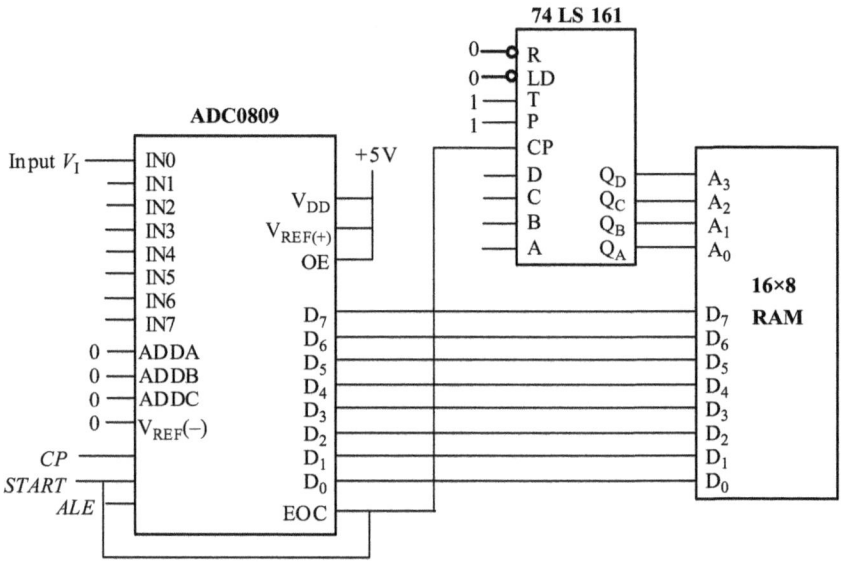

Figure P11.12

12 Integrated gate circuit

12.1 Introduction

Gate circuit is the unit circuit that can implement both the basic logic operation and combinational logic operation. Basic logic gates include AND, OR and inverter, and combinational logic gates involve NAND, NOR, XOR, XNOR, etc. Gate circuits can be constructed with multiple diodes, transistors and other components. In 1960s, the invention of the *integrated circuit* (IC) allowed multiple diodes, transistors, resistors and other components to be fabricated on a single chip. Also the first integrated-circuit logic families were introduced.

A *logic family* is a collection of different integrated-circuit chips that have similar inputs, outputs, and internal circuit characteristics, but they perform different logic functions [20]. Now the most successful logic families are bipolar logic family and metal-oxide semiconductor field-effect transistor (MOSFET) logic family. Bipolar logic family is one based on bipolar junction transistors called *transistor-transistor logic* (*TTL*). TTL now is actually a family of logic families that are compatible with each other but differ in speed, power consumption, and cost. With TTL, small- and medium-scale applications can be implemented with the relative high speed. In this chapter, the basic operation characteristics and parameters of TTL are introduced in Section 12.2. After that, we briefly introduce the basic structures of diode logic in Section 12.3 and then focus on the introduction of TTL families in Section 12.4.

Metal oxide semiconductor (MOS) includes n-channel MOS (NMOS), p-channel MOS (PMOS) and complemented MOS (CMOS). Among them, CMOS logic family is the most commonly used to construct large-scale integrated circuits, such as microprocessors and memories due to their lower power consumption and higher integration level. Now CMOS circuits occupy a vast majority of the worldwide IC market. So, we focus on introducing the basic structures of CMOS logic gate circuits and the most commonly used commercial CMOS logic families in Section 12.5.

Chips from different logic families may not be compatible; they may use different power supply voltages or different input or output conditions to represent logic values. Lots of circuit systems include chips of TTL families and CMOS families. In Section 12.6, we introduce how TTL and CMOS families can be mixed within a single system.

The objectives of this chapter are to
– Understand the performance characteristics of integrated logic families.
– Describe how basic TTL and CMOS gates operate at the component level.
– Recognize the difference between TTL totem-pole outputs and TTL open-collector outputs
– Connect circuits in a wired-AND configuration.
– Explain how tristate gate operates.

https://doi.org/10.1515/9783110614916-012

– Properly deal with unused gate inputs.
– Compare the performance of TTL and CMOS families.
– Describe how TTL and CMOS families can be mixed within a single system.

12.2 Basic operational characteristics and parameters

When you work with digital integrated circuits, you should understand not only their operations but also some important operational properties such as voltage levels, noise immunity, power dissipation, fan-out, and propagation delay time [47]. This section introduces basic operational characteristics and parameters. These operational properties and parameters guarantee you use digital ICs correctly.

The objectives of this section are to
– Explain the logic levels for CMOS and TTL
– Discuss noise immunity
– Determine the power dissipation of a logic circuit
– Define the propagation delay time of a logic gate
– Discuss speed-power product and explain its significance
– Explain what the fan-out of a gate means

12.2.1 DC supply voltage

Every digital IC has DC supply voltage and ground distributed internally to all elements within the package, as shown in Figure 12.2.1. The nominal value of the DC supply voltage for TTL devices is +5 V. For CMOS devices, there are different supply voltages, for example, +5 V, +3.3 V, +2.5 V, and +1.2 V. For simplicity, DC supply voltage and ground are usually omitted from logic diagram.

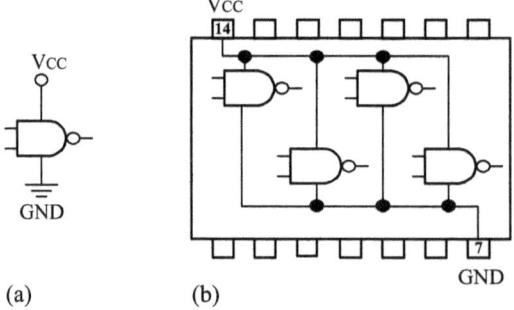

(a) (b)

Figure 12.2.1: Single gate (a), V_{CC} and ground connection and distribution (b) in an IC dual in-line package. Other pin connections are omitted for simplicity.

12.2.2 Logic levels

Logic elements process binary digits, 0 and 1. But real logic circuits process electrical signals such as voltage levels. In any logic circuit, it is a range of voltages that is interpreted as a logic 0, and a non-overlapping range that is interpreted as logic 1. Usually there are four different logic-level specifications i.e. U_{IL}, U_{IH}, U_{OL} and U_{OH} for a logic circuit. U_{IL} and U_{IH} are input low voltage and input high voltage, respectively. U_{OL} is output low voltage and U_{OH} is output high voltage. Different logic families accept different range of voltage levels as the input and the output voltages.

Figure 12.2.2 shows the range of input and output voltages for +5 V CMOS family and +5 V TTL family, respectively.

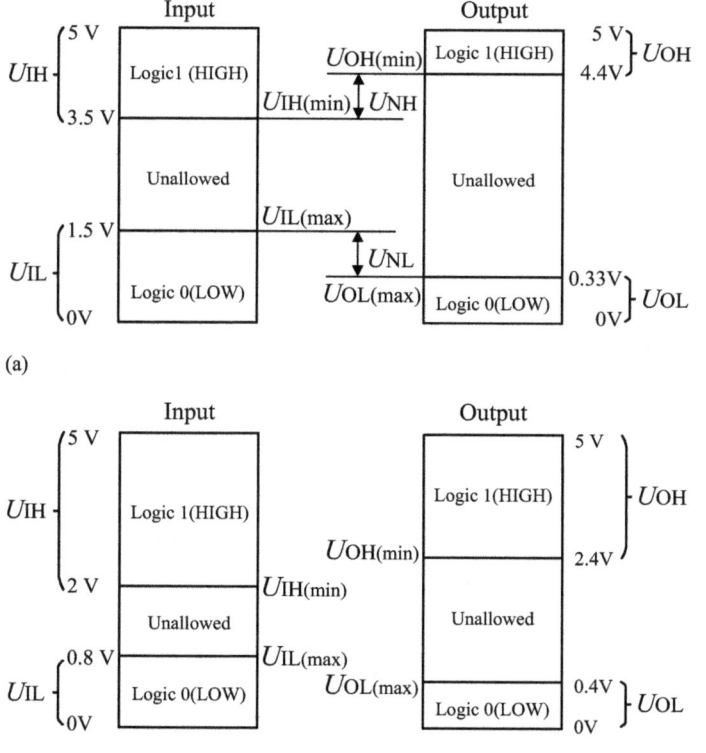

Figure 12.2.2: Input and output logic levels for + 5 V CMOS (a) and + 5 V TTL (b).

For +5 V CMOS circuit, the range of input low voltage (U_{IL}) that represents a valid LOW (logic 0) is from 0 V to 1.5 V. The range of input high voltage (U_{IH})

that represents a valid HIGH (logic 1) is from 3.5 V to 5 V. When an input voltage is in one of these ranges, it can be interpreted as either HIGH or LOW by the logic circuit. The range of values from 1.5 V to 3.5 V is an unallowed region. CMOS gates cannot be operated reliably when the input voltage is in the unallowed range. The ranges of CMOS output voltages (U_{OL} and U_{OH}) for 5 V logic are shown in Figure 12.2.2(a). Notice that the minimum high level output, $U_{OH(min)}$ is greater than the minimum high level input, $U_{IH(min)}$. Also, notice that the maximum low level output, $U_{OL(max)}$, is less than the maximum low level input, $U_{IL(max)}$.

For +5 V TTL family, the ranges of input and output voltages, as shown in Figure 12.2.2(b), are different from that of CMOS family. Generally, the range of input low voltages (U_{IL}) is from 0 V to 0.8 V and input high voltages (U_{IH}) range from 2 V to 5 V.

12.2.3 Noise immunity

Noise is the unwanted voltage level induced in electrical circuits and may bring a trouble to the proper operation of the circuit. In order to resist the effect from noise, a logic circuit should have a certain amount of *noise immunity*. That represents the ability to tolerate a certain amount of unwanted voltage level fluctuation on the input without changing the output.

Noise margin is used to measure a circuit's noise immunity and assure that the highest LOW output is always lower than the highest voltage value that can be reliably interpreted as LOW, and the lowest HIGH output is always higher than the lowest voltage value that can be interpreted as HIGH. There are two values of noise margin specified for a given logic circuit, i.e. the HIGH-level noise margin (U_{NH}) and the LOW-level noise margin (U_{NL}). These two parameters are defined as

$$U_{NH} = U_{OH(min)} - U_{IH(min)} \tag{12.2.1}$$

$$U_{NL} = U_{IL(max)} - U_{OL(max)} \tag{12.2.2}$$

U_{NH} is the difference between the lowest possible HIGH output from a driving gate ($U_{OH(min)}$) and the lowest possible HIGH input that the load gate can tolerate ($U_{IH(min)}$). Noise margin, U_{NL}, is the difference between the maximum possible LOW input ($U_{IL(max)}$) that a gate can tolerate and the maximum possible LOW output ($U_{OL(max)}$) of the driving gate. Noise margins are illustrated in Figure 12.2.2.

Example 12.1 Determine the HIGH-level and LOW-level noise margins for +5 V CMOS and for +5 V TTL circuits by using the information in Figures 12.2.2.

Solution

For 5V CMOS,

$U_{IH(min)} = 3.5V$

$U_{IL(max)} = 1.5V$

$U_{OH(max)} = 4.4V$

$U_{OL(max)} = 0.33V$

$U_{NH} = U_{OH(min)} - U_{IH(min)} = 4.4V - 3.5V = 0.9V$

$U_{NL} = U_{IL(max)} - U_{OL(max)} = 1.5V - 0.33V = 1.17V$

For TTL,

$U_{IH(min)} = 2V$

$U_{IL(max)} = 0.8V$

$U_{OH(min)} = 2.4V$

$U_{OL(max)} = 0.4V$

$U_{NH} = U_{OH(min)} - U_{IH(min)} = 2.4V - 2V = 0.4V$

$U_{NL} = U_{IL(max)} - U_{OL(max)} = 0.8V - 0.4V = 0.4V$

It can be seen from the results of Example 12.1, the noise margin of a CMOS gate is greater than that of a TTL gate. This means that CMOS family has stronger noise immunity than TTL family. That is to say, CMOS family is more suitable for a high-noise environment than TTL family.

12.2.4 Power dissipation

The power consumed by a logic gate depends on a number of factors, including its internal structure, the input signals, the devices that it drives, and how often its output changes between LOW and HIGH. Figure 12.2.3 shows the current drawn from the DC supply voltage in a NAND gate. When the output of a logic gate is a HIGH, the DC supply voltage source offers an amount of current designated by I_{CCH}. At that time, the power dissipation is the product of the DC supply voltage, V_{CC}, and I_{CCH}. When the output of a logic gate is a LOW, the DC supply voltage source offers a different amount of current designated by I_{CCL}. At this situation, power dissipation is the product of V_{CC} and I_{CCL}.

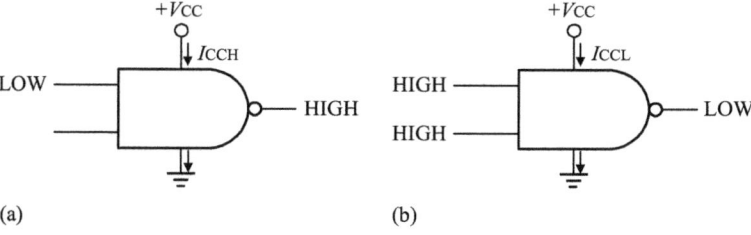

Figure 12.2.3: Current from the DC supply in a HIGH (a) and a LOW output state (b)

When a pulse signal is applied to the input, its output switches back and forth between HIGH and LOW, and the amount of supply current varies between I_{CCH} and I_{CCL}. The average power dissipation depends on the duty cycle of the pulse signals. If the duty cycle is 50%, then the average supply current is therefore

$$I_{CC} = \frac{I_{CCH} + I_{CCL}}{2} \tag{12.2.3}$$

The average power dissipation is

$$P_D = V_{CC}I_{CC} \tag{12.2.4}$$

The operating frequency also affects the power dissipation of a logic gate. Figure 12.2.4 shows the power dissipation versus frequency. For a TTL circuit, the power dissipation is essentially constant over its operating frequency range. But for a CMOS circuit, the power dissipation is frequency dependent. Power dissipation is extremely low under static (DC) conditions and increases as the frequency increases under dynamic operation. For example, the power dissipation of a low-power Schottky (LS) TTL gate is a constant of 2.2 mW. The power dissipation of an HCMOS gate is 2.75 μW under static conditions and 170 μW at 100 kHz.

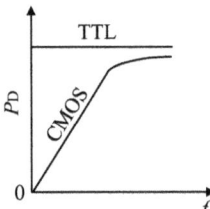

Figure 12.2.4: Power versus frequency curves for TTL and CMOS.

12.2.5 Propagation delay time

When a signal propagates through a logic circuit, it always experiences a time delay, as illustrated in Figure 12.2.5. When the input level has a transition from HIGH to LOW

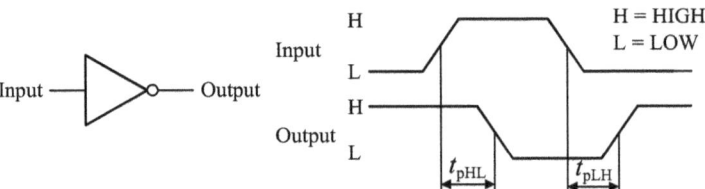

Figure 12.2.5: A basic illustration of propagation delay time.

for an inverter, the output level changes from LOW to HIGH after a short delay time called the *propagation delay time* and denoted by t_{pd}.

There are two propagation delay times specified for logic gates:

t_{pHL}: The delay time between an input change and the corresponding output change from HIGH to LOW.

t_{pLH}: The delay time between an input change and the corresponding output change from LOW to HIGH.

These propagation delay times are determined by the 50% points on the pulse edges used as references. The propagation delay time t_{pd} of a logic circuit can be determined by the average time of t_{pHL} and t_{pLH}

$$t_{pd} = \frac{t_{pHL} + t_{pLH}}{2}$$

(12.2.5)

The propagation delay time of a gate limits its operating frequency. Generally, the greater the propagation delay time is, the lower the maximum frequency is. Thus, a higher speed circuit is one that has a smaller propagation delay time. For example, a gate with a delay of 10 ns is faster than one with a delay of 40 ns.

12.2.6 Speed-power product

Propagation delay time and power dissipation are two important parameters for logic circuits. Speed-power product is defined as the product of propagation delay time and power dissipation. It provides a basis for the comparison of logic circuits and allows you to simultaneously consider the effect of the propagation delay time and the power dissipation with only one parameter in the selection of the type of logic gates to be used in a certain application. The lower the speed-power product is, the better the performance of logic circuits is. The unit of speed-power product is the picojoule (pJ). For example, HCMOS has a speed-power product of 1.2 pJ at 100 kHz while LS TTL has a value of 22 pJ.

12.2.7 Loading and fan-out

This refers to the number and the type of inputs that are connected to a given output. As shown in Figure 12.2.6, the output of a NAND gate G_1 is connected to the same type inputs of three NAND gates. Gate G_1 is called as a *driving gate*, and Gates G_2, G_3, G_4, are called as the *load gates*. Generally, there is a limit to the number of load gate inputs that a given gate can drive. This limit is called the fan-out of the gate. *Fan-out* is the maximum number of load gate inputs that can be connected to a certain output without affecting the specified operational characteristics of the gate.

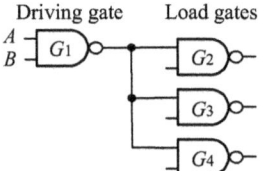

Driving gate   Load gates

Figure 12.2.6: Illustration of a driving gate and load gates.

1. CMOS Loading

Loading in CMOS differs from that in TTL because MOSFET in CMOS logic presents a predominantly capacitive load to the driving gate, as illustrated in Figure 12.2.7. In this case, the limitations are the charging and discharging times associated with the output resistance of the driving gate and the input capacitance of the load gates. When the output of the driving gate is HIGH, the input capacitance of the load gate is charging through the output resistance of the driving gate, as indicated in Figure 12.2.7(a). When the output of the driving gate is LOW, the capacitance is discharging, as indicated in Figure 12.2.7(b).

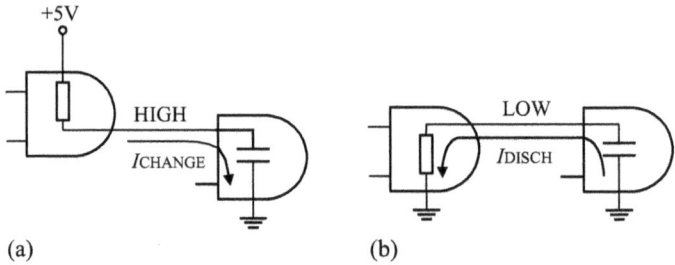

(a)   (b)

Figure 12.2.7: Capacitive loading of a CMOS gate: (a) charging; (b) discharging.

When more load gate inputs are added to the driving gate output, the total capacitance increases because the input capacitances effectively appear in parallel. The increase in capacitance results in the long charging and discharging times, thus reduces the maximum frequency at which the gate can be operated. Therefore, the fan-out of a CMOS gate depends on the operation frequency. The fewer the load gate input is, the greater the maximum operation frequency is.

2. TTL Loading

Loading in TTL presents a resistive load to the driving gate, as illustrated in Figure 12.2.8.

As more load gates are connected to the driving gate, the loading on the driving gate increases. For a TTL driving gate in a HIGH output state, each load gate input need a current I_{IH} and thus the total source current I increases with the number of

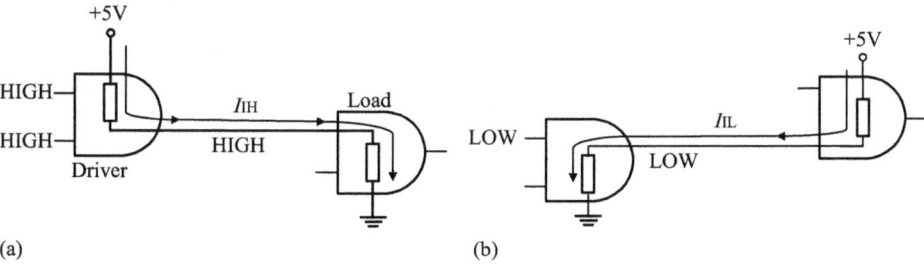

Figure 12.2.8: Illustration of current sourcing (a) and current sinking (b) in logic gates.

load gate inputs, as illustrated in Figure 12.2.9(a). With the increase of total source current I, the voltage drop on the internal resistor of driving gate becomes high, causing the output U_{OH} drops below $U_{OH(min)}$. Due to the output U_{OH} out of high level noise margin, thus the logic operation may be wrong. In addition, as the total source current increases, the power dissipation of the driving gate increases, accordingly.

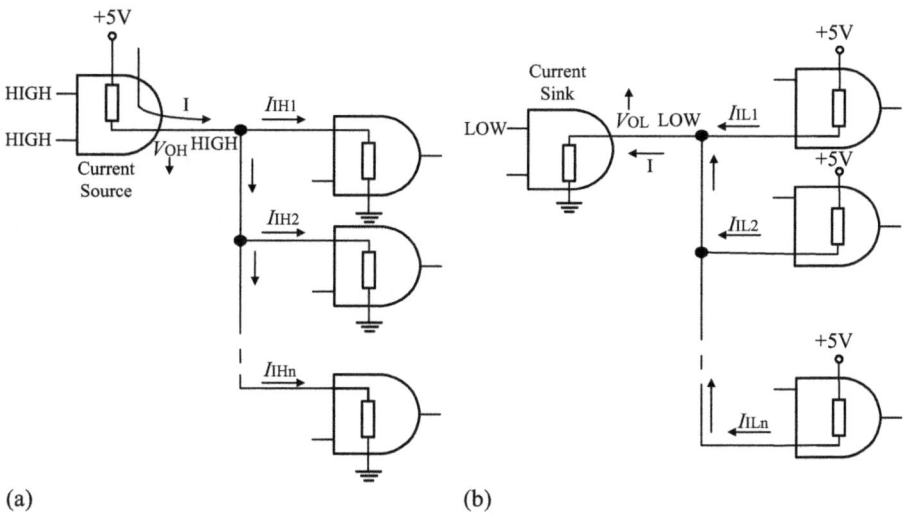

Figure 12.2.9: Illustration of a TTL gate driving more load gates in a HIGH (a) and LOW output state (b).

For a TTL driving gate in a LOW output state, each load gate input injects a current I_{IL} sinking into the driving gate and thus the total sink current I rises with the increase of load gates, as illustrated in Figure 12.2.9(b). As the total sink current increases, the voltage drop on the internal resistor of the driving gate increases, causing U_{OL} increase. If an excessive number of load gates are added, U_{OL} exceeds $U_{OL(max)}$. This may cause logic wrong.

One input of the same logic family for the driving gate is called a *unit load*. For example, the low-power Scottky (LS) TTL has a fan-out of 20 unit loads. In TTL, the current-sinking capability (LOW state) is the main limit factor in determining the fan-out.

12.3 Diode logic

Bipolar logic families use semiconductor diodes and bipolar junction transistors as the basic building blocks of logic circuits. The simplest bipolar logic elements use diodes and resistors to complete logic functions, therefore, called *diode logic*. This section introduces the basic operation of the diode logic.

The objectives of this section are to
-   Understand the current switch characteristics of a diode.
-   Explain how AND logic function can be implemented with diode logic.

12.3.1 Diode

In analogic circuit, you have learned the transfer characteristic of a diode. If the anode-to-cathode voltage, V, is negative, the diode is said to be reverse biased and the current I through the diode is near zero. At this moment, the diode acts like an open circuit. If the V is positive, the diode is said to be forward biased and the current I can be a large positive value. The diode acts like a short circuit. This means that the diode acts like a switch element that is controlled by the anode-to-cathode voltage.

A real diode has a resistance that is less than infinity when reverse biased and greater than zero when forward biased. Therefore, the forward-biased voltage is usually greater than or equal to 0.7 V and the reverse-biased voltage is considered with the value less than 0.7 V, as shown in Figure 12.3.1.

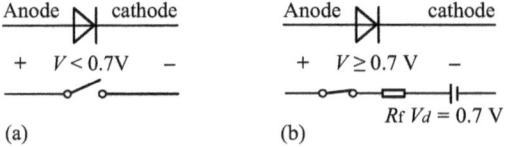

Figure 12.3.1: Model of a real diode: (a) reverse biased; (b) forward biased.

12.3.2 Diode logic

Diodes can be used to perform the logic operation. Figure 12.3.2 shows a diode AND gate with two diodes and one resistor. In this circuit, there are two inputs A and B, one output P with a 5 V power supply voltage Vcc.

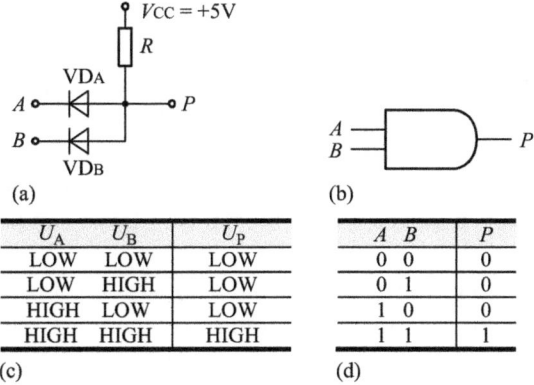

(a)

(b)

U_A	U_B	U_P
LOW	LOW	LOW
LOW	HIGH	LOW
HIGH	LOW	LOW
HIGH	HIGH	HIGH

A	B	P
0	0	0
0	1	0
1	0	0
1	1	1

(c)

(d)

Figure 12.3.2: AND gate constructed with Diode: (a) circuit diagram; (b) logic symbol; (c) function table; (d) truth table.

Suppose that the diode voltage drop is 0.7 V at the forward biased state. When the input A or B is connected, the LOW voltage of 0.3 V, diode VD_A or VD_B is forward biased and thus the voltage of the output P is limited to 1 V. When the input A and B are all connected to the HIGH voltage of 5 V, diode VD_A and VD_B are reverse biased and thus the voltage of the output P is approximate to the supply voltage.

Assume that 5 V is a HIGH that is represented by logic 1, and the voltage less than 1 V is considered to be LOW denoted by logic 0, the function table and truth table can be summarized in Figure 12.3.2 (c) and (d). It can be found that the output P is the result of AND operation of input A and input B and the corresponding logic symbol is given in Figure 12.3.2 (b).

12.4 Transistor-transistor logic

The most commonly used bipolar logic family is transistor-transistor logic (TTL). Actually, there are several TTL families with different speed, power consumption, and other characteristics. This section introduces some typical TTL gate circuits, which include TTL NAND gates, TTL gates with open-collector output and tristate TTL gates. Also TTL families are briefly introduced.

The objectives of this section are to

- Describe the switch action of a bipolar junction transistor.
- Describe how basic TTL gates operate at the component level
- Recognize the difference between TTL totem-pole output and TTL open-collector output
- Connect circuits in a wired-AND configuration
- Describe the operation of a gate with tristate output
- Properly deal with unused gate inputs

12.4.1 Transistor logic

In digital switching application, bipolar transistors are often operated at either cutoff or saturation acting like a current-controlled switch. If a small current is put into the base, the switch is "on" and current flows between the collector and the emitter. If no current is put into the base, the switch is "off" and no current flows between the collector and the emitter. The switching characteristics of bipolar transistors can be used to construct a simple logic.

Figure 12.4.1 shows a transistor logic circuit with a *npn* transistor using a *common-emitter configuration*. Except for the single *npn* transistor, there are two resistors R_b and R_c in the configuration. The switch characteristics of the circuit are explained as the following.

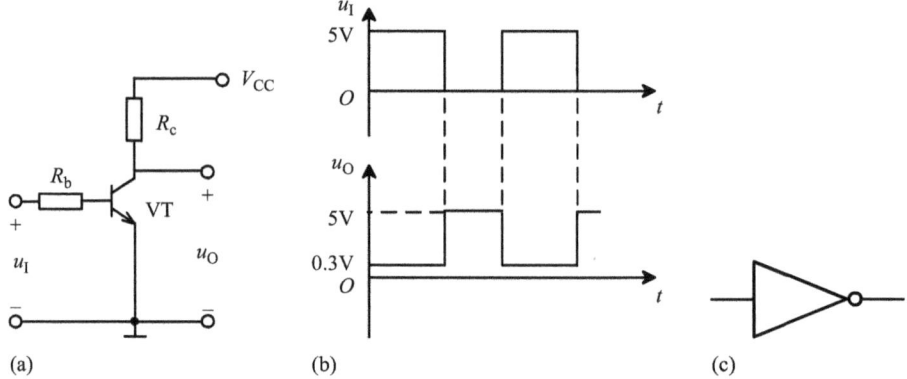

Figure 12.4.1: Transistor logic with common-emitter configuration of a npn transistor: (a) circuit diagram; (b) waveform diagram; and (c) logic symbol.

If u_I is LOW with the voltage less than 0.7 V, then the base-to-emitter diode junction is reverse biased. Therefore, no base current (I_b) and thus no collector current (I_c) flows between the collector and the emitter. At this moment, the transistor is said to be cut off (OFF); the emitter and the collector is disconnected i.e. open. And the output voltage u_O is approximate to the supply voltage (V_{CC}); the output is in a HIGH state.

If u_I is HIGH with the voltage of 5 V, then the base-to-emitter diode junction is forward biased. As a result, a biased current is injected into the base and thus the transistor operates in saturation region. By choosing suitable values of resistor R_b and R_c, the transistor is said to be saturated (ON) and hence the output voltage $u_O \approx U_{CES}$ i.e. 0.3 V. Thus the output u_O is LOW. At this situation, the saturated threshold of collector current I_{CS} can be calculated by

$$I_{CS} = \frac{V_{CC} - U_{CES}}{R_C} \approx \frac{V_{CC}}{R_C}.$$ (12.4.1)

The corresponding saturated threshold of base current I_{BS} is

$$I_{BS} = \frac{I_{CS}}{\beta} \approx \frac{V_{CC}}{\beta R_c}$$ (12.4.2)

where β is the gain coefficient of the transistor.

For the common-emitter configuration, the transistor starts to enter into the saturation region when the base current i_B increases to the saturated threshold of base current I_{BS}.

In terms of the relation between the output u_O and the input u_I, the waveform of output u_O can be drawn as shown in Figure 12.4.1 (b). When the input voltage is LOW, the output voltage is HIGH, and vice versa. Obviously, the transistor logic in Figure 12.4.1 is an inverter.

12.4.2 Basic TTL NAND gate

1. Circuit
The circuit diagram of a 2-input TTL NAND gate, part of number 7400-series in standard TTL family, is shown in Figure 12.4.2. The circuit is consisted of three stages i.e. input stage, phase splitter and output stage.

The input stage includes a multiple-emitter transistor VT_1, a resistor R_1 and two clamp diodes VD_1 and VD_2. The multiple-emitter transistor contains two base-emitter junctions and one base-collector junction. It can be compared to the diode arrangement, as shown in Figure 12.4.3. Obviously, the input stage forms a diode AND gate and implement AND operation of input A and input B. Clamp diodes VD_1 and VD_2 do nothing in normal operation, but limit undesirable negative excursion on two inputs A and B.

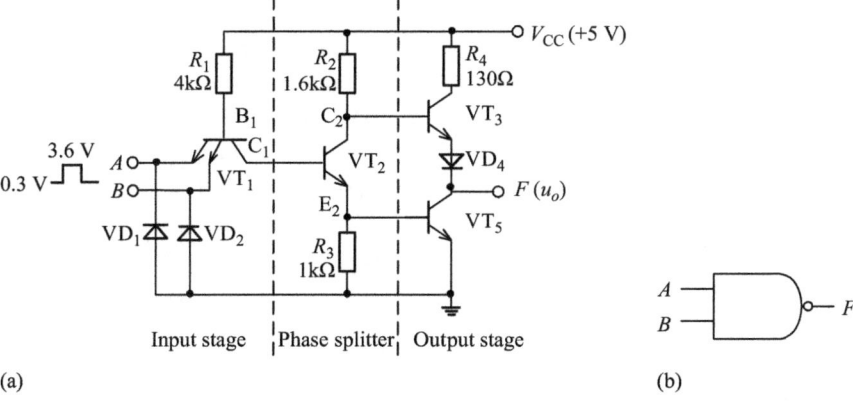

(a)

(b)

Figure 12.4.2: Two-input standard TTL NAND gate: (a) circuit diagram; (b) logic symbol.

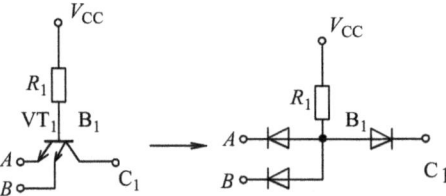

Figure 12.4.3: Multiple-emitter transistor and its analogical structure of diode arrangement.

Transistor VT_2 and its surrounding resistors R_2, R_3 form a phase splitter that offers two opposite voltage levels to control the output stage. The phase splitter is a basic common-emitter amplifier as an inverting buffer amplifier. Depending on whether V_{C1} is at a "LOW" or a "HIGH" voltage, VT_2 is either cutoff or turned on.

The output stage has two transistors VT_3 and VT_4, only one of which is on at any time. This TTL output stage is called a *totem-pole or push-pull output*, which is helpful for improving the switching speed and the load capacity. Diode VD_4 ensures that VT_3 will turn off when VT_2 is on.

2. Operational Principle

For TTL AND gate, the ideal voltage level of LOW is 0.3 V and that of HIGH is 3.6 V. When all inputs are HIGH, two base-emitter junctions of VT_1 are reverse biased, and the base-collector junction of VT_1 is forward biased. The supply voltage V_{CC} can offer a current i_{B2} through R_1 and the base-collector junction of VT_1 into the base of VT_2. This current can drive VT_2 into saturation. Then VT_2 goes on driving VT_5 into saturation and thus the output u_O is LOW ($u_O = U_{CES} \approx 0.3V$). At the same time, the potential of the collector of VT_2 to the ground is equal to the sum of the potential of the emitter of VT_2 to ground and the saturated voltage drop U_{CES2} of collector-emitter of VT_2 i.e. $u_{C2} = u_{E2} + U_{CES2} \approx 0.7 + 0.3 = 1V$. The voltage level of 1 V is too low to simultaneously keep VT_3 and VD_4 on. So VT_3 and VD_4 are cut off.

When at least one input is LOW (0.3 V), the corresponding base-emitter junction of VT_1 is forward biased and the base-collector junction of VT_1 is reverse biased. The supply voltage V_{CC} can offer a current I_{IL} through R_1 and the base-emitter junction of VT_1 to the LOW input. Then a LOW input offers a current path to ground. So the potential of the base of VT_1 to ground is 1 V i.e. $u_{B1} = 0.3 + 0.7 = 1V$. The voltage level of 1 V is too low to simultaneously keep the base-collector of VT_1, the base-emitter of VT_2 and VT_5 on. There is no current into the base of VT_2, so VT_2 is off. The supply voltage V_{CC} can offer a current through R_2 into the base of VT_3. This current can drive VT_3 into saturation. A saturated VT_3 provides a low-resistance path from V_{cc} to the output. The output voltage level u_o can be deduced as below:

$$u_o \approx V_{cc} - u_{BE3} - u_{D4} = 5 - 0.7 - 0.7 = 3.6V \tag{12.4.3}$$

We therefore have a HIGH output for at least one LOW on the inputs. At the same time, the emitter of VT2 is at ground potential, keeping VT5 off.

The function operation of the TTL NAND gate is summarized in Table 12.4.1. The corresponding truth table is given in Table 12.4.2. It can be seen from Table 12.4.2 that the gate does indeed perform the NAND function. The corresponding logic symbol is shown in Figure 12.4.2(b). TTL NAND gate can be designed with any desired number of inputs simply by increasing the number of base-emitter junction of VT1.

Table 12.4.1: Function table.

A	B	V_A	V_B	VT_2	VT_3	VT_5	V_P	P
L	L	0.3V	0.3 V	off	on	off	3.6 V	H
L	H	0.3V	3.6 V	off	on	off	3.6 V	H
H	L	3.6V	0.3 V	off	on	off	3.6 V	H
H	H	3.6V	3.6 V	on	off	on	0.3 V	L

Note: LOW(L), HIGH(H)

Table 12.4.2: Truth table.

A	B	P
0	0	1
0	1	1
1	0	1
1	1	0

3. Special characteristics and parameters

Section 12.2 already introduces basic operational properties and parameters, such as voltage levels, noise immunity, power dissipation, fan-out and propagation delay time. Here, we further emphasize several special characteristics and parameters of TTL NAND gate.

(1) Transfer characteristics

Transfer characteristics refer to the relation between the output voltage and input voltage. The relation can be measured with the circuit as shown in Figure 12.4.4(a) and the measured result is shown in Figure 12.4.4(b).

Transfer characteristics curve in Figure 12.4.4 can be divided into four segments. **Segment AB:** $u_{B1} < 1.3$ V when $u_I < 0.6$ V, then VT_2 and VT_5 cutoff. The output is HIGH, i.e. $u_O = 3.6$ V. Therefore, segment AB is called cutoff region or NAND gate off region.
Segment BC: 1.3 V $\leq u_{B1} < 2.1$ V when 0.6 V $\leq u_I < 1.3$ V, then VT_2 on and VT_5 off. At this condition, part of input current begins to flow into the base of VT_2. This causes VT_2 entering into amplifier region and thus u_{C2} falls with u_{B1} rises. Therefore, segment BC is called linear region.

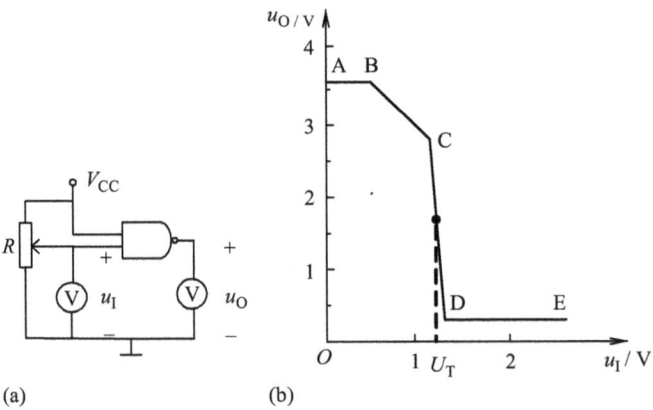

(a) (b)

Figure 12.4.4: Measurement circuit (a) and transfer characteristics curve (b) of a TTL NAND Gate.

Segment CD: When $u_I \geq 1.3$ V, the output voltage falls rapidly even if input voltage slightly rises. At this moment, VT_5 starts turning on and VT_2 does not enter into saturation. That result in VT_2, VT_3 and VT_5 all entering the amplifier region. Hence, with slight rising of u_I, u_O will fall rapidly to low-level. Segment CD is usually called the transition region.

Segment DE: When u_I continues rising, both VT_2 and VT_5 are on, and VT_3 and VD_4 are off. Therefore, the output is in a LOW state i.e. $u_O = 0.3$ V. At this situation, the output voltage would keep constant in spite of the rise in input voltage.

From the transfer characteristics curve, you can obtain some key voltage parameters, such as $U_{IL(max)}$, $U_{IH(min)}$, U_{NH} and U_{NL}. For a standard TTL NAND gate, $U_{OH(MIN)} = 2.4$ V and $U_{OLMAX} = 0.4$ V. $U_{IL(max)}$ is the maximum of input low level voltage which can ensure the output in a HIGH state, i.e, $u_O > U_{OH(MIN)}$. $U_{IH(min)}$ is a minimum of input high level voltage which can allow the output in a LOW state, i.e. $u_O < U_{OLMAX}$. From the transfer characteristics curve, you can also get the parameter values $U_{IL(max)} = 0.8$ V and $U_{IH(min)} = 1.8$ V. From the practical effects of engineering, a redundancy should be considered to guarantee the correct implementation of logic function. Parameters of commercial products are $U_{IL(max)} = 0.8$ V and $U_{IH(min)} = 2$ V. So

$$U_{NH} = U_{OH(MIN)} - U_{IH(min)} = 0.4V$$

$$U_{NL} = U_{IL(max)} - U_{OLMAX} = 0.4V.$$

Another key voltage is threshold voltage, U_T, which generally refers to the input voltage corresponding to the middle point of transition region i.e. segment CD. For standard TTL NAND gate, $U_T = 1.3 \sim 1.4$ V. Threshold voltage can be used to approximately analyze operation of the gate. If $u_I > U_T$, VT_5 is on and the output is a LOW level. If $u_I < U_T$, VT_5 is off and the output is a HIGH level.

(2) Input characteristics

Input characteristics refer to the relation between input current (i_i) and input voltage (u_i), which can be described by the function equation of $i_i=f(u_i)$ as shown in Figure 12.4.5. The relation is very important for considering TTL loading. The current flowing into a TTL input is defined to be positive while the current flowing out of a TTL input is defined to be negative.

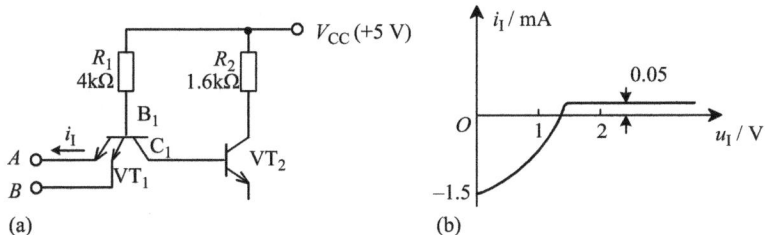

Figure 12.4.5: input circuit part (a) and characteristics curve (b) of the TTL NAND gate.

The amount of current required in a TTL input depends on which the input is LOW, HIGH, or connected to ground (short circuit), and is specified by the following three parameters.

I_{IS} refers to the input current when an input is connected to ground i.e. $u_I = 0$V. If the input A in Figure 12.4.4 (a) is connected to the ground, there is a current (I_{IS}) from V_{cc}, through R_1, through base-emitter of VT$_1$, out of input A into the ground. Due to the current flowing out of the input lead, I_{IS} is a negative value, which can be calculated by

$$I_{IS} = -\frac{V_{CC} - U_{BE1}}{R_1} = -\frac{5-0.7}{4} \approx -1.1\text{mA} \tag{12.4.4}$$

I_{IL} is the input current when an input is in a LOW state i.e. $u_I = 0.3$ V. If the input A in Figure 12.4.4 (a) is a LOW. I_{IL} is from V_{cc}, through R_1, through base-emitter of VT$_1$, out of input A, and then through the driving gate to ground. Since the current flows out of the input lead, I_{IL} is a negative value which can be deduced as follows:

$$I_{IL} = -\frac{V_{CC} - U_{BE1} - 0.3}{R_1} = -\frac{5-0.7-0.3}{4} = -1.0\text{mA} \tag{12.4.5}$$

Obviously, the absolute value of I_{IL} is slightly less than that of I_{IS}. In general case, I_{IS} can be instead of I_{IL} for an approximate analysis. When the TTL gate works as a load gate, I_{IL} is the current sinking to the driving gate.

I_{IH} refers to the input current when the inputs are HIGH i.e. $u_I = 3.6$ V. I_{IH} comes from the driving gate, through the input leads and flows into VT$_1$. Generally, I_{IH} has a very small value (~40 µA for a standard TTL gate). When the TTL gate works as a load gate, I_{IH} is the current sourcing from the driving gate.

(3) Fan-out

As we defined it in Section 12.2.5, *fan-out* is the maximum number of load gate inputs that can be connected to and driven by a single output without affecting the specified operational characteristics of the gate.

TTL driving gate can source or sink a certain amount of current depending on the output state, HIGH or LOW. I_{OLmax} is the maximum current which a driving gate can sink in LOW state and maintain an output voltage no more than U_{OLmax}. Since current flows into the output, I_{OLmax} has a positive value. I_{OHmax} is the maximum current which an output can source in HIGH state and maintain an output voltage no more than U_{OHmin}. Since current flows out of the output, I_{OHmax} has a negative value. Therefore, fan-out is denoted by N_{OH} when a driving gate is in a HIGH output state and by N_{OL} when a driving gate is in LOW output state. They can be deduced from

$$N_{OH} = \left| \frac{I_{OHMAX}}{I_{IHMAX}} \right| \tag{12.4.6}$$

$$N_{OL} = \left| \frac{I_{OLMAX}}{I_{ILMAX}} \right| \tag{12.4.7}$$

where I_{ILmax} is the maximum of I_{IL} and I_{IHmax} is the maximum of I_{IH}.

In order not to affect the specified operational characteristics of the gate, the *overall fan-out* should be the minimum integer of N_{OH} and N_{OL} i.e. $N_O = \min\{N_{OH}, N_{OL}\}$. Since I_{ILmax} is two orders of I_{IHmax}, the fan-out of a TTL driving gate is mainly determined by N_{OL}. Generally, the fan-out of standard TTL circuit is ten.

Example 12.2 Determine the fan-out of a TTL gate with the following current parameters: $I_{IS} = -1.4$ mA, $I_{IHMAX} = 0.02$ mA, $I_{OLMAX} = 15$ mA and $I_{OHMAX} = -0.4$ mA

Solution

Due to $I_{IL} \approx I_{IS} = -1.4$ mA, so

$$N_{OL} = \left| \frac{I_{OLMAX}}{I_{ILMAX}} \right| = \frac{15\text{mA}}{1.4\text{mA}} = 10.7$$

The integer of 10 should be reserved.

$$N_{OH} = \left| \frac{I_{OHMAX}}{I_{IHMAX}} \right| = \frac{0.4}{0.02} = 20$$

$$N_O = \min\{N_{OH}, N_{OL}\} = 10$$

Loading a TTL output with more than its rated fan-out has deleterious effects. Noise margins may be reduced, transition times and delays may increase, and the device may overheat.

---

(4) Input load characteristics

It is a common thing to connect a resistor between the input and the ground for practical engineering. This would result in the change of input voltage due to the introduction of input resistor, as shown in Figure 12.4.6 (a). When the input current flows through R_i, the input voltage (u_i) is formed. Within a certain range, u_i would

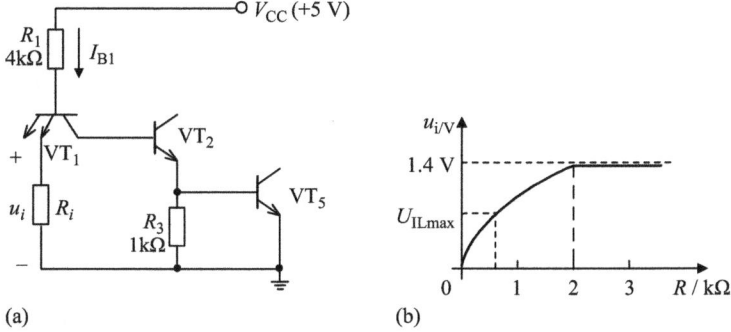

(a)   (b)

Figure 12.4.6: Input load characteristics of standard TTL NAND gate: (a) input load circuit; (b) characteristics.

increase with increase in R_i. Figure 12.4.6(b) shows the curve of the input voltage as the function of the input resistor i.e. $u_i = f(R_i)$, which is called the curve of input load characteristics.

It can be found that u_i is approximately proportional to R_i for small R_i. When $u_i \leq U_{ILmax}$, the output of TTL NAND gate stay steadily in a HIGH state. VT_2 and VT_5 are off, thus the requirement of R_i can be deduced from

$$u_i = \frac{R_i}{R_i + R_1}(V_{CC} - u_{BE1}) \leq U_{ILmax} \tag{12.4.8}$$

Therefore,

$$R_i \leq \frac{U_{ILmax}}{V_{CC} - u_{BE1} - U_{ILmax}}R_1 \tag{12.4.9}$$

The resistor calculated from eq. (12.4.9) is written as R_{off}. For example, if $U_{ILmax} = 0.8$ V, then $R_{off} \leq 0.9$ kΩ. That is to say, only if $R_i \leq R_{off}$, TTL NAND gate can stay steadily at a HIGH output state.

If the value of R_i increases further, u_i continues increasing. When u_i increases to 1.4 V, VT_2 and VT_5 are both into saturation and thus the potential of the base of VT_1 (u_{B1}) is confined to 2.1 V. Even if R_i continues increasing, u_i stays at the fixed value of 1.4 V. When $u_i = 1.4$ V, the corresponding R_i is written as R_{on}. For standard TTL NAND gate in Figure 12.4.5, R_{on} can be deduced from

$$u_i = \frac{R_i}{R_i + R_1}(V_{CC} - u_{BE1}) = 1.4\text{V} \tag{12.4.10}$$

Therefore, $R_{on} = 2$ kΩ. Only if $R_i \geq R_{on}$, TTL NAND gate can stay at a LOW output state.

If the inputs of TTL NAND gate are empty, the current flows from the power supply V_{CC}, through R_1 into the VT_2 and VT_5. That makes VT_2 and VT_5 into saturation and thus the output is LOW. At the same time, the potential of the base of VT_1 is confined to 2.1 V. If you measure the voltage of empty input with a multimeter, which is equivalent to connecting a high resistor between the empty input and ground, $u_i = 2.1-0.7 = 1.4$ V which is equivalent to a HIGH input level.

Example 12.3 Determine the output expression of F_1, F_2 and F_3 for TTL NAND gate circuits as shown in Figure 12.4.7.

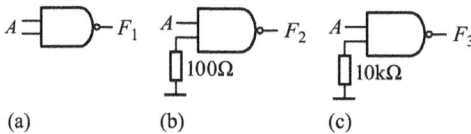

(a)   (b)   (c)

Figure 12.4.7: TTL NAND gate circuits

Solution

For the TTL NAND gate as shown in Figure 12.4.7(a), there is an empty input. This is equivalent to a HIGH input level, thus $F_1 = \overline{A \cdot 1} = \overline{A}$.

For the TTL NAND gate as shown in Figure 12.4.7(b), a resistor of 100Ω, less than R_{off} (900Ω), is connected between an input and the ground. This is equivalent to input a LOW level, so $F_2 = \overline{A \cdot 0} = 1$.

For the TTL NAND gate as shown in Figure 12.4.7(c), $R_i = 10$ kΩ, greater than R_{on} (2 kΩ). This input is equivalent to a HIGH level, thus $F_3 = \overline{A \cdot 1} = \overline{A}$.

12.4.3 TTL family

Since 1963, the original TTL family of logic gates was introduced by Sylvania. It was popularized by Texas Instruments, whose 7400-series for gates and other TTL components quickly became an industry standard. TTL families have evolved over the years in response to the demands of digital designers for better performance [20].

The earliest TTL family is *74-series* TTL. Through changing the resistor values in 74-series TTL, 74H (*high-speed TTL*) and 74L (*Low-power TTL*) were developed. The 74H family uses lower resistor values to reduce propagation delay at the expense of an increase in power consumption. The 74L family uses higher resistor values to reduce power consumption at the expense of an increase of propagation delay. Due to the contradiction between propagation delay and power consumption, three TTL families have gone. Today, there have been five surviving families which are compatible as they use the same power supply voltage and logic levels, but each family has its own advantages in terms of speed, power consumption and cost.

Aiming at solving the contradiction between propagation delay and power consumption, the first family to make use of Schottky transistors was 74S (*Schottky TTL*). With Schottky transistors and low resistor values, this family has much higher speed, but higher power consumption, than the original 74-series TTL. After 74S, 74LS (*Low power Schottky TTL*) was developed. By combining Schottky transistors with higher resistor values, 74LS TTL matches the speed of 74series TTL but has about one-fifth of its power consumption.

With the development of IC processing, circuit innovations gave rise to two more Schottky logic families. The 74AS (*Advanced Schottky TTL*) family offers speeds approximately twice as fast as 74S with approximately the same power consumption. The 74ALS (*Advanced Low-power Schottky TTL*) family offers both lower power and higher speed. Its delay time is about one third of 74-series TTL and power consumption is one eighth of 74-series TTL. The 74F (*Fast TTL*) family is positioned between 74AS and 74ALS in the speed/power tradeoff and is probably the most popular choice for high-speed requirements in new TTL design. Till date, 74LS is still a preferred logic family for TTL designs. Table 12.4.3 lists delay time, power consumption and speed-power product of five TTL families.

Table 12.4.3: Parameters of the typical TTL gates.

Family	Delay time/ns	Power dissipation/mW	Speed-power product/pW·s
74	10	10	100
74S	3	20	60
74LS	9	2	18
74ALS	4	1	4
74F	2.7	4	11

Table 12.4.4, 12.4.6 list some product parameters of several common used TTL NAND gates so that you can overall understand the parameters of TTL circuits. 7400: four two-input NAND gates; 7404: six inverts; 7410: three three-input NAND gates; 7420: two four-input NAND gates; 7430: one eight-input NAND gate. The 54-series TTL listed in the table is identical to the 74-series, except that it is specified to operate over the full military temperature and voltage range, and it is more expensive than 74-series.

12.4.4 Open-collector gate

TTL gate circuits introduced in the previous sections all have the totem-pole output. Another type of output available in TTL integrated circuits is the *open-collector* output.

Table 12.4.4: Product parameters of TTL gates.

parameter	Symbol	54/74-series	Standard 54/74 TTL 00,04,20,30			54/74LS TTL 00,04,10,20,30			Unit
			MIN	NOM	MAX	MIN	NOM	MAX	
Supply voltage	U_{CC}	54	4.5	5	5.5	4.5	5	5.5	V
		74	4.75	5	5.25	4.75	5	5.25	
high-level output current	I_{OH}	54/74		-400				-400	µA
Low-level output current	I_{OL}	54			16		8	4	mA
		74		16			8		
Operating temprature	T_A	54	-55		125	-55		125	°C
		74	0		70	0		70	
Low-level input voltage	U_{IL}	54			0.8			0.7	V
		74			0.8			0.8	
High-level input voltage	U_{IH}	54/74	2			2			V
Low-level output voltage	U_{OL}	54		0.2	0.4		0.25	0.4	V
		74		0.2	0.4		0.25	0.5	
High-level output voltage	U_{OH}	54	2.4	3.4		2.5	3.4		V
		74	2.4	3.4		2.7	3.4		
Low-level input current	I_{IL}	54/74			-1.6			-0.4	mA
High-level input current	I_{IH}	54/74			40			20	µA
Short-circuit output current	I_{OS}	54	-20		-55	-20		-100	mA
		74	-20		-55	-20		-100	

Table 12.4.5: Current parameters of TTL gates (V_{CC} = 5 V, T_A = 25 °C)

model	I_{CCH}(mA)		I_{CCL}(mA)		I_{CC}(mA) 50% duty circle TYP
	TYP	**MAX**	**TYP**	**MAX**	
00	4	8	12	22	2
04	6	12	18	33	2
20	2	4	6	11	2
30	1	2	3	6	2
LS00	0.8	1.6	2.4	4.4	0.4
LS04	1.2	2.4	3.6	6.6	0.4
LS10	0.6	1.2	1.8	3.3	0.4
LS20	0.4	0.8	12	2.2	0.4
LS30	0.35	0.5	0.6	1.1	0.48

Table 12.4.6: Propagation delay time of TTL gates (V_{CC}=5V, T_A=25°C).

model	Test conditions	t_{PLH}(ns)		t_{PHL}(ns)	
		TYP	**MAX**	**TYP**	**MAX**
00	C_L=15pF R_L=400Ω	11	22	7	15
04,20		12	22	8	15
30		13	22	8	15
LS00,LS04		9	15	10	15
LS10,LS20	C_L=15pF R_L=2kΩ	91	15	10	15
LS30		8	15	13	20

1. Circuit

A standard TTL NAND gate with an open collector output is shown in Figure 12.4.8(a).

Notice that the output is the collector of transistor VT_5 with nothing connected to it, hence getting the name *open collector*. When VT_5 is on, the output is LOW. While VT_5 is off, there is no path to provide a HIGH output due to nothing connected to the collector of VT_5. Therefore, in order to get a proper HIGH logic level out of the circuit, an external resistor must be connected to V_{cc} or an external power supply voltage from the collector of VT_5, as shown with dotted line part in Figure 12.4.8. When VT_5 is off, the output is pulled up to the power supply through the external resistor and then gets a HIGH output state. This external resistor is also called *pull-up resistor*. The logic symbol that designates a NAND gate with open-collector output is shown in Figure 12.4.8(b).

Notice that the pull-up resistor can be also connected to an external supply voltage V_c which can be up to +30 V. Therefore, compared to TTL gates with totem-pole output, TTL gates with open-collector output have a higher power supply voltage and thus stronger current-handling capability. And, they are generally used for driving the

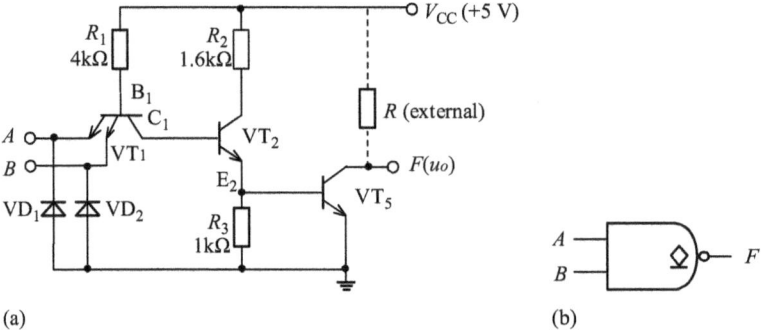

(a)   (b)

Figure 12.4.8: TTL NAND gate with open-collector output: (a) circuit diagram; (b) logic symbol.

heavy current loads, such as LEDs, lamps, and relays. Typically, the output current of an open-collector gate can reach to 40 mA. While the amount of current sinking in the LOW state (I_{OLmax}) is limited to 16 mA for standard TTL and 8 mA for LS TTL. Typical ICs of open-collector gates include the inverter 7406 and the buffer 7407.

2. Wired-AND operation

The outputs of open-collector gates can be wired together to form a *wired AND* configuration. Figure 12.4.9 illustrates that the outputs of two open-collector gates are wired together to implement a wired AND operation. A single external pull-up resistor, R_c, and an external power supply voltage, V_C (or V_{CC}), are required in all wired AND circuits.

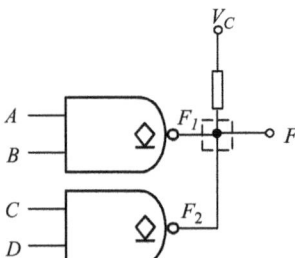

Figure 12.4.9: A wired AND operation of two open-collector gates.

Individual outputs of two open-collector gates are $F_1 = \overline{AB}$ and $F_2 = \overline{CD}$, respectively. When the outputs of two open-collector gates are wired together, $F = F_1 \cdot F_2 = \overline{AB} \cdot \overline{CD}$.

Notice that totem-pole outputs cannot be wired together because such a connection might produce excessive current and result in damage to the devices. For example, the outputs of two standard TTL gates (G_1 and G_2) are wired together in Figure 12.4.10.

When the F_1 output of G_1 is a HIGH state and the F_2 output of G_2 is LOW, there is a large current from V_{CC}, through R_4, VT_3 and VD_4 of G_1, and then through VT_5 of G_2 to ground. Such a large current might result in both low-level output of G_2 beyond the specification logic level and damage to the devices.

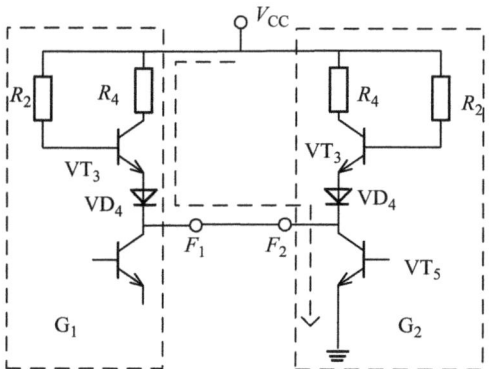

Figure 12.4.10: Totem-pole outputs wired together.

Example 12.4 Three open-collector AND gates are connected in a wired-AND configuration as shown in Figure 12.4.11. Assume that the wired-AND circuit is driving four standard TTL inputs with $I_{iL} = -1.6$ mA (each).

(a) Write the logic expression for X.

(b) Determine the minimum value of R if $I_{OL(max)}$ for each open-collector gate is 30 mA and $U_{OL(max)}$ is 0.4 V.

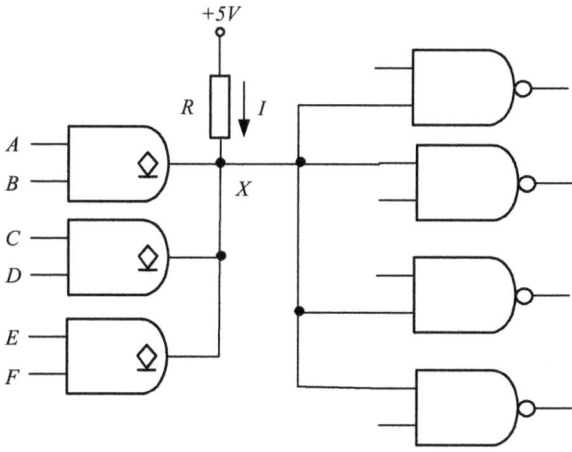

Figure 12.4.11: Logic diagram.

Solution

(a) Since three open-collector AND gates are connected in a wired-AND configuration, the final output, X, can be expressed by

$$X = ABCDEF$$

(b) When an open-collector AND gate is in a LOW output state, every load input sinks a I_{iL} into this driving gate and thus four load inputs sink $4I_{iL}$ into this driving gate.

$$4I_{IL} = 4 \times 1.6 \text{ mA} = 6.4 \text{ mA}$$
$$I = I_{OL(max)} - 4I_{IL} = 23.6 \text{ mA}$$

Therefore,

$$R = \frac{V_{CC} - U_{OL(max)}}{I} = 195\Omega$$

12.4.5 Tristate gates

The tristate gates are different from general gates. Besides a HIGH state and a LOW state, the outputs of the tristate gates include the third state called high-impedance state (**high-Z**). The tristate gates combine the advantage of totem-pole and open-collector output.

1. Circuit

Figure 12.4.12 illustrates a TTL tristate NAND gate. In fact, this tristate NAND gate is composed of a TTL inverter, a TTL NAND gate and a diode VD. The TTL inverter in dashed frame is the control circuit of the tristate gate. Its input, $\overline{EN}$, is called the enable input. The right part out of the dashed frame is a standard TTL NAND gate with three inputs.

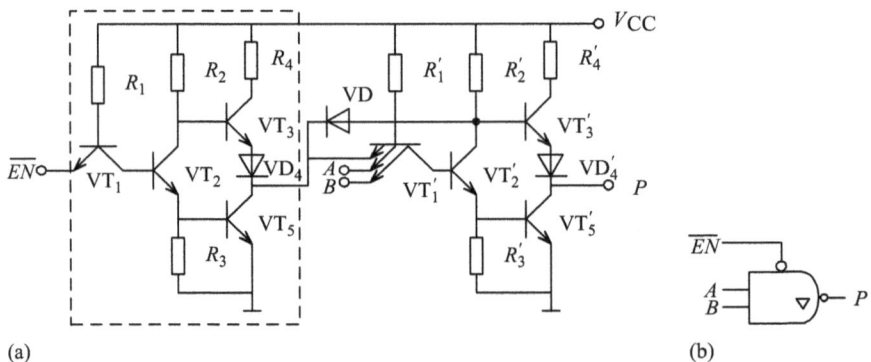

(a)   (b)

Figure 12.4.12: A tristate NAND gate with a LOW-level enable input: (a) circuit diagram; (b) logic symbol.

If a LOW is applied to $\overline{EN}$ input, the output of the inverter is a HIGH level, which is one input of the next NAND gate. At this situation, diode VD is reverse biased. The circuit implements the normal NAND operation i.e. $F = \overline{AB}$.

When a HIGH is applied to $\overline{EN}$ input, the output of the inverter is a LOW level (0.3 V). That causes VT'$_2$ and VT'$_5$ to be turned off and VD is forward biased. The potential of the base of VT'$_3$ is 1V, which is not enough to keep VT'$_3$ and VD'$_4$ on. Thus, VT'$_3$ and VD'$_4$ is also off. When both totem-pole output transistors are off, they are effectively open, and the output is completely disconnected from the internal circuitry. The output is in a high-impedance state written as **high-Z**. The logic symbol for the NAND gate with a tristate output is shown in Figure 12.4.13(b). The inverted triangle (∇) designates a tristate output.

The above tristate gate has a low-level enable input. It is possible to use high-level as the enable input. By omitting the inverter part in dashed frame, a tristate gate with a high-level enable input can be constructed as shown in Figure 12.4.13.

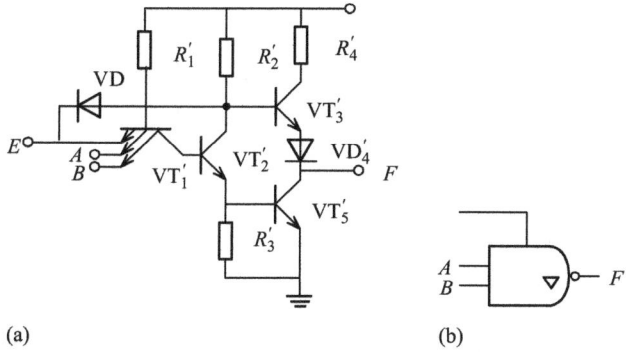

(a)   (b)

Figure 12.4.13: A tristate NAND gate with a high-level enable input: (a) circuit Diagram; (b) logic symbol.

2. Applications of the tristate gates

The tristate gate is an important device in digital circuits. It can be widely applied for computer system as bidirectional data transfer and data bus, as shown in Figure 12.4.14.

Figure 12.4.14 (a) shows the application of two tristate gates in the bidirectional data transfer. When the enable input, C, is a LOW, the tristate gate, G_1, is enabled and the tristate gate, G_2, is disabled. Data can be transmitted from the input A to the output B. While the enable input, C, is a HIGH, the tristate gate, G_2, is enabled and the tristate gate, G_1, is disabled. Data can be transmitted from the input B to the output A.

Figure 12.4.14 (b) shows the application of a group of tristate gates connected to data bus for transferring data between different digital devices. A group of tristate gates are connected to the same data bus. By controlling the sequence of each enable input, only one device can transfer data to data bus while other devices are in high impedance state. Therefore, all devices connected to data bus can operate and transfer data to data bus at time division mode by controlling the enable input of each tristate gate. This prevents data collision from different devices with each other on data bus.

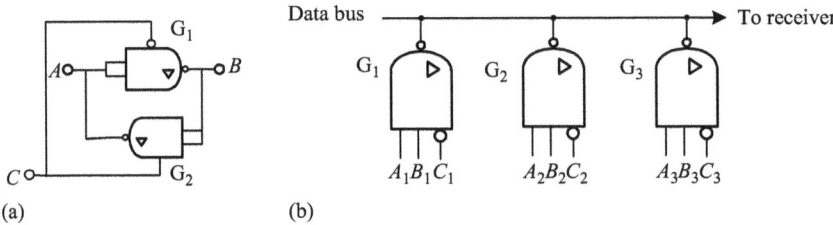

(a)   (b)

Figure 12.4.14: Application of the tristate gates: (a) bidirectional data transfer;
(b) application on data bus.

12.4.6 Unused TTL inputs

The rule of dealing with unused inputs is that the logic operation of gate circuits could not be affected. For a TTL gate, there are several ways to deal with unused inputs. The most commonly used method is to connect them to a used input of the same gate. For AND gates and NAND gates, all tied-together inputs count as a unit load in the LOW state. This is because the NAND gate uses a multiple-emitter input transistor; so no matter how many inputs are low, the total LOW-state current is limited to a fixed value. While for OR gates and NOR gates, each input tied to another input counts as a separate unit load in LOW state; this is because the OR gates and NOR gates use a separate transistor for each input; therefore, the LOW-state current is the sum of the currents from all the tied-together inputs. In the HIGH state, each tied-together input counts as a separate unit load for all types of TTL gates.

Also, unused inputs of AND gates and NAND gates can be connected to V_{CC} through a pull-up resistor. The connection pulls the unused inputs to a HIGH level. While unused inputs of OR gates and NOR gates can be connected to a LOW level or ground.

For TTL NAND gates, unused inputs can be left open since an unconnected input on a TTL gate acts as a HIGH level. This is already analyzed in the part of 12.4.2. However, due to noise sensitivity, it is best not to leave unused TTL inputs open. Another alternative method is to connect the unused input with any used input of the gate together.

12.5 CMOS logic circuits

MOS integrated circuits include three kinds of integrated circuits i.e. PMOS, NMOS and CMOS. Due to the higher speed and the lower power consumption, CMOS integrated circuits have already become the current mainstream of digital integrated circuits and analog-digital hybrid integrated circuits. They are widely applied to memory and microprocessor.

Basic internal CMOS circuitry and its operation are discussed in this section. The abbreviation CMOS stands for complementary MOS. The term complementary refers to two types of MOS transistors, that is, N-channel MOSFET (MOS field-effect transistor) and p-channel MOSFET that are used in pairs.

The objectives of this section are to
- Discuss the switching action of a MOSFET
- Explain the basic operation of a CMOS inverter, CMOS NAND and NOR gate
- Describe the operation of a CMOS gate with an open-drain output
- Explain the operation of tristate CMOS gate
- List the precautions required when handling CMOS devices

12.5.1 Switching action of MOSFET

In digital circuit, only enhancement MOSFET is used to play a role of a voltage-controlled switch. There are two types of enhancement MOSFET, n-channel and p-channel, as shown in Figure 12.5.1. The control voltage is applied to the gate, the resistance between drain and source can be controlled.

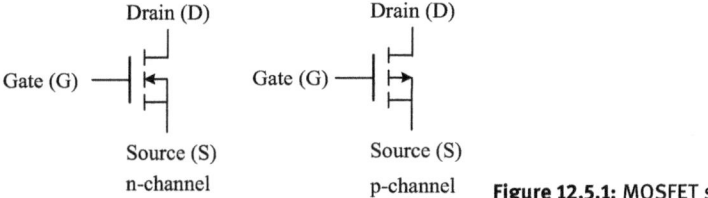

Figure 12.5.1: MOSFET symbols.

In digital applications, a MOSFET operates whether its resistance is always either very high (and the MOSFET is "off") or very low (and the MOSFET is "on"). The switch circuit with n-channel MOSFET is shown in Figure 12.5.2(a). When the gate voltage of an n-channel MOSFET is positive and enough higher than the source i.e. $U_{GS} > U_T$ (threshold voltage), the MOSFET is on; this is equivalent to an ideally closed switch between the drain and the source. When the gate-to-source voltage is zero or $U_{GS} < U_T$, the MOSEFT

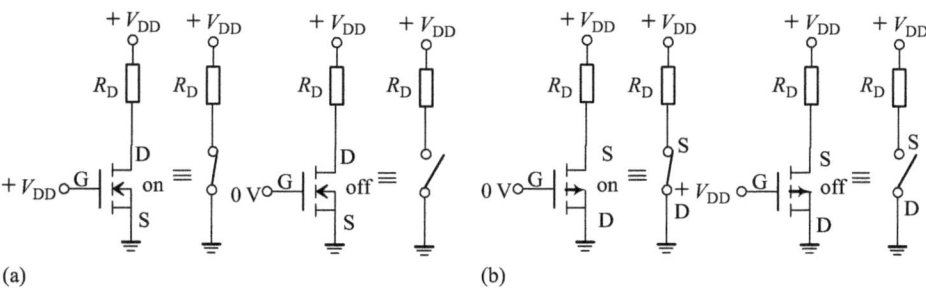

(a)   (b)

Figure 12.5.2: Switching action of a MOSFET: (a) n-channel; (b) p-channel.

is off (cutoff); this is equivalent to an ideally open switch between the drain and the source. So does the switch circuit with p-channel MOSFET. The p-channel MOSFET operates with opposite voltage polarities, as shown in Figure 12.5.2 (b).

Notice that only enhancement MOSFETs can be used as a switch in digital circuits.

12.5.2 CMOS invertor

1. Circuit and operating principle

Complementary MOS (CMOS) logic uses the MOSFET in complementary pairs as its basic elements. A complementary pair uses both p-channel and n-channel enhancement MOSFETs, as shown in CMOS inverter in Figure 12.5.3. For two MOSFETs, their gates are connected together as an input and drains are linked together as an output. The source and substrate of PMOS are connected to the positive supply voltage, $+V_{DD}$, and those of NMOS are linked with the ground. In order to assure CMOS inverter in the normal operation, the supply voltage should be greater than the sum of absolute values of two threshold voltage i.e. $V_{DD} > |U_{TP}| + |U_{TN}|$.

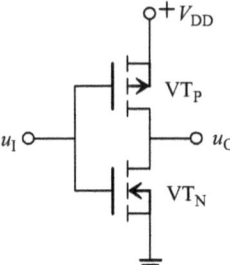

Figure 12.5.3: A CMOS inverter.

When a HIGH level $(+V_{DD})$ is applied to the input (u_I), as shown in Figure 12.5.4(a), n-channel MOSFET VT$_N$ is on since $U_{GSN} > U_{TN}$ and p-channel MOSFET VT$_P$ is off since $|U_{GSP}| < |U_{TP}|$. Thus the output is connected to ground through the on resistance of VT$_N$, resulting in a LOW output. When a LOW level (0 V) is applied to the input (u_I), as shown in Figure 12.5.4(b), VT$_N$ is off and VT$_P$ is on. Thus the output is connected

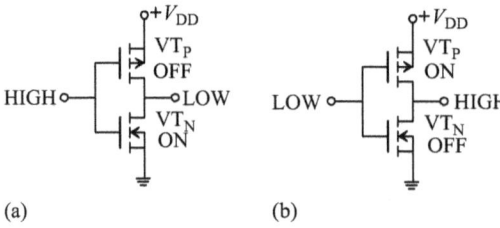

(a)   (b)

Figure 12.5.4: Operation of a CMOS inverter: (a) HIGH input; (b) LOW input.

to + V_{DD} through the on resistance of VT$_P$, resulting in a HIGH output. Therefore, an inverter operation is implemented by the circuit in Figure 12.5.3.

Since at least one MOS transistor is turned on in the normal operation of CMOS inverter and the on resistance is very small, the charge and the discharge time to load capacitance are efficiently decreased. This results in accelerating the switching speed of CMOS inverter, which can even be compared to that of the TTL gate. Moreover, because at least one of VT$_P$ and VT$_N$ is turned off in the normal operation and the off resistance is very high, current from V_{DD}, through VT$_P$ and VT$_N$ to ground is nearly zero and thus static power dissipation is very low. Generally, static power dissipation is in the magnitude of order of nW. In addition, the input impedance of CMOS gate is extremely high and thus fan-out is very large when CMOS gate is cascaded.

2. Transfer characteristics

For the CMOS inverter in Figure 12.5.3, VT$_P$ and VT$_N$ have the same parameters. That is to say, $|U_{TP}|=U_{TN}$, on resistance and off resistance of VT$_P$ and VT$_N$ are identical. The transfer characteristics curve is shown in Figure 12.5.5, which can be divided into three parts: segment AB, BC and CD.

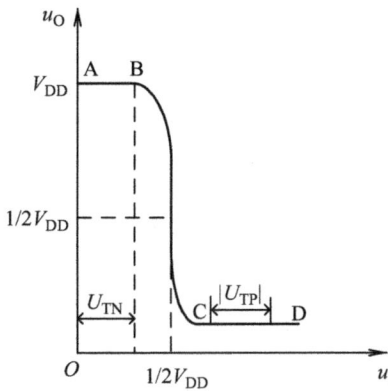

Figure 12.5.5: Transfer characteristics of CMOS inverter.

Segment AB: Since $u_I = U_{GSN} < U_{TN}$ and $|U_{GSP}|> |U_{TP}|$, VT$_N$ is off and VT$_P$ is on. So the output is a HIGH, i.e $U_{OH} \approx V_{DD}$.

Segment CD: Since $u_I = U_{GSN} > U_{TN}$ and $|U_{GSP}|< |U_{TP}|$, VT$_N$ is on and VT$_P$ is off. So the output is a LOW, i.e $U_{OL} \approx 0$ V.

Segment BC: Since $U_{TN} < u_I < V_{DD} - |U_{TP}|$, $U_{GSN} > U_{TN}$ and $|U_{GSP}|>|U_{TP}|$. So VT$_N$ and VT$_P$ are both on. If $u_I = 1/2\, V_{DD}$, then $u_O = 1/2\, V_{DD}$ since VT$_P$ and VT$_N$ have the same parameters. It is corresponding to the middle point of the transition region i.e. threshold voltage $U_{TH} =1/2\, V_{DD}$. Therefore, CMOS inverter has a higher noise margin than TTL counterpart.

3. Noise margin

Compared with TTL gate, CMOS inverter has the stronger noise immunity. Noise margin of the CMOS inverter increases with an increase in power supply voltage V_{DD}. Usually, $U_{NL} = U_{NH}$ for a given power supply voltage. For CMOS 4000-series, the measurement result shows that $U_{NL} = U_{NH} \geq 30\% \ V_{DD}$.

4. Propagation delay time

Although CMOS inverter does not exist the carrier accumulation and diffusion like the TTL gate, it still occurs propagation delay coming from the IC internal resistor and load capacitance.

Since the output resistor of CMOS inverter is higher than TTL gate, the load capacitances have a distinct effect on propagation delay time.

For CMOS inverter, $t_{pHL} = t_{pLH}$ due to its complementary circuit structure. The average propagation delay time is about tens of ns.

12.5.3 CMOS NAND gate and NOR gate

CMOS inverter is the basic element of CMOS gate. Based on CMOS inverter, you can construct CMOS NAND gate and CMOS NOR gate. Generally, CMOS gate circuit can be constructed with NMOS logic block circuit and PMOS logic block circuit substituting single NMOS and PMOS in a CMOS inverter. For NMOS logic block circuit, follow the rules of "*NMOS in serial for AND logic operation, and in parallel for OR logic operation*". For PMOS logic block circuit, follow the rules of "*PMOS in parallel for AND logic operation, and in serial for OR logic operation*" [48]. You should remember that the number of PMOS are equal to that of NMOS in CMOS gate circuit.

1. CMOS NAND gate

Figure 12.5.6 shows a CMOS NAND gate with two inputs. Each input is connected to a pair of PMOS and NMOS. Two NMOS transistors are in serial and PMOS transistors are in parallel.

The operations of CMOS NAND gate are as follows:
- When both inputs are LOW, VT_{N1} and VT_{N2} are off, and VT_{P1} and VT_{P2} are on. The output is pulled up to HIGH through the on resistance of VT_{P1} and VT_{P2} in parallel to the power supply voltage.
- When input A is LOW and input B is HIGH, VT_{P1} and VT_{N2} are on, and VT_{N1} and VT_{P2} are off. The output is pulled up to HIGH through the on resistance of VT_{P1}.
- When input A is HIGH and input B is LOW, VT_{N1} and VT_{P2} are on, and VT_{P1} and VT_{N2} are off. The output is pulled up to HIGH through the on resistance of VT_{P2}.

- Finally, when both inputs are HIGH, VT_{N1} and VT_{N2} are on, and VT_{P1} and VT_{P2} are off. The output is pulled down to LOW through the on resistance of VT_{N1} and VT_{N2} to ground.

It can be found from the function table in Figure 12.5.6(b) that circuit in Figure 12.5.6 (a) implements NAND operation i.e. $F = \overline{AB}$

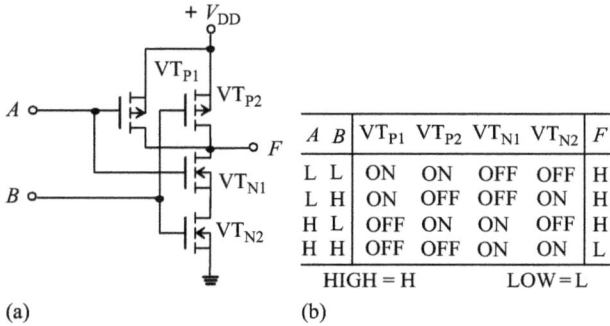

A B	VT_{P1}	VT_{P2}	VT_{N1}	VT_{N2}	F
L L	ON	ON	OFF	OFF	H
L H	ON	OFF	OFF	ON	H
H L	OFF	ON	ON	OFF	H
H H	OFF	OFF	ON	ON	L

HIGH = H   LOW = L

(a)   (b)

Figure 12.5.6: A CMOS NAND gate with two inputs: (a) circuit diagram; (b) function table.

2. CMOS NOR gate

Figure 12.5.7 shows a CMOS NOR gate with two inputs. Each input is connected to a pair of PMOS and NMOS. Two NMOS transistors are in parallel and two PMOS transistors are in serial.

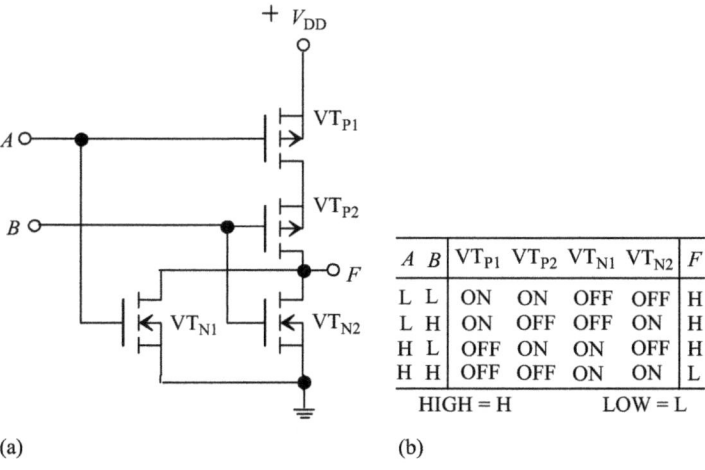

A B	VT_{P1}	VT_{P2}	VT_{N1}	VT_{N2}	F
L L	ON	ON	OFF	OFF	H
L H	ON	OFF	OFF	ON	H
H L	OFF	ON	ON	OFF	H
H H	OFF	OFF	ON	ON	L

HIGH = H   LOW = L

(a)   (b)

Figure 12.5.7: A CMOS NOR gate with two inputs: (a) circuit diagram; (b) function table

The operations of CMOS NOR gate are as follows:
- When both inputs are LOW, VT_{N1} and VT_{N2} are off, and VT_{P1} and VT_{P2} are on. The output is pulled up to HIGH through the on resistance of VT_{P1} and VT_{P2} in serial to the power supply voltage.
- When input A is LOW and input B is HIGH, VT_{P1} and VT_{N2} are on, and VT_{N1} and VT_{P2} are off. The output is pulled down to LOW through the on resistance of VT_{N2} to ground.
- When input A is HIGH and input B is LOW, VT_{N1} and VT_{P2} are on, and VT_{P1} and VT_{N2} are off. The output is pulled down to LOW through low on resistance of VT_{N1}.
- Finally, when both inputs are HIGH, VT_{N1} and VT_{N2} are on, and VT_{P1} and VT_{P2} are off. The output is pulled down to LOW through the on resistance of VT_{N1} and VT_{N2} to ground.

Therefore, the circuit in Figure 12.5.7 implements a NOR logic operation i.e. $F = \overline{A + B}$

For a CMOS NAND gate, if the number of inputs increases, the number of NMOS in serial and PMOS in parallel increases. When all inputs are HIGH, all NMOS in serial are on and the equivalent on resistance to ground increases. As a result, low-level output voltage rises, which might exceed the U_{OLmax} and thus makes logic operation wrong. In order to control voltage level in the allowed region, input buffers for each input and output buffer can be used to avoid voltage level inconsistence when input terminals increase. A CMOS NAND gate with the inverters as input and output buffers is shown in Figure 12.5.8. Due to using the inverters as input and output buffers, the original structure of NAND gate must be replaced with NOR gate to implement NAND operation. The corresponding effective logic circuit is also shown in Figure 12.5.8(b). So does a CMOS NOR gate with the inverters as input and output buffers in as shown Figure 12.5.9.

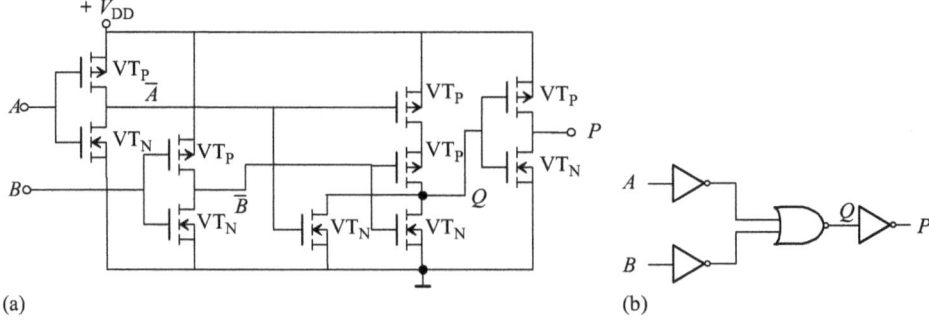

(a)   (b)

Figure 12.5.8: A CMOS NAND gate with buffers (a) and its effective logic diagram (b)

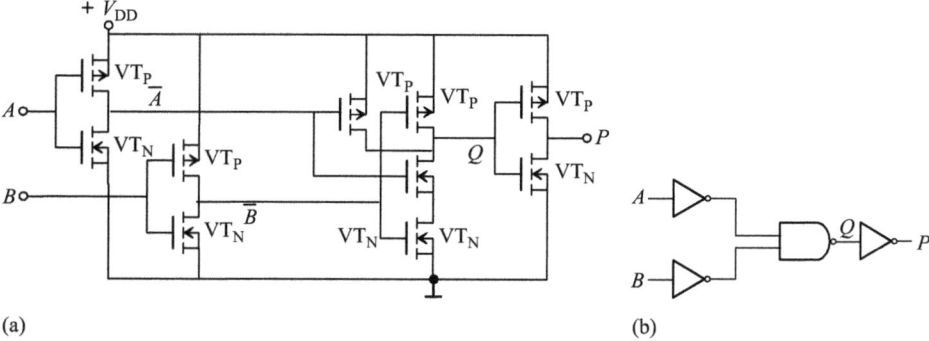

Figure 12.5.9: A CMOS NOR gate with buffers (a) and its effective logic diagram (b)

Example 12.5 Write the output logic expression and explain logic function of the CMOS logic circuit shown in Figure 12.5.10.

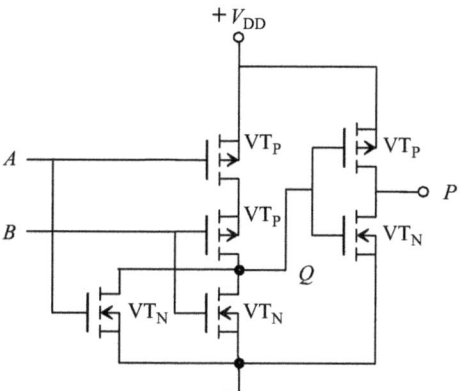

Figure 12.5.10: Circuit diagram.

Solution

It can be found that the circuit consists of a 2-input NOR gate and an inverter. The output expression of the 2-input NOR gate is $Q = \overline{A + B}$. The output Q is the input of the inverter, and thus the output expression of the inverter is $P = \overline{Q} = A + B$. So the final output expression of the circuit is

$$P = A + B$$

Therefore, the circuit in Figure 12.5.10 is OR gate and perform the OR operation of input A and input B.

12.5.4 CMOS transfer gate

Based on the switching action of single NMOS and PMOS, a CMOS transfer gate is constructed with a NMOS and a PMOS in parallel, as shown in Figure 12.5.11. Two sources are connected together as an input (u_I) and two drains are linked as an output (u_O). A pair of complementary signals, C and $\bar{C}$, control the gate of a NMOS and a PMOS, respectively. Figure 12.5.11(c) shows the operation schematic diagram of transfer gate. Transfer gate can be equivalent to a voltage-controlled switch, in which the state of the switch is controlled by the voltage level of input C.

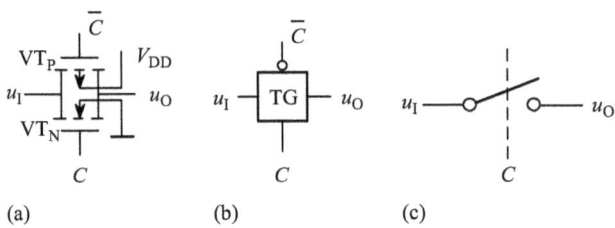

(a) (b) (c)

Figure 12.5.11: CMOS transfer gate: (a) circuit diagram; (b) logic symbol; (c) operational schematic diagram.

CMOS transfer gate requires that $V_{DD} \geq |U_{TP}| + |U_{TN}|$ and $0 \leq u_I \leq V_{DD}$. Assume that $V_{DD} = 10V$, $|U_{TP}| = |U_{TN}| = 3V$.

A 10 V voltage is applied to the control input C, and the complementary input $\bar{C}$ is 0V, as shown in Figure 12.5.12 (a). When u_I varies in the range from 0 V to 3 V, only VT_N is on; when u_I varies in the range from 3 V to 7 V, both VT_N and VT_P are on; when u_I varies in the range from 7 V to 10 V, only VT_P is on. Therefore, there is at least one MOS transistor turned on when u_I varies in the range from 0 V to 10 V. This is equivalent to the closed switch. As a result, $u_O = u_I$. This means that the data can be transferred from the input to the output.

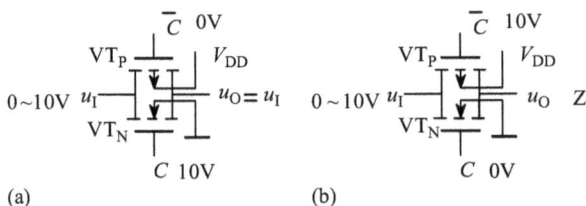

(a) (b)

Figure 12.5.12: CMOS transfer gate as a closed switch for transmitting the signal (a) and an open switch to cutoff the transmission (b).

A 0 V voltage is applied to the gate of VT_N and 10 V to the gate of VT_P, as shown in Figure 12.5.12 (b). When u_I varies in the range from 0 V to 10 V, both VT_N and VT_P is off, which is equivalent to the open switch. The input signal u_I could not be transferred to the output u_O. The output is in high impedance state.

In a word, the voltage level of the control input C determines whether CMOS transfer gate is turned on or off. The transfer gate is turned on when a HIGH is applied on the input C, and off when a LOW is applied on the input C. CMOS transfer gate is approximate to an ideal switch since it has very low on resistance (several hundreds of ohm) and very high off resistance greater than $10^7\Omega$. In addition, source and drain of MOS transistor have the same structure so that the input of CMOS transfer gate can be used as the output and the output can be used as the input. Therefore, CMOS transfer gate is also called as bidirectional switch.

12.5.5 Open-drain gate

An open-drain gate is the CMOS counterpart of an open-collector TTL gate. The term *open-drain* means the drain of the output MOS transistor with nothing connected to it. When you use an open-drain gate, an external load must be connected from the drain terminal to power supply voltage. An open-drain gate with a single n-channel MOSFET is shown in Figure 12.5.13(a). An external pull-up resistor must be used, as shown in Figure 12.5.13(b), to provide a HIGH output state. Also open-drain outputs can be connected in a wired-AND configuration that is discussed in the previous section.

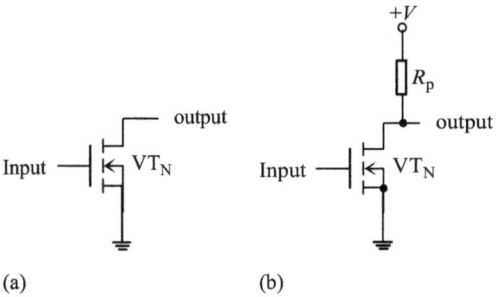

Figure 12.5.13: Open-drain CMOS gates: (a) unconnected output; (b) with pull-up resistor.

12.5.6 Tristate CMOS gate

Tristate outputs are also available in CMOS gates. As you recall, three output states are HIGH, LOW, and high-Z [47]. A tristate CMOS inverter is shown in Figure 12.5.14.

The operation of tristate CMOS inverters is illustrated in Figure 12.5.15.
- When enable input is a LOW (logic 0), the device is enabled for normal logic operation. The output is NOT A.
- When enable input is a HIGH (logic 1), VT_P and VT_N are both cutoff. The output is in a high-Z state.

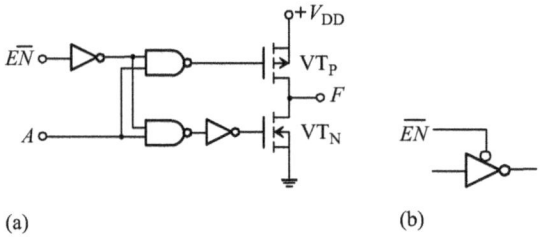

(a)　　　　　　　　　　　　　　　　(b)

Figure 12.5.14: A tristate CMOS inverter: (a) circuit diagram; (b) logic symbol.

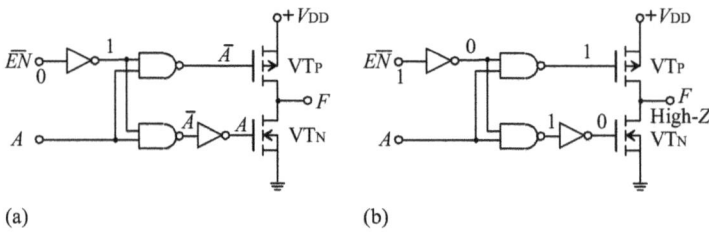

(a)　　　　　　　　　　　　　　　　(b)

Figure 12.5.15: Illustration of tristate CMOS inverter when $\overline{EN}$ is a LOW (a) and a HIGH (b).

Therefore, the logic expression of the tristate CMOS inverter is described as follows:

$$F = \begin{cases} \bar{A} & \overline{EN} = 0 \\ Z & \overline{EN} = 1 \end{cases}$$

12.5.7 CMOS family

The first commercial CMOS family is 4000-series CMOS. Although 4000-series circuits have low power dissipation and strong noise immunity, their switching speed and driving capacity need to be improved. To improve the switching speed and the driving capacity, a series of upgrade products come out from 4000B to high-speed 74HC and 74HCT; in succession, from low supply voltage 74LVC to ultra-low supply voltage 74AUC. In the recent years, BiCMOS is developed to push CMOS gate toward high speed and strong driving capacity. With CMOS circuits as the logic part and TTL totem structure as the output part, BiCMOS has not only the advantages of high speed and strong driving capacity of TTL gates, but also the advantages of high integrated level, low power dissipation and low cost of CMOS gates. These advantages make it a potential application of developing integrated digital circuit and analogic circuit. Table 12.5.1 lists the parameters of the commonly used typical series [47].

Table 12.5.1: Parameters of typical CMOS circuit series.

Series	Propagation delay time/ns	Power dissipation/mW	Speed-power product/pW·s
4000B	105	1(1MHz)	105
74HC	10	1.5(1MHz)	15
74BCT	2.9	0.0003~7.5	0.00087~22

12.5.8 Comparison of CMOS and TTL Circuits

There are some differences between CMOS circuits and TTL circuits. The main differences are described as below.

1. TTL gate circuits are consisted of bipolar junction transistors and CMOS gate circuits constructed with single polar MOSFETs.
2. CMOS gate circuits have a wide supply voltage range from 1.2 V to 18 V while the supply voltage of TTL gate circuits is usually 5 V.
3. Empty input is equivalent to high level input for TTL gate circuits. But empty input is not allowed for CMOS gate circuits. Since the input impedance of CMOS circuits is extremely high, the induced charge in gate capacitance discharge slowly, which results in the output in an unknown state. Due to the existence of induced charge in gate capacitance, MOSFET is easy to be damaged.
4. For TTL circuits, when there is an input resistor, R, connected between the input and ground, the input voltage changes with the value of resistor. If $R \geq R_{on}$, the input is a high level. If $R \leq R_{off}$, the input is equivalent to a low level. While input resistor is accessed between the input and ground for CMOS circuit, the input is a low level due to near zero input current.
5. Fan-out of a CMOS circuit can be evaluated with a different method from that of a TTL circuit. If the evaluating method of a TTL circuit is adopted for a CMOS circuit, fan-out of a CMOS circuit would be a very large number. However, the more the number of load gates is, the higher the equivalent load capacitance is. That affects the operation frequency of CMOS circuits. Therefore, fan-out of a CMOS circuit is determined by the operational frequency. Fan-out can be increased for low operational frequency and decreased for high operational frequency for a CMOS circuit. Usually, fan-out of a TTL circuit is smaller than that of a CMOS circuit.
6. CMOS circuits have very small static power dissipation, but dynamic power dissipation increases with the operational frequency increasing. When the operational frequency reaches around 1MHz, the power dissipation of a CMOS circuit is

approximate to that of a TTL circuit. CMOS circuits are suitable for large-scale integrated circuit.

7. Noise margin of CMOS circuit is higher than that of TTL circuit. Thus, CMOS circuit has the stronger noise immunity.
8. CMOS circuits have the better thermal stability than TTL circuits.

12.6 CMOS/TTL interfacing

In previous sections, various logic circuits in different family are introduced. These offer the freedom for a digital designer to select the requiring logic gates for designing a digital system based on general requirements of speed, power dissipation and cost. It is possible to select the logic gates from the same or different logic family. If logic gates in a system are from different logic family i.e. some from TTL family and the others from CMOS family, it is important for a designer to understand the implications of connecting TTL outputs to CMOS inputs, and vice versa. Since logic circuits from different family have different requirement of voltage and current, the interface must consider the matches of voltage and current between TTL and CMOS.

There are two main factors to be considered in TTL/CMOS interfacing. The first factor is that output voltage of a driving gate should be in the range of the requiring voltage of the load gate. Logic level between a driving gate and a load gate must satisfy the following requirement.

$$\begin{cases} U_{OH(min)} \geq U_{IH(min)} \\ U_{OL(max)} \leq U_{IL(max)} \end{cases} \tag{12.6.1}$$

The other factor to be considered is if a driving gate owns the ability of meeting the current requirement to the load gates. The current requirements between a driver gate and the load gates are expressed as follows:

$$\begin{cases} I_{OH(max)} \geq mI_{IH(max)} \\ I_{OL(max)} \geq nI_{IL(max)} \end{cases} \tag{12.6.2}$$

where m and n are the number of load gates that a driving gate is sourcing current to and sinking current from, respectively. Voltage and current parameters of logic gates are main parameters for a designer to consider the interface between TTL and CMOS. The above four equations must be satisfied to ensure the logic correctness of the design system. Table 12.6.1 lists the main characteristic parameters for the design reference.

Table 12.6.1: Main parameters of TTL and CMOS.

Series \ Parameter	TTL 74 series	TTL 74LS	CMOS 4000 series	CMOS 74HC	CMOS 74HCT
$U_{IH(min)}/V$	2	2	3.5	3.5	2
$U_{IL(max)}/V$	0.8	0.8	1.5	1	0.8
$I_{IH(max)}/\mu A$	40	20	0.1	0.1	0.1
$I_{IL(max)}/mA$	−1.6	−0.4	-0.1×10^{-3}	-0.1×10^{-3}	-0.1×10^{-3}
$U_{OH(min)}/V$	2.4	2.7	4.6	4.4	4.4
$U_{OL(max)}/V$	0.4	0.5	0.05	0.1	0.1
$I_{OH(max)}/mA$	−0.4	−0.4	−0.51	−4	−4
$I_{OL(max)}/mA$	16	8	0.51	4	4

12.6.1 TTL driving CMOS interface

When a TTL gate as a driving gate drives a CMOS gate as a load gate, the interfacing circuit can be divided into two categories in terms of supply voltage. One is for the same supply voltage and the other is for different supply voltage [21].

1. Interface for the chips with the same supply voltage

TTL 74LS series and CMOS 74HC series have the same supply voltage of 5 V. If a TTL gate in 74LS series drives a CMOS gate in 74HC series, two gates can be directly connected together. Through checking the match condition in eqs. (12.6.1), (12.6.2) and parameters in Table 12.6.1, you can find that only one required condition could not be satisfied. The minimum of high-level output voltage of the TTL gate is 2.7 V while the minimum of high-level input voltage of the CMOS gate is 3.5 V. This means that the output high-level voltage of the TTL gate could not satisfy the requirement of the input high-level voltage of the CMOS gate if two gates are directly connected.

One method of solving the above problem is to connect a pull-up resistor from the output of the TTL gate to supply voltage, as shown in Figure 12.6.1. This pull-up resistor can increase the high-level output voltage of TTL gate so that the high-level

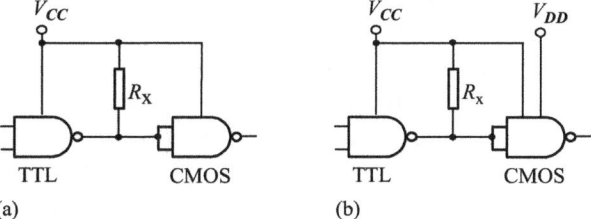

(a)   (b)

Figure 12.6.1: A TTL gate driving a CMOS gate: (a) with the same supply voltage; (b) with different supply voltages.

output voltage of TTL gate is compatible with the high-level input voltage of CMOS gate. The pull-up resistor can be determined by the following equation.

$$R_{x(min)} \geq \frac{U_{CC} - U_{OL(max)}}{I_{OL(max)} - nI_{IL}}; \quad R_{x(max)} \geq \frac{U_{CC} - U_{IH(min)}}{nI_{IH} + I_{OH(max)}}. \tag{12.6.3}$$

where $U_{OL(max)}$, $I_{OL(max)}$ and $I_{OH(max)}$ are the parameters for TTL driving gate, I_{IL} and I_{IH} are the parameters for CMOS load gate, and n is the number of the load gates.

If a TTL gate in 74LS series is used for a driving gate and the CMOS gates in 74HCT series for the load gate, the parameters of two series is completely compatible for each other. Therefore, no additional interfacing circuit is required. In digital system, CMOS gates in 74HCT series are generally selected to eliminate the demand of pull-up resistor.

Therefore, another optimum method is that a CMOS gate in 74HCT series is used as the transition interface between TTL 74LS series and CMOS 74HC series. A TTL gate in 74LS series drives a CMOS gate in 74HCT series first. Then a CMOS gate in 74HCT continues to drive a CMOS gate in 74HC series.

2. Interface for the chips with different supply voltage

There are two methods of designing the interface when the supply voltage of CMOS load gate is higher than that of TTL driving gate.

One method is to use the CMOS gate with voltage level offset. For example, CD40109 has two supply voltage inputs, V_{CC} and V_{DD}. When V_{CC} = 5 V and V_{DD} = 10 V, the input of CD40109 can receive the TTL logic level of 1.5 V/3.5 V and the output logic level is 9 V/1 V which can satisfy the requirement of CMOS load gate. The connection between TTL gate and CMOS gate of CD40109 is shown in Figure 12.6.1(b).

Another method is to use open-collector gate as a driving gate. Since the supply voltage of open-collector gate is allowed to be higher than the supply voltage V_{DD} of CMOS gate, the external pull-up resistor can be directly connected to the supply voltage V_{DD} of CMOS gate.

12.6.2 CMOS driving TTL interface

When a CMOS gate drive a TTL gate , the characteristics of the driving gate and the load gate in series should be considered.

For the case of CMOS 4000 series driving TTL 74 series with the same supply voltage, the interface should consider not only the match of the voltage level but also the match of the current. It can be found from Table 12.6.1 that CMOS driving gate can accept the sinking current of 0.51 mA while the low-level input current of TTL load gate is 1.6 mA. Obviously, the sinking current of the CMOS driving gate is not able to satisfy the requirement of TTL load gate. If you still use this CMOS gate to drive TTL gate, the low-level output of CMOS gate would be rising and thus the logic operation error might occur.

In order to improve the current sinking capacity of the CMOS driving gate, there are several schemes for selection. The simple method is to increase the number of driving gates in parallel for lifting current sinking capacity. The other common used method is to insert a CMOS driver between a CMOS driving gate and a TTL load gate to increase the current sinking capacity. For example, an in-phase driver of cc4010, which has the same supply voltage with the CMOS driving gate, is inserted to complement current sinking capacity for the CMOS driving gate. The connection of the circuit is shown in Figure 12.6.2(a). Also an open-drain CMOS gate, for example, CC40107 can be inserted in Figure 12.6.2(b). CC40107 can drive ten TTL load gates in 74 series. In addition, the output current of the CMOS driving gate can be amplified by using a transistor amplifier to drive the TTL load gate, as shown in Figure 12.6.2(c).

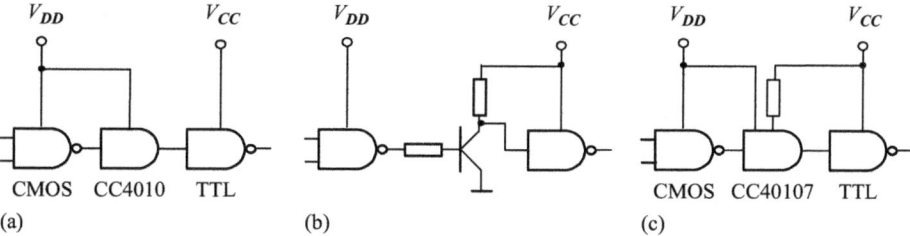

Figure 12.6.2: CMOS gate driving TTL gate: (a) with an in-phase driver; (b) with an open drain gate; (c) with a transistor amplifier.

12.6.3 Gate circuit loading interface

In digital application, it is a common thing to use a gate circuit to drive a heavy current load, such as motor, relay, and LED. Thus the interface between a gate circuit and a heavy current load should be designed according to the parameters of data sheet of IC.

Take a common example of a gate circuit driving LED. Usually, there are two kinds of connection for a gate circuit driving LED, as shown in Figure 12.6.3.

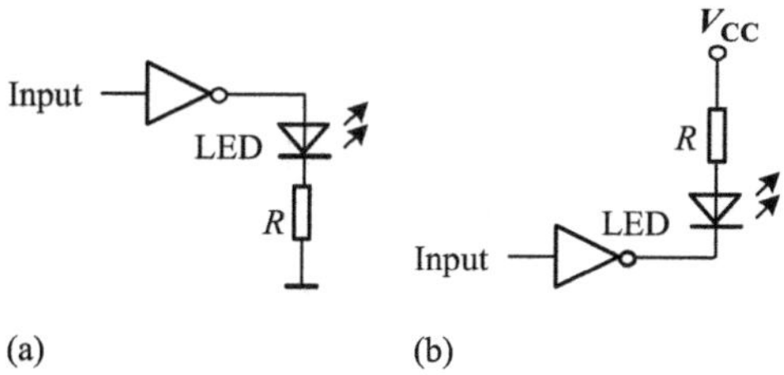

Figure 12.6.3: Circuits of a gate driving LED.

Assume that the operating current of LED is denoted by I_D and the voltage drops for LED forward biased is V_D. Current limiting resistor, R, is selected as follows.

When the output of the gate circuit is a HIGH output, current limiting resistor is selected by the following equation

$$R = \frac{V_{OH} - V_D}{I_D} \tag{12.6.4}$$

When the output of gate circuit is a LOW output, current limiting resistor is determined as follows.

$$R = \frac{V_{CC} - V_D - V_{OL}}{I_D} \tag{12.6.5}$$

12.7 Summary

1. The most successful logic families are bipolar logic family and metal-oxide semiconductor field-effect transistor (MOSFET) logic family.
2. Bipolar logic family is one based on bipolar junction transistors called *transistor-transistor logic* (*TTL*). The TTL family of logic devices offers many SSI logic gates, and MSI devices.
3. Metal oxide semiconductor (MOS) includes n-channel MOS (NMOS), p-channel MOS (PMOS) and Complemented MOS (CMOS). CMOS logic family is most commonly used to construct large-scale integrated circuits and captured the market due to its very low power and competitive speed.
4. To correctly use digital integrated circuits, you should understand some important operational properties such as voltage levels, noise immunity, power dissipation, fan-out, and propagation delay time.
5. When you are connecting devices together, it is vital to know how many inputs a given output can drive without compromising reliability. This is referred to as *fan-out*.
6. Noise margins are the parameters related to voltage parameter, which represent noise immunity ability of gate circuit.
7. Totem-pole outputs of TTL cannot be directly connected for implementing wired-AND.
8. Open-collector and open-drain outputs can be wired together to implement a wired-AND function.
9. Tristate outputs can be wired together to allow numerous devices to share a group of data bus. At any one time, only one device is allowed to assert a logic level on the data bus.

10. CMOS transfer gate is an ideal bidirectional switch which can transfer not only digital signal but also analogue signal. These devices can pass or block an analog signal, depending on the digital logic level that controls it.
11. Logic devices that use various technologies cannot always be directly connected together and operated reliably. The voltage and current characteristics of inputs and outputs must be considered and precautions need to be taken.

Key terms

Gate circuit: The unit circuit who can implement basic logic operation and combinational logic operation.

Logic family: A collection of different integrated circuits that have similar input, output, and internal circuit characteristics, but that perform different logic function.

TTL: Transistor-transistor logic; a type of integrated circuit that uses bipolar junction transistors.

CMOS: A type of integrated circuit that uses pairs of complementary MOS transistors.

Noise immunity: The ability to tolerate a certain amount of unwanted voltage fluctuation on its inputs without changing its output.

Noise margin: The parameters for measuring noise immunity of gate circuits.

Propagation delay time: The amount of time that a gate takes for a change to produce a change in the output signal.

Speed-power product: The product of propagation delay time and power dissipation.

Current sourcing: The action of a logic circuit in which it sends current from its output to a load.

Current sinking: The action of a logic circuit in which it accepts current into its output from a load.

Fan-out: the maximum number of load gate inputs that can be connected to a certain output without affecting the specified operational characteristics of the driving gate.

Totem-pole or push-pull: a type of output in TTL circuits. The output stage has two transistors, only one of them is on at any time.

Open-collector: A type of output for a TTL circuit in which the collector of the output transistor is left internally disconnected and is available for connection to an external load that requires relatively high current or voltage.

Pull-up resistor: A resistor with one end connected to the dc supply voltage used to keep a given point in a logic circuit HIGH when in the inactive state.

Tristate:A type of output in logic circuits that have three operational states: HIGH, LOW, high-Z.

Self-test

12.1 Which of the following is not a TTL circuit?
(a) 74F00　(b) 74AS00　(c) 74HC00　(d) 74ALS00

12.2 If two unused inputs of a LS TTL gate are connected to an input being driven by another LS TTL gate, the total number of remaining unit loads that can be driven by this gate is
(a) seven　(b) eight　(c) eighteen　(d) unlimited

12.3 In a TTL circuit, if an excessive number of load gate inputs are connected,
(a) $U_{OH(min)}$ drops below U_{OH}
(b) U_{OH} drops below $U_{OH(min)}$
(c) U_{OH} exceeds $U_{OH(min)}$
(d) U_{OH} and $U_{OH(min)}$ are unaffected

12.4 An open-collector output requires _____.
(a) a pull-down resistor
(b) a pull-up resistor
(c) no output resistor
(d) an output resistor

12.5 When the frequency of the input signal to a CMOS gate is increased, the average power dissipation _____.
(a) increases
(b) decreases
(c) does not change
(c) decreases exponentially

12.6 CMOS operates more reliably than TTL in a high-noise environment because of its _____.
(a) lower noise margin
(b) input capacitance
(c) higher noise margin
(d) smaller power dissipation

12.7 Which factor does not affect CMOS loading?
(a) Charging time associated with the output resistance of the driving gate
(b) Discharging time associated with the output resistance of the driving gate
(c) Output capacitance of the load gates
(d) Input capacitance of the load gates

12.8 Proper handling of a CMOS device is necessary because of _____.
(a) fragile construction
(b) high noise margin

(c) susceptibility to electrostatic discharge

(d) low power dissipation

12.9 It is best not to leave unused TTL inputs unconnected (open) because of TTL's _____.

(a) noise sensitivity

(b) low-current requirement

(c) open-collector outputs

(d) tristate construction

12.10 One output structure of a TTL gate is often referred to as a _____.

(a) totem-pole arrangement

(b) diode arrangement

(c) JBT arrangement

(d) base, emitter, collector arrangement

12.11 A certain gate draws 1.8 μA when its output is HIGH and 3.3 μA when its output is LOW. V_{CC} is 5 V and the gate is operated on a 50% duty cycle. What is the average power dissipation (P_D)?

(a) 2.55 μW   (b) 1.27 μW   (c) 12.75 μW   (d) 5 μW

12.12 If I_{CCH} is specified as 1.1 mA when V_{CC} is 5 V and if the gate is in a static (noncharging) HIGH output state, the power dissipation (P_D) of the gate is

(a) 5.5 mW   (b) 5.5 W   (c) 5 mW   (d) 1.1 mW

Problems

12.1 A certain logic gate has a $U_{OH(min)}$ = 2.2 V, and it is driving a gate with a $U_{IH(min)}$ = 2.5 V. Are these gates compatible for HIGH-state operation? Why?

12.2 A certain logic gate has a $U_{OL(max)}$ =0.45 V, and it is driving a gate with $U_{IL(max)}$ = 0.8 V. Are these gates compatible for LOW-state operation? Why?

12.3 Voltage specifications for three types of logic gates are given in the following table. Which gates would you select for use in a high-noise industrial environment?

	$U_{OH(min)}$	$U_{OL(max)}$	$U_{IH(min)}$	$U_{IL(max)}$
Gate A	4.4 V	0.2 V	3.5 V	0.8 V
Gate B	2.4 V	0.4 V	2 V	0.8 V
Gate C	3.5 V	0.3 V	2.7 V	0.8 V

12.4 A certain gate draws 1.8 µA when its output is HIGH and 3.3 µA when its output is LOW. V_{CC} is 5 V and the gate operates on a 50% duty cycle. What is the average power dissipation (P_D)?

12.5 Parameters for three types of gates are listed in the following table. Considering the speed-power product, which one would you select for best performance? If you wanted the gate to operate at the highest possible frequency, which gate would you select?

	t_{PLH}	t_{PHL}	P_D
Gate A	1 ns	1.2 ns	1.5 mW
Gate B	5 ns	4 ns	8 mW
Gate C	10 ns	10 ns	0.5 mW

12.6 For a given circuit with five NAND gates in Figure P12.1, if the propagation of gates G1, G2, G3 and G4 is 30 ns and the frequency of output F is 3.2 MHz, determine the average propagation delay time, t_{pd5}, of gate G5.

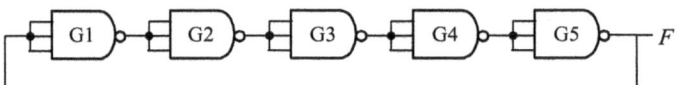

Figure P12.1

12.7 Write the logic expression for each circuit in Figure P12.2.

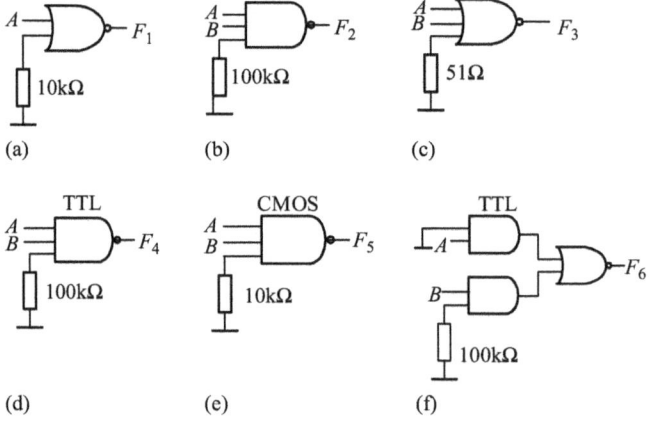

Figure P12.2

12.8 Which TTL circuits in Figure P12.3 can implement NOT operation?

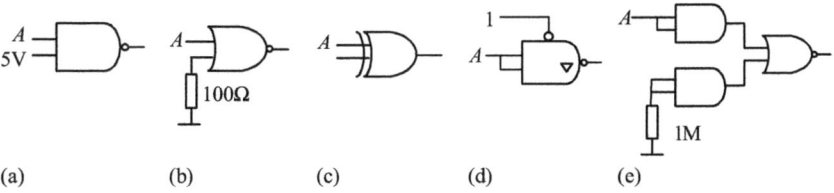

(a)   (b)   (c)   (d)   (e)

Figure P12.3

12.9 Which CMOS circuits in Figure P12.4 can implement NOT operation?

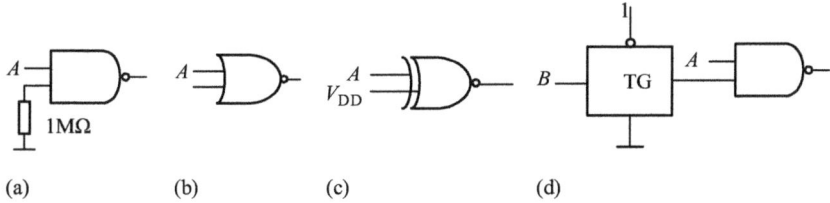

(a)   (b)   (c)   (d)

Figure P12.4

12.10 For each part of the circuits in Figure P12.5, determine whether each part can implement the required logic expression. If not, indicate the change that should be made to implement the required logic. All gates are standard TTL.

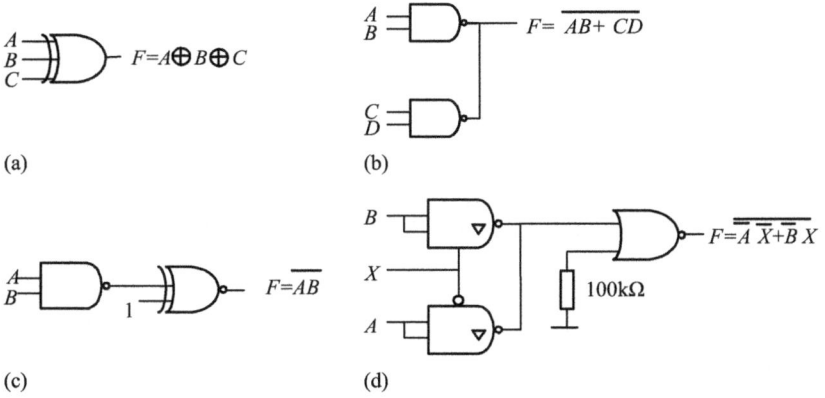

(a)   (b)

(c)   (d)

Figure P12.5

12.11 The input waveforms are applied to the inputs of the TTL tristate circuits, as shown in Figure P12.6. Show the output waveform in proper relation to the inputs with timing diagram.

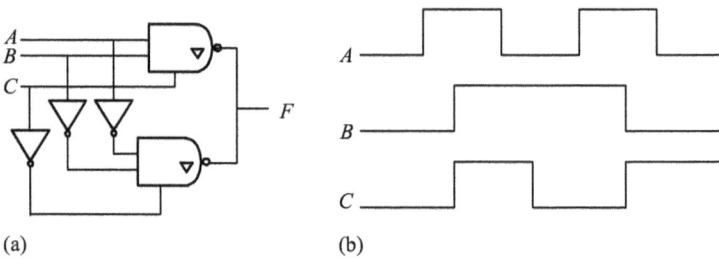

(a) (b)

Figure P12.6

12.12 For a CMOS circuit in Figure P12.7 (a), the input waveforms in Figure P12.7 (b) are applied to inputs A, B and C, and $R = 10$ kΩ. Show the output waveform in proper relation to the inputs with timing diagram.

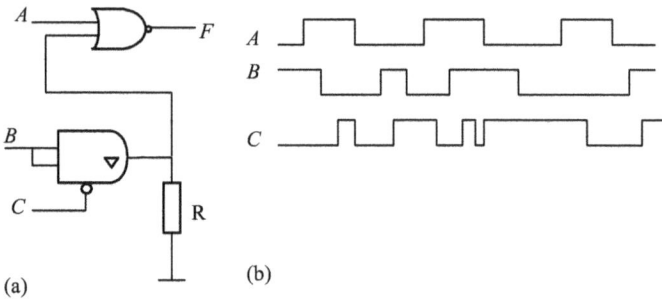

(a) (b)

Figure P12.7

12.13 Figure P12.8 (a) shows a circuit consisting of two transfer gates and an inverter. $u_{I1} = 10$ V and $u_{I2} = 5$ V. If the input waveform in Figure P12.8 (b) is applied to the input C, draw out the waveform of the output u_O in proper relation to the inputs with a timing diagram.

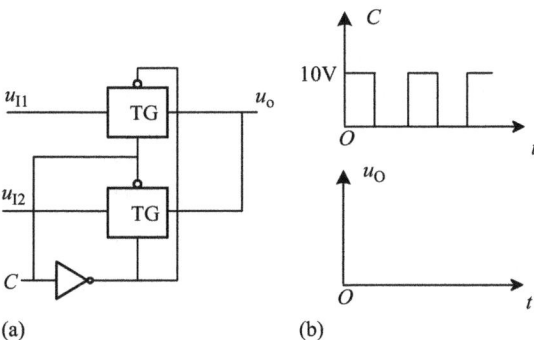

Figure P12.8

12.14 Write the logic expression of each part of logic circuit in Figure P12.9.

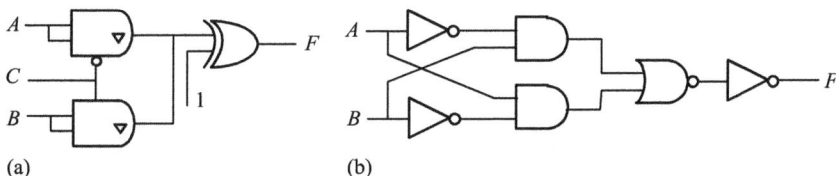

Figure P12.9

12.15 Write the logic expression for each of TTL circuits in Figure P12.10 (a) and (b). If the input waveforms in Figure P12.10 (c) are applied to inputs A, B, and C, show the output waveforms in proper relation to the inputs with a timing diagram.

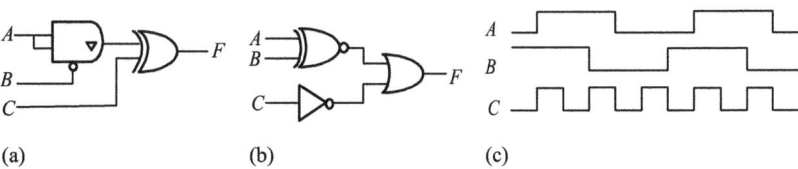

Figure P12.10

12.16 For a circuit in Figure P12.11 (a) to (g), input waveforms in figure (h) are applied to input A and B. Determine the corresponding output waveforms in proper relation to inputs with timing diagram.

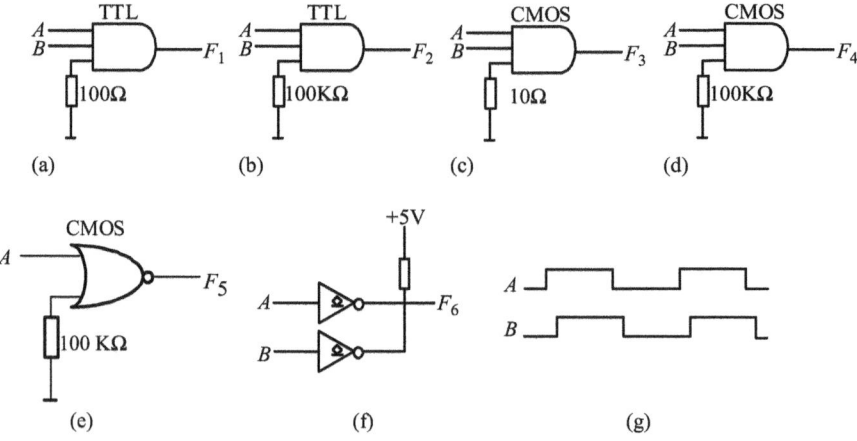

Figure P12.11

12.17 Write the logic expression for each of CMOS circuits in Figure P12.12, and explain what the similarities are and what the difference is between the two circuits.

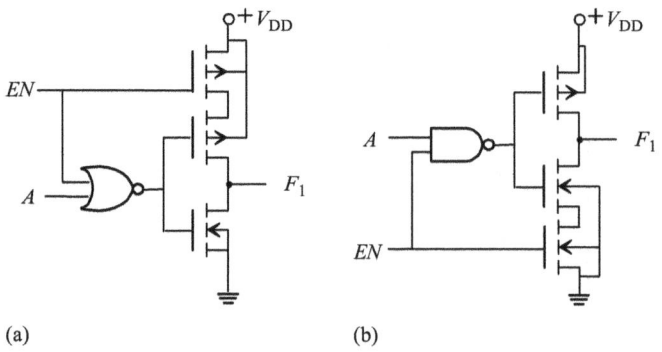

Figure P12.12

12.18 Design a CMOS circuit for implementing the logic expression as follows:

$$F = AB + C$$

12.19 In viewpoint of voltage level match, explain the function of R_P for the TTL/ CMOS interface in Figure P12.13.

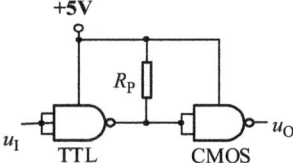

Figure P12.13

12.20 Explain the reason why two CMOS inverters are in parallel in Figure P12.14.

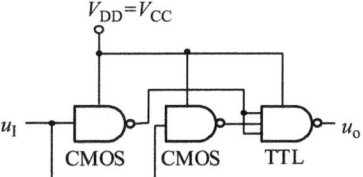

Figure P12.14

12.21 Determine which driving circuits in Figure P12.15 is wrong. Indicate the change that should be made to get the correct driving circuit.

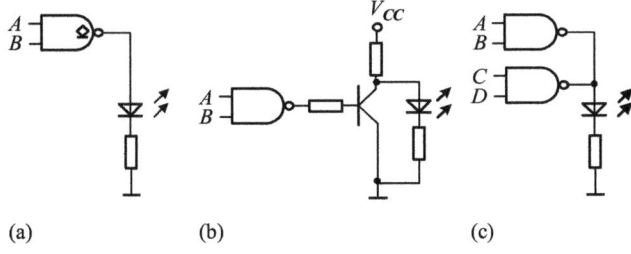

(a)   (b)   (c)

Figure P12.15

Appendix I: Quartus II software guide

Quartus II is an Electronic design automation (EDA) software for implementing a digital circuit design, which is a software development package of Alrera Inc. The programming hardware adopts Altera DE2 Development and Education Board. Taking a 4-bit adder design as an example, this appendix introduces how to use Quartus II software for implementing a digital circuit design. Figure A1 shows the block diagram of a 4-bit adder, in which $A3A2A1A0$ and $B3B2B1B0$ are two 4-bit binary addends, $C0$ is a carry input, $S3S2S1S0$ are 4-bit sum outputs, and $C1$ is carry output.

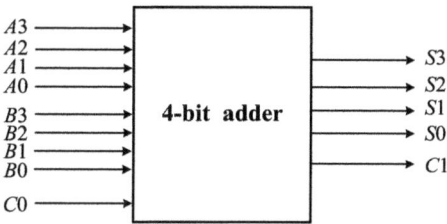

Figure A1: Block diagram of a 4-bit adder.

Step 1 Creat a New Project

Run Quartus II 7.2 software and enter into the initial interface, as shown in Figure A2.

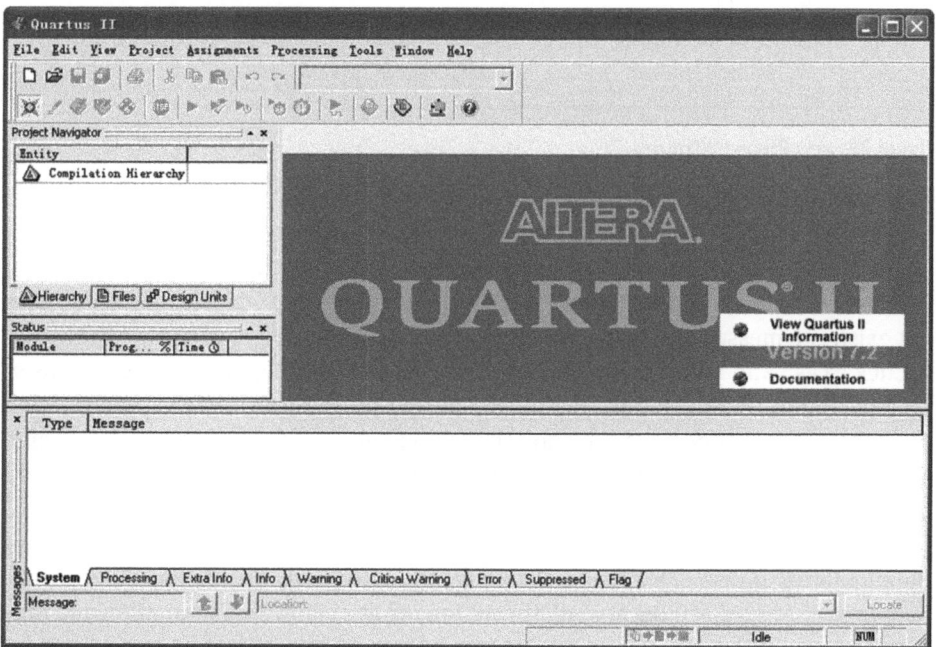

Figure A2: Initial interface of Quartus II.

https://doi.org/10.1515/9783110614916-013

Select **File>New Project Wizard** and appear a dialog box of New Project Wizard, as shown in Figure A3. This dialog box lists all steps to help you creat a new project.

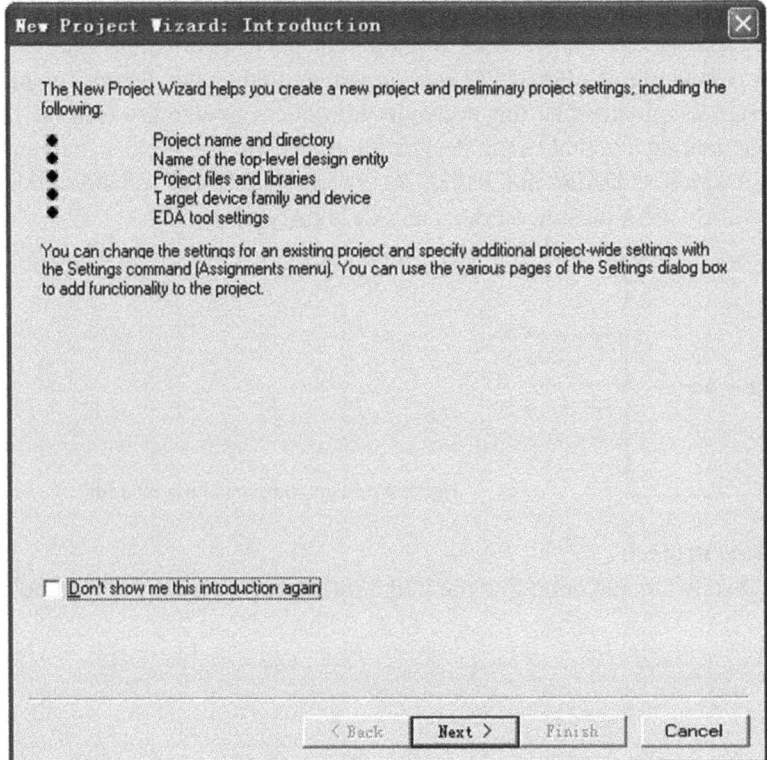

Figure A3: New Project Wizards.

Click **Next** and appear a dialog box as shown in Figure A4.

Type the working directory of the project: **"d:\DE2\example"** or other directory name. The project name must be the same as the name of top-level design entity, here **"adder"** is used and click **"Next."** If the assigned directory "d:\DE2\example" does not exist, the message box pops out, as shown in Figure A5.

Click **"Yes"** and a new dialog box pops out, as shown in Figure A6. Then the user-specified design files can be added to the current project.

Click **Next** and appear a dialog box shown in Figure A7. Select the target programmable device **'EP2C35F672C6'** in available device list, which is a FPGA device in DE2 board.

Click **"Next"** and appear a dialog box as shown in Figure A8. The dialog box give the prompting message for the user to select the third party design tools that will be used in the new project. Because all the design inputs, synthesis, simulation and time sequence analysis tools provided by Quartus II are used in this design, no the third party design tool is used in this project. Then click **"Next"** to go to the next step.

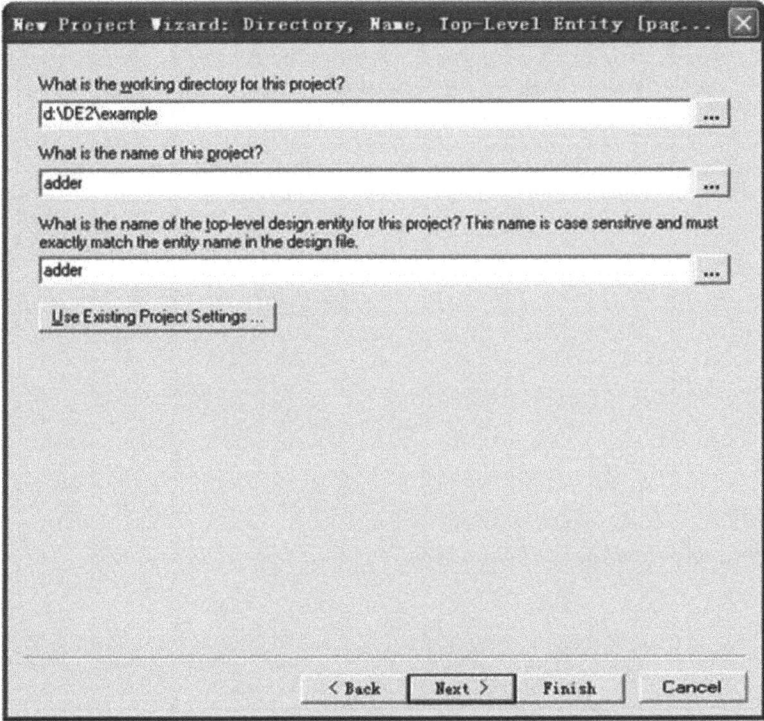

Figure A4: Window for entering the working directory, the project name, and the top-level design entity.

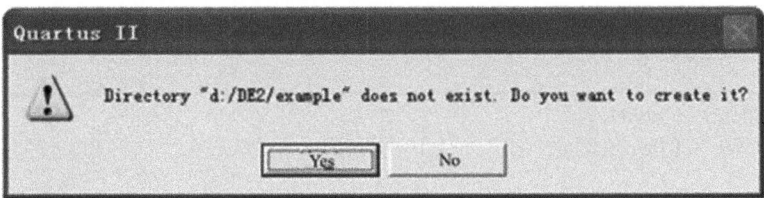

Figure A5: Create a new directory for the project.

Figure A9 show a summary dialog box, and click **"Finish"** to complete new project creation.

Step 2 Design Entry

In the current project directory, select **File > New** to open a dialog box shown in Figure A10.

Select **Verilog HDL File** and click **OK** to open Text Editor. Then select **File>Save as** to open a dialog box as shown in Figure A11.

Enter the file name **"adder"** and save as the type of **"Verilog HDL File"**. Select the tick of **"Add file to current project"** and add the file **"adder.v"** to the current project.

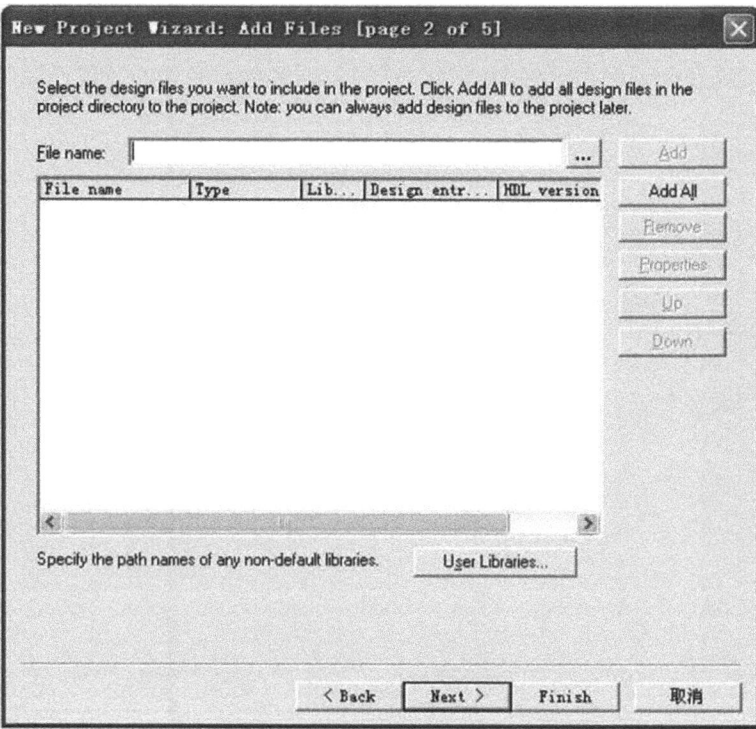

Figure A6: Adding the user-specified design files for the project.

Then in the **Text Editor**, you can input the program code of 4-bit adder in Example A1 into the file "adder.v."

–**Example A1** Edit input file (adder.v)
```verilog
module adder (A, B, C0, S, C1);

    input [3:0] A, B;
    input C0;
    output [3:0] S;
    output C1;reg [3:0] S;
    reg C1;
    always@(A or B or C0)
    begin
        S=A+B+C0;
        if(A+B+C0>15)
            C1=1;
        else
            C1=0;
    end
end module
```

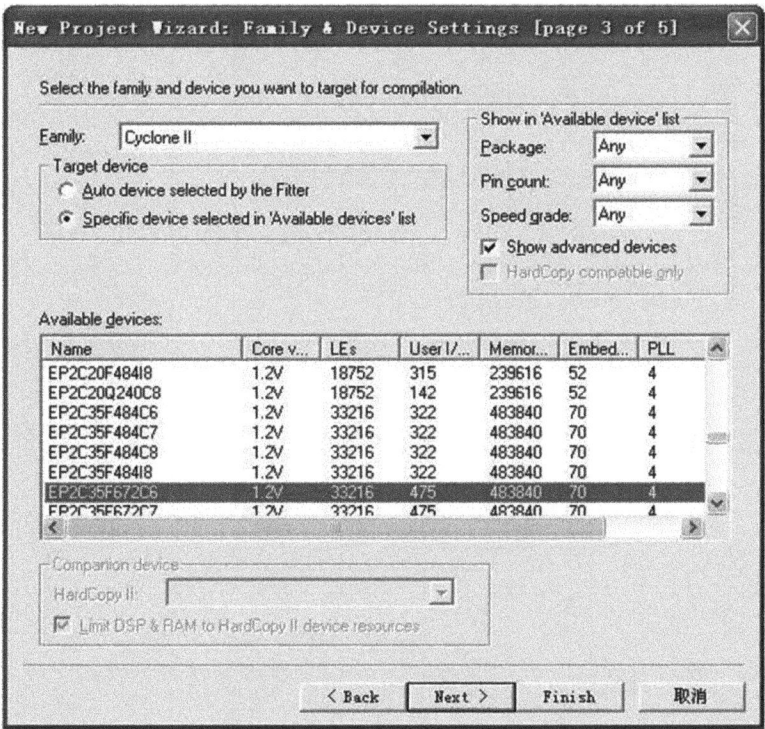

Figure A7: Selecting the target programmable logic device.

The next is to compile the input file of Verilog HDL description. Select **Processing>start compilation**, or click the toolbar button ▶ and enter into the process of the code analysis, synthesis and implementation for the target chip. The successful compilation will pop out a message box to tell the success message. After pressing the **"OK"** button, Quartus II displays the interface shown in Figure A12. At the bottom of the interface, various compilation messages are displayed. Once errors occur during the compiling process, the compilation quit and the error message will be displayed in the "Message" window. When the compilation is finished, the system will open a **"Compilation Report"** window automatically and report the chip's resource occupancy of the compiled design for the target FPGA chip.

Step 3 Assigning Pins

Select **Assignments>Pins** to assign pins as the inputs and outputs of the design circuit. Table A1 shows the corresponding relation between input/output variables of the file adder.v, input control switches and output LED display on the board DE2, and the pins of target FPGA devices.

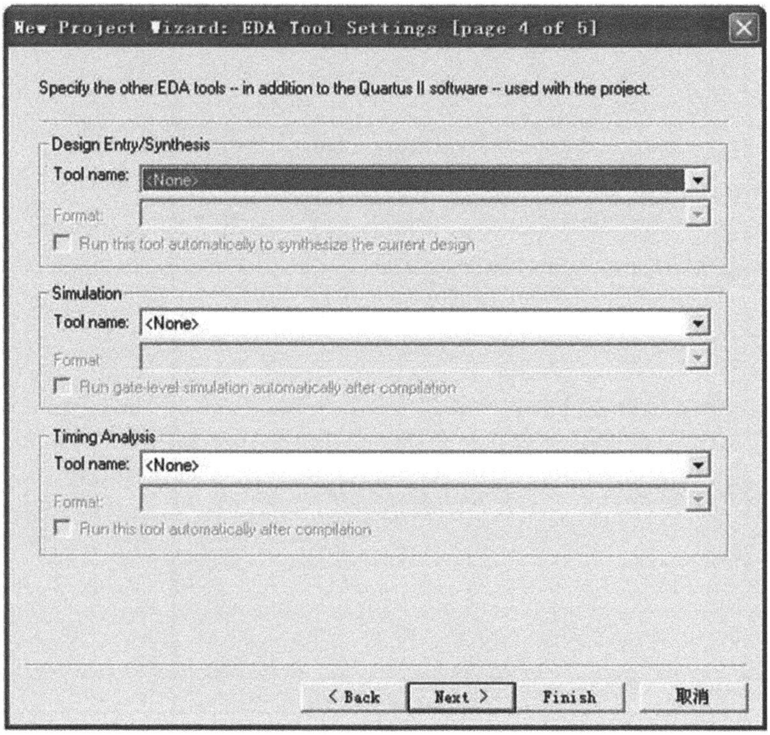

Figure A8: Select the third part tools.

In terms of Table A1, assign pins of FPGA to input/output variables, and the final result is shown in Figure A13.

Step 4 Simulation

To check the correctness of the design, the function implemented by the program need be simulated. Select **File>New** and appear a dialog box shown in Figure A14. Click **"Other File"** item and select **"Vector Waveform File."**

Click **OK** and go into the Waveform editor window shown in Figure A15.

Select **File > Save as** and enter the filename **"adder"** to save the file as "adder.vwf." Choose **Edit>End Time** to set the simulation end time of 200 ns, and select **View>Fit in Window** to display the entire simulation time range in the menu window.

Next, the input/output nodes need to be added into the waveform. Select **Edit> Insert>Insert Node or Bus** and open a dialog box shown in Figure A16.

Click **"Node Finder"** and select Filter item as **"Pins: all"** in the Node Finder window shown in Figure A17. Then click **"List"** and all nodes and buses are displayed in the Nodes Found window.

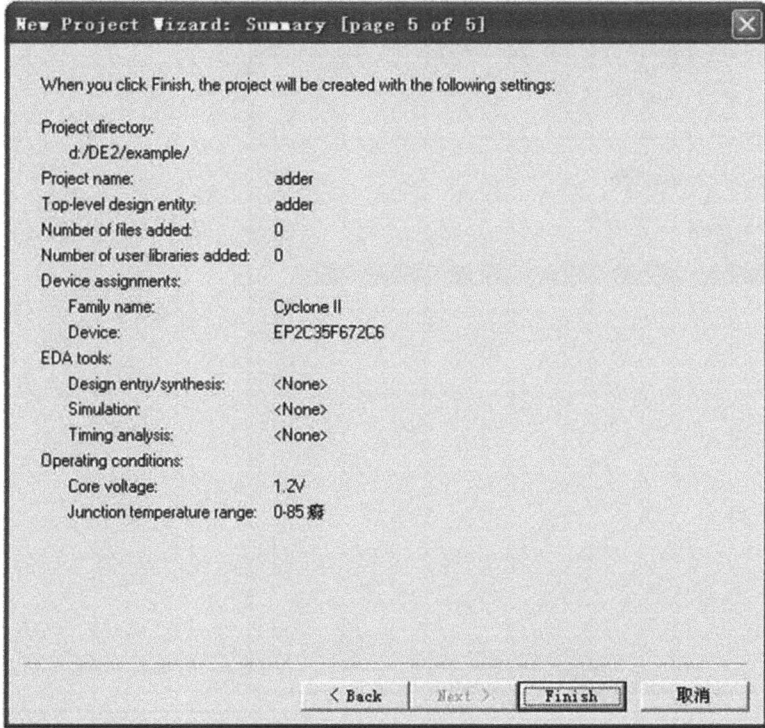

Figure A9: Summary dialogue message of the new project.

Select all nodes and then click ≫ button. This makes all nodes added into **"Selected Nodes"** window. Click OK and return to waveform editor window. Use the select icon ▷ and the waveform edit icon ⌘ to edit the input waveform. Figure A18 shows the waveform of A[3:0]=[0000], B[3:0]=[0000] to [1111], and C0 = 0.

The following step is to perform functional simulation. Select **Assignments>Settings** to open the "Setting" window shown in Figure A19.

Select **"Simulator Settings"** and set **"Simulation mode"** as **"Functional"**. Click "OK" to finish the settings. Then, select **Processing>Generate Functional Simulation Netlist** to generate the netlist for functional simulation, and select **Processing>Start Simulation** or press the icon ▶ to start functional simulation. The result is shown in Figure A20.

Step 5 Configuration and Programming
Configuration and programming is to configure the 4-bit adder design into the FPGA chip. Select the **ALTERA's DE2 board,** which contains a Cyclone II chip.

Quartus II offer two kinds of configuration and programming mode. One is JTAG (Join Test Action Group) mode, in which FPGA device is in-system programming (ISP)

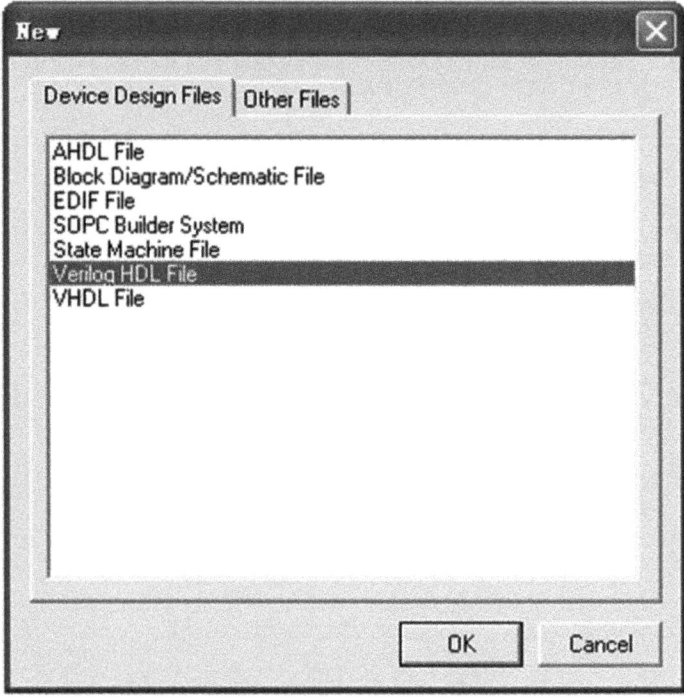

Figure A10: Creating a new Verilog file.

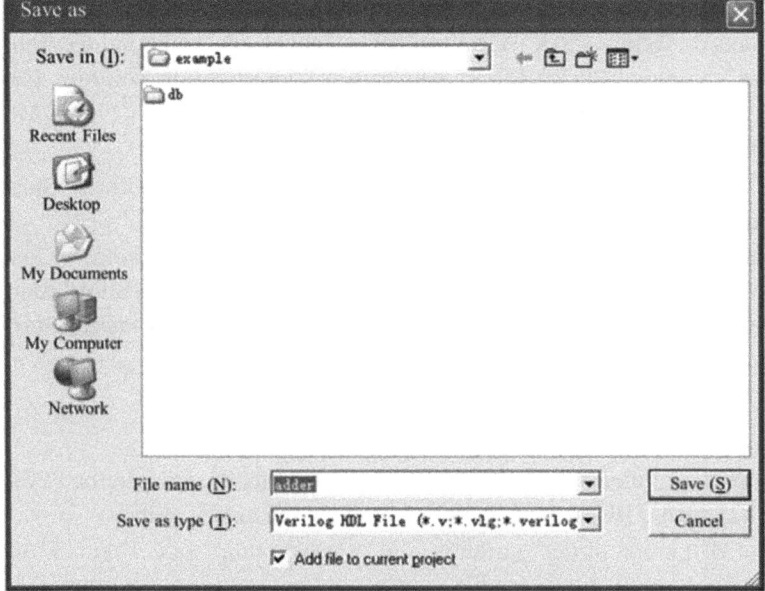

Figure A11: Dialog box of "Save as" and creating the file name.

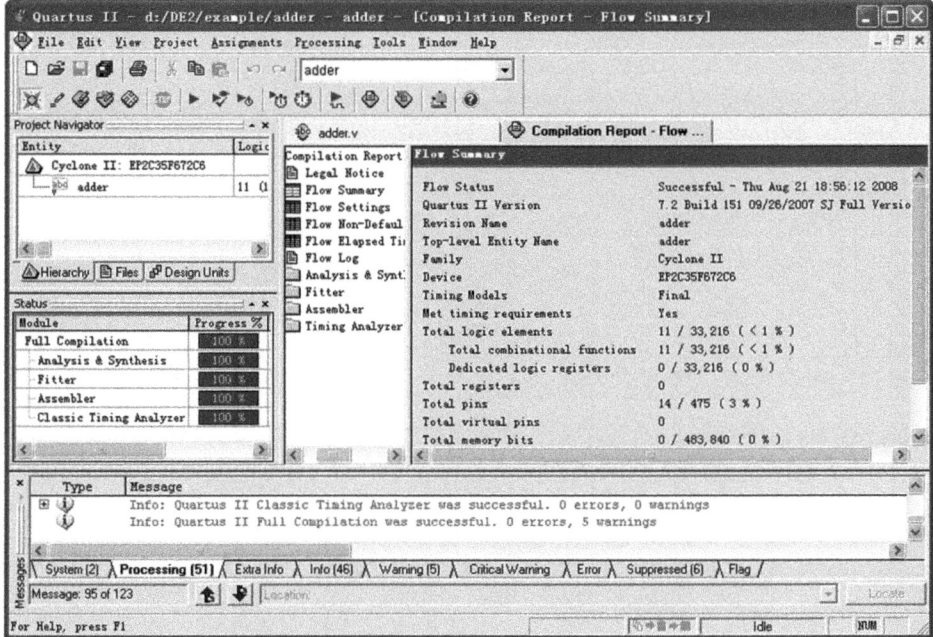

Figure A12: Compilation report.

Table A1: Pins assignment.

I/Os	Variables	Signal name of DE2 board	FPGA pin No.
Inputs	A[3]	SW4	PIN_AF14
	A[2]	SW3	PIN_AE14
	A[1]	SW2	PIN_P25
	A[0]	SW1	PIN_N26
	B[3]	SW8	PIN_B13
	B[2]	SW7	PIN_C13
	B[1]	SW6	PIN_AC13
	B[0]	SW5	PIN_AD13
	C0	SW0	PIN_N25
Outputs	S[3]	LEDG3	PIN_V18
	S[2]	LEDG2	PIN_W19
	S[1]	LEDG1	PIN_AF22
	S[0]	LEDG0	PIN_AE22
	C1	LEDG4	PIN_U18

	Node Name	Direction	Location	I/O Bank	Vref Group	I/O St
1	A[3]	Input	PIN_AF14	7	B7_N1	3.3-V LVTTL
2	A[2]	Input	PIN_AE14	7	B7_N1	3.3-V LVTTL
3	A[1]	Input	PIN_P25	6	B6_N0	3.3-V LVTTL
4	A[0]	Input	PIN_N26	5	B5_N1	3.3-V LVTTL
5	B[3]	Input	PIN_B13	4	B4_N1	3.3-V LVTTL
6	B[2]	Input	PIN_C13	3	B3_N0	3.3-V LVTTL
7	B[1]	Input	PIN_AC13	8	B8_N0	3.3-V LVTTL
8	B[0]	Input	PIN_AD13	8	B8_N0	3.3-V LVTTL
9	C0	Input	PIN_N25	5	B5_N1	3.3-V LVTTL
10	C1	Output	PIN_U18	7	B7_N0	3.3-V LVTTL
11	S[3]	Output	PIN_V18	7	B7_N0	3.3-V LVTTL
12	S[2]	Output	PIN_W19	7	B7_N0	3.3-V LVTTL
13	S[1]	Output	PIN_AF22	7	B7_N0	3.3-V LVTTL
14	S[0]	Output	PIN_AE22	7	B7_N0	3.3-V LVTTL
15	<<new node>>					

Figure A13: Pin assignment.

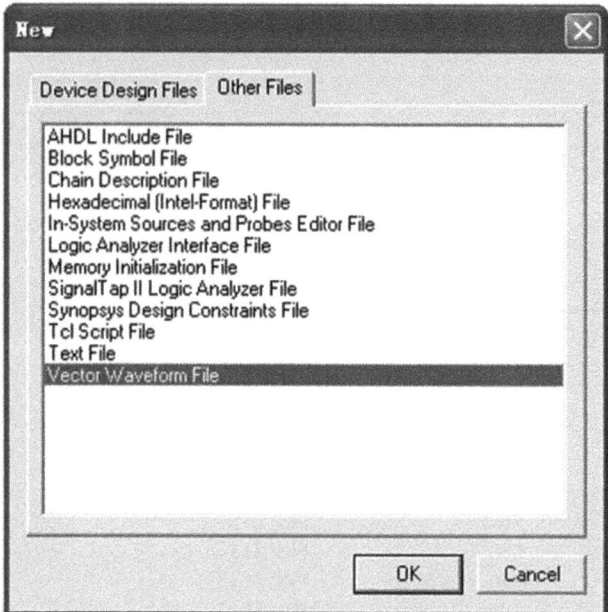

Figure A14: Creating a new Vector Waveform File.

through the "USB Blaster." Since FPGA belong to volatile device, the configuring data will be lost when the power is turned off. Therefore, FPGA device must be reprogrammed each time the power is turned on. Another is AS (active serial configuration) mode, in which a serial EPROM chip (EPCS16) is used to configure FPGA device on the DE2 board. In AS mode, the programming data is written into EPCS16 through the USB

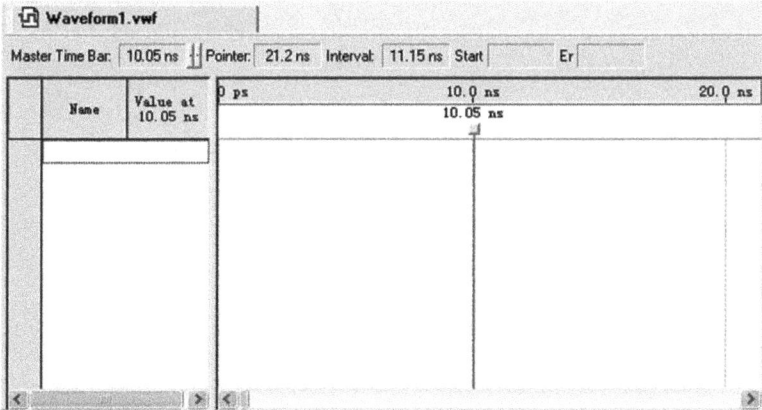

Figure A15: Waveform editor window.

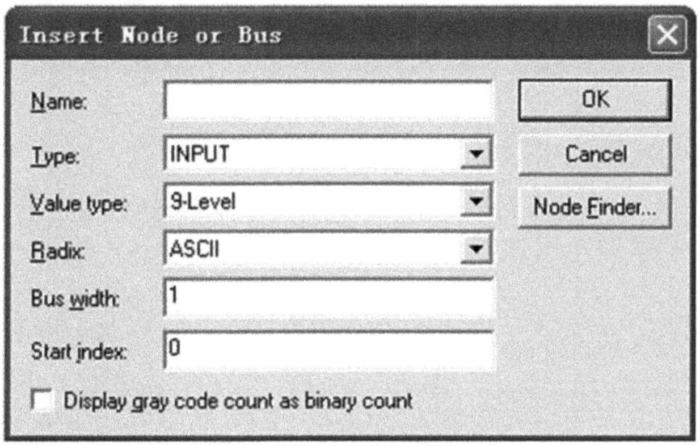

Figure A16: Dialog box of Insert Node or Bus.

Blaster. Then the data in EPCS16 is configured into FPGA device when the power is turned on. SW19 on DE2 board is used to select the configuration mode. When SW19 is switched to the RUN position, JTAG mode is selected; when SW19 is switched to the PROG position, AS mode is selected.

The detailed steps to configure or program a device with JTAG mode as fellows.

(1) Connect USB port of the host computer to BLASTER port on the DE2 board, and turn on the power of DE2 board.

(2) Switch SW19 to RUN position.

(3) Select **Tool>Programmer** or click 🖐 to open the programming window shown in Figure A21. Double Click **"USB Blaster"** and then Click **"Close"** to configure the hardware.

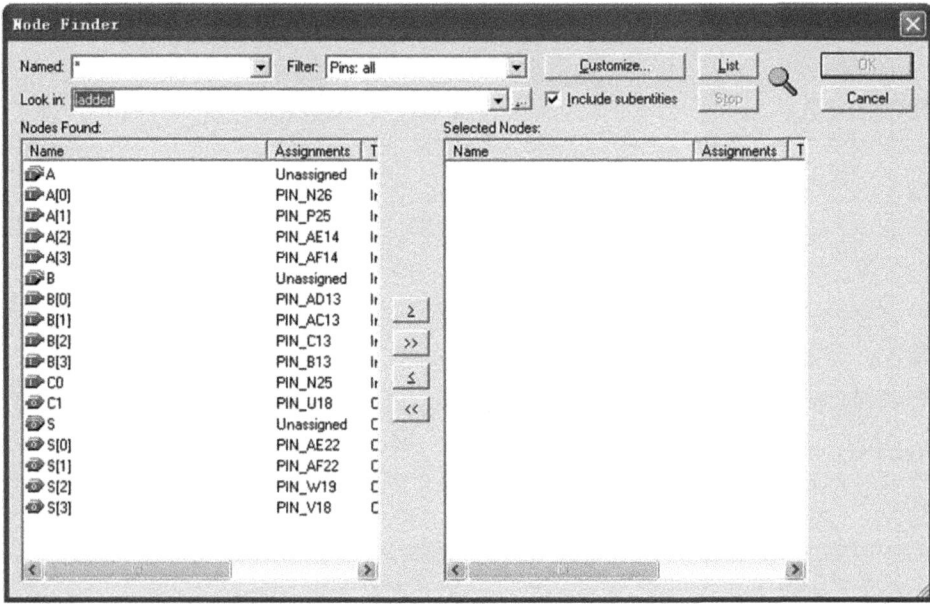

Figure A17: Node Finder dialog box.

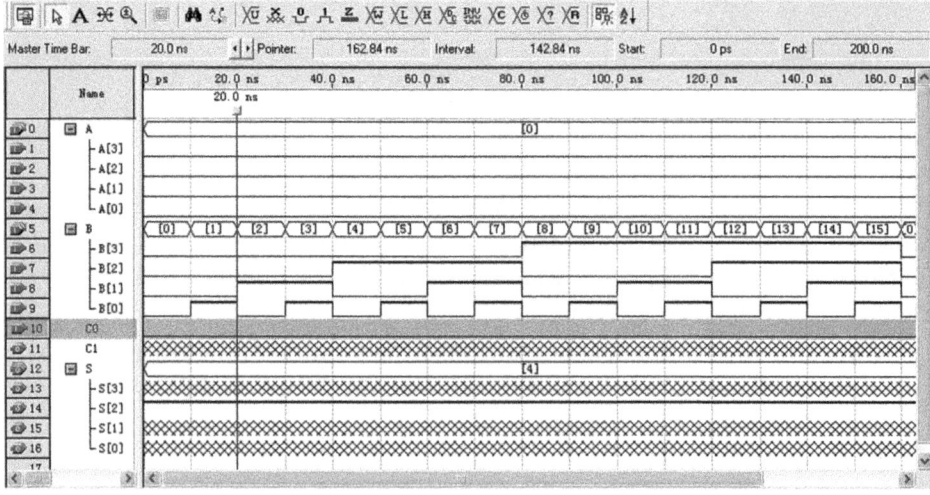

Figure A18: Waveform editor window with all nodes.

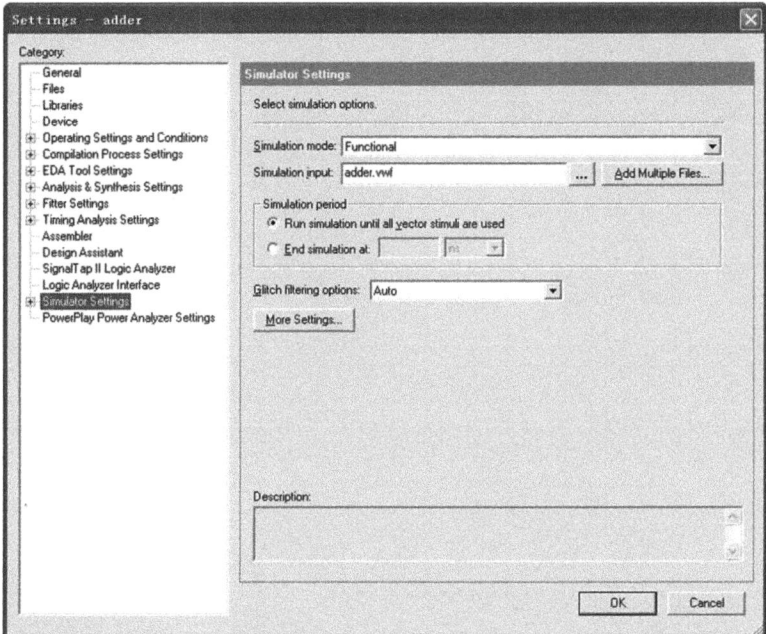

Figure A19: Simulation Settings.

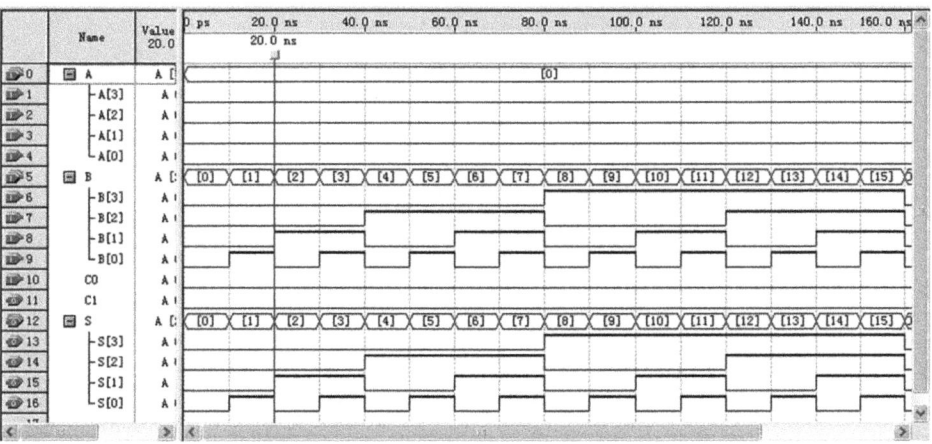

Figure A20: Simulation waveforms.

(4) If "No Hardware" is displayed, click **"Hardware Setup"** to open the Hardware Setup window in Figure A22.

(5) Click **Add File** to add the "add.sof" file to the programmer. Notice that this step can be omitted if the file is already displayed. Set Device as EP2C35F672 and select **"Program/Configure."**

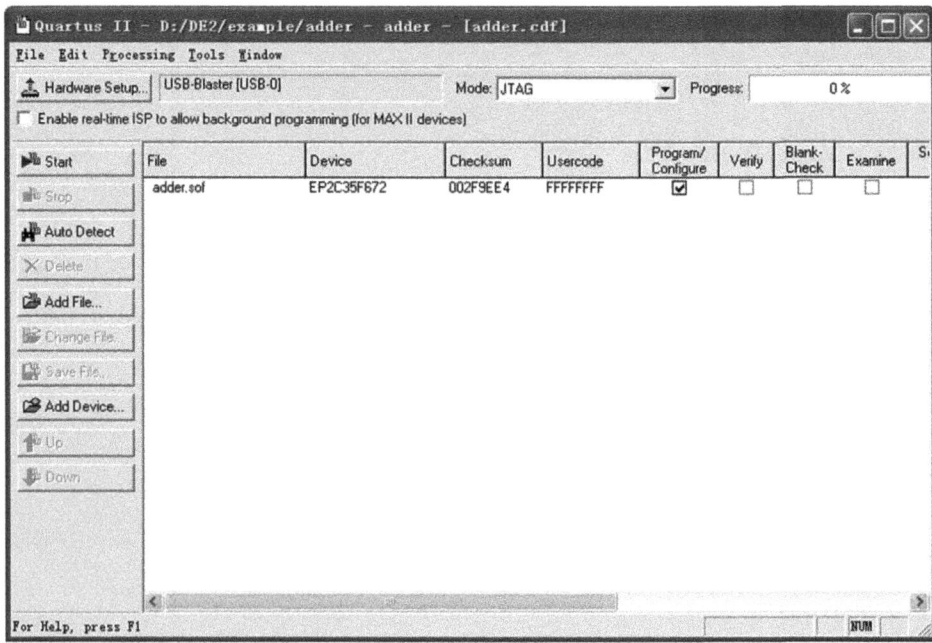

Figure A21: Programming window.

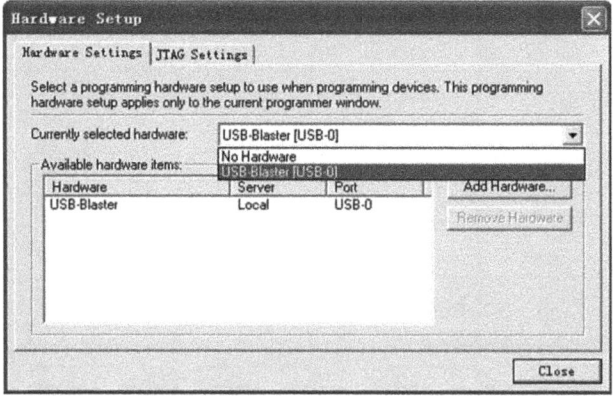

Figure A22: Hardware setup window.

(6) Click **"Start"** button and start programming. After programming, the LED "GOOD" on DE2 is turned on.

The detailed steps to configure or program a device with AS mode as fellows.

(1) Select **Settings>Device** to open the setting window in Figure A23.

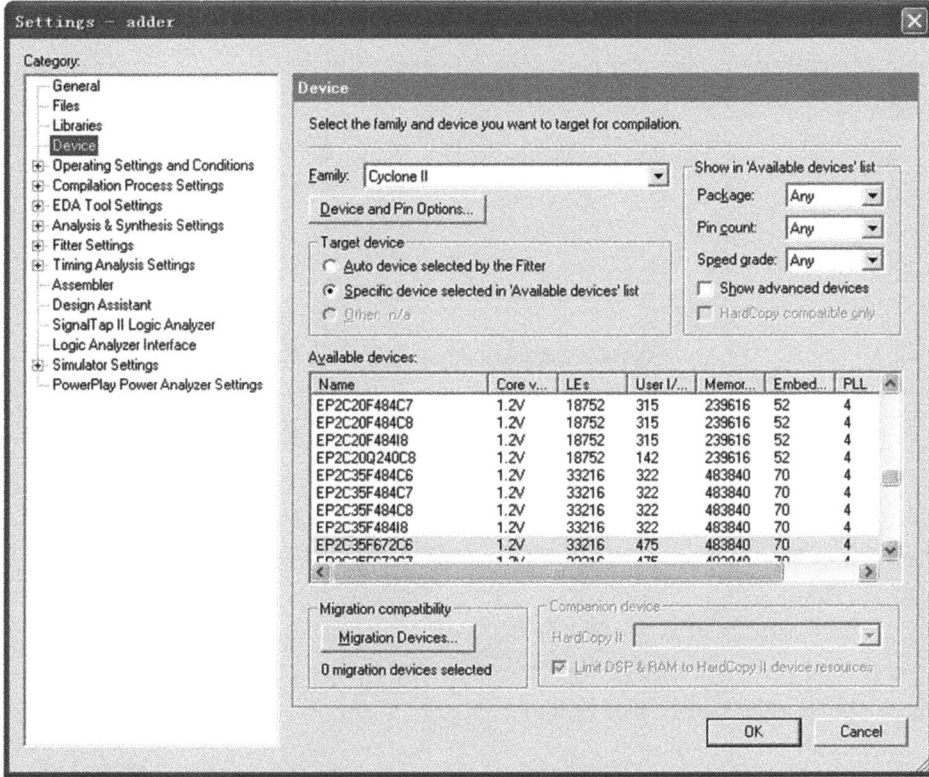

Figure A23: Device configuration window.

(2) Click **"Device & Pin Options"** in Figure A23 and select "Configuration" item and choose EPCS16 in the "Configuration Device" window as shown in Figure A24. Click "OK" and return to the previous interface. Then click "OK" to end the configuration setting and recompile.

(3) If the power of DE2 board is on, turn it off. Switch SW19 to PROG position and connect DE2 board to the host computer by USB cable, and then turn on the power of DE2 board;

(4) Select **Tools>Programmer** or Click 🔌 to open the programming window in Figure A21 and choose Mode as **"Active Serial Programmer."**

(5) Click **"Add File"** to add the "adder.sof" file to the programmer and select **"Program/Configure"** item shown in Figure A25.

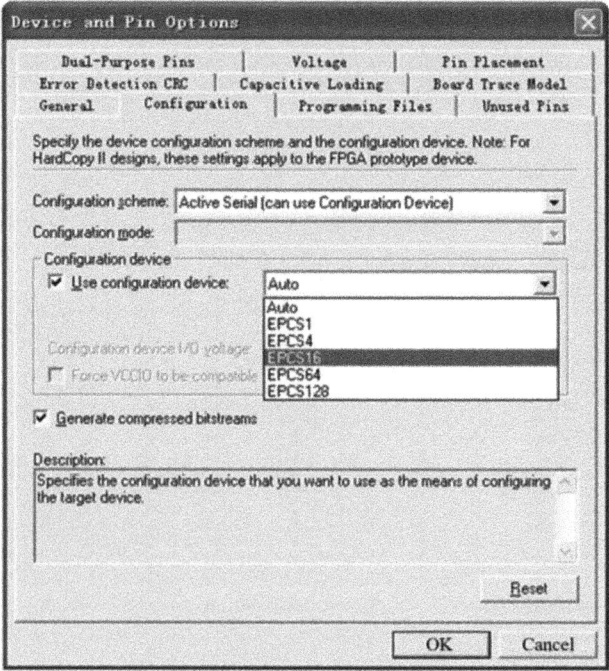

Figure A24: Selecting configuration device.

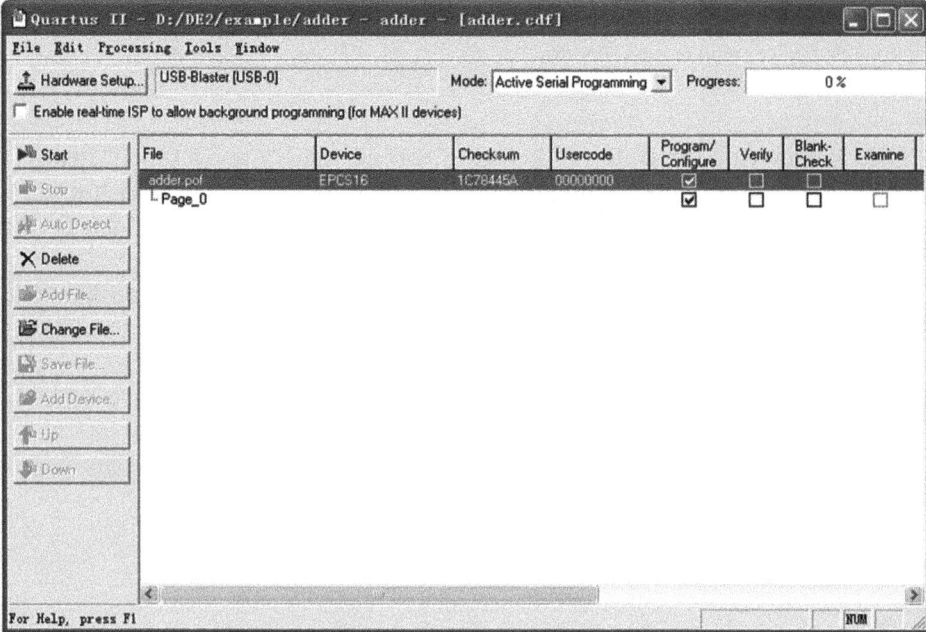

Figure A25: Programming and configuration window in AS mode.

(6) Click **"Start"** button to start programming. After programming, the LED "GOOD" on DE2 board is turned on.

Step 6 Circuit Test

Switch SW19 to RUN position. The circuit function can be tested by observing the variation of LEDG0 to LEDG4 with the state of SW0 to SW8

Appendix II: Introduction to Altera DE2 board

The purpose of the Altera DE2 Development and Education board is to provide the ideal vehicle for learning about digital logic, computer organization and FPGAs. It uses the state-of-the-art technology in both hardware and CAD tools to expose students and professionals to a wide range of topics. The board offers a rich set of features that make it suitable for use in a laboratory environment for university and college courses, for a variety of design projects, as well as for the development of sophisticated digital systems. Altera provides a suite of supporting materials for the DE2 board, including tutorials, "ready-to-teach" laboratory exercises, and illustrative demonstrations. You can download from the website of Altera Inc [49].

1. DE2 Board Features

The DE2 board features a state-of-the-art Cyclone® II 2C35 FPGA in a 672-pin package, as shown in Figure A26.

Figure A26: The DE2 board.

All important components on the board are connected to pins of this chip, allowing the user to control all aspects of the board's operation. For simple experiments, the DE2 board includes a sufficient number of robust switches (of both toggle and push-button type), LEDs, and 7-segment displays. For more advanced experiments, there are SRAM, SDRAM, and Flash memory chips, as well as a 16 × 2 character display. For experiments that require a processor and simple I/O interfaces, it is easy to instantiate Altera's Nios II processor and use interface standards such as RS-232 and PS/2. For experiments that involve sound or video signals, there are standard connectors for microphone, line-in, line-out (24-bit audio CODEC), video-in (TV Decoder) and VGA (10-bit DAC); these features can be used to create CD-quality audio applications and professional-looking video. For larger design projects, the DE2 provides USB 2.0 connectivity (both host and device), 10/100 Ethernet, an infrared (IrDA) port, and an SD memory card connector. Finally, it is possible to connect other user defined boards to the DE2 board by means of two expansion headers.

2. Main Pin Table

Table A2 lists part of main pin of Altera DE2 board for reference.

Table A2: Main pin table.

Signal Name	FPGA Pin No.	Description	Signal Name	FPGA Pin No.	Description
SW[0]	PIN_N25	Toggle switches Up: HIGH Down: LOW	LEDR[0]	PIN_AE23	
SW[1]	PIN_N26		LEDR[1]	PIN_AF23	
SW[2]	PIN_P25		LEDR[2]	PIN_AB21	
SW[3]	PIN_AE14		LEDR[3]	PIN_AC22	
SW[4]	PIN_AF14		LEDR[4]	PIN_AD22	
SW[5]	PIN_AD13		LEDR[5]	PIN_AD23	
SW[6]	PIN_AC13		LEDR[6]	PIN_AD21	
SW[7]	PIN_C13		LEDR[7]	PIN_AC21	
SW[8]	PIN_B13		LEDR[8]	PIN_AA14	RED LED
SW[9]	PIN_A13		LEDR[9]	PIN_Y13	
SW[10]	PIN_N1		LEDR[10]	PIN_AA13	
SW[11]	PIN_P1		LEDR[11]	PIN_AC14	
SW[12]	PIN_P2		LEDR[12]	PIN_AD15	
SW[13]	PIN_T7		LEDR[13]	PIN_AE15	
SW[14]	PIN_U3		LEDR[14]	PIN_AF13	
SW[15]	PIN_U4		LEDR[15]	PIN_AE13	

Table A2 (continued)

Signal Name	FPGA Pin No.	Description	Signal Name	FPGA Pin No.	Description
SW[16]	PIN_V1		LEDR[16]	PIN_AE12	
SW[17]	PIN_V2		LEDR[17]	PIN_AD12	
KEY[0]	PIN_G26		LEDG[0]	PIN_AE22	
KEY[1]	PIN_N23	4 Push-button switches, which can be uses as manual clock.	LEDG[1]	PIN_AF22	
KEY[2]	PIN_P23		LEDG[2]	PIN_W19	
KEY[3]	PIN_W26		LEDG[3]	PIN_V18	
CLOCK_27	PIN_D13	27MHz clock	LEDG[4]	PIN_U18	GREEN LED
CLOCK_50	PIN_N2	50MHz clock	LEDG[5]	PIN_U17	
EXT_CLOCK	PIN_P26	External clock	LEDG[6]	PIN_AA20	
HEX0[0]	PIN_AF10		LEDG[7]	PIN_Y18	
HEX0[1]	PIN_AB12		LEDG[8]	PIN_Y12	
HEX1[2]	PIN_AC12		HEX5[0]	PIN_T2	
HEX0[3]	PIN_AD11	7-Seg display LED 0	HEX5[1]	PIN_P6	
HEX0[4]	PIN_AE11		HEX5[2]	PIN_P7	
HEX0[5]	PIN_V14		HEX5[3]	PIN_T9	7-Seg display LED 5
HEX0[6]	PIN_V13		HEX5[4]	PIN_R5	
HEX1[0]	PIN_V20		HEX5[5]	PIN_R4	
HEX1[1]	PIN_V21		HEX5[6]	PIN_R3	
HEX1[2]	PIN_W21		HEX6[0]	PIN_R2	
HEX1[3]	PIN_Y22	7-Seg display LED 1	HEX6[1]	PIN_P4	
HEX1[4]	PIN_AA24		HEX6[2]	PIN_P3	7-Seg display LED 6
HEX1[5]	PIN_AA23		HEX6[3]	PIN_M2	
HEX1[6]	PIN_AB24		HEX6[4]	PIN_M3	

(continued)

Table A2 (continued)

Signal Name	FPGA Pin No.	Description	Signal Name	FPGA Pin No.	Description
HEX2[0]	PIN_AB23		HEX6[5]	PIN_M5	
HEX2[1]	PIN_V22		HEX6[6]	PIN_M4	
HEX2[2]	PIN_AC25		HEX7[0]	PIN_L3	
HEX2[3]	PIN_AC26	7-Seg display LED 2	HEX7[1]	PIN_L2	
HEX2[4]	PIN_AB26		HEX7[2]	PIN_L9	
HEX2[5]	PIN_AB25		HEX7[3]	PIN_L6	7-Seg display LED 7
HEX2[6]	PIN_Y24		HEX7[4]	PIN_L7	
HEX3[0]	PIN_Y23		HEX7[5]	PIN_P9	
HEX3[1]	PIN_AA25		HEX7[6]	PIN_N9	
HEX3[2]	PIN_AA26				
HEX3[3]	PIN_Y26	7-Seg display LED 3			
HEX3[4]	PIN_Y25				
HEX3[5]	PIN_U22				
HEX3[6]	PIN_W24				
HEX4[0]	PIN_U9				
HEX4[1]	PIN_U1				
HEX4[2]	PIN_U2				
HEX4[3]	PIN_T4	7-Seg display LED 4			
HEX4[4]	PIN_R7				
HEX4[5]	PIN_R6				
HEX4[6]	PIN_T3				

Appendix III: Abbreviations

ABEL	Advanced Boolean Expression Language
ADC	Analog-to-Digital Converter
AIM	Advanced Interconnect Matrix
ALE	Address Latch Enable
ASCII	American Standard Code for Information Interchange
ASIC	Application-Specific Integrated Circuit
BCD	Binary Coded Decimal
BEDO DRAM	Burst Extended Data Out Dynamic Random Access Memory
BGA	Ball Grid Array
BIOS	Basic Input/Output System
BJT	Bipolar Junction Transistor
BST	Boundary-Scan Test
CAD	Computer-Aided Design
CD	Compact Disc
CLB	Configurable Logic Blocks
CLR	Clear
CMOS	Complemented Metal Oxide Semiconductor
CP	Clock Pulse
CPLD	Complex Programmable Logic Device
CPU	Central Processing Unit
CV	Control Voltage
DAC	Digital-to-Analog Converter
DC	Direct Current
DEMUX	Demultiplexer
DIP	Dual Inline Package
DIS	Discharge
DPRAM	Dual-Ported Random Access Memory
DSP	Digital Signal Process
ECAD	Electronic Computer-Aided Design
EDA	Electronic Design Automation
EDO DRAM	Extended Data Out Dynamic Random Access Memory
EEPROM	Electrically Erasable Programmable Read-Only Memory
EN	Enable Port
EOC	End of the Conversion
EPROM	Erasable Programmable Read-Only Memory
FF	Flip-Flop
FIFO	First In First Out
FPGA	Field Programmable Gate Array
FPM DRAM	Fast Page Mode Dynamic Random Access Memory
FRAM	Ferroelectric Random Access Memory
GAL	Generic Array Logic
GDA	Gateway Design Automation
GND	Ground
HDL	Hardware Description Language
I/O	Input/Output
I/OB	Input/Output Block

IC	Integrated Circuit
ICT	In-Circuit Testers
IEEE	Institute of Electrical and Electronic Engineers
ISO	International Standard Organization
ISP	In-System Programmability
JTAG	Joint Test Action Group
K-map	Karnaugh map
LAB	Logical Array Block
LCD	Liquid Crystal Display
LED	Light Emitting Diode
LFSR	Linear Feedback Shift Register
LQFP	Low-Profile Quad Flat Package
LS	Low Power Scottky
LSB	Least Significant Bit
LSI	Large-Scale Integrated Circuits
LUT	Look-Up Table
MOD	Modulus
MOS	Metal Oxide Semiconductor
MOSFET	Metal Oxide Semiconductor Field Effect Transistor
MROM	Mask Read-Only Memory
MSB	Most Significant Bit
MSI	Medium-Scale Integration
MTP ROM	Multiple Times Programmable Read-Only Memory
MUX	Multiplexer
NAND	Not And (Electronic Logic Gate)
NMOS	N-Channel Metal Oxide Semiconductor
NPN	Negative-Positive-Negative Transistor
OLMC	Output Logic Macro Cell
OTP	One-Time Programmable
OTP ROM	One-Time Programmable Read-Only Memory
PAC	Pad Array Carrier
PAL	Programmable Array Logic
PCB	Printed Circuit Board
PI	Programmable Interconnection
PLA	Programmable Logic Array
PLCC	Plastic Leaded Chip Carrier
PLD	Programmable Logic Device
PMOS	P-Channel Metal Oxide Semiconductor
POS	Product of Sum
PRE	Preset
PROM	Programmable Read-Only Memory
PWM	Programmable Switch Matrix
QFP	Quad Flat Package
RAM	Random Access Memory
RC	Resistance–Capacitance Circuits
RCLK	Read Clock
RLC	Resistance–Inductor–Capacitance Circuits
ROM	Read-Only Memory
S/H	Sample and Hold Circuit

SAR	Successive Approximation Register
SDRAM	Synchronous Dynamic Random Access Memory
SMT	Surface-Mount Technology
SOIC	Small Outline Integrated Circuit
SOP	Sum of Products
SPLD	Simple Programmable Logic Devices
SSC	Synchronous Sequential Circuits
SSI	Small-Scale Integration
TB	Terabyte
TH	Threshold
TQFP	Thin Quad Flat Package
TSOP	Thin Small Outline Plastic
TTL	Transistor–Transistor Logic
ULSI	Ultra Large-scale Integration
USB	Universal Serial Bus
VHSIC	Very High-Speed Integrated Circuit
VHDL	Hardware Description Language
VLSI	Very Large-Scale Integration
VRAM	Video Random Access Memory
V-T	Voltage-to-Time
VT	Vacuum Tube
WCLK	Write Cloc
XNOR	Exclusive-NOR
XOR	Exclusive-OR

References

[1] Robert K. Dueck. 2000. Digital Design with CPLD Applications and VHDL. Delmar: Thomson Delmar Learning.

[2] Proakis, John G., Manolakis, Dimitris G. 2007. Digital Signal Processing. Upper Saddle River: Pearson Prentice Hall.

[3] Thomas L. Floyd. 2013. Digital fundamental: A system approach. Upper Saddle River: Pearson Prentice Hall.

[4] Paul Horowitz, Winfield Hill. 2015. The Art of Electronics. Third Edition. Cambridge: Cambridge University Press.

[5] Linda Null, Julia Lobur. 2006. The essentials of computer organization and architecture. Sudbury: Jones & Bartlett Publishers.

[6] Chunling Yang, Shujuan Wang. 2017. Digital Electronic Technology Fundamental. Second Edition. Beijing: Higher Education Press.

[7] Mohammed Ferdjallah. 2011. Introduction to digital system: modeling, synthesis and Simulation using VHDL. Hoboken: John Wiley & Sons, Inc.

[8] D. Jansen. 2003. The Electronic Design Automation Handbook. Dordrecht:Kluwer Academic Publishers.

[9] Michael D. Ciletti. 2005. Advanced Digital Design with Verilog HDL. New Delhi: Prentice Hall of India.

[10] Institute of Electrical and Electronics Engineers (IEEE). 1993. IEEE Standard VHDL Language Reference Manual (LRM), IEEE Std. 1076–1987, 1988, 1993. Piscataway: IEEE.

[11] Volnei Pedroni. 2008. Digital Electronics and Design with VHDL. Burlington: Morgan Kaufmann Publishers.

[12] Institute of Electrical and Electronics Engineers. 2001. IEEE Standard Verilog Hardware Description language Reference Manual. Piscataway: IEEE.

[13] The Institute of Electrical and Electronics Engineers. 2005. IEEE Standard Hardware Description Language Based on the Verilog Hardware Description Language, Language Reference Manual (LRM), IEEE Std. 1364–1995, 1996, 2001, 2005. Piscataway: IEEE.

[14] Randy Katz, Gaetano Borriello. 2005. Contemporary Logic Design. Upper Saddle River: Pearson Prentice Hall.

[15] David Patterson, John Hennessy. 2014. Computer Organization and Design: The Hardware/Software Interface. Fifth edition. San Francisco: Morgan Kaufmann.

[16] Shi Yan, Hong Wang. 2016. Digital Electronic Technology Fundamental. Sixth Edition. Beijing: Higher Education Press.

[17] Huaguang Kang. 2014. Electronic Technology Fundamental: Digital. Sixth Edition. Beijing: Higher Education Press.

[18] George Boole. 1854. An Investigation of the Laws of Thought. New York: Dover Publications.

[19] A.K. Maini. 2007. Digital Electronics Principles, Devices and Applications. Chichester: John Wiley & Sons Ltd.

[20] John F. Wakerly. 2007. Digital Design Principles and Practices (4th Edition, Photocopy version). Beijing: Higher Education Press.

[21] Jianjun Hou. 2015. Digital Electronic Technology Fundamental. Third Edition. Beijing: Higher Education Press.

[22] Jerry Daniels, 1996. Digital Design from Zero to One. New York: John Wiley and Sons, Inc.

[23] M Karnaugh. 1953. The map method for synthesis of combinational logic circuits. Transactions of the American Institute of Electrical Engineers, Part I: Communication and Electronics, 72 (5):593–599.

https://doi.org/10.1515/9783110614916-014

[24] Randy Katz, Gaetano Borriello. 2005. Contemporary Logic Design. Upper Saddle River: Pearson Prentice Hall.

[25] R.P. Jain. 2010. Modern Digital Electronics. Fourth Edition. New Delhi: Tata McGraw-Hill Education.

[26] Alan Marcovitz. 2010. Introduction to Logic Design. Third Edition. New York: McGraw-Hill.

[27] Parag Lala. 2007. Principles of Modern Digital Design. Hoboken: John Wiley & Sons, Inc.

[28] Brian Holdsworth and Clive Woods 2002 Digital Logic Design. Fourth Edition. Oxford Elsevier Science & Technology.

[29] N. Zheludev. 2007. The life and times of the LED: a 100-year history. Nature Photonics. 1 (4): 189–192.

[30] Hiroshi Kawamoto. 2002. "The History of Liquid-Crystal Displays". Proceedings of the IEEE. 90 (4): 460–500.

[31] M. Morris Mano, Michael D. Ciletti. 2015. Digital Design with an Introduction to the Verilog HDL. Firth Edition. Singapore: Pearson Education South Asia Pte Ltd.

[32] Don Thomas, Philip Moorby. 2002. The Verilog Hardware Description Language. Fifth edition. Norwell: Kluwer.

[33] M. Rafiquzzaman. 2005. Fundamentals of Digital Logic and Microcomputer Design. Fifth Edition. Hoboken: John Wiley & Sons, Inc. (Verilog)

[34] Stephen Brown, Zvonko Vranesic. 2014. Fundamentals of Digital Logic with Verilog Design. Third Edition. New York City: McGraw-Hill Education.

[35] Zainalabedin Navabi. 2006. Verilog Digital System Design. Second edition. New York: McGraw-Hill.

[36] Samir Palnitkar. 2003. Verilog HDL: A Guide to Digital Design and Synthesis. Second Edition. Upper Saddle River: Prentice Hall PTR.

[37] Victor Nelson, H. Troy Nagle, Bill Carroll, J David Irwin. 1995. Digital Logic Circuit Analysis and Design. Englewood Cliffs: Prentice Hall.

[38] Uday Bakshi, Atul Godse. 2009. Analog and Digital Electronics. Pune: Technical Publications Pune.

[39] Anil Maini. 2007. Digital Electronics: Principles, Devices and Applications. Chichester: John Wiley & Sons Ltd.

[40] Myke Predko. 2005, Digital Electronics Demystified: a Self-teaching Guide. New York City: McGraw-Hill Education.

[41] William Kleitz. 2003. Digital and Microprocessor Fundamentals: Theory and Application. Fourth Edition. Upper Saddler Reviver: Pearson Prentice Hall.

[42] Carl Hamacher, Zvonko Vranesic, Safwat Zaky, Naraig Manikian. 2012. Computer Organization and Embedded Systems. Sixth Edition. New York: McGraw-Hill.

[43] Thomas M. Coughlin 2018 Digital Storage in Consumer Electronics: The Essential Guide. Basel: Springer International Publishing AG.

[44] Ronald Tocci, Neal Widmer, Gregory Moss. 2007. Digital Systems: Principles and Applications. Tenth Edition. Upper Saddle River: Pearson Education, Inc.

[45] Steven T. Karris. 2007. Digital Circuit Analysis and Design: with Simulink®Modeling and Introduction to CPLDs and FPGAs. Fremont: Orchard Publications.

[46] Richard Tinder. 2000. Engineering Digital Design. Second Edition. Washington: Cambridge: Academic Press.

[47] Thomas L. Floyd 2015 Digital Fundamentals (11th Edition) London:Pearson Education Inc.

[48] Anant Agarwal, Jeffrey Lang. 2005. Foundations of Analog and Digital Electronic Circuits. San Francisco: Morgan Kaufmann Publishers.

[49] Altera Inc. DE2_UserManual_1.5. 2012. Https://www.altera.com.

Index

https://doi.org/10.1515/9783110614916-015